International Conference on Material Strength and Applied Mechanics (MSAM 2018)

IOP Conference Series: Materials Science and Engineering
Volume 372

Kitakyushu, Japan
10 – 13 April 2018

ISBN: 978-1-5108-6485-6
ISSN: 1757-8981

Printed from e-media with permission by:

Curran Associates, Inc.
57 Morehouse Lane
Red Hook, NY 12571

Some format issues inherent in the e-media version may also appear in this print version.

Printed by Curran Associates, Inc. (2018)

For permission requests, please contact the Institute of Physics at the address below.

Institute of Physics
Dirac House, Temple Back
Bristol BS1 6BE UK

Phone: 44 1 17 929 7481
Fax: 44 1 17 920 0979

techtracking@iop.org

Additional copies of this publication are available from:

Curran Associates, Inc.
57 Morehouse Lane
Red Hook, NY 12571 USA
Phone: 845-758-0400
Fax: 845-758-2633
Email: curran@proceedings.com
Web: www.proceedings.com

Table of contents

Volume 372

2018 International Conference on Material Strength and Applied Mechanics (MSAM 2018)

10–13 April 2018, Kitakyushu City, Japan

Accepted papers received: 22 May 2018
Published online: 12 June 2018

Preface

Papers

IOP Conf. Series: Materials Science and Engineering **372** (2018) 011001 doi:10.1088/1757-899X/372/1/011001

MSAM2018 Preface

International Conference on Material Strength and Applied Mechanics (MSAM 2018) was held from April 10th to 13th, 2018 in Kitakyushu City, Japan. This conference represented a broad forum for all aspects of material strength to promote our fundamental understanding of material strength. MSAM 2018 also discussed recent analytical, experimental and computational results in the field of Applied Mechanics to bridge the gap between physical theory and its application to technology. The conference had parallel sessions with dedicated selected subjects related to Material Strength and Applied Mechanics. The technical program consisted of 4 international keynote speakers, 30 invited speakers, 53 oral presentations, and several poster presentations. MSAM 2018 was supported by several local companies, universities and academic association close to Kitakyushu. They are Kyusyu Institute of Technology, Fukuoka University, University of the Ryukyus, Oita University, The Japan Society of Materials Science, Kyushu Branch Hinode, Ltd., Yaskawa Electric Corporation, Shiromizu Iron Works Co., Ltd., Japan Casting & Forging Corp., Toyota Motor Kyushu, Inc., Shin-Nippon Nondestructive Inspection Co., Ltd., Kumagaigumi Co., Ltd., Takagi Co., Ltd., Kitakyushu Convention & Visitors Association and Kitakyushu Convention Bureau.

The present issue of IOP Conference Series: Materials Science and Engineering (MSE) records the proceedings of MSAM2018 and contains 49 specially selected manuscripts submitted to the MSAM2018 conference. The electronic submission and handling of manuscripts via the conference website, including the selection of reviewers and evaluation of manuscripts, were identical to the procedures applied to manuscripts submitted as regular contributions for international per reviewed publications.

The organization of this conference and the preparation of proceedings volume would have been impossible without the tremendous efforts and dedication of many individuals, especially from Ms. Yajun Ye, who oversaw the organization of the conference and the program, and a large team of reviewers with their timely submission of quality reports. We express our sincere thanks to all authors and presenters for their contributions. We also thank very much our sponsors for their generous support.

MSAM 2019 is scheduled to be held in Kiev, capital of Ukraine. All participants are welcome to MSAM2019! More information will be available at the conference website: http://www.msamconf.org.

Prof. Nao-Aki Noda
General Chairman MSAM2018
Kitakyushu, Japan, May 2018

IOP Conf. Series: Materials Science and Engineering **372** (2018) 011002 doi:10.1088/1757-899X/372/1/011002

Peer review statement

All papers published in this volume of *IOP Conference Series: Materials Science and Engineering* have been peer reviewed through processes administered by the proceedings Editors. Reviews were conducted by expert referees to the professional and scientific standards expected of a proceedings journal published by IOP Publishing.

IOP Conf. Series: Materials Science and Engineering **372** (2018) 012001 doi:10.1088/1757-899X/372/1/012001

An evaluation of fatigue limit of notched specimen of a C/C composite

C Makabe[1], M Fujikawa[1] and M S Ferdous[2]

[1]Faculty of Engineering, University of the Ryukyus, Okinawa, 903-0213, Japan
[2]Department of Mechanical Engineering, Bangladesh Army University of Science and Technology, Saidpur, 5310, Bangladesh

E-mail: makabe@tec.u-ryukyu.ac.jp

Abstract. The fatigue strengths in notched specimens of carbon fiber reinforced carbon composites (C/C composites) were investigated. The fatigue limit was measured by *S-N* curves and load increase tests. The value of fatigue limit obtained by those methods was almost the same. Slits of several sizes were cut on both sides of a test section, and different sizes of slit length were chosen. Also, specimens with blunt-notches were used to compare the fatigue strength. The weakest fatigue limit was obtained in the case of specimens with blunt-notches. However, the stress concentration factor of those is smaller than that of slit specimens. The relationships between fatigue strengths and specimen shapes were analyzed by stress distribution. The effect of slit configuration on fatigue strength was then discussed regarding the experimental consequences. Consequently, it was discussed that the fatigue strength of the present specimens was determined depending on the damage conditions in the vicinity of the notch and on the crack initiation behavior. It is expected that the tendency of the *S-N* curve and fatigue limit was related to the shear damage and shear mode of the crack growth.

1. Introduction

Composites made by carbon fibers are widely used as engineering materials, and there are many kinds of carbon composites [1]. For example, there are carbon fiber reinforced aluminium composites, carbon fiber reinforced epoxy composites, carbon fiber reinforced thermoplastic composites, and so on. The carbon fiber including resin was sintered for carbonizing the resin. That was called the carbon fiber reinforced carbon compound composites or C/C composites [1-6].

The fatigue limit of slit specimens of C/C composites was investigated in the previous study by an experimental method [5]. Also, the effects of notch root radius on fatigue behavior of some notched specimens was investigated [6]. In this paper, we review the results of those studies and some discussion about the qualitative tendency of a fatigue limit of a C/C composite of machine-able type was performed. The fatigue limits of the notched specimens were discussed in this study from the viewpoints of a local shear deformation of carbon fiber, fracture constraint of composites, stress concentration, etc. In the case of metals, fatigue limit was evaluated by the *S-N* curve. However, in the same cases of carbon composite, the fatigue limit can be approximated by the load increase test in which single specimen was used for testing. The present study examined how the damage of the matrix and crack growth mode is related to the fatigue limit and its evaluation method.

2. Material and experimental procedure

The materials used are C/C composites of machine-able type (Grade CX-31 in Toyo Tanso Co. Ltd.). Bulk density and ash content (mass %) of that are 1.61 g/m^3 and 0.02%, respectively. Flexural strength (perpendicular to laminate), compressive strength (perpendicular to laminate) and tensile strength (parallel to laminate) are 90 MPa, 249 MPa and 98 MPa, respectively.

The phenol resin flew into the sheet stitched by carbon fiber. Several of those sheets were piled up and compressed under high temperature at 2000°C to make the plate for the specimen. The nominal thickness of a plate consisting of 26 sheets is 6.2 mm and the ultimate tensile strength is 98 MPa. Figure 1 shows the macroscopic appearance of the plate. The thickness b and length h of specimen cut from the plate were 4.2 mm and 130 mm, respectively. The width $2W$ of the specimen was changed depending on the length of blunt-notch or slit. The shapes of the test section of the specimen are shown in figure 2. Two types of specimens were made. One of them was a slit specimen with ρ = 0.5 mm of the root radius of the ends cut, and the other was a notched specimen with ρ = 8 mm of notch root radius. The loading direction was shown in figure 2 by arrow. Now, in the case of the slit specimen (figure 2(a)), slit or notch length t and width $2W$ are determined to be set as the same size as the ligament $2(W\text{-}t)$. So, $2(W\text{-}t)$ was set 20 mm, the combinations of t and $2W$ were 0.5 mm and 19 mm, 1.0 mm and 18 mm, 2.5 mm and 15 mm, and 5.0 mm and 10 mm.

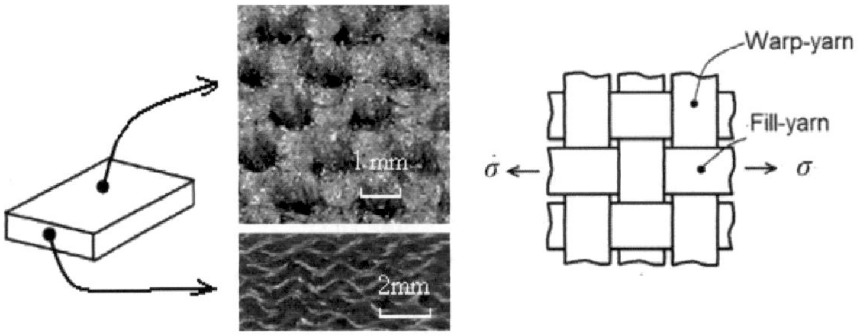

Figure 1. Macroscopic appearance of the carbon composite used in the study.

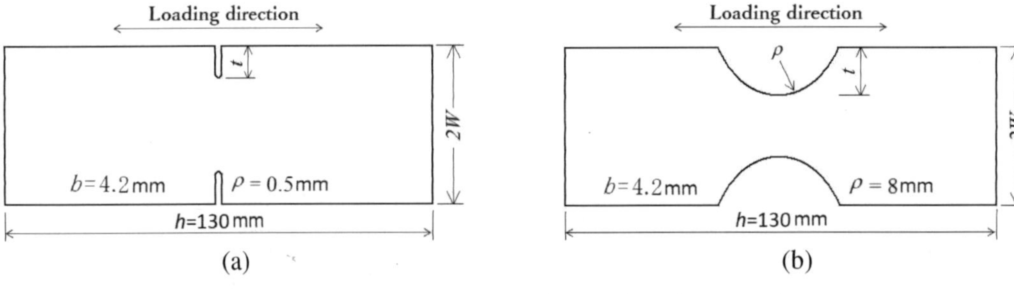

Figure 2. Geometry of the specimens, (a) Slit specimens (t = 0.5, 1.0, 2.5, 5.0 mm; $2(W\text{-}t)$ = 20 mm; ρ=0.5 mm), (b) Blunt-notch specimens (t = 7.5 mm; $2(W\text{-}t)$ = 15 mm; ρ = 8 mm).

The net stress σ defined by equation (1) was used for the evaluation of fatigue strength.

$$\sigma = P/(2(W-t)b) \tag{1}$$

where, P is applied load and $2(W\text{-}t)b$ is area of net section.

The tendency of fatigue limit was evaluated by the two methods and whether the fatigue limit can be approximated by a simple method in the present material was discussed. One of them is the normal method of using several specimens while applying constant stress amplitude. The fatigue limit σ_w is determined by the maximum stress where the specimen endures the 10^7 cycles of stress repetition. Another method is called the load increase test (LIT). In this method, the approximated fatigue limit σ_w' is obtained by using only one specimen. As shown in figure 3, the stress amplitude was increased by 2 MPa after every application of 2x10^4 cycles. We can approximate the fatigue limit by the stress amplitude of one stage before the specimen breaks within 10^4 cycles of stress application. According to the reference [2], a notched specimen of C/C composites did not break under critical stress that is the maximum value of endurance stress of 10^4 times of stress repetition when the fatigue test was performed under constant stress amplitude. That method was applied in the present study.

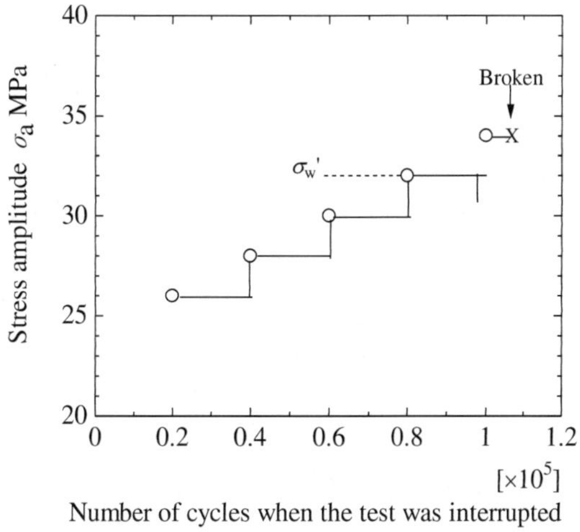

Figure 3. An example of measurement of approximated fatigue limit by the load increase test (LIT).

The carbon fibers are stitched so that they may intersect perpendicularly in the present material. In this study, the axial direction of the specimen coincides with one of the fiber directions, so the angles between the specimen axis and fiber direction β were 0°/90°. The fatigue tests were performed with push pull by using a hydraulic testing machine with a capacity of 9.8 kN. Stress ratio (= minimum stress amplitude σ_{min} / maximum stress amplitude σ_{max}) was -1, and frequency was 10 Hz. The crack initiation and growth were observed by a microscope with 100 times of magnification.

3. Experimental results and discussion

3.1. Measurement results of fatigue limit σ_w and approximated fatigue limit σ_w'

Usually, the fatigue limit of a metal was determined by using plural specimens. However, in the case of carbon composites, the fatigue limit was approximated with LIT by using single specimen [2]. We measured the fatigue limit of notched specimens and the tendency of fatigue behavior was discussed.

The approximated fatigue limit σ_w' was compared with the fatigue limit σ_w measured by *S-N* curve which is the relationship between the stress amplitude σ_a and the number of stress cycles to failure N_f. Then, the results were discussed from the viewpoint of the relationship between the fatigue limit and fracture behavior.

Figure 4 shows the *S-N* curves of the notched specimen. The results for the slit specimens and blunt notched specimen are shown. The arrow in figure 4 means that the specimen did not break by 10^7 times of stress repetition. The stress at that point was regarded as the fatigue limit σ_w. In the case of a slit specimen, the fatigue limit σ_w became lower when slit length was longer, because the stress concentration at the slit edge affected the fatigue limit. This is the same tendency as in the case of a metal [7]. In the case of ρ = t = 0.5 mm, the specimen was broken at the section without notch or slit. In the other cases, the broken area of specimen developed from notched or slit sections. Therefore, the fatigue limit in the case of ρ = t = 0.5 mm is almost the same as that of a smooth specimen because of small effect of stress concentration on fatigue behavior. Also, it is shown in figure 4 that fatigue limit of slit specimen of t = 1.0 mm was almost the same as that of smooth specimen.

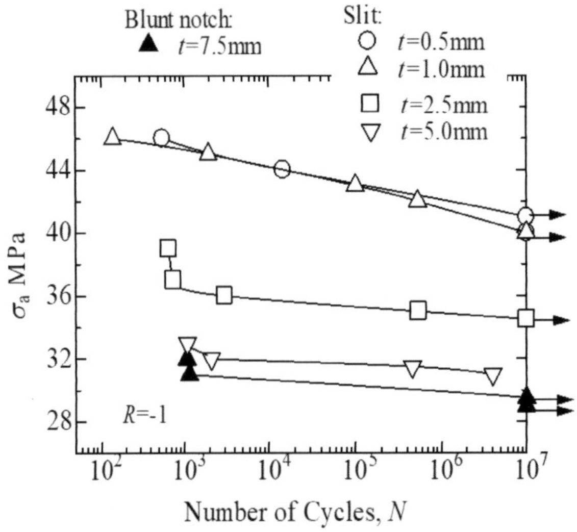

Figure 4. *S-N* curves of the present study.

Besides the cases where the notch length t was shorter than 1.0 mm, the fatigue limit σ_w was not so much different from the stress σ_4, that is, the endurance stress of 10^4 cycles of stress application. Therefore, it is expected that the fatigue limit of the notched specimen in the present study can roughly be approximated by the stress σ_4. This is related to the possibility of the approximation of σ_w by the approximated fatigue limit σ_w'. The fatigue limit σ_w is compared with the approximated fatigue limit σ_w' in the cases of the blunt notched specimen and the slit specimen of ρ =0.5 mm, t = 5.0 mm. Within the present experiment, the fatigue limits of those specimens are almost the same as shown in table 1.

Table 1. Comparison of fatigue strength measured from *S-N* curve and approximated method.

Notch (mm)	σ_w (MPa)	σ_w' (MPa)	σ_w / σ_w'
Slit ρ=0.5, t=5.0	31*	32	0.97
Blunt notch ρ=8.0, t=7.5	29	32	0.91

* Estimated value from figure 4

3.2. Relationship between fatigue limit and stress distribution

In the case of the slit specimen, the fatigue limit decreased when the slit length became longer. However, the relationship between notch radius and fatigue limit cannot be explained by the variation of maximum stress at the notch bottom.

Figure 5 shows the stress distribution calculated from the notch bottom along the axis perpendicular to the specimen axis. Two dimensional calculation was performed by the body force method [8]. We can get solution by applying boundary conditions and load to the program software of that. The symbol σ_0 shows the applied net stress and σ_y the distributed stress in the loading direction. The fatigue limit of the slit specimen with ρ = 0.5 mm, t = 5.0 mm was almost the same as that of a blunt notched specimen with ρ = 8.0 mm, t = 7.5 mm. The stress distribution of those two cases are compared in figure 5. Line 1 in figure 5 shows the distance x (= 0.5 mm) where the same value of stress σ_y is applied to cases of slit specimens with ρ = 0.5 mm, t = 5.0 mm and notched specimen with ρ = 8.0 mm, t = 7.5 mm. Within a shorter distance than that (x = 0.5 mm), the stress σ_y is higher in the case of slit specimens with ρ = 0.5 mm, t = 5.0 mm. Inversely, for a longer distance than that (x=0.5 mm), the stress σ_y is higher in the case of a notched specimen with ρ = 8.0 mm, t = 7.5 mm. From those stress distributions, following situation can be expected. Even if the initial damage or cracking area in front of the notch in the case of ρ = 0.5 mm, t = 5.0 mm was created earlier, the damage expanded more slowly than in the case of ρ = 8.0 mm, t = 7.5 mm. Line 2 shows the other case where smaller initial damage was created than in the case of ρ = 0.5 mm, t = 5.0 mm. In this case, the fatigue limit becomes higher than in the case of ρ = 0.5 mm, t = 5.0 mm. Therefore, the fatigue limit is related not only to maximum stress at notch bottom but also stress distribution. Now, line 2 in figure 5 shows the distance x where the same value of stress σ_y is applied to cases of slit specimens with ρ = 0.5 mm, t = 2.5 mm and blunt notched specimens with ρ = 8.0 mm, t = 7.5 mm. From those results and the results of figure 4 (*S-N* curves), it is clear that the maximum stress is not the only parameter for controlling the fatigue life.

The fatigue limit can be approximated by the load increase test. Damage to matrix and crack growth behavior can be related to that reason. So, we investigated the relationship between the damage to the matrix and crack growth behavior in the next section.

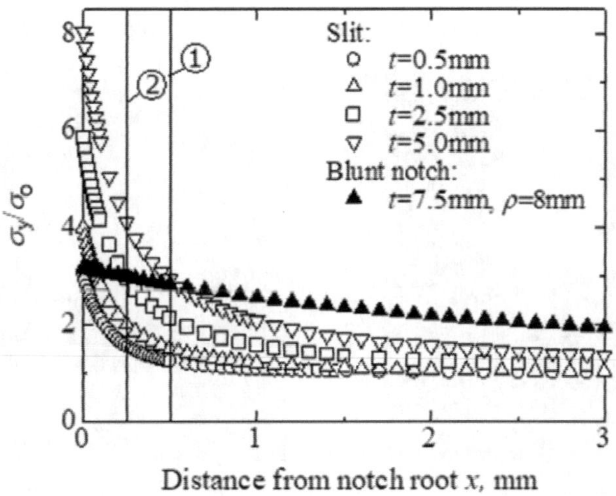

Figure 5. The distribution of non-dimensional stress along the vertical of loading direction from the notch bottom.

3.3. Failure mechanism of present notched specimen

Figure 6 shows the example of the growth of cracks in two cases where the fatigue life of the specimens was almost the same, that is shown in figure 4 with solid triangle symbol and open inverse triangle symbol. The shear deformation of slit specimens with ρ = 0.5 mm, t = 5.0 mm was concentrated in a narrow band which developed from the notch root. Initiation of damage was observed in the vicinity of the slit bottom, and the plural cracks were initiated in the narrow band as shown in figure 6(a). In the case of the blunt notched specimen with ρ = 8.0 mm and t = 7.5 mm, the damaged area spread in front of the notch. However a small number of main cracks grew from the notch bottom as shown in figure 6(c). Consequently, the damaged area was related to the crack expansion, and shear deformation of the matrix was limited to a narrow band in the cases of the slit specimen and blunt notched specimen.

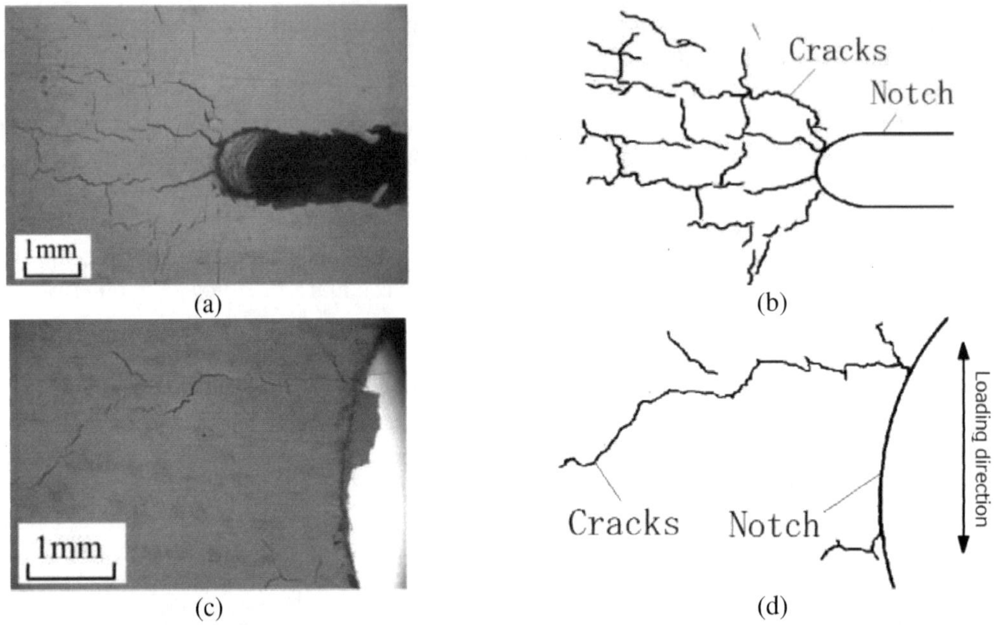

Figure 6. Observation of cracks in the vicinity of the crack tips, (a) Slit (σ_a=31.8 MPa), (b) Schematic representation of (a), (c) Blunt notch (σ_a=32 MPa), (d) Schematic representation of (c).

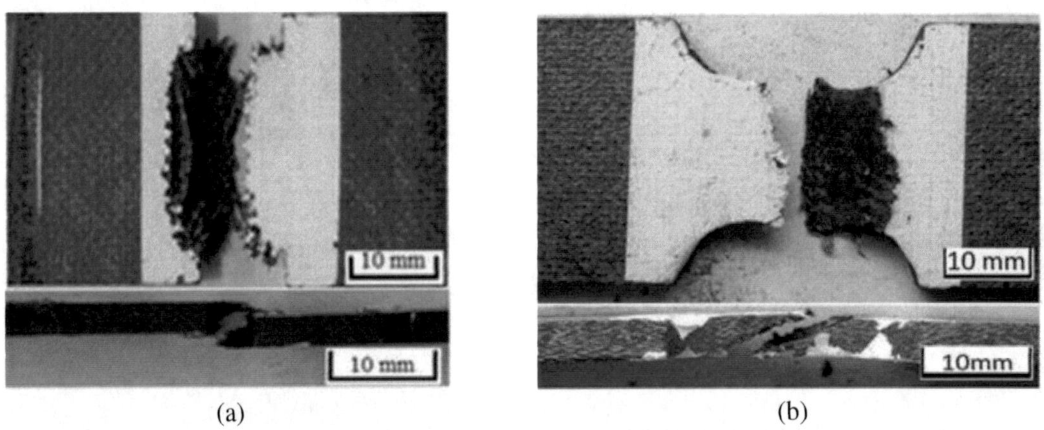

Figure 7. Observation of the broken specimen (a) Slit specimen, (b) Blunt notched specimen, A thin film was stuck on the surface of test-section to ease the observation of crack.

Figure 7 shows the features of broken specimens. Figure 7(a) is the case of slit specimen with ρ = 0.5 mm and t = 5.0 mm (σ_a= 31.8 MPa, N_f = 2x10^3) and figure 7(b) blunt notched specimen with ρ = 8.0 mm and t = 7.5 mm (σ_a = 32 MPa, N_f = 1x10^3). The cracks in these specimens were observed within a limited area as shown in figure 6. In the final stage of fracture, the crack grew with the shear deformation along the 45 degree inclined plane in the thickness direction. Figure 8 represents the schematic representation of failure pattern in a closed part of the center axis of the specimen. The fracture process was not observed during the fatigue tests, but followings are expected from the initiation behavior of cracks (figure 6) and situation of the inclined surface of the fracture part (figure 7). During the fatigue process, the shear damage accumulated in a limited area of the matrix. Then slipping of the matrix cut fibers, and the main cracks grew by the shear mode. Finally, the specimen was broken by the shear mode in the present notched specimen.

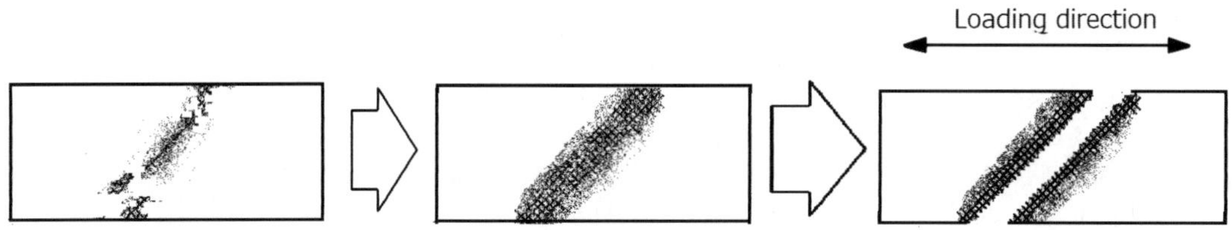

Figure 8. Schematic representation of the damage expansion and fracture behavior at the center section of the specimen.

In the case of a slit specimen, the fatigue life became longer when the slit length was shorter. Within the case of slit specimens, the stress concentration is strongly related to the stress concentration at slit bottom. However, the tendency of fatigue limit cannot be explained in the total cases of the present specimens. The fatigue limit can be related to the crack initiation patterns and crack growth mode. As shown in figures 6 and 7, the density of the crack initiation was different depending on the notch root radius. In the stage of the crack growth, the matrix deformation by shear mode in the thickness direction was expected to occurred, and the fibers were cut by the shear deformation of the matrix. That mechanism can be related to the fatigue limits of the present specimens. Thus, critical stress and deformation are expected to be related to unstable crack growth.

In the present case, the fatigue life of notched specimens or slit specimens can be approximated by the load increase test (LIT). It is expected that this reason is related to the fracture behavior of the material. The crack initiation and growth are strongly related the shear deformation of the matrix. When the crack began to grow unstably, the damage to the matrix led to the sliding direction of cracked faces in the thickness direction. It is also expected that there was a critical stress level for starting unstable shear deformation in each notch shape. The fatigue limit can be related to such behavior of cracks.

4. Conclusions

The investigation results on fatigue limit of notched specimens of the C/C composites was reviewed. Two types of specimens were used: one of them had a narrow notch which was called the slit specimen. The other one had a blunt notch with a notch radius of 8 mm. The fatigue limit can be approximated by the load increased test by using a single specimen. It is expected that this reason is related to the tendency of the fracture behavior of the material. In the case of the slit specimen, the higher fatigue limit was obtained where the notch length or slit length was shorter. However, the fatigue limit was not evaluated by only the maximum stress value at notch root, because the fatigue life of the specimen with a blunt notch was lower than that of the slit specimen in which maximum stress at notch root was low. The local shear deformation was related to the crack initiation and growth. It is expected that the fatigue limit is related to the critical level of local shear stress and shear deformation.

Acknowledgments

This work was performed as a part of a Project of Okinawa innovation system building business of science and technology, which was supported by the Okinawa Science and Technology Promotion Center.

References

[1] Hull D and Clyne T W 2003 An introduction to composite materials *Cambridge Solid State Science Series* (UK: Cambridge University Press)
[2] Goto K, Hatta H, Katsu D and Machida T 2003 Tensile fatigue of a laminated carbon- carbon composite at room temperature *Carbon* **41** 1249-55

[3] Xu G, Li H, Bai R, Wei R and Zhai Y 2008 Influence of the matrix texture on the fracture behavior of 2D carbon/carbon *Compost. Mat. Sci. Eng. A* **478** 319-23

[4] Mahfuz H, Maniruzzaman M, Krishnagopalan J, Haque A, Ismail M and Jeelani S 1995 Effects of stress ratio on fatigue life of carbon-carbon composites *Theor. Appl. Fract. Mec.* **24** 23-31

[5] Murdani A, Makabe C and Fujikawa M 2009 Fatigue and fracture behavior in notched specimens of C/C composite with fine-woven carbon fiber laminates *Carbon* **47** 3355-64

[6] Ferdous M, Sofyan A, Makabe C and Fujikawa M 2013 Fatigue behavior and relationship between crack propagation and the slit configuration of C/C composites *Mod. Phys. Lett. B* **27** 1-11

[7] Murakami Y and Usuki H 1989 Quantitative evaluation of effects of non-metallic inclusions on fatigue strength of high strength steel *Int. J. Fatigue* **11** 299-307

[8] Nisitani H and Saimoto A 2003 Effectiveness of two-dimensional versatile program based on body force method and its applications to crack problems *Key Eng. Mat.* **251-252** 97-102

IOP Conf. Series: Materials Science and Engineering **372** (2018) 012002 doi:10.1088/1757-899X/372/1/012002

Practical method for analyzing singular index and intensity of singular stress field for three dimensional bonded plate

Tatsujiro Miyazaki[1,4], Takuma Inoue[2] and Nao-Aki Noda[3]

[1]University of the Ryukyus, 1 Senbaru, Nishihara-cho, Nakagami-gun, Okinawa 903-0213, Japan

[2]Graduate School of Engineering and Science, University of the Ryukyus, 1 Senbaru, Nishihara-cho, Nakagami-gun, Okinawa 903-0213, Japan

[3]Kyushu Institute of Technology, 1-1 Sensui-cho, Tobata-ku, Kitakyushu-shi, Fukuoka 804-8550, Japan

E-mail: t-miya@tec.u-ryukyu.ac.jp

Abstract. In this paper, the practical analysis methods are proposed for analyzing the singular index and the intensity of singular stress field (ISSF) at the vertex of the interface in the three dimensional (3D) bonded plate. The analysis methods focus FEM stresses at and around the vertex. The singular index is determined from the FEM stress ratio at the vertex obtained by performing FEM analyses on the finely and coarsely meshed models. Then, the ISSF is determined by the ratio of the average FEM stresses at and around the vertex obtained by performing the FEM analyses on the reference and unknown models under the same mesh pattern. The validity of the present methods was examined by the plane strain bonded plate and 3D bonded plate in the literature. It was found that the present methods for the singular index have the same accuracy as the FEM eigenvalue analysis. The asymptotic solutions with the singular index and ISSF by the present method correspond to FEM stress distributions. Since the ISSF by the body force method (BFM) is used as the reference solution, the present method for ISSF has the same accuracy as BFM. Moreover, the critical ISSF values were calculated in the experimental results of the butt joints with various adhesive thicknesses. In the case of the ductile epoxy adhesive, it was shown that the critical ISSF at the vertex by 3D model was more constant against the thickness than that by 2D model. The result is quite different from that of the brittle epoxy adhesive and can be never obtained by 2D model.

1. Introduction

The intensity of singular stress field (ISSF) is useful for evaluating the debonding strength [1–4]. Generally, the finite element method (FEM) is often used to evaluate the strength in various industries [5, 6]. However, the ISSF cannot be calculated directly by FEM [7–10]. The authors proposed the method for calculating the ISSF easily and accurately by the FEM [3, 4, 11, 12]. The method does not require the complex calculation and can be applied to various bonded structures. In the previous studies, the ISSF for the butt joint was analyzed by 2D model [3, 4]. It was found that the debonding fracture criterion can be expressed with the ISSF [11, 13]. Actual butt joint has the vertex and its debonding fracture is caused from the vertex of the interface. However, it is difficult to analyze the ISSF at the vertex and the debonding fracture

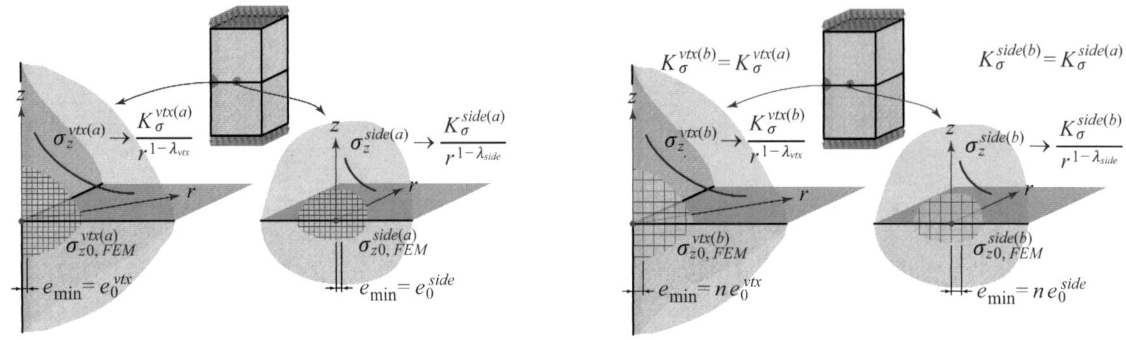

(a) Fine mesh model with minimum element size $e_{\min} = e_0^{vtx}$

(b) Coarse mesh model with $e_{\min} = ne_0^{vtx}$

Figure 1. Schematic illustration of 3D bonded plate models for analyzing the singular index

criterion is not discussed by the ISSF at the vertex in 3D model.

In this paper, the practical analysis methods are proposed for analyzing the singular index and the ISSF at the vertex of the interface in 3D bonded plate. The critical ISSF values are calculated in the experimental results of the butt joints with various adhesive thicknesses and are compared with those by 2D model.

2. Analysis method for singular index and intensity of singular stress field

2.1. Analysis method for singular index

Figure 1 shows the schematic illustrations of the 3D bonded plate models. The model (a) is subdivided by the minimum element size $e_{\min} = e_0^{vtx}$. The FEM stress in the z direction and the ISSF at the vertex of the interface are denoted with $\sigma_{z0,FEM}^{vtx(a)}\big|_{e_{\min}=e_0^{vtx}}$ and $K_\sigma^{vtx(a)}$, respectively. The model (b) is as large as the model (a) and subdivided by $e_{\min} = ne_0^{vtx}$. The FEM stress in the z direction and the ISSF at the vertex of the interface are denoted with $\sigma_{z0,FEM}^{vtx(b)}\big|_{e_{\min}=n\,e_0^{vtx}}$ and $K_\sigma^{vtx(b)}$, respectively.

The real singular stress in the z direction at the vertex of the interface in 3D bonded plate is given by the following equation [9].

$$\sigma_z^{vtx} = \frac{K_\sigma^{vtx}}{r^{1-\lambda_{vtx}}} \tag{1}$$

Here, r is the distance from the vertex, λ is the singular index at the vertex, K_σ is the ISSF at the vertex. Since the minimum element size of the model (b) is n times as large as that of the model (a), the relation between the FEM stresses $\sigma_{z0,FEM}^{vtx(a)}\big|_{e_{\min}=e_0^{vtx}}$ and $\sigma_{z0,FEM}^{vtx(b)}\big|_{e_{\min}=n\,e_0^{vtx}}$ can be expressed as follows [11].

$$\sigma_{z0,FEM}^{vtx(a)}\big|_{e_{\min}=e_0^{vtx}} \cdot e_0^{1-\lambda_{vtx}} = \sigma_{z0,FEM}^{vtx(b)}\big|_{e_{\min}=ne_0^{vtx}} \cdot (ne_0^{vtx})^{1-\lambda_{vtx}} = \text{constant} \tag{2}$$

Solving Eq. (2) on λ_{vtx}, the following equation is obtained.

$$\lambda_{vtx} = 1 - \frac{\ln\left(\sigma_{z0,FEM}^{vtx(a)}\big|_{e_{\min}=e_0^{vtx}} \Big/ \sigma_{z0,FEM}^{vtx(b)}\big|_{e_{\min}=ne_0^{vtx}}\right)}{\ln n} \tag{3}$$

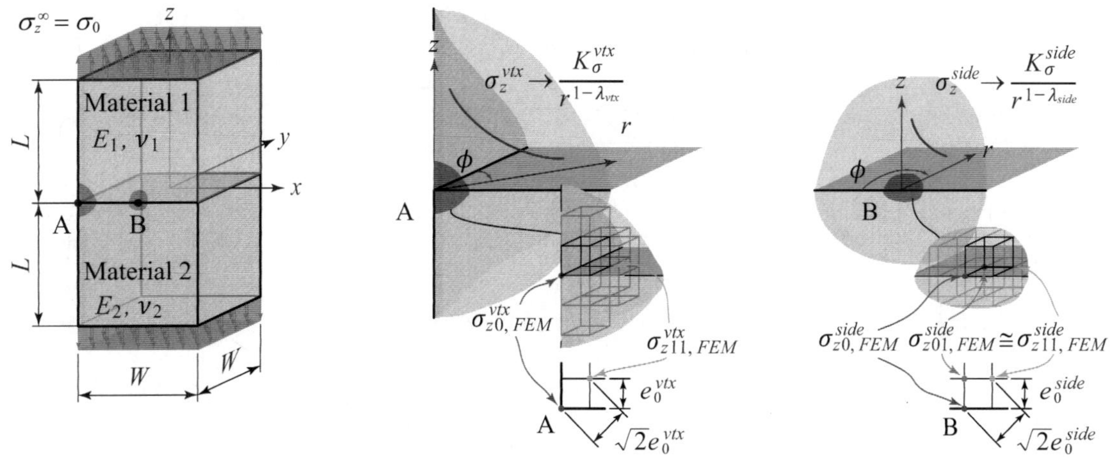

(a) 3D bonded plate model.

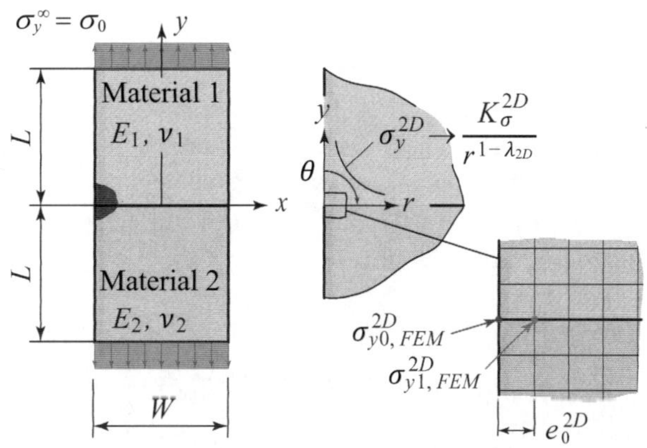

(b) 2D bonded plate model.

Figure 2. Schematic illustration of 2D and 3D bonded plate models.

Generally, the finite element eigen analysis method is often used for the analysis of λ_{vtx}. λ_{vtx} can be determined by Eq. (3) much more easily than the finite element eigenanalysis method.

When the FEM stresses $\sigma_{z0,FEM}^{side(a)}\big|_{e_{\min}=e_0^{side}}$ and $\sigma_{z0,FEM}^{side(b)}\big|_{e_{\min}=ne_0^{side}}$ are used instead of the $\sigma_{z0,FEM}^{vtx(a)}\big|_{e_{\min}=e_0^{vtx}}$ and $\sigma_{z0,FEM}^{vtx(b)}\big|_{e_{\min}=ne_0^{vtx}}$, respectively, the λ_{side} is determined by the following equation.

$$\lambda_{side} = 1 - \frac{\ln\left(\sigma_{z0,FEM}^{side(a)}\big|_{e_{\min}=e_0^{side}} \Big/ \sigma_{z0,FEM}^{side(b)}\big|_{e_{\min}=ne_0^{side}}\right)}{\ln n} \tag{4}$$

Here, $\sigma_{z0,FEM}^{side(a)}\big|_{e_{\min}=e_0^{side}}$ and $\sigma_{z0,FEM}^{side(b)}\big|_{e_{\min}=ne_0^{side}}$ are the FEM stresses at the center $(0, -W/2, 0)$ on the interface end $y = -W/2$ in the finely and coarsely meshed 3D bonded plate models, respectively.

IOP Conf. Series: Materials Science and Engineering **372** (2018) 012002 doi:10.1088/1757-899X/372/1/012002

2.2. Mesh-independent technique useful for 3D bonded plate

The authors proposed the mesh-independent technique useful for 2D butt joint [3, 4]. The method is extended so that the ISSF at the vertex in 3D bonded plate can be analyzed. Figure 2 shows (a) unknown 3D bonded plate model and (b) reference 2D bonded plate model. The method for analyzing the ISSF K_σ^{vtx} at $(x, y, z) = (-W/2, -W/2, 0)$ in Fig. 2(b) by using the ISSF K_σ^{2D} at $(x, y) = (-W/2, 0)$ as the reference solution is mentioned below.

As shown in Fig. 2(a), the local polar coordinate (r, ϕ) is set at the vertex $(x, y, z) = (-W/2, -W/2, 0)$in the 3D model. ϕ is the angle between the interface end $y = -W/2$ and r axis. When $\phi = \pi/4$, the singular stress field at $(-W/2, -W/2, 0)$ is expressed with the following equation [9].

$$\sigma_z^{vtx}(r, \phi = \pi/4) = \frac{K_\sigma^{vtx}|_{\phi=\pi/4}}{r^{1-\lambda_{vtx}}} \tag{5}$$

As shown in Fig. 2(b), the local polar coordinate (r, θ) is set at the interface end $(x, y) = (-W/2, 0)$ in the 2D model. θ is the angle between the interface end $y = 0$ and r axis. When $\theta = \pi/2$, the singular stress field at $(-W/2, 0)$ is expressed with the following equation [3, 4].

$$\sigma_y^{2D}(r, \theta = \pi/2) = \frac{K_\sigma^{2D}}{r^{1-\lambda_{2D}}} \tag{6}$$

The analysis methods focus average real/FEM stresses at and around the vertex. The FEM analyses are performed on the reference and unknown models under the similar mesh pattern. Since the mesh size dependency in the FEM stress is canceled by using a ratio, the average FEM stress ratio $\bar{\sigma}_{z,FEM}^{vtx}/\bar{\sigma}_{y,FEM}^{2D}$ corresponds to the average real stress ratio $\bar{\sigma}_z^{vtx}/\bar{\sigma}_y^{2D}$. When the stress in 3D model is averaged from $r = 0$ to $\sqrt{2}e_0^{vtx}$ and the stress in 2D model is averaged from $r = 0$ to e_0^{2D}, the following equation is obtained.

$$\frac{K_\sigma^{vtx}|_{\phi=\pi/4}}{K_\sigma^{2D}} = \frac{\lambda_{vtx}}{\lambda_{2D}} \cdot \frac{\sigma_{z0,FEM}^{vtx} \cdot (\sqrt{2}e_0^{vtx})^{1-\lambda_{vtx}} + \sigma_{z11,FEM}^{vtx} \cdot (\sqrt{2}e_0^{vtx})^{1-\lambda_{vtx}}}{\sigma_{y0,FEM}^{2D} \cdot (e_0^{2D})^{1-\lambda_{2D}} + \sigma_{y1,FEM}^{2D} \cdot (e_0^{2D})^{1-\lambda_{2D}}} \tag{7}$$

Here, $\sigma_{z11,FEM}^{vtx}$ is the FEM stress at $(r, \phi) = (\sqrt{2}e_0^{vtx}, \pi/4)$ in Fig. 2(a), $\sigma_{y0,FEM}^{2D}$ and $\sigma_{y1,FEM}^{2D}$ are the FEM stresses at $(r, \theta) = (0, \pi/2)$, $(e_0^{2D}, \pi/2)$ in Fig. 2(b), respectively.

The local polar coordinate (r, ϕ) is set at the center $(0, -W/2, 0)$ in Fig. 2(b). ϕ is the angle between the interface end $y = -W/2$ and r axis. When $\phi = \pi/2$, the singular stress field at the center of interface end is expressed with the following equation [9]

$$\sigma_z^{side}(r, \phi = \pi/2) = \frac{K_\sigma^{side}|_{\phi=\pi/2}}{r^{1-\lambda_{side}}} \tag{8}$$

The following equation is obtained as well as Eq. (7).

$$\frac{K_\sigma^{side}|_{\phi=\pi/2}}{K_\sigma^{2D}} = \frac{\lambda_{side}}{\lambda_{2D}} \cdot \frac{\sigma_{z0,FEM}^{side} \cdot (e_0^{side})^{1-\lambda_{side}} + \sigma_{z01,FEM}^{side} \cdot (e_0^{side})^{1-\lambda_{side}}}{\sigma_{y0,FEM}^{2D} \cdot (e_0^{2D})^{1-\lambda_{2D}} + \sigma_{y1,FEM}^{2D} \cdot (e_0^{2D})^{1-\lambda_{2D}}} \tag{9}$$

Because the ISSF for 2D bonded plate by the body force method (BFM) [8, 10] is used, the ISSF for 3D bonded plate can be obtained by Eqs. (7) and (9) accurately.

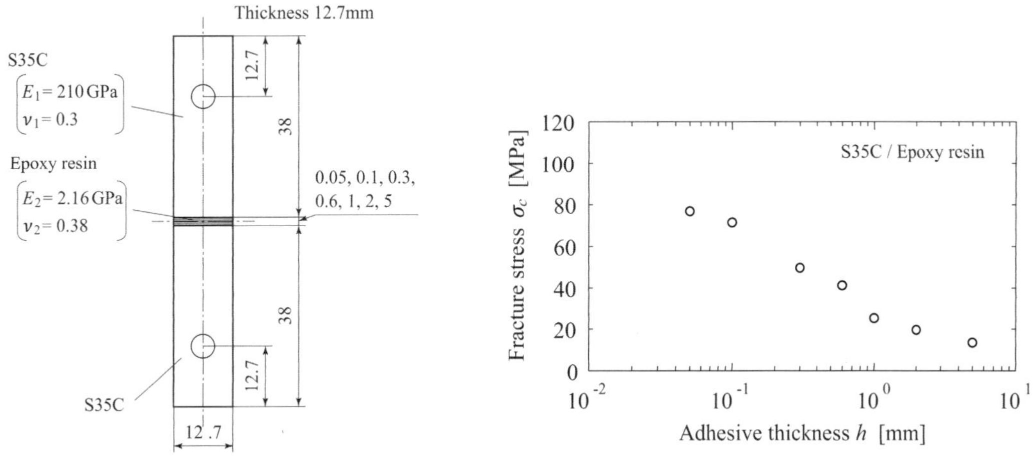

(a) Butt joint used in the tensile experiment (b) Experimental adhesive strength σ_c

Figure 3. Schematic illustration of butt joint specimen and experimental adhesive strength [13, 14]

Table 1. Material properties used in the present analyses [13, 14]

Adherend (S35C)		Adhesive (Epoxy resin)		α	β	λ_{2D}	F_{σ}^{2D*}
E_1 [GPa]	ν_1	E_2 [GPa]	ν_2				
210.0	0.3	3.14	0.37	0.969	0.199	0.684	0.405

3. Debonding strength evaluation for butt joint by 3D model

The debonding strength for butt joint by 3D model is discussed by using the experimental result by Suzuki [14]. Figure 3 shows the schematic illustration of the butt joint specimen and the fracture stress. In the experiment, the adhesive thickness h is varied from 0.05 mm to 5 mm. The adherend is the low cabon steel JIS S35C. The adhesives are the epoxy resin. The fracture stress decreases with increasing h. Table 1 shows the material constants of the adherend and adhesives. Then, λ_{2D} and the dimensionless ISSF F_{σ}^{2D} for 2D bonded plate, which are used in the evaluation of the ISSF as the reference solution, are also shown in Table 1.

Analysis results in the case of $h = 5$ mm are mentioned as example below. Table 2 shows λ_{vtx} by Eq. (3) and λ_{side} by Eq. (4). λ_{side} nearly equals λ_{2D}. Table 3 shows the FEM stresses of unknown 3D butt joint model and reference 2D bonded plate model. All FEM stresses depend on the element size. However, it is found that the mesh size dependency can be overcome by multiplying the FEM stress by $(e_{\min})^{1-\lambda}$. Therefore, the left-hand sides of Eqs. (7) and (9) have mesh size independency. Table 4 shows the ISSFs at the vertex by Eq. (7) and at the center by Eq. (9). It is found that K_{σ}^{vtx} and K_{σ}^{side} are independent of the element size. Figure 4 shows the singular stress distributions at the vertex and the center in the 3D butt joint model. The solid lines are the asymptotic solutions which are obtained by substituting the singular indexes in Table 2 and the ISSFs in Table 4 into Eqs. (5) and (8). The lines are good agreement with the FEM stress distributions. The differences between the asymptotic solution and FEM stress are less than about 1% in $4.4 \times 10^{-6} < r/(W/2) < 7.89 \times 10^{-4}$ at the vertex and the center except for the nodes near the singular point which is strongly influenced by the singularity. It can be confirmed that the singular indexes and the ISSFs are determined accurately by the present method.

Figure 5 shows the critical ISSFs at the vertex and the center by 3D model and the critical

Table 2. Singular indexes λ_{vtx} by Eq. (3) and λ_{side} by Eq. (4)

(a) λ_{vtx}

Fine mesh model		Coarse mesh model		λ_{vtx}
e_0^{vtx}	$\sigma_{z0,FEM}^{vtx}$	e_0^{vtx}	$\sigma_{z0,FEM}^{vtx}$	
4.360×10^{-6}	89.49	1.744×10^{-5}	51.12	0.5961

(b) λ_{side}

Fine mesh model		Coarse mesh model		λ_{side}	λ_{2D}
e_0^{side}	$\sigma_{z0,FEM}^{side}$	e_0^{side}	$\sigma_{z0,side}^{side}$		
4.360×10^{-6}	36.02	1.744×10^{-5}	22.93	0.6741	0.6735

Table 3. Mesh-dependent FEM stresses of 3D butt joint model with $h = 5\,\text{mm}$ in Fig. 3 and reference 2D model under plane strain in Fig. 2(b)

(a) FEM stress at the vertex of unknown 3D butt joint model in Fig. 3

e_0^{vtx}	$\sigma_{z0,FEM}^{vtx}$	$\sigma_{z0,FEM}^{vtx} \cdot (e_0^{vtx})^{1-\lambda_{vtx}}$	$\sigma_{z11,FEM}^{vtx}$	$\sigma_{z11,FEM}^{vtx} \cdot (e_0^{vtx})^{1-\lambda_{vtx}}$
4.360×10^{-6}	89.49	0.6118	64.92	0.4438
1.744×10^{-5}	51.12	0.6118	37.09	0.4438

(b) FEM stress at the center of unknown 3D butt joint model in Fig. 3

e_0^{side}	$\sigma_{z0,FEM}^{side}$	$\sigma_{z0,FEM}^{side} \cdot (e_0^{side})^{1-\lambda_{side}}$	$\sigma_{z01,FEM}^{side}$	$\sigma_{z01,FEM}^{side} \cdot (e_0^{side})^{1-\lambda_{side}}$
4.360×10^{-6}	36.02	0.6453	27.87	0.4993
1.744×10^{-5}	22.93	0.6453	17.74	0.4993

(c) FEM stress at the interface end of reference 2D bonded plate model in Fig. 2(b)

e_0^{2D*}	$\sigma_{y0,FEM}^{2D*}$	$\sigma_{y0,FEM}^{2D*} \cdot (e_0^{2D*})^{1-\lambda_{2D}}$	$\sigma_{y1,FEM}^{2D*}$	$\sigma_{y1,FEM}^{2D} \cdot (e_0^{2D})^{1-\lambda_{2D}}$
3^{-12}	57.88	0.4399	44.79	0.3404
3^{-9}	19.73	0.4399	15.27	0.3404

Table 4. ISSFs for 3D butt joint with $h = 5\,\text{mm}$

(a) K_σ^{vtx}

e_0^{2D*}	e_0^{vtx}	K_σ^{vtx}
3^{-12}	4.360×10^{-6}	0.392
3^{-12}	1.744×10^{-5}	0.392
3^{-9}	4.360×10^{-6}	0.392
3^{-9}	1.744×10^{-5}	0.392

(b) K_σ^{side}

e_0^{2D*}	e_0^{side}	K_σ^{side}
3^{-12}	4.360×10^{-6}	0.418
3^{-12}	1.744×10^{-5}	0.418
3^{-9}	4.360×10^{-6}	0.418
3^{-9}	1.744×10^{-5}	0.418

ISSFs by 2D model [13]. $K_{\sigma c}^{side}$ almost equals $K_{\sigma c}^{2D}$. The $K_{\sigma c}^{vtx}$ is more constant against the thickness than that by 2D model and has the smallest scatter. The result can be never obtained

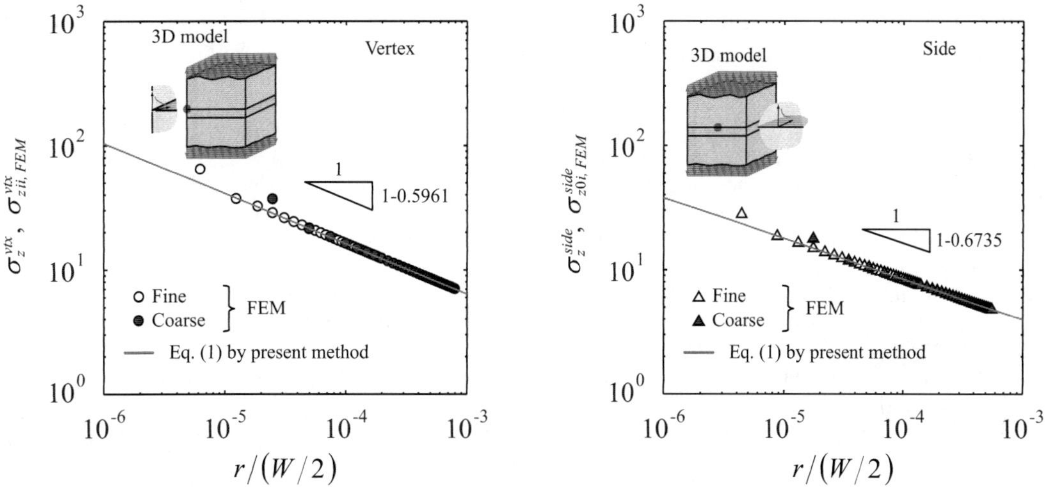

(a) Vertex (b) Center

Figure 4. Singular stress distributions at the vertex and the center of butt joint with $h = 5\,\mathrm{mm}$

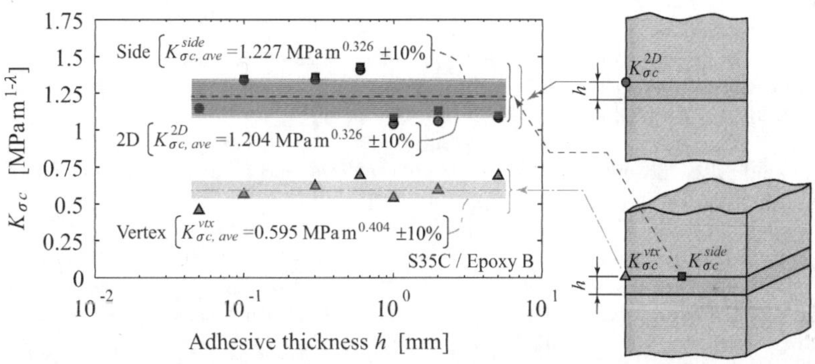

Figure 5. Relation between critical ISSF $K_{\sigma c}$ and adhesive layer thickness h

by 2D model.

4. Conclusion

In this paper, the practical analysis methods were proposed for analyzing the singular index and the intensity of singular stress field (ISSF) at the vertex of the interface in 3D bonded plate. The debonding strength for butt joint by 3D model is discussed by using the experimental result in the earlier study.

(i) The critical ISSFs of the butt joints with various adhesive thicknesses in the earlier study were calculated by 3D and 2D models. The critical ISSFs at the center on the side in 3D model almost equaled those by 2D model. The critical ISSFs at the vertex in 3D model was less scattered than those at the center on the side in 3D model.

(ii) The asymptotic solutions with the singular indexes and the ISSFs obtained by the present method were good agreement with the FEM stress distributions. The differences between the asymptotic solution and FEM stress were less than about 1% in $4.4\times10^{-6} < r/(W/2) <$

7.89×10^{-4} at the vertex and the center except for the nodes near the singular point which is strongly influenced by the singularity. It was confirmed that the singular indexes and the ISSFs are determined accurately by the present method.

References

[1] Qian, Z. and Akisanya, A. R., An experimental investigation of failure initiation in bonded joints, Acta Materialia, Vol. 46, No. 14 (1998), pp. 4895 - 4904.

[2] Mintzas, A. and Nowell, D., Validation of an H_{cr}-based fracture initiation criterion for adhesively, Engineering Fracture Mechanics, Vol. 80 (2012), pp. 13 - 27.

[3] Zhang, Y., Noda, N. -A., Takaishi, K. -T., Lan, X., Effect of adhesive thickness on the intensity of singular stress at the adhesive dissimilar joint, Journal of Solid Mechanics and Materials Engineering, Vol. 4, No. 10 (2010), pp. 1467 - 1479.

[4] Zhang, Y., Noda, N. -A., Wu, P. and Duan, M., A mesh-independent technique to evaluate stress singularities in adhesive joints, International Journal of Adhesion and adhesives, Vol. 57 (2015), pp. 105 - 117.

[5] Noda, N. -A., Suryadi, D., Kumasaki, S., Sano, Y., Takase, Y, Failure analysis for coming out of shaft from shrink-fitted ceramic sleeve, Engineering Failure Analysis, Vol. 57 (2015), pp. 219 - 235.

[6] Noda, N. -A., Chen, X, Sano, Y., Wahab, M. A., Maruyama, H., Fujisawa, R., Effect of pitch difference between the bolt-nut connections upon the anti-loosening performance and fatigue life, Materials & Design, Vol. 96 (2016), pp. 476 - 489.

[7] Carpenter, W. C. and Byers, C., A path independent integral for computing stress intensities for V-notched cracks in a bi-material, International Journal of Fracture, Vol. 35 (1978), pp. 245 - 268.

[8] Chen, D. H. and Nisitani, H., Intensity of singular stress field near the interface edge point of a bonded strip, Transactions of the Japan Society of Mechanical Engineers, Series A, Vol. 59,No. 567 (1993),pp. 2682 - 2686 (in Japanese).

[9] Koguchi, H., Distribution of displacements and stresses near stress singularity field at a vertex in three-dimensional bonded structures, Transactions of the Japan Society of Mechanical Engineers, Series A, Vol. 66, No. 648 (2000), pp. 1597 - 1605 (in Japanese).

[10] Noda, N. -A., Shirao, R. and Li, J., Sugimoto J. -S. Intensity of singular stress fields causing interfacial debonding at the end of a fiber under pullout force and transverse tension, International Journal of Solids and Structures, Vol. 44 (2007), pp. 4472 - 4491.

[11] Miyazaki, T., Noda, N. -A., Evaluation of debonding strength of single lap joint by the intensity of singular stress field, Journal of Physics: Conference Series, Vol. 842 (2017), pp. 012078.

[12] Miyazaki, T. , Noda, N. -A., Ren, F., Wang, Z., Sano, Y., Iida, K., Analysis of intensity of singular stress field for bonded cylinder and bonded pipe in comparison with bonded plate, International Journal of Adhesion and Adhesives, Vol. 77, (2017), pp. 118 - 137

[13] Noda, N. -A., Miyazaki, T., Li, R., Uchikoba, T. and Sano, Y., Debonding strength evaluation in terms of the intensity of singular stress at the interface coner with and without fictitious crack, International Journal of Adhesion and adhesives, Vol. 61 (2015), pp. 46 - 64.

[14] Suzuki, Y., Adhesive tensile strengths of scarf and butt joints of steel plates (Relation between adhesive layer thicknesses and adhesive strengths of joints), JSME International Journal, Vol. 30, No. 265 (1987), pp. 1042 - 1051.

IOP Conf. Series: Materials Science and Engineering **372** (2018) 012003 doi:10.1088/1757-899X/372/1/012003

Fatigue and fracture mechanical behavior for Chinese A508-3 steel at room temperature

K K Shi[1,2], H Xie[1], B Zheng[1] and X L Fu[1]
[1]Science and Technology on Reactor System Design Technology Laboratory, Nuclear Power Institute of China, Chengdu, 610213, China

E-mail: shikai1000@163.com

Abstract. Material, A508-3 steel, has been used in nuclear reactor vessels. In the present study, fatigue and fracture mechanical behavior of Chinese A5083 steel at room temperature are studied by mechanical material testing machine (MTS). Test data of material's mechanical behavior including uniaxial tension, low cycle fatigue (LCF), threshold value of stress intensity factor (SIF) range, fatigue crack growth (FCG), and fracture toughness is generated and given for further study. It is worth noting that the model in predicting FCG of material from LCF parameters is verified and discussed.

1. Introduction

The basic material experimental data is important design input parameters for the structural integrity assessment of engineering components. Reactor pressure vessel (RPV) is an important part of the reactor coolant for the second and third generation nuclear power plants (NPPs). A set of operation, emergency, and accident transients could exist that allow cold water to come into contact with the inner surface of the vessel wall, producing high thermal stress. The low alloy ferritic steel, A508-3 steel, have been used to make RPVs. Because the A508-3 steel has good weld ability and radiation resistance than other low carbon ferritic steel. The A508-3 steel, firstly, is developed from A533-B steel in American for pressure vessel with the higher safety standards of nuclear. The A533-B steel's mechanical properties were widely studied [1-4] and the steel has been used extensively in heavy section pressure vessels for nuclear applications. During the operating lifetime of these vessels, neutron bombardment is sufficient to affect substantial changes in the metallurgical and mechanical properties of the vessel structural materials. Neutron irradiation increase strength properties and decrease ductility.

For the Chinese A508-3 steel, at room temperature, the uniaxial tension, low cycle fatigue (LCF), threshold value of stress intensity factor (SIF), fatigue crack growth (FCG), and fracture toughness is studied, analyzed, and discussed in the present study.

2. Mechanical behavior of ChineseA508-3 steel

In this section, test data comes from the hydraulic mechanical testing machines (MTS).

2.1. Uniaxial tension properties

Figure 1 givens the uniaxial tension engineer stress and strain curves. Before necking stage, the true stress and strain can be calculated by using Bridgman's correction factor [5].

Based on the arc funnel specimen (Radius=10 mm), the full-range stress and strain curve (figure 2) is obtained by finite aided testing method [6].

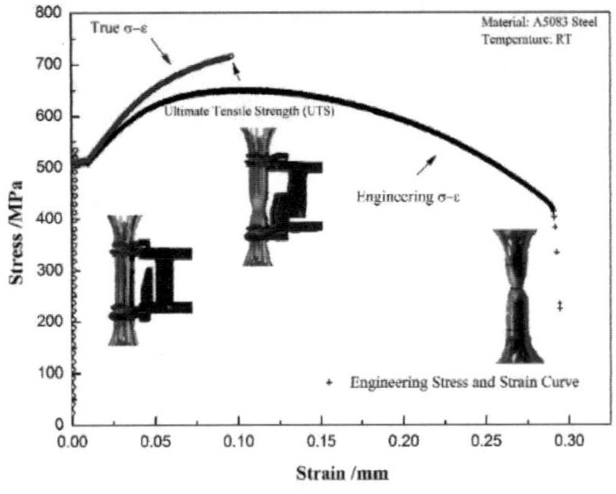

Figure 1. Test specimen of equal diameter round bar, Diameter=8 mm.

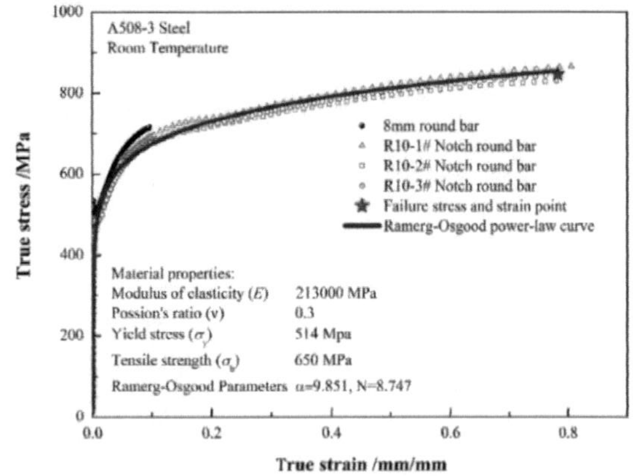

Figure 2. Test specimen of equal diameter round bar (Diameter=8 mm) and arc funnel specimen (Radius=10 mm).

2.2. Low cycle fatigue (LCF) behavior

There exist many methods available to analyze the crack nucleation and failure life. For complex structures and environments, especially, finite element analyses combined with the local strain theory are often used in design phase analysis [7]. Previous studies including Basquin, Manson, and Coffin have obtain the relationship between strain range and cyclic failure life for testing materials. In addition, the average stress σ_m parameter is considered in Manson-Coffin equation which is a classic relation in order to describe the low cycle fatigue behavior of materials under different load environments. Figure 3 shows the typically low cycle fatigue curve of A508-3 steel.

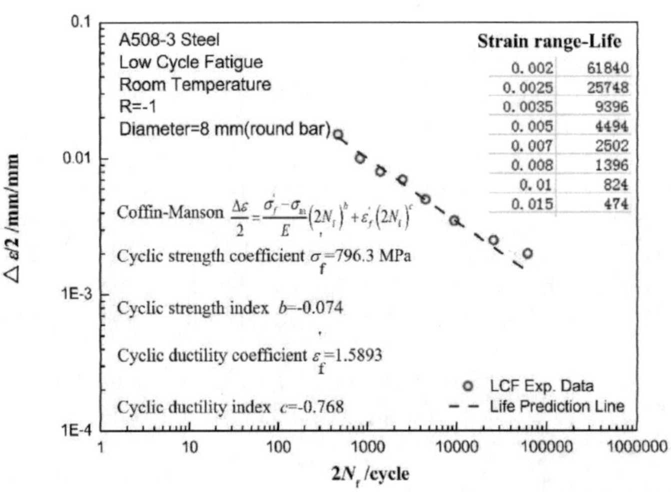

Figure 3. Low cycle fatigue properties of A508-3 steel.

2.3. Threshold value of SIF

Considering the simplicity and accuracy of elastic unloading compliance theory, the amount of crack propagation is detected and calculated using elastic unloading compliance method by computer codes. Based on the definition of threshold value of stress intensity factor for metal materials, test data is obtained and analyzed when no crack propagation is found in 10^7 cyclic life. For a material, commonly, the linear relation between the load stress ratio and the threshold value of SIF does not exist. However, a conservative calculation result would be obtained based on assuming the establishment of a linear dependency for servicing engineering analysis [8]. Figure 4 givens only the threshold value of stress intensity factor (SIF) range under load stress ratio R=0.1.

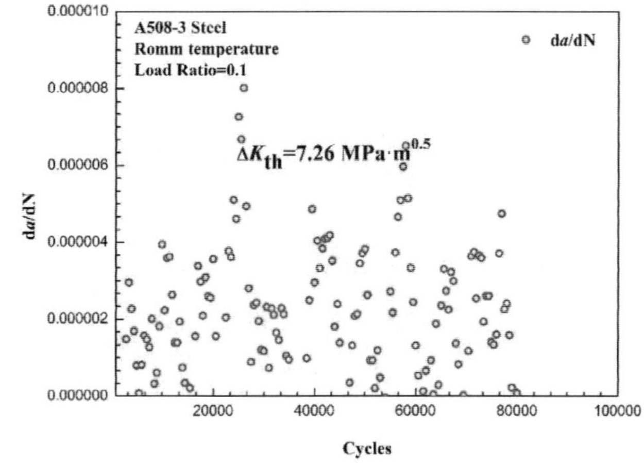

Figure 4. Threshold value of stress intensity factor (SIF) range of A508-3 steel.

Combination the threshold value of SIF and the low cycle fatigue (LCF) properties, material's fatigue crack growth (FCG) of second stage is estimated by kinds of prediction model of mode tension type failure cyclic life.

2.4. Fatigue crack growth (FCG)

Considering previous research results, fatigue crack growth would be studied from the following two scales, macroscopic and microscopic. Reference [9] has also studied the driving force, stress intensity factor K, near crack tip based on the stress criterion. The load stress ratio can affect the result of FCG rate of material. But, in the present study, the load stress ration is not considered and investigated.

Figure 5 shows only the fatigue crack growth (FCG) rate under second stage corresponding to the load stress ratio R=0.1.

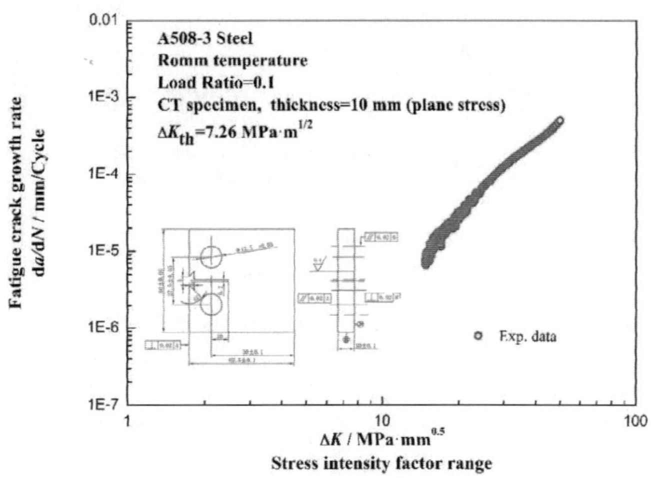

Figure 5. Fatigue crack growth rate of A508-3 steel.

The conception that there is the highly strained zone near the crack tip is like a small low cycle fatigue (LCF) test specimen is assumed by many studies. In Pandey's model, for example, the process zone is viewed as the highly strained zone in the front of crack tip. Considering the energy of process zone equals to the energy of whole plastic zone [10]. Using the essential relationship between low cycle fatigue (LCF) and fatigue crack growth (FCG) rate, various prediction models are applied for analyzing behavior of Chinese A508-3 steel. Many prediction results including SHI-CAI model [11], Glinka model [12], and Ellyin-Li model [13] are showed in figure 6.

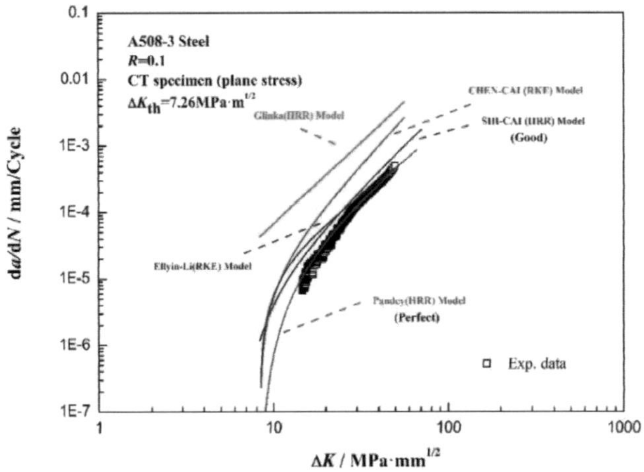

Figure 6. Prediction results based on estimating models of A508-3 steel.

2.5. *Fracture toughness*

The three primary factors which affect the fracture toughness of structural steels are temperature, loading rate, and constraint. Generally, the fracture toughness of structural steels increases with increasing temperature and decreases with increasing loading rate.

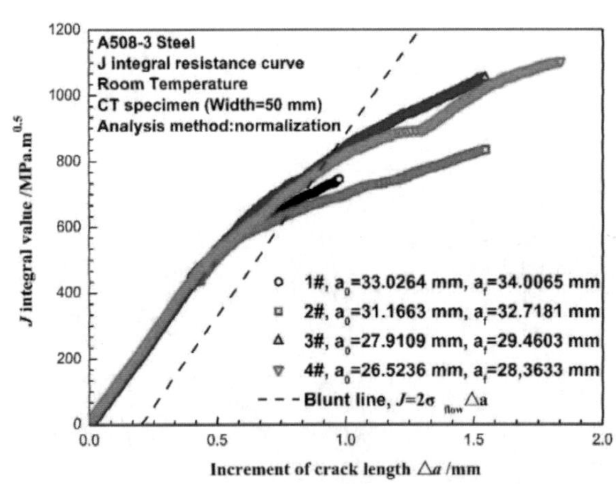

Figure 7. *J* integral resistance curve of A508-3 steel.

For real fracture, the crack failure degree, test specimen thickness, and experimental loading environment have influence on the constraint in front of the crack tip [14]. Figure 7 obtains the result of *J* integral resistance curve with different initial crack length a_0. Results show the different *J* integral resistance curves for compact tensile specimens with 50mm thickness. A so-called constraint effect would be introduced into explaining above phenomenon. The constraint have defined in measuring of the resistance to plastic flow. Basically, a low *J* integral value corresponds to a high level of constraint near the crack tip. The reason is that the plastic deformation and associated energy dissipation are restricted by constraint. Therefore, the resistance to fracture is low. The stress triaxiality can often been used to described the level of constraint in the front of the crack tip. In other words, a high value of stress triaxiality is equivalent to a high level of constraint [15].

3. Conclusion

For Chinese A508-3 steel, the mechanical properties at room temperature are given, which include uniaxial tension, low cycle fatigue (LCF), threshold value of stress intensity factor (SIF), fatigue crack growth (FCG), and fracture toughness. These basic behaviors is analyzed and discussed, in the present study, in order to service the nuclear power plant safety.

The mechanical properties in different temperature environments will be studied and discussed in the future.

Acknowledgments

The authors gratefully acknowledge the support of Lixun CAI professor of applied mechanics and structure safety key laboratory of Sichuan Province, school of mechanics and engineering, southwest JIAOTONG University, Chengdu, China.

References

[1] Yu M, Chao Y J and Luo Z 2015 An assessment of mechanical properties of A508-3 steel used in Chinese nuclear reactor pressure vessels *J Press Vessel-T ASME* **137**(3) 031402
[2] Fang Y 2011 Fracture toughness prediction of domestic A508-III steel based on master curve approach and its application to reactor pressure vessel P-T curve *Master's thesis* (Shanghai: East China University) (in Chinese)
[3] Yu M, Luo Z and Chao Y J 2015 Correlations between Charpy V-notch impact energy and fracture toughness of nuclear reactor pressure vessel (RPV) steels *Eng Fract Mech* **147** 187-202
[4] Richard Wright 2014 Creep of A508/533 pressure vessel steel United States doi: 10.2172/1166040
[5] Bridgman P 1944 The stress distribution at the neck of a tension specimen *J Press Vess-T ASME* **32** 553-74
[6] Yao D, Cai L and Bao C 2016 A new approach on necking constitutive relationships of ductile materials at elevated temperatures *Chinese J Aeronaut* **29**(6) 1626-34
[7] Ricotta M 2015 Simple expressions to estimate the Manson–Coffin curves of ductile cast irons *Int J Fatigue* **78**(0) 38-45
[8] Davenport R and Brook R 1979 The threshold stress intensity range in fatigue *Fatigue Fract Eng M* **1**(2) 151-8
[9] Tang K K, Berto F and Wu H 2016 Fatigue crack growth in the micro to large scale of 7075-T6 Al sheets at different R ratios *Theor Appl Fract Mec* **83** 93-104
[10] Pandey K and Chand S 2003 An energy based fatigue crack growth model *International Journal of Fatigue* **25**(8) 771-8
[11] Shi K K *et al* 2015 Structural fatigue crack growth on a representative volume element under cyclic strain behavior *Int J Fatigue* **74**(0) 1-6
[12] Glinka G 1982 A cumulative model of fatigue crack growth *Int J Fatigue* **4**(2) 59-67
[13] Li D, Nam W and Lee C 1998 An improvement on prediction of fatigue crack growth from low cycle fatigue properties *Eng Fract Mech* **60**(4) 397-406
[14] Petti J P and Dodds R H Jr 2004 Constraint comparisons for common fracture specimens: C(T)s and SE(B)s *Eng Fract Mech* **71**(18) 2677-83
[15] Huang Y, Zhou W and Wang E 2014 Constraint−corrected J–R curve based on three−dimensional finite element analyses *Fatigue & Fract Eng M* **37** 1101-15

Mechanical and physical properties of selected magnesium base nanocomposites

Z Drozd[1,3], Z Trojanová[1], P Lukáč[1], M Kučeráková[2] and T Václavů[2]

[1]Faculty of Mathematics and Physics, Charles University, Ke Karlovu 3, CZ-121 16 Prague 2, Czech Republic
[2]Department of Structure Analysis, Institute of Physics of the Czech Academy of Sciences, Cukrovarnická 10, CZ-162 00 Prague 6, Czech Republic

E-mail: zdenek.drozd@mff.cuni.cz

Abstract. The effect of various nanoparticles (BN, Al_2O_3, ZrO_2, Gr) on the mechanical and physical properties of magnesium was studied in the present paper. Microcrystalline magnesium reinforced with nanoparticles was prepared by ball milling and hot extrusion. Microhardness of samples was measured at room temperature. The linear thermal expansion of the nanocomposites was measured over a wide temperature range from RT up to 400°C. Dynamic modulus and amplitude dependent internal friction were measured at room temperature. Although all nanocomposites were prepared with the same technology substantial differences in mechanical and physical properties were estimated among various nanocomposites. Bonding between magnesium matrix and ceramic and graphite nanoparticles plays important role for the resulting nanocomposites properties.

1. Introduction

Magnesium alloys and composites are very attractive materials due to their good properties: low density, high specific stiffness and good creep resistance. These properties nominate Mg materials predominantly for applications where low weight is required. On the other hand, there is a big competition of aluminum alloys in this field. One possibility of further improvement of Mg is adding of ceramic particles or fibres. By this process composite microstructure – metal matrix composite (MMC) can be obtained. MMCs provide promising solution for various structural applications. Such composites have much better mechanical properties in comparison with unreinforced materials. The special types of MMCs are so called nanocomposites. Nanocomposites (nc) combine two or more components where at least one of them has typical dimension in the nanometer region. Nanocomposites with the magnesium matrix exhibit improved strength, better creep resistance and they still have low density and good machinability [1,2].

It is known that addition of the second phase (e.g. SiC, MgO, Al_2O_3) with a micrometer size improves properties of the matrix itself (yield strength, elastic moduli, hardness, damping, thermal properties and so on) [1-6]. Using of nano-reinforcements can bring many interesting improvements as it has been reported in [7-9].

Magnesium composites with the micrometer-size reinforcements have usually lower ductility caused by the crack of the reinforcements which occurs at the particle-matrix interface and also by the voids in the vicinity of the interfaces. Presence of the voids at the interface play role also in the process of crack initiation. The improvement of the composites properties can be caused by the hot extrusion manufacturing process. Certain consolidation of the composite and homogenization of the reinforcing phase is usually achieved using this technology. On the other hand, hot extrusion can increase the probability of particle fracture.

Recently reported experimental results indicated, that in the case of nanometer-scaled reinforcing phase some improvement of the mechanical properties occurred without a loss of ductility [10-13]. It has been also reported that a small volume fraction of nanoparticles (np) resulted in the comparable or better properties comparing the case of composites with the higher volume fraction of micro scaled reinforcements. Quite effective reinforcing effect of nanoparticles was found in the nanocomposites with Al_2O_3 particulates; as it was reported by Hassan et al [14]. Authors estimated that the nano-reinforcements improve the dimensional stability of pure Mg. The thermally stable alumina (Al_2O_3) nanoparticles (np) cause a significant increase in the microhardness at room temperature, enhance of dynamic elastic modulus as well as the yield strength, ultimate tensile strength and ductility of the material. This effect was observed up to 150 °C. There are several possible mechanisms describing the effect of nanoparticles on the plastic deformation of the nanocomposites [15]. Following process can be taken into account: an increase of the grain boundary area caused by a certain grain refinement, an increase of the dislocation density which accommodates the thermal stresses generated at the particle/matrix interface – such stresses occur due to the difference in the thermal expansion coefficients between the reinforcement phase and the matrix metal and good transfer of the applied stress in the case of effective particle-matrix binding. The Orowan strengthening as well as generation of geometrically necessary dislocations which accommodate the reinforcement/matrix interface during the plastic deformation should also play not negligible role. The distribution of particles in nanocomposites is usually not uniform. The clusters of nanoparticles are often formed. Clusters of various sizes can be found at the vicinity of grain boundaries. This fact represents a serious problem for modeling of processes in the nanocomposites. Despite great effort, no satisfying theoretical model describing the mechanical and physical properties of nanocomposites was developed. This is caused by the very complex character of this matter. Too many of factors and mechanisms play role and no satisfying complex research of these mechanisms has been done yet.

The properties of nanocomposites are influenced by the size, volume fraction and distribution of np and manufacturing technology. In this study, we reinforced pure magnesium with 3 vol.% of various np. All samples were prepared with the same technology. Mechanical and selected physical properties were estimated and compared in individual nanocomposites. Different strengthening effects of various np indicate that there is/are some other effect(s) influencing resulting properties of nanocomposites. We tried to find these hidden parameters which are important for the nc behavior.

2. Experimental

2.1. Materials preparation

Nanocomposites presented in this study - pure magnesium with 3 vol.% of various ceramic and graphite nanoparticles (Al_2O_3, ZrO_2, Gr and hexagonal BN-hBN) were prepared by the powder metallurgical route. In the first step magnesium powder was mixed and milled for 6 hr together with nanoparticles in a planetary ball mill with a speed of 150-200 rpm. Milled powder was then compacted under a pressure of 65 MPa. Hot extrusion (1:5.8 extruded ratio) of pre-compacted materials was next step of the preparation process. Rods of 12 mm in diameter were prepared using a horizontal 600 t extrusion press. Extrusion temperature was 350 °C; the extrusion speed exhibited value of 1.5 mm/s. In this paper the materials studied will be depicted as Mg+Al_2O_3 (Mg+ 3vol% alumina np) Mg+ZrO_2 (Mg+3vol% zirconia np), Mg+Gr (Mg+3vol.% graphite np) and Mg+BN (Mg+3vol.% hBN np).

TEM showed that the distribution of the nanoparticles was not homogenous. Particles were situated mainly in grain boundaries, only few particles were found within the grains.

2.2. Experimental methods

Microhardness was measured with the Vickers pyramid at room temperature in a Qness 10a apparatus with the load of 0.1 kg. The dwell time was 10 s. Microhardness was estimated at the polished surface perpendicular to the extrusion direction. The indents were automatically evaluated along concentric circles as it is shown in figure 1. For the samples preparation the conventional mechanical polishing and etching were done using Glycol solution (1 ml HNO_3, 24 ml water, 75 ml ethylene glycol).

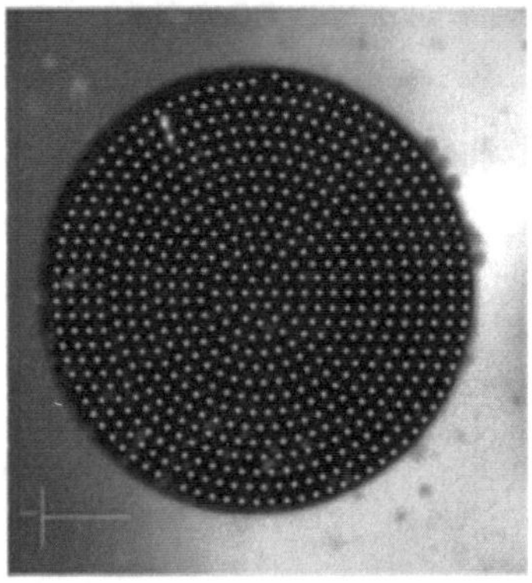

Figure 1. Microhardness measurements map.

Texture of samples was studied using scanning electron microscope ZEISS Auriga Compact equipped with EDAX EBSD camera; OIM software was utilized for EBSD observations. Samples were first mechanically grinded then polished with a diamond suspension of the grade 3.1 and ¼ μm and alumina suspension of the grade 0.05 μm. Finally, the samples surface was ion-polished by Leica EM RES102 device in order to get a high quality surface for EBSD measurements.

Characteristic stresses – offset compressive yield stress (CYS) and ultimate compression strength (UCS) were estimated from the true stress-true strain curves performed at room temperatures.

The thermal expansion was measured in an argon protective atmosphere using Netzsch 410 dilatometer in the temperature interval from room temperature up to 400 °C. The elongation of cylindrical samples with a diameter of 6 mm and 50 mm length were measured in an argon protective atmosphere. Four thermal cycles (heating-cooling) were performed with every specimen; the heating/cooling rate exhibited 0.9 K/min. The thermal expansion coefficient (CTE), α, was estimated as a derivative $\alpha=(1/\ell)\cdot(d\ell/dT)$, where, ℓ, is the sample length.

Young's modulus (YM) was estimated in a Resonant Frequency and Damping Analyser (RFDA). The cantilever beams were used for the modulus measurements. Samples were excited into resonant frequency using a small striker. Free vibrations of the sample were registered by a microphone. Fast Fourier transform was used to estimate the resonant frequency which served for the Young's modulus calculations. Amplitude dependent internal friction was measured at room temperature as damping of the sample free vibrations in a bending mode. Electromagnetic excitation and recording of vibrations was used. The resonant frequency exhibited ~ 40 Hz. The logarithmic decrement, δ, was used as a measure of the internal friction.

3. Results and discussion

Transmission electron micrograph of the Mg+Gr is reported in figure 2. The Gr nanoparticles form agglomerates which are situated mainly in the grain boundaries. Resulting materials after the hot extrusion exhibited very small grain size - units of micrometers. Grains were slightly elongated into the extrusion direction. Texture of Mg+BN and Mg+ZrO_2 nanocomposites is shown in figures 3(a) and 3(b), respectively. A typical ring texture was observed with the (0001) basal planes of the hexagonal cell parallel and <c> axis perpendicular to the extrusion direction. Such texture is typical for the hexagonal magnesium materials submitted to the hot extrusion.

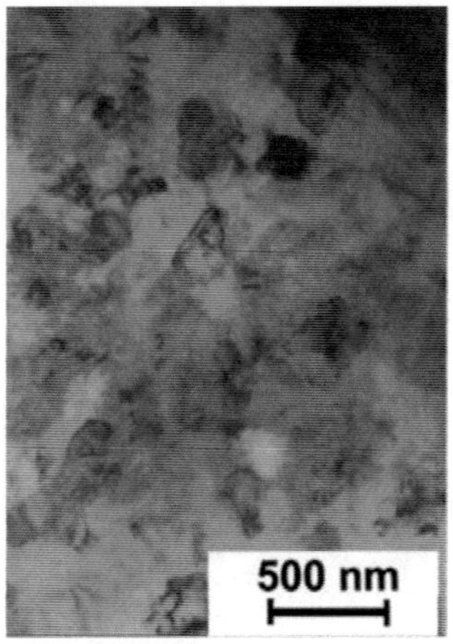

Figure 2. TEM micrograph of Mg+Gr nanocomposite.

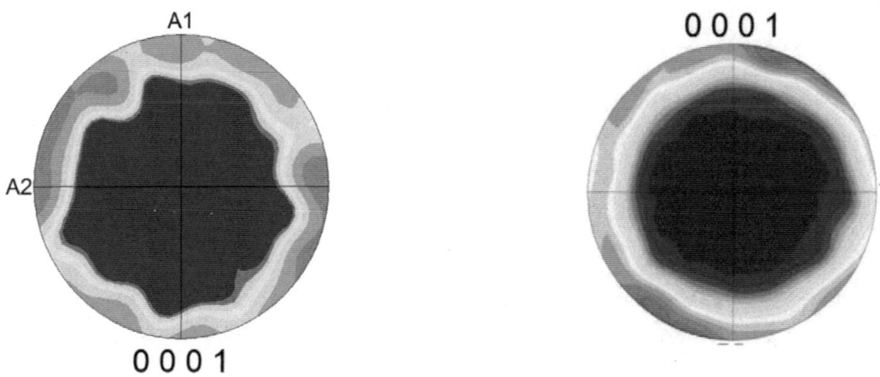

(a) (b)

Figure 3. (a) Pole figure of Mg+ZrO_2 nanocomposite and (b) Pole figure of Mg+BN nanocomposite.

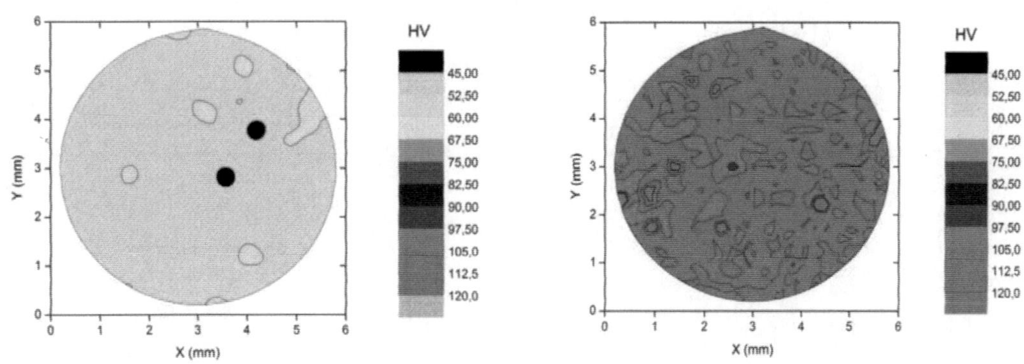

(a) (b)

Figure 4. Microhardness estimated for Mg+ZrO_2 (a), Mg+BN (b).

Vickers microhardness (HV) measured at room temperature in the samples cross-section is shown in figures 4(a) and 4(b) for Mg+ZrO_2 (a) and Mg+BN (b). Various colours of pictures indicate that microhardness of samples substantially differs. The mean HV values were: HV(Mg+ZrO_2) = 51.09±4.66 and HV(Mg+BN) = 110.2±4.66 i.e. practically two times higher compared with Mg+ZrO_2 sample.

Comparing HV values estimated for all nanocomposites, it can be seen that there is a big difference between both materials although they were prepared with the same method and contain 3 vol.% of ceramic nanoparticles.

Offset compressive yield stress (CYS) and ultimate compression strength (UCS) estimated at room temperature are given in table 1. Mechanical characteristics substantially vary for the materials studied.

Table 1. Compressive mechanical properties of Mg reinforced with different nanoparticles.

	$Mg+Al_2O_3$	$Mg+ZrO_2$	Mg+Gr	Mg+BN
CYS (MPa)	189	162	259	399
UCS (MPa)	413	236	293	406

Hardness, microhardness and mechanical properties of magnesium nanocomposites reinforced with alumina, zirconia and yttria nanoparticles were reported by several authors [1,6,13]. Comparing of experimental results is rather problematic because various preparation techniques and various contents of nanoparticles were used.

The temperature dependences of thermal expansion coefficient (CTE) estimated for microcrystalline Mg (μMg) and Mg+BN are shown in figure 5. The CTE values extrapolated to 20 °C are reported in table 2. It can be seen that the ceramic np decreased the thermal expansion. The values of CTE of the Mg+BN sample obtained from experiment were compared with the analytical prediction of the composite sample, α_C, by the rule of mixtures (RoM).

The rule of mixture may be written as

$$\alpha_c = (1 - V_p)\alpha_m + V_p\alpha_p \quad (1)$$

where α_c, α_m and α_p are the CTEs of the composite, matrix, and particle, respectively, and V_p is the volume fraction of the reinforcing phase particles. Comparing calculated CTE-value according to rule of mixtures good agreement with the experimentally estimated value can be seen (table 2).

Table 2. Thermal expansion coefficient (CTE) and Young`s modulus (E) estimated at room temperature.

	μMg	Mg+BN	pkMg	hBN
$CTEx10^{-6}K^{-1}$	24.38	23.55	25.38	4.56
$CTE(RoM)x10^{-6}K^{-1}$		23.79		
E(GPa)	44.15	48.80	43.1	675

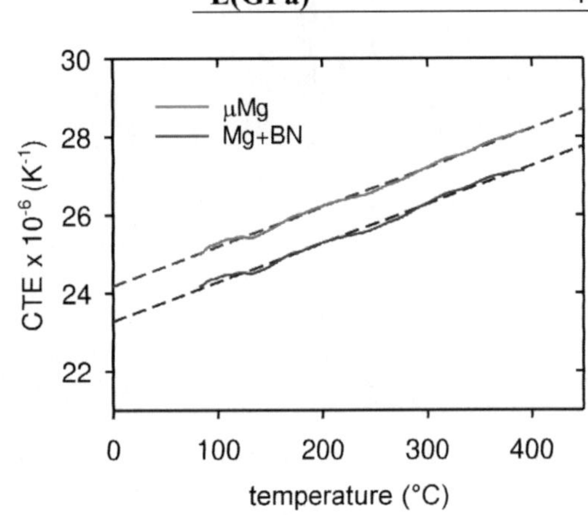

Figure 5. Temperature dependence of CTE.

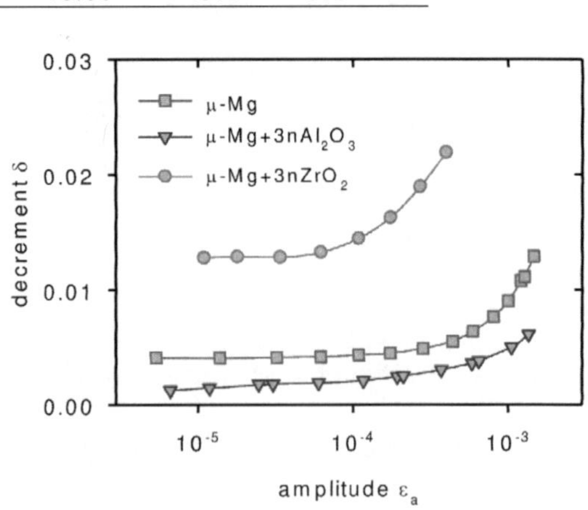

Figure 6. Strain amplitude dependence of decrement.

Experimental values of Young`s moduli for polycrystalline magnesium (pkMg) and hBN were taken from literature [16-18]. Good agreement between measured values of CTE and calculated according rule of mixture is very probably a consequence of the good bonding between matrix and the reinforcing phase nanoparticles. This hypothesis was confirmed by the internal friction measurements. The amplitude dependence of the logarithmic decrement is shown in figure 6.

High values of decrement estimated for Mg+ZrO_2 are due to sliding at the interface between ZrO_2 np and matrix. The interface sliding is the effective damping source. The weak bonding between the magnesium matrix and zirconia nanoparticles is the reason for high values of decrement. On the other hand, low values of decrement estimated for Mg+Al_2O_3 indicate a good bonding at the interface, which was also manifested by a higher deformation stresses. Mg reacts strongly with oxide producing MgO or spinels such as $MgAl_2O_4$. On the other hand, bonding between Mg and zirconia particles is weaker. Sliding in the interface between Mg and ZrO_2 is a reason for a high decrement value and consequently also for worse mechanical properties compared with Mg-Al_2O_3 nanocomposite.

Reinforcing particles reported in this study can be divided into two groups. While δ-Al_2O_3 exhibits tetragonal and ZrO_2 the monoclinic structure; Gr and hBN exhibit the hexagonal structure where well bonded basal planes are stacked in the hexagonal cell, with a weak bonding in the <c> axis direction. On the other hand, the effect of Gr and hBN nanoparticles is different. Reischer et al [19] have reported that the graphite-like nanostructure and chemical inertness of the BN may serve as interlayer material in metal matrix composites with focus on the possibilities of an interlayer texturing. We consider that the BN (and/or Gr) coating of Mg grains created during milling may screens the stress transfer inside grains. Hard hBN interlayers influence the deformation behaviour and contribute to the high strengthening of the Mg+BN nc. In materials with the ultrafine grains (lower than 1 μm), grain boundary sliding may contribute to the total strain, especially at elevated temperatures. This process is in the Mg+BN and Mg+Gr nc very probably restricted due to strengthening of grain boundaries with the nanoparticles and their agglomerates. Hard hBN layers well bonded to the Mg grains hinder sliding of the grain boundary dislocations and grain boundary diffusion.

All materials contained 3 vol.% of nanoparticles and were prepared with the same technology. Results show that properties of individual nanocomposites are different. Models calculating stresses necessary for deformation of composites which take into account yield stress in the matrix, particles size and geometry, thermal expansion coefficient of the reinforcing phase are not able to explain difference between composites with various nanoparticles. There are other factors which should be taken into account constructing models characterise the strengthening of nanocomposites.

4. Conclusions

Mechanical and physical properties of various magnesium based nanocomposites were studied. Composites containing 3 vol% of the reinforcing phase nanoparticles were prepared with the same technology. Significant differences between various composites were found. Large – scale application of nanocomposites requires better understanding of processes influencing resulting mechanical and physical properties of nanocomposites. Binding between particles and matrix plays an important role for the resulting properties. Graphite and graphite like structure of hBN serve as interlayer material which screens the stress transfer between the individual grains and contribute to the high deformation stresses. Models calculating the stress necessary for deformation of composites may take into account these circumstances.

Acknowledgments

Authors gratefully acknowledge the financial support from the Charles University under grant PROGRES Q47.

References

[1] Dieringa H 2011 Properties of magnesium alloys reinforced with nanoparticles and carbon nanotubes: A review *J. Mater. Sci.* **46** 289-306

[2] Chen Y, Tekumalla S, Guo Y B, Shabadi R, Shim V P W and Gupta M 2017 The dynamic compressive response of a high-strength magnesium alloy and its nanocomposite *Mater. Sci. Eng. A* **702** 65-72
[3] Ferkel H and Mordike B L 2001 Magnesium strenghthened by SiC nanoparticles *Mater. Sci. Eng. A* **298** 193-9
[4] Trojanová Z, Lukáč P, Ferkel H, Mordike B L and Riehemann W 1997 Stability of microstructure in magnesium reinforced by nanoscaled alumina particles *Mater. Sci. Eng.* **234** 798-801
[5] Haasen S F and Gupta M 2008 Effect of submicron-sized Al_2O_3 particulates on microstructural and tensile properties of Mg *J. Alloy Compd.* **457** 244-50
[6] Gupta M and Wong W L E 2015 Magnesium-based nanocomposites: Lightweight materials of the future *Mater. Charact.* **105** 30-46
[7] Thostenson E T, Li C and Chou T W 2005 Nanocomposites in contex *Compos. Sci. Technol.* **65** 491-516
[8] Jayakumar J, Raghunath B K and Rao T H 2012 Recent development and challenges in synthesis of magnesium matrix nano composites - A review *Int. J. Latest Res. Sci. Technol.* **1** 164-71
[9] Zhao C, Wu H, Ni J, Zhang S and Zhang X 2017 Development of PLA/Mg composite for orthopedic implant: Tunable degradation and enhanced mineralization *Compos. Sci. Technol.* **147** 8-15
[10] Gupta M and Wong W L E 2014 An insight into processing and characteristic of magnesium based composites *Magnes. Technol.* (John Willey & Sons, Inc.) p 419-28
[11] Tjong S C 2007 Novel nanoparticle-reinforced metal matrix composites with enhanced mechanical properties *Adv. Eng. Mater.* **9** 639-52
[12] Cole G D 2003 Magnesium *Chem. Eng. News archive* **36** 52
[13] Casati R and Vedani M 2014 Metal matrix composites reinforced by nano particles – A review *Metal (Basel)* **4** 65-83
[14] Hassan S F, Tan M J and Gupta M 2008 High temperature tensile properties of Mg/Al_2O_3 nanocomposite *Mater. Sci. Eng. A* **486** 56-62
[15] Trojanová Z, Lukáč P, Száraz Z and Drozd Z 2014 Mechanical and accoustic properties of magnesium based alloys (nano)composites prepared by powder metallurgical routes Waldemar A Monteiro ed. *Light Metal Alloys Applications InTech* pp 163-97
[16] Köster W and Rosenthal K 1940 Der Verlauf des elastizitätsmoduls in den systemen des magnesiums mit zink, aluminium, zinn, blei, antimon und wismut *Z. Metallkd.* **32** 163-4
[17] http://konsys-t.tanger.cz/files/proceedings/metal_00/papers/135.pdf
[18] https://www.ceramicindustry.com/ext/resources/pdfs/2013-CCD-Material-Charts.pdf
[19] Reischer F, Pippel E, Woltersdorf J, Stockel S and Marx G 2007 Carbon fibre-reinforced magnesium, Improvement of bending strength by nanodesign of boron nitride interlayers *Mater. Chem. Phys.* **104** 83-7

IOP Conf. Series: Materials Science and Engineering **372** (2018) 012005 doi:10.1088/1757-899X/372/1/012005

Influence of crack-like surface defects on the fatigue limit of nitrocarburized carbon steel

Y Yamada[1,3], H Eto[1], J Konya[2] and K Takahashi[2]

[1]Isuzu motors LTD, 8 Tsuchidana, Fujisawa-shi, Kanagawa, 252-0881 Japan
[2]Yokohama National University, 79-5 Tokiwadai, Hodogaya-ku, Yokohama, 240-8501 Japan

E-mail: yoshitomi_yamada@notes.isuzu.co.jp

Abstract. The aim of this study was to evaluate the influence of surface crack on the bending fatigue limit of nitrocarburized steel. Semicircular slits with depths of 0.040, 0.075, and 0.100 mm were introduced on the surface of nitrocarburized medium carbon steel specimens to simulate the presence of small cracks. Bending fatigue tests were performed at room temperature in air with the stress ratio $R = -1$. It was observed that the fatigue limit decreased as the depth of the slit increased. The acceptable crack size was estimated based on fracture mechanics. The predicted acceptable crack size was consistent with the experimental result. Therefore, the acceptable crack size of nitrocarburized carbon steel could be predicted via analysis based on fracture mechanics.

1. Introduction

A crankshaft is an automobile engine part, and is often subjected to nitrocarburizing treatment to improve its fatigue limit and wear resistance. Nitrocarburizing treatment has an advantage in that deformation after this type of heat-treatment is considerably smaller than that after other heat-treatments, such as carburizing, because the heat-treatment temperature for nitrocarburizing treatment is lower than the transformation temperature of metals. However, when the nitrocarburizing treatment is applied to a long part, such as a crankshaft, deformation tends to exceed the control value. In this case, reworking is performed within the prescribed reworking limit range. However, if the deformation of a part is too large, the part is discarded because surface cracks are induced in the reworking process.

The nitrocarburizing process of steels creates a compound layer consisting mainly of ε-$Fe_{2\text{-}3}N$ on the surface. A diffusion layer is created below the compound layer. Compound layers of nitrocarburized parts are more brittle compared to the base metals. Thus, micro surface cracks tend to initiate in the compound layers if the part is subjected to plastic deformation beyond the reworking limit range. It is well known that, if the size of a surface defect is smaller than that of the critical defects (a_c), the fatigue strength of steels is not reduced by the defect [1,2]. If the value of a_c in nitrocarburized steels is clarified, we can expand the reworking limit range of nitrocarburized parts.

Takahashi *et al* showed that the value of a_c could be increased by the compressive residual stress induced by shot peening [3,4]. The values of a_c depend on the distribution of residual stress. After the nitrocarburizing process, compressive residual stress is introduced near the surface. It is presumed that the value of a_c in nitrocarburized steels depends on the residual stress distribution. However, the effects of compressive residual stress on a_c in nitrocarburized steels have not yet been studied.

The aim of this study was clarifying the effects of surface crack on the fatigue limit of nitrocarburized steel. Plane bending fatigue tests were performed for the nitrocarburized medium carbon steel specimens containing a small semi-circular slit. Furthermore, the value a_c was estimated based on fracture mechanics.

2. Experimental procedure

The test material used in the study was medium carbon steel (JIS-S53C). The chemical composition of JIS-S53C was shown in table 1. Figure 1 shows a flowchart of the preparation of the test specimens. The plates were oil quenched from 1123 K followed by tempering at 873 K. After the heat treatment, the plated were machined into bending fatigue test specimens shown in figure 2(a). The maximum bending stress at the minimum section of the test specimen were evaluated. After machining of the test specimens, nitrocarburizing treatments were carried out. A semicircular slit, as shown in figure 2(b), was introduced perpendicular to the longitudinal direction by electric discharge machining. The depth (a) of the semi-circular slit is 0.040, 0.075, and 0.100 mm. The slit width w is 0.03 mm.

Table 1. Chemical composition of JIS-S53C (wt %).

C	Si	Mn	P	S	Cr
0.54	0.26	0.78	0.019	0.017	0.12

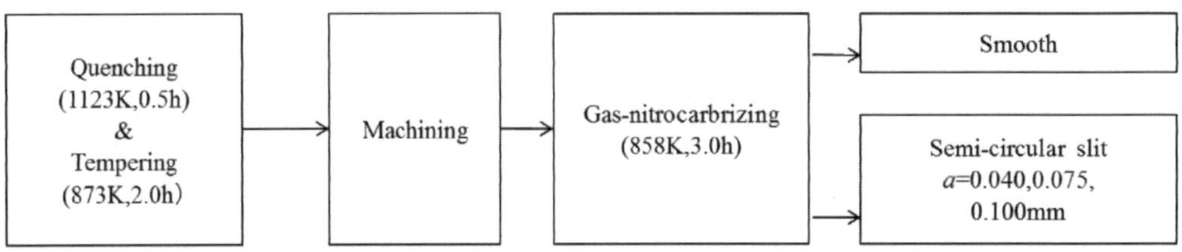

Figure 1. Flowchart of specimen preparation.

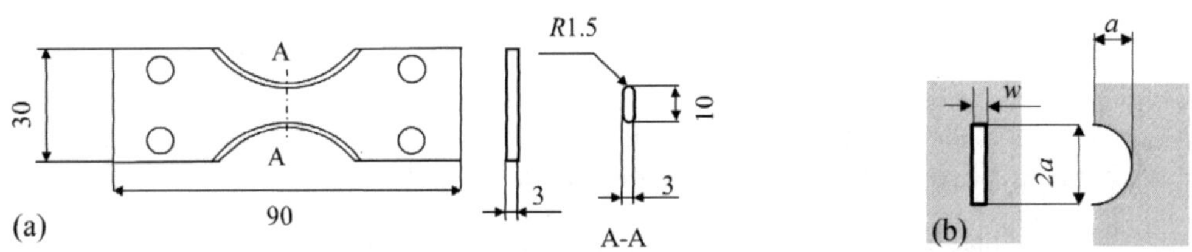

Figure 2. (a) Shape and dimension of a test specimen. (b) Shape of semi-circular slit. (Unit: mm).

The residual stress distributions of the specimen were evaluated using X-ray diffraction. Table 2 lists the conditions of measurement. The X-ray stress constants used were −611 MPa/deg for ε phase and −318 MPa/deg for α-Fe phase [5]. The residual stress distributions of the specimens were evaluated after removing the surface layers by chemically etching.

Table 2. Residual stress measurement condition.

	ε-Fe_3N	α-Fe
Characteristic X-ray	Cr-Kα	
Diffraction plane	211	103
Diffraction angle (deg)	156.4	134-137
Diameter of incident collomator (mm)	1.0	
Tube current (kV)	30	
Tube current (mA)	30	

The surface hardness at the compound layer and hardness distribution at the diffusion layer were measured by using a micro Vickers hardness tester with a holding time of 10 s and at an indenter load of 0.24 N and 0.98 N, respectively.

Plane bending fatigue test were performed by using a bending fatigue test machine. The conditions of fatigue test included a stress ratio $R = -1$ and a cyclic frequency of 25 Hz. After the fatigue tests, the fractured surfaces were observed by using a stereoscopic microscope and scanning electron microscope (SEM).

3. Experimental results

3.1. Vickers hardness and residual stress

The microstructure of the cross-section of the test specimens is shown in figure 3. A compound layer with a depth of approximately 0.033 mm was observed, and a porous layer of approximately 0.015 mm was formed on the outermost surface of the compound layer. Figure 4 shows the distributions of Vickers hardness. The hardness was measured at the center of the gauge section where the gauge width was 10 mm. The hardness of the compound layer was 724 HV, the maximum hardness of the diffusion layer was 374 HV (0.050 mm from the surface), and the depth of the diffusion layer from the surface of the specimen was 0.8 mm.

Figure 3. Microstructure of the test specimen.

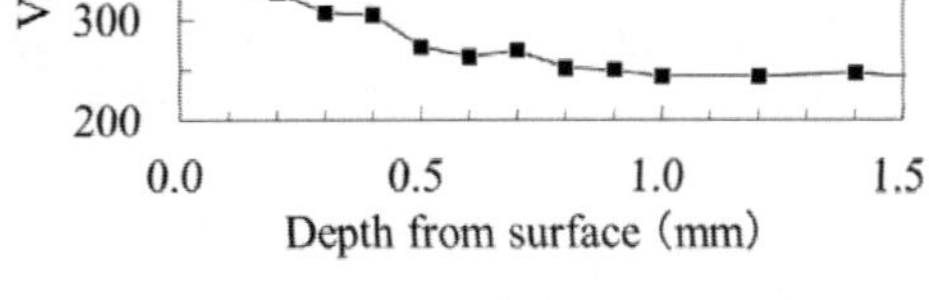

Figure 4. Vickers hardness of the compound and diffusion layers.

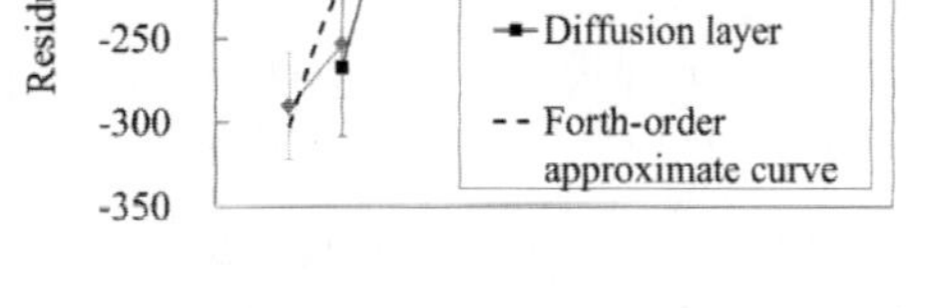

Figure 5. Residual stress distributions and fourth-order polynomial fitting curve.

Figure 5 shows the measurement results of the residual stress distribution. The residual stress at the porous layer was not measured owing to difficulty of measurement. The maximum compressive residual stress of 300 MPa was measured at the compound layer. However, the compressive residual stress sharply decreased at the interface between the compound layer and diffusion layer at the depth of 0.03–0.04 mm. The compressive residual stress gradually decreased at a depth greater than 0.05 mm.

3.2. Fatigue test results and observation of fracture surface

Figure 6 shows the *S–N* curves. Figure 7 shows the relationship between the stress amplitude and the semi-circular slit depth. The open and solid symbols indicate the fractured and non-fractured specimens under 10^7 cycles, respectively. The maximum stress amplitude in the non-fractured specimen corresponds to the fatigue limit. The results prove that the fatigue limit of the specimens decreased with an increase in slit depth.

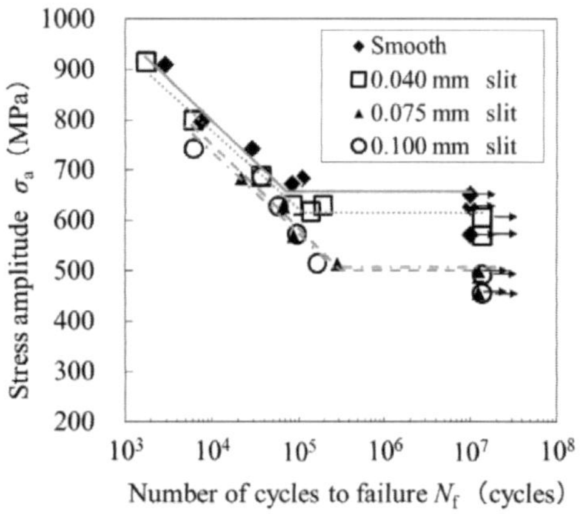

Figure 6. *S–N* curves.

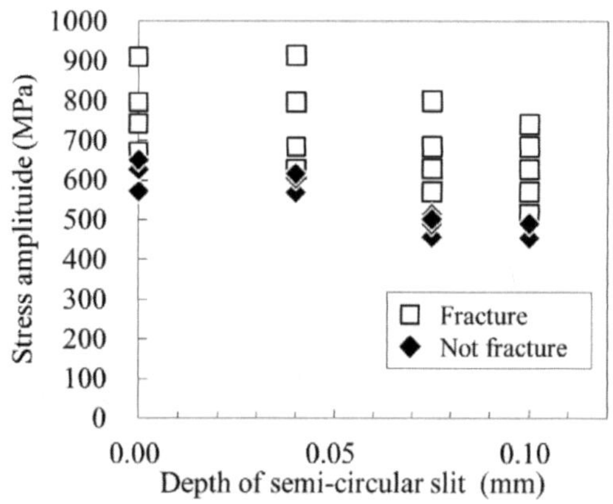

Figure 7. Stress amplitude as a function of depth of semi-circle slit.

Figure 8 shows the typical fracture surface of the specimen with a semi-circular slit. All the test specimens with a semi-circular slit fractured from the slit. From the results of the fatigue test and fracture surface observation, it was observed that a_c of this material is smaller than $a = 0.04$ mm.

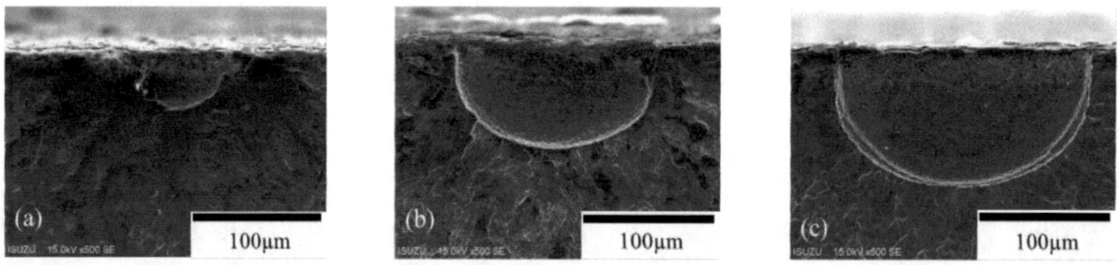

Figure 8. Typical fracture surface of specimen with a semi-circular slit. (a) a = 0.040 mm, N_f = 141800; (b) a = 0.075 mm, N_f = 88200; (c) a = 0.100 mm, N_f = 96900.

3.3. Evaluation of crack size which does not affect the fatigue limit

To expand the limit range for reworking, it is necessary to clarify the micro crack size which does not affect the fatigue limit (a_c). The values of a_c were evaluated based on fracture mechanics by assuming that the semicircular slit is equivalent to a semicircular crack. We used the following equation in order to evaluate the effective stress intensity factor range (ΔK_T) [6].

$$\Delta K_T = K_{max} + K_r \quad (1)$$

where K_{max} is defined as the stress intensity factor under maximum applied stress σ_{max}. To determine the values of K_{max}, σ_{max} = 662 MPa corresponding to the fatigue limit of the smooth specimens was substituted into the equation proposed by Newman and Raju [7]. K_r is the stress intensity factor owing to residual stress, which is a negative value in the compressive residual stress field. The equations of the American Petroleum Institute Recommended Practice (API RP 579) were used to determine the value of K_r for the surface cracks [8]. The K_r for a semielliptical crack is indicated as follows.

$$K_r = [G_0\sigma_0 + G_1\sigma_1\left(\frac{a}{t}\right) + G_2\sigma_2\left(\frac{a}{t}\right)^2 + G_3\sigma_3\left(\frac{a}{t}\right)^3 + G_4\sigma_4\left(\frac{a}{t}\right)^4]\sqrt{\frac{\pi a}{Q}}f_W \quad (2)$$

$$Q = 1 + 1.464\left(\frac{a}{c}\right)^{1.65} \quad (3)$$

$$f_W = \left[sec\left(\frac{\pi c}{2W}\right)\sqrt{\frac{a}{t}}\right]^{\frac{1}{2}} \quad (4)$$

In the above equation, G_0 to G_4 represent the influence coefficients for a semielliptical crack according to API RP 579. a and c indicate the depth and half-length of the semielliptical surface crack, respectively. t and W are the thickness and width of the specimens, respectively. The coefficients, σ_0 to σ_4, are obtained using the fourth-order polynomial fitting curve of the residual stress distribution by using the following equation (see figure 5):

$$\sigma(x) = \sigma_0 + \sigma_1\left(\frac{x}{t}\right) + \sigma_2\left(\frac{x}{t}\right)^2 + \sigma_3\left(\frac{x}{t}\right)^3 + \sigma_4\left(\frac{x}{t}\right)^4 \quad (5)$$

where x indicates the distance from the specimen surface in the depth direction.

Figure 9 shows ΔK_T, K_r, K_{max}, and the effective threshold stress intensity factor range ($\Delta K_{eff,th}$) as a function of the crack depth (a). The value of the stress intensity factor depends on the position of the crack front. The values of ΔK_T are the highest at the deepest point of the crack. Thus, the values of ΔK_T at the deepest point are plotted in figure 9. It is reported that the value of $\Delta K_{eff,th}$ for steel is 3.0 MPa·m$^{1/2}$[9]. Thus, it is presumed that $\Delta K_{eff,th}$ is 3.0 MPa·m$^{1/2}$ in this study.

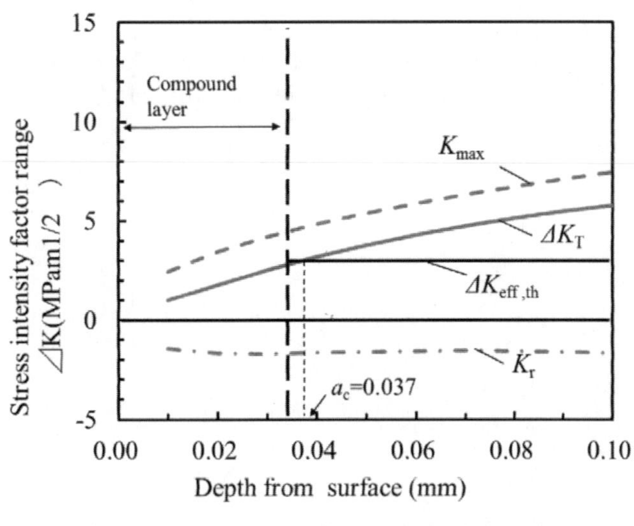

Figure 9. Estimation of acceptable defect size.

The surface slit is considered to be harmless if ΔK_T is less than $\Delta K_{eff,th}$. Thus, the intersection between ΔK_T and $\Delta K_{eff,th}$ yields a_c. The value of a_c is estimated to be 0.037 mm. The value of a_c is consistent with the experimental result, which indicated that its value is less than a = 0.04 mm.

As shown in figure 5, the compressive residual stresses at the compound layer were much higher than that those at the diffusion layer. Thus, surface cracks that were less in dimension compared to the depth of the compound layer were acceptable in the nitrocarburized specimen used in this study. To increase the value of a_c, it is effective to increase the value of the compressive residual stress or deepen the regions with high compressive residual stress.

4. Conclusion

The objective of the present study was clarifying the effect of the surface crack on the fatigue limit of nitrocarburized steel. Plane bending fatigue tests were performed using the test specimens of nitrocarburized medium carbon steel containing a small semicircular slit. Furthermore, the value of the maximum acceptable crack size (a_c) was predicted based on the theory of fracture mechanics. Following conclusions were obtained:

- The results of the fatigue test and fracture surface observation indicated that a_c of this material is smaller than $a = 0.04$ mm.
- The value of a_c estimated based on fracture mechanics was determined to be 0.037 mm, which was consistent with the experimental result.
- In order to increase the value of a_c, it is effective to increase the value of the compressive residual stress or deepen the regions with high compressive residual stress.

References

[1] Kitagawa H and Takahashi S 1979 *Trans. Jpn. Soc. Mech. Eng. Ser.* A **45**(399) 1289–303
[2] Murakami Y 2002 *Metal Fatigue: Effects of Small Defects and Nonmetallic Inclusions* (Amsterdam: Elsevier) pp 35–55
[3] Takahashi K, Amano T, Hanaori K, Ando K and Takahashi F 2009 *J. Soc. Mater. Sci.* **58**(12) 1030–6
[4] Fueki R and Takahashi K 2018 *Int. J. Struct. Integrity* **9(1)** pp 50–64
[5] Tanaka K, Gu Q P, Mikuriya T and Akiniwa Y 1996 *Trans. Jpn. Soc. Mech. Eng. Ser.* A **62**(604) 2734–40
[6] Takahashi K, Hayashi T, Ando K and Takahashi F 2010 *Trans. Jpn. Soc. Spring. Eng.* **2010**(55) 25–30
[7] Newman Jr J C and Raju I S 1981 *Eng. Fract. Mech.* **15** 185–92
[8] American Petroleum Institute 2000 Recommended practice 579, fitness for service p C3
[9] Liaw P K, Lea T R and Logsdon W A 1983 *Acta Metall.* **31**(10) 1581–7

3D finite element analysis of tightening process of bolt and nut connections with pitch difference

X Liu[1], N-A Noda[1,2], Y Sano[1], Y T Huang[1] and Y Takase[1]

[1]Mechanical Engineering Department, Kyushu Institute of Technology, 1-1 Sensui-cho Tobata-ku Kitakyushu-shi Fukuoka 804-8550, Japan

E-mail: noda@mech.kyutech.ac.jp

Abstract. In a wide industrial field, the bolt-nut joint is unitized as an important machine element and anti-loosening performance is always required. In this paper, the effect of a slight pitch difference between a bolt and nut is studied. Firstly, by varying the pitch difference, the prevailing torque required for the nut rotation, before the nut touches the clamped body, is measured experimentally. Secondly, the tightening torque is determined as a function of the axial force of the bolt after the nut touches the clamped body. The results show that a large value of pitch difference may provide large prevailing torque that causes an anti-loosening effect although a very large pitch difference may deteriorate the bolt axial force under a certain tightening torque. Thirdly, a suitable pitch difference is determined taking into account the anti-loosening and clamping abilities. Furthermore, the chamfered corners at nut ends are considered, and it is found that the 3D finite element analysis with considering the chamfered nut threads has a good agreement with the experimental observation. Finally, the most desirable pitch difference required for improving anti-loosening is proposed.

1. Introduction

The bolt-nut connections can be regarded as one of the most important material joining techniques. They are widely used in various engineering fields, including aerospace, automotive and mechanical/civil engineering constructions. To ensure the structures safety, high fatigue strength has been required, as well as, anti-loosening performance. Many previous studies are focusing on the anti-loosening performance for newly developed bolt-nut connections [1-3], and many studies contribute toward improving fatigue strength [4-6]. This paper, therefore, focuses on the effect of pitch difference in a connection on the anti-loosing performance. Here, several pitch differences between the bolt and nut are designed as shown in figures 1 and 2, the nut pitch is larger than the bolt pitch. However, the previous studies on pitch difference were limited to fatigue strength improvement, and the effect of pitch difference on the anti-loosening performance has not been investigated yet. There is no three-dimensional model is available. This paper, a three-dimensional FEM simulation is used to calculate the tightening torque and the tightening force during the tightening and loosening process. Taking the anti-loosening performance into account, the most desirable pitch difference will be proposed.

2. Experimental and analytical method of tightening and loosening mechanism

2.1. *Specimen and experiment equipment*

In this study, the Japanese Industrial Standard (JIS) M12 bolt-nut connections with strength grade 8.8

are employed as shown in figure 2. The bolt material is chromiummolybdenum steel SCM435, and the nut material is medium carbon steel S45C quenched and tempered. The standard M16 bolt-nut connection has the same pitch dimension of 1750 μm, here, the nut pitch is assumed to be equal or slightly larger than the bolt pitch (figure 3) [7]. Three types of pitch differences, namely α=30 μm, α=40 μm and α=50 μm, are considered in this study. In addition, the horizontal clearance of a bolt and a nut (Cx) is 59 mm. when tightening a nut with pitch difference, torque occurs before the nut contact with the clamped body, and the torque occurred in this period is called prevailing torque [8,9] as shown in figure 4.

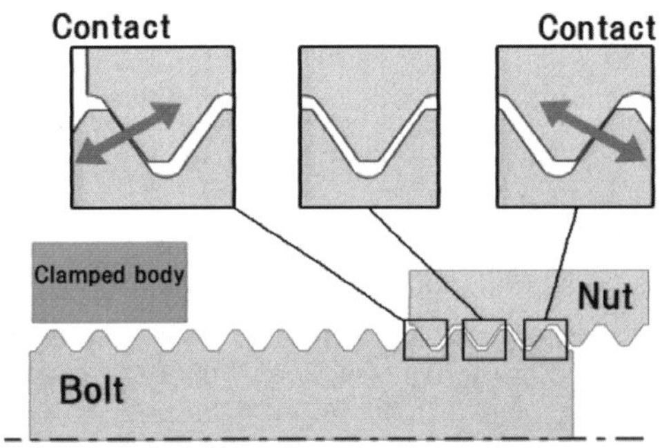

Figure 1. Contact status between bolt and nut, when the nut pitch is slightly larger than the bolt pitch.

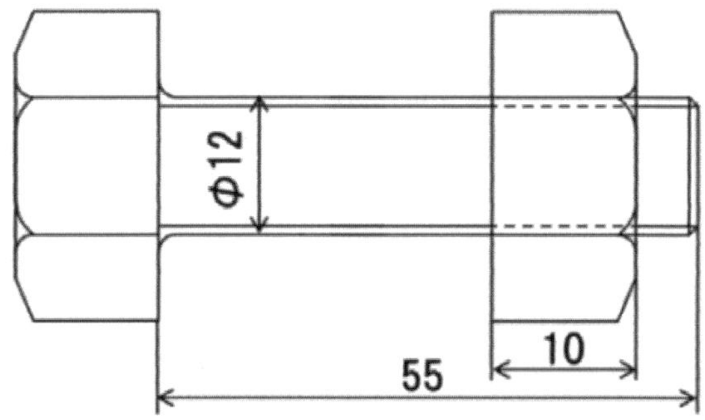

Figure 2. M12 Bolt-nut specimen (dimensions in mm).

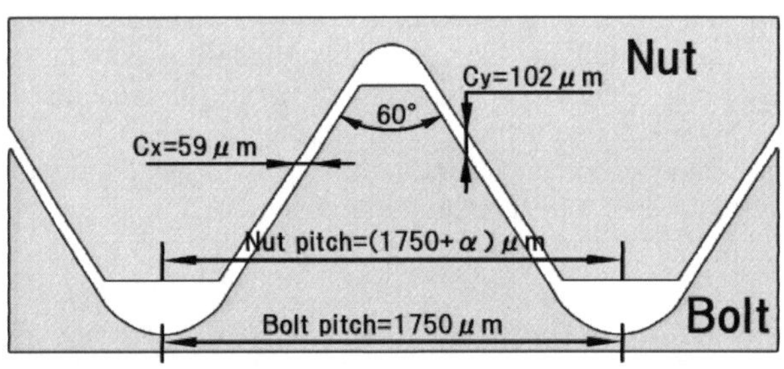

Figure 3. Pitch difference and clearance between bolt and nut.

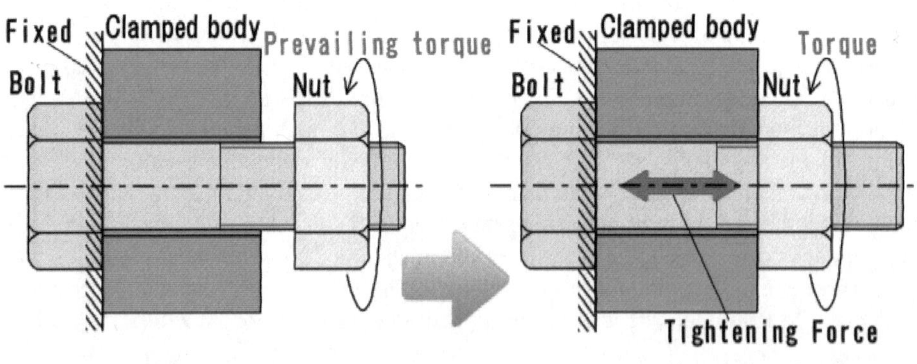

Figure 4. FEM model and boundary conditions for prevailing torque analysis.

2.2. Experimental conditions

The method to measure prevailing torque is shown in figure 5. The model of the torque wrench is TOHNICHI DB 50 N, and lubricating oil is MoS_2. Start from the position where the blot and nut begin to connect (0 turns), record the prevailing torque every 45° the nut is tightened. After the whole nut (5.7 turns) are tightened in the bolt, continue tighten the nut 2.3 turns, thus the nut is tightened in the bolt 8 turns in total. Then, loosen the nut until the bolt and the nut separate from each other, and record the prevailing torque every 45° the nut is tightened.

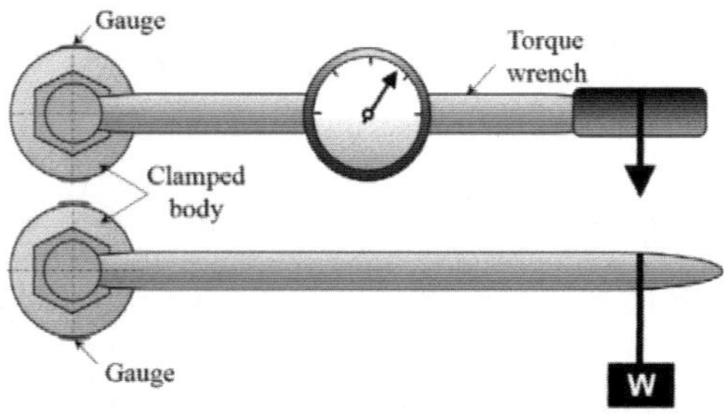

Figure 5. Torque measuring device.

2.3. Analytical method

A precise three-dimensional model as shown in figure 6 is used for finite element analysis. To simplify the calculation, the hexagonal part of the bolt and the nut is replaced by two cylinders. Especially, the helical thread of the bolt and nut were subdivided into smaller elements compared with other parts [10]. Contact types and material non-linearity are considered in the analysis. Boundary conditions are set as follows, one side of the bolt is fixed, a set of tightening angels are applied on the side surface of the nut, and then loosen the nut. The connecting way of the nut and the bolt is frictional contact. The start position of the analysis is where the prevailing torque begins to occur and the end position of the analysis is where 2.3 cycles tightened after all the nut thread is tightened into bolt.

Figure 6. FEM model of M12 bolt and nut.

Table 1. Material properties of bolt and nut.

	Young's modulus (GPa)	Poison'sratio	Yield strength (MPa)	Tensile strength (MPa)
SCM435 (Bolt)	206	0.3	800	1200
S45C (Nut)	206	0.3	530	980

3. Comparison of experimental result and analytical result of prevailing torque

3.1. Tightening process

Figure 7(d) is a schematic diagram of nut tightening process. From A to E is the process of tightening, and from E to A is the process of loosening. In figure 7(d), A is the position where the nut begins to contact with the bolt, B is the position where prevailing torque begins to occur, C is in the period when prevailing torque is increasing, D is the position where the nut is completely tightened into the bolt, E is the position where to continue to tighten the nut for 2.3 turns until the nut contact with clamped body. Figures 7(a) to 7(c) shows the prevailing torque in the tightening and loosening process obtained by experiment and FEM simulation. The red line from A to E shows the process of the tightening, and the blue line from E to A shows the process of loosening. The FEM results of the solid line are in good agreement with the dotted line of the experimental results. It is found that prevailing torque remains steady after the nut was completely screwed into the bolt (from position D to position E), and the positions from D to E are called convergent position. And the average of those data of the convergent positions was calculated to represent the convergent positions.

3.2. Comparison of experimental result and analytical result of prevailing torque

Figure 7(a) shows the measured results when α=30 μm, and it can be seen that prevailing torque begins to occur when the nut is tightened into the bolt for 4.2 turns, the average of the convergent prevailing torque is 4.7 Nm. And analytical result can be seen that prevailing torque begins to occur when the nut is tightened into the bolt for 4.25 turns, the average of the convergent prevailing torque is 7 Nm.

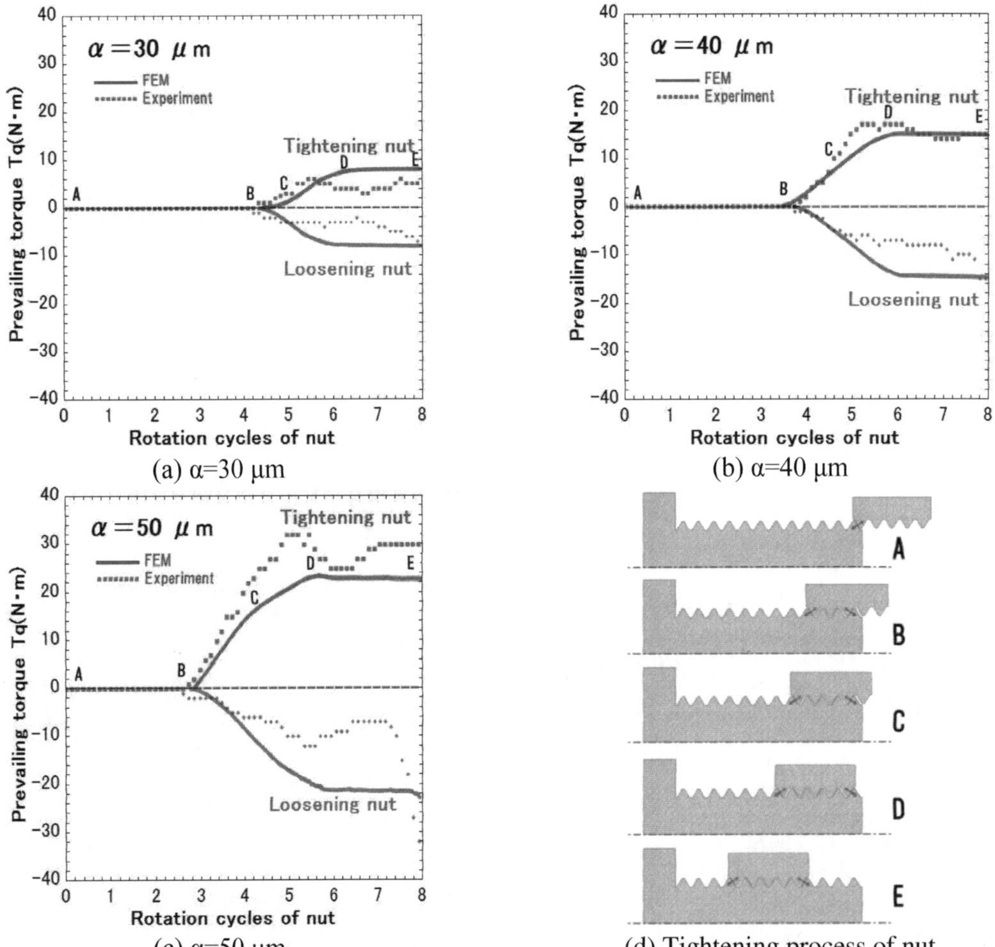

(a) α=30 μm (b) α=40 μm (c) α=50 μm (d) Tightening process of nut

Figure 7. Results of prevailing torque simulation by 3D FEM in comparison with experimental measurement.

Figure 7(b) shows the measured results when α=40 μm, and it can be seen that prevailing torque begins to occur when the nut is tightened into the bolt for 3.9 turns, the average of the convergent prevailing torque is 14.6 Nm. And analytical result can be seen that prevailing torque begins to occur when the nut is tightened into the bolt for 2.8 turns, the average of the convergent prevailing torque is 16 Nm.

Figure 7(c) shows the measured results when α=50 μm, and it can be seen that prevailing torque begins to occur when the nut is tightened into the bolt for 2.6 turns, the average of the convergent prevailing torque is 29.9 Nm. And analytical result can be seen that prevailing torque begins to occur when the nut is tightened into the bolt for 2.8 turns, the average of the convergent prevailing torque is 23 Nm.

4. FEM analysis of tightening process of the nut

4.1. Analytical method

Figure 5 shows the FEM model, a clamped body was added in the connection compared with the model in previous sections. The boundary conditions are shown in figure 6. The head of the bolt and the bottom of the clamped body were fixed. The simulation begins at position B at which the nut began to be tightened, and the nut finally reaches position E at where the nut begins to contact with the clamped body after it passes through position C and position D. Then, continues to tighten a certain angle, during which period a tightening force to compress the clamped body will be generated. The contact type of the thread surface of the bolt and the nut was set as frictional contact, and so were the bottom surface of the nut and the top surface of the clamped body. Here, our special interest focus on the relationship between the tightening torque and tightening force from position E to position F (figure 8).

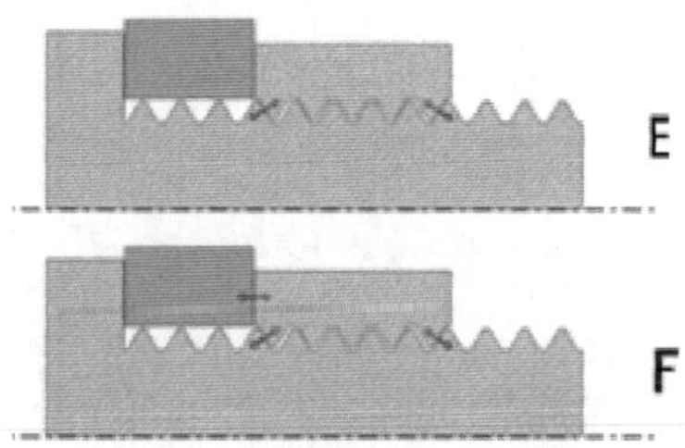

Figure 8. Tightening process of nut.

4.2. Comparison of torque formula and analysis in normal bolt · nut (α = 0)

When tighten the common nut into the bolt, the generated tightening torque consists of three parts, the screw surface friction torque $F/2 \cdot d_2/\cos\beta \cdot \mu$, the axial force torque $F/2 \cdot P/\pi$, and the seating surface friction torque $F/2 \cdot d_w \cdot \mu_w$, and it is shown in the following equation called Motosh equation.

$$d_w = \frac{2(d_o^3 - d_h^3)}{3(d_o^2 - d_h^2)} \quad T = \frac{F}{2}\left(\frac{d_2}{\cos\beta}\mu + \frac{p}{\pi} + d_w\mu_w\right) \ldots \qquad \ldots (1)$$

Here, tightening torque *T*, tightening force *F*, effective diameter *d2*, half angle of thread *β*, outside diameter of bolt seat surface *d0*, bolt hole diameter *dh*. The relationship between the tightening torque and the tightening force of M12 obtained by this evaluation formula and analysis shows in figure 8. The evaluation formula and the analysis result agreed well within an error of 5%. From this result, it is considered that the analysis result of α=0 is high in accuracy.

4.3. 3-D analysis during the tightening process of pitch difference nut

Since the bolt and nut are used for connecting components or structures, the tightening ability to produce enough tightening force is essential. Therefore, after the nut touches the clamped body, the relationship between the tightening torque and the tightening force was investigated. Note that tightening torque T is different from prevailing torque Tp, which is defined only before the nut touches the clamped body. To obtain the relationship between torque and tightening force, Three-dimensional finite element analysis is applied.

Figure 9 shows the relationship between the tightening torque and the tightening force obtained by FEM analysis of M12 bolt and nut with pitch difference. Although the tightening force of pitch difference α=30 is large, insufficient prevailing torque can cause the nut to loosen. The prevailing torque of pitch difference α=50 is large, but a sufficient tightening force cannot be imparted. The pitch difference α=40 is considered to be the most suitable for anti-loosening performance because the prevailing torque and tightening force are both sufficient.

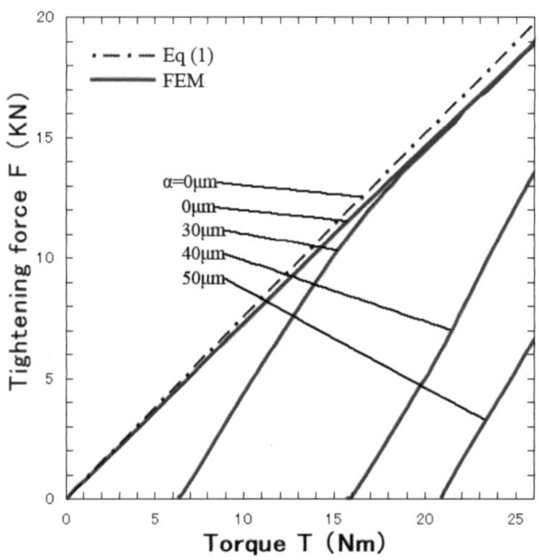

Figure 9. Relationship between torque and tightening force.

When a common nut is tightened, the tightening forces increase from 0. However, if there is a pitch difference between the nut and bolt, the initial points of tightening forces will no longer be 0. The reason for this is that when tighten pitch a difference nut, there will be prevailing torque between the bolt-nut connections before the nut contacts with the clamped body. Thus, when tighten a pitch difference nut with a certain torque T, a part of the torque needs to be used to offset the prevailing torque, and the other part of the torque, which equals to T-Tp, will convert into tightening force. Therefore, when the tightening torque is fixed, the tightening force decreases with the increase of pitch difference, meanwhile, the prevailing torque increases with the increase of pitch difference.

5. Conclusion

In this paper, the following results were obtained by tightening and loosening experiments on bolts and nuts with different pitch differences and analysis by three-dimensional finite element method.

- When pitch difference is 30 μm, the corresponding prevailing torque equals to 4.7 Nm; when pitch difference is 40 μm, the corresponding prevailing torque equals to 14.6 Nm; when pitch difference is 50 μm, the corresponding prevailing torque equals to 29.9 Nm, in the experiment.
- Suitable pitch difference can maintain the good anti-loosening performance in the whole loosening process, and this has been proved by FEM simulation.
- The relationship between tightening force and tightening torque can be predicted by FEM simulation. In order to obtain the appropriate tightening force and tightening torque, the optimum pitch difference should be 40 μm.

References

[1] Noda N A, Sano Y, Takase Y, Chen X, Maruyama H, Wang H and Fujisawa R 2015 Anti-Loosening performance of special bolts and nuts having enhanced fatigue life by introducing pitch difference *Transactions of Society of Automotive Engineers of Japan* **46**(1) 121-6 (In Japanese)

[2] Hard Lock Industry KK, Looseness Preventing Nut, Patent Abstracts of Japan, Publication No.: 2002–195236, 2002-7-10

[3] Ishimura M, Yamanaka H, Syoji Y, Kobayashi T and Sawa T 2011 *Jpn. Soc. Mech. Eng. A* **77**(2011) 1444

[4] Akaishi Y I, Chen X, Yu Y, Tamasaki H, Noda N A, Sano Y and Takase Y 2013 Fatigue strength analysis for bolts and nuts which have slightly different pitches considering clearance *Transactions of Society of Automotive Engineers of Japan* **44**(4) 1111-7 (In Japanese)

[5] Chen X, Noda N A, Wahab M A, Akaishi Y I, Sano Y, Takase Y and Fekete G 2015 Fatigue failure analysis in bolt-nut connection having slight pitch difference using experiments and finite element method *Acta Polytechnica Hungarica* **12**(8) 61-79

[6] Noda N-A, Chen X, Sano Y, Wahab M A, Maruyama H, Fujisawa R and Takase Y 2016 Effect of pitch difference between the bolt–nut connections upon the anti-loosening performance and fatigue life *Materials & Design* **96** 476–89

[7] Noda N-A, Sano Y, Takase Y, Chen X, Maruyama H, Wang H and Fujisawa R 2015 Anti-loosing performance of special bolts and nuts having enhanced fatigue life by introducing pitch difference *Transactions of Society of Automotive Engineers of Japan* **46** 121-6 (in Japanese)

[8] Noda N-A, Noma S, Saito K, Nagawa M, Yumoto A and Xiao Y 2006 Stress Reduction Effect and Anti-loosening Performance of Outer Cap Nut by Finite Element Method *J-STAGE Volume 72 (2006)* **721** 1318-1325

[9] Izumi S, Yokoyama T, Teraoka T, Iwasaki A, Sakai S, Saito K, Nagawa M and Noda H 2005 Verification of anti-loosening performance of super slit nut by finite element method *Transaction of the Japan Society of Mechanical Engineers, Series A* **71** 380-6 (in Japanese)

[10] Izumi S, Yokoyama T, Iwasaki A and Sakai S 2005 Three-dimensional finite element analysis on tightening and loosening mechanism of bolted joint *Transaction of the Japan Society of Mechanical Engineers, Series A* **71** 204-12 (in Japanese)

IOP Conf. Series: Materials Science and Engineering **372** (2018) 012007 doi:10.1088/1757-899X/372/1/012007

Suitable pitch difference to realize anti-loosening performance for various bolts-nuts diameter

S Kubo[1], K Tateishi[1], N-A Noda[1,2], Y Sano[1], Y Takase[1] and K Honda[1]

[1]Mechanical Engineering Department, Kyushu Institute of Technology, 1-1 Sensui-cho Tobata-ku, Kitakyushu-shi, 804-8550, Japan

E-mail: noda@mech.kyutech.ac.jp

Abstract. In bolt-nut connection, the anti-loosening performance and high fatigue strength are always required with low cost to ensure the connected structure's safety. In the previous study, a suitable pitch difference between the bolt-nut was obtained as $\alpha = 33\ \mu m$ for M16 JIS bolt-nut through loosening experiment and FEM simulation for tightening process. However, other bolt-nut diameters have not been considered yet. In this paper, therefore, suitable pitch difference is considered for various diameters to realize anti-loosening performance. Since bolt-nut thread geometries are different depending on the diameter, they are expressed as approximate formula. Then, loosening force and anti-loosening force are considered by varying the diameter. Finally, suitable pitch difference $\alpha_{min}^{suit} < \alpha < \alpha_{max}^{suit}$ was determined from mechanical condition.

1. Introduction

In a wide industrial field, bolt-nut joint is unitized as a machine element and is of great importance. However, the anti-loosening and anti-damage performances are always required. When vibration is added, the rotational loosening force on the spiral shape of the screw becomes larger than the friction between the screw surfaces, then the loosening occurs. Most previous studies have been mainly focusing on anti-loosening performance [1-4]. Only a few studies contribute toward improving fatigue strength [5-7]. To improve both the fatigue life and the performance of anti-loosening, a slight pitch difference between bolt and nut was previously proposed [8-12]. Then, the anti-loosening performance was confirmed for M16 bolt-nut connection by using FEM simulation as well as experiment [13-15]. However, other bolt-nut diameters have not been considered yet. Therefore, in this paper, the anti-loosening will be studied for different bolt-nut diameter to clarify the suitable pitch difference theoretically and experimentally. This special nut may improve the fatigue strength [13-15] without using additional components. Moreover, this special nut can be manufactured just by changing feed speed of a numerical control (NC) lathe.

2. Effect of the pitch difference on the rotation

2.1. Bolt-nut specimens

Japanese Industrial Standard (JIS) M16 bolt-nut connections with strength grade 8.8 are employed. The bolt material is chromium- molybdenum steel SCM435, and the nut material is medium carbon steel S45C quenched and tempered, whose properties are indicated in table 1, and whose stress–strain curves are shown in figure 1, respectively. The normal M16 bolt and nut have the same pitch dimension as 2000 μm. Herein, the nut pitch is assumed to be (2000+α) μm. Figure 2 shows a contact status between bolt and nut threads during the tightening process. Notation $\boldsymbol{C_x}$ is the horizontal clearances between bolt and nut.

Table 1. Material properties of bolt and nut.

	Young's modulus (GPa)	Poison's ratio	Yield strength (MPa)	Tensile strength (MPa)
SCM435 (Bolt)	206	0.3	800	1200
S45C (Nut)	206	0.3	530	980

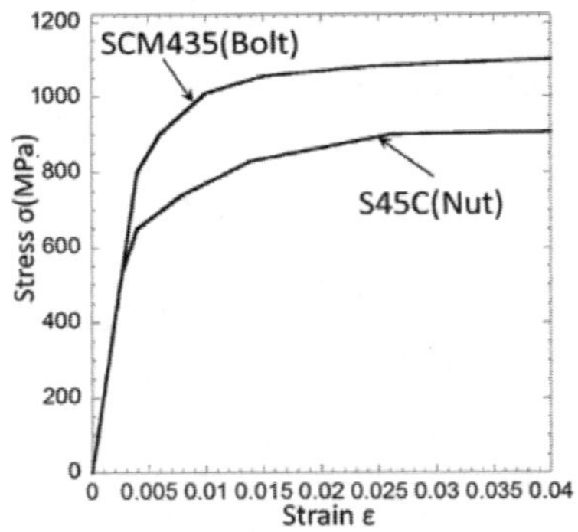

Figure 1. Stress strain relation for SCM435 and S45C.

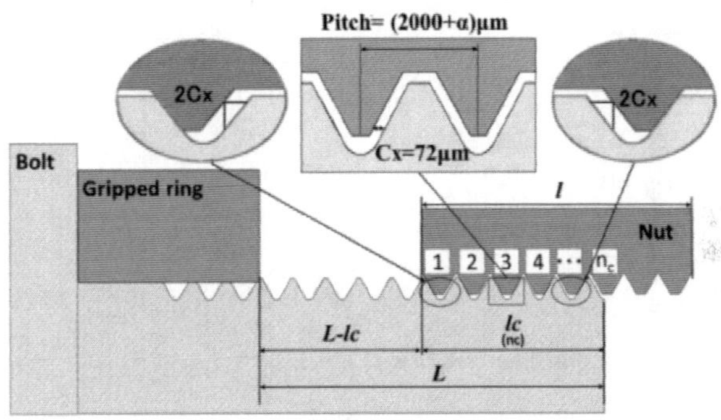

Figure 2. Schematic diagram of bolted joint.

2.2. Occurrence condition of Tp

Notation l_c is defined as the distance where the bolt-nut threads contact takes place, which can be obtained geometrically as shown in equation (1)

$$(n_c - 1)\alpha = 2C_x$$

$$C_x = \frac{C_y}{\tan\theta}, \qquad l_c = n_c p \tag{1}$$

In equation (1), p is the pitch of bolt (for example, p=2 mm for M16), α is the pitch difference, n_c is the number of nut thread in contact in figure 2 except for n=1, θ is the thread angle (=60°) and C_x and C_y are the horizontal and vertical clearances between bolt and nut threads. Note that equation (1) is valid when $n_c < 8$ which is the total nut threads number 8 and the nut length < 16 mm.

2.3. Pitch difference α vs prevailing torque

To obtain the relationship between torque and clamping force, the torque was controlled by using an electric torque wrench, and the clamping force was measured by using the strain gauge attached to the clamped body surface as shown in figure 3. The uniaxial strain gauge with a length of 2 mm KFG-2 (Kyowa Electronic Instruments Co., Ltd.) was used in this measurement. Before the experiments, calibration tests were performed by compressing the clamped body to obtain the relationship between the clamping force and surface strain as shown in figure 3. Similar tests were performed to calibrate the torque wrench as shown in figure 4.

IOP Conf. Series: Materials Science and Engineering **372** (2018) 012007 doi:10.1088/1757-899X/372/1/012007

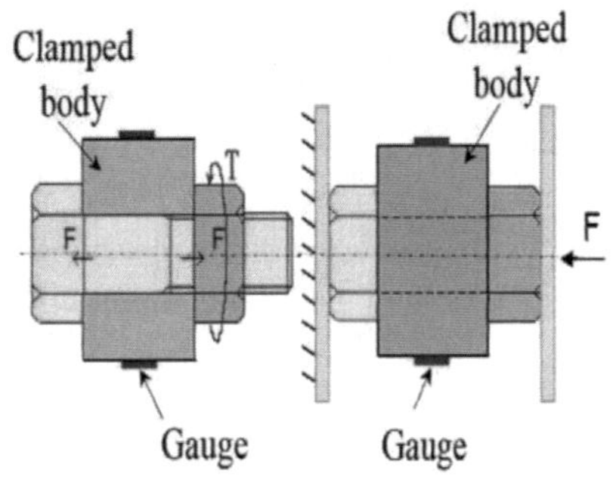

Figure 3. Calibration method for bolt axial force measurement.

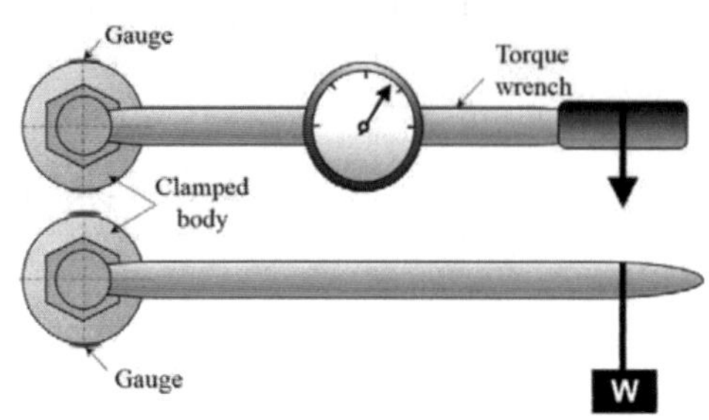

Figure 4. calibration method for torque wrench.

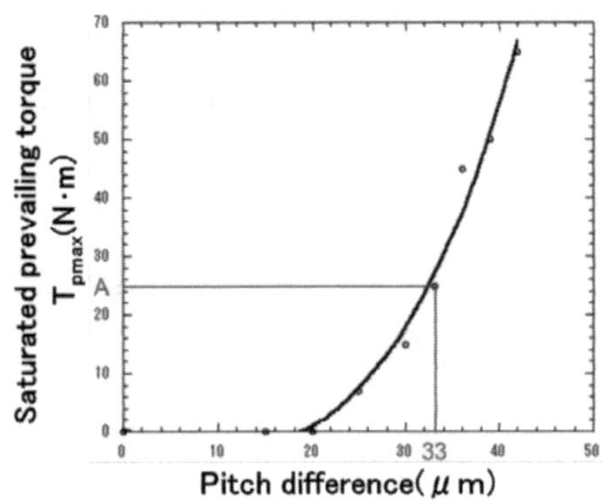

Figure 5. Relationship between torque and Pitch difference (M16).

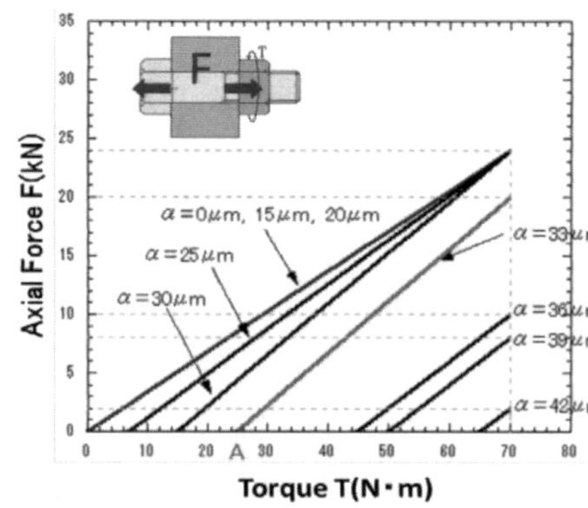

Figure 6. Relationship between torque and axial force (M16).

Figure 5 shows the prevailing torque T_p when the nut rotation is large enough by varying the pitch difference. From figure 5, it is seen that saturated prevailing torque T_p is nearly proportional to the pitch difference α in the range of $15\ \mu m < \alpha < 42\ \mu m$. Figure 6 shows the tightening torque vs clamping force experimentally obtained. When $\alpha = 0{\sim}30\ \mu m$, the torque–clamping force relationship is the same as the one of $\alpha = 0\ \mu m$. When $\alpha = 33\ \mu m$, the prevailing torque of 25 N·m is required before the nut touches the clamped plate. Under the same tightening torque of 70 N·m, the clamping force was reduced to 20 kN. When α = 39 μm, under a torque of 70 N·m the axial force decreased significantly to 8 kN, which was only 1/3 of the axial force of α=0 μm.

3. Loosening experiment

3.1. Device and conditions

The bolt-axial force 24 kN was considered for the standard bolt–nut, and the corresponding tightening torque was 70 N·m. It is tighten ring that insert to bolt at 70 N·m for nut. And experiment machine is built up. Also an extra length is 10mm. As shown in figure 7, the experimental device was an impact-vibration testing machine based on NAS3350 (National Aerospace Standard), whose vibration frequency was about 30 Hz, and vibration acceleration is 20 g. The maximum vibration cycle of NAS3350 is 30 000, therefore, if the number of vibration cycles was over 30 000, the anti-loosening performance may be considered to be good enough.

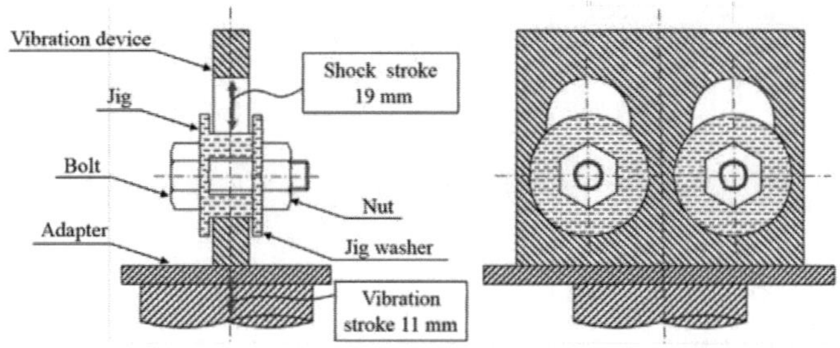

Figure 7. Loosening experimental device based on NAS3350.

3.2. Results

Table 2 lists the number of cycles for the start loosening and the nut dropping. Table 2 also lists the prevailing torque measured in the loosening experiments and the bolt axial forces estimated from figure 6. For each pitch difference α, two specimens were tested. For α=0 μm and $\alpha = 15$ μm, the nuts dropped at about 1000 cycles. For $\alpha = 33$ μm, no loosening was observed until 30 000 cycles and prevailing torque was 25 N · m. For $\alpha = 42$ μm, no loosening was observed until 30 000 cycles although the axial force was estimated to be only1~4 kN. It maybe concluded that if α is too small, the anti-loosening cannot be expected and if α is too large, the clamping ability is not good enough. By considering both the anti-loosening and clamping abilities, α=33~42 μm can be selected as the most suitable pitch difference.

Table 2. Anti-loosening Performance (M16).

Pitch difference α(μm)	Cycles for dropping	Prevailing torque (N·m)	Axial force(kN)
0	751	0	24
	876		
15	813	0	24
	1528		
33	30000	25	14~24
	30000		
42	30000	67	1~4
	30000	57	

4. Suitable pitch difference for various bolt-nut diameter

4.1. How to estimate suitable pitch difference

In this section, suitable pitch difference will be considered for various thread nominal diameters. First bolt-nut geometries are different depending on the diameter. Figure 8 shows JIS clearance C_x between the bolt and nut threads by varying nominal thread diameter d. It is seem that the clearance C_x can be expressed as equation (2)

$$C_x = 1.4882d + 41.239 \tag{2}$$

Similarly, figure 9 shows JIS nut thickness h by varying nominal thread diameter d. In this study, we have to estimate loosening force F_l which is closely related to the volume of the nut. Then, h is an important dimension. It is seen that nut thickness can be expressed as equation (3).

$$h = 0.3099d + 0.4253 \tag{3}$$

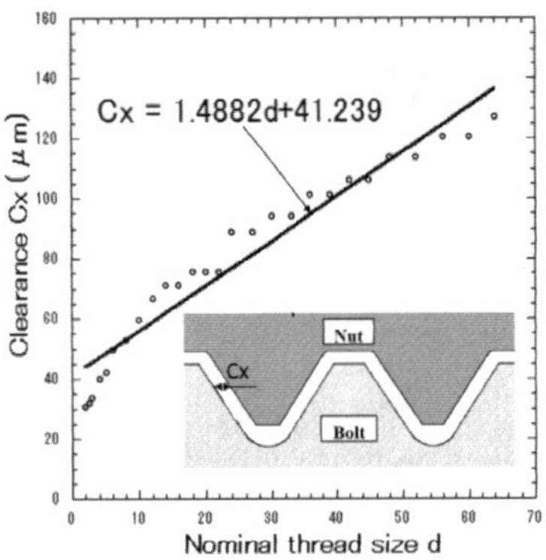

Figure 8. Relationship between Nominal thread size d and Clearance C_x.

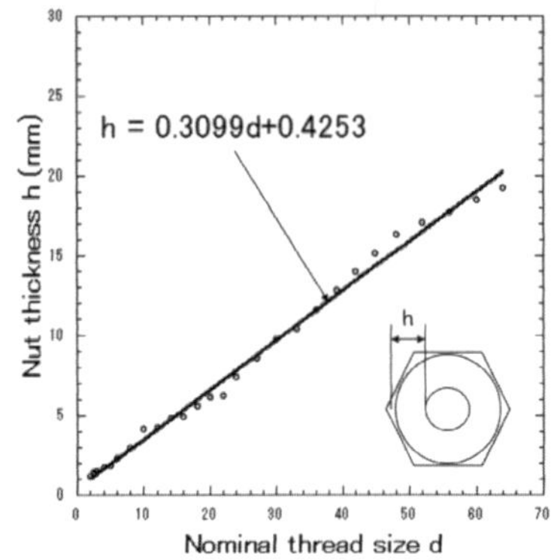

Figure 9. Relationship between Nut size d and Nut thickness h.

In the following discussion, loosening force F_l and anti-loosening force F_a are assumed. Figure 10 shows the relationship between F_l and F_a. On the one hand, loosening force F_l can be proportional as equation (4) since F_l should be proportional to the nut weight.

$$F_l \propto \gamma V \cong \gamma(\pi d)hl \tag{4}$$

Here, γ= specific gravity, V= the volume of nut, l= the length of nut.

On the other hand, anti-loosening force F_a should be proportional to be contact surface area between the nut and bolt threads.

$$F_a \propto \tau_s(\pi d)l_r \tag{5}$$

In equation (6) τ_s = average shear stress between the nut and bolt threads, and l_r is the length of the nut threads in contact in figure 10. Therefore, loosening condition can be expressed as equation (6) assuming loosening and anti-loosening forces should be balanced.

$$F_l = KF_a \tag{6}$$

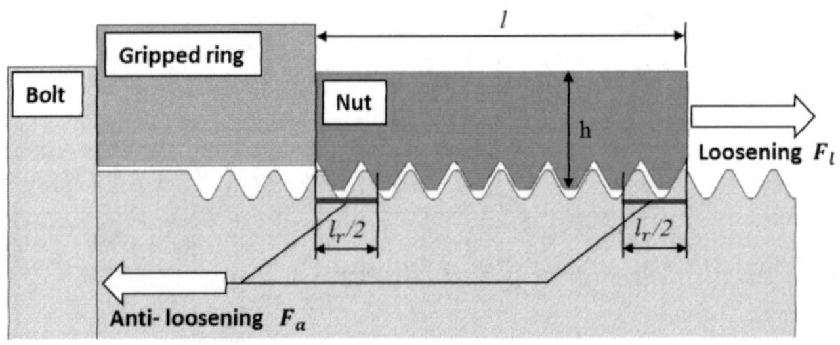

Figure 10. Relationship between loosening force F_l and anti-loosening force F_a.

Therefore, we have

$$\gamma(\pi d)hl = K\tau_s(\pi d)l_r, = \frac{\gamma hl}{\tau_s l_r} = \frac{\gamma}{\tau_s}K' \ , \ K' = \frac{lh}{l_r} \tag{7}$$

Here, K and $K'$are proportional constants. From figure 2 and equation (1) by using equation (7), the number of nut thread in contact in figure 2 will be expressed since n_c is important for anti-loosening.

$$l_r = \frac{lh}{K'}$$

From figure 2 with equation (1)

$$l_r = l - l_c = l - n_c p \tag{8}$$

$$n_c = n\left(1 - \frac{h}{K\prime}\right) \tag{9}$$

From table 2, suitable prevailing torque range to prevent loosening can be expressly as 30<T_p<60 for M16. Similarly, suitable pitch difference can be expressed as $\alpha_{min}^{suit} = 34\mu m < \alpha^{suit} < \alpha_{max}^{suit} = 41\ \mu m$. From equation (1), we have $\alpha = 2C_x/(n_c - 1)$. Assume $n_{c\ max}$ is the number of nut threads for α_{min}^{sut} and $n_{c\ min}$ is the number of nut threads for α_{max}^{sut}. Then,

$$n_{c\ max} = n\left(1 - \frac{h}{K\prime_{max}}\right) \tag{10}$$

$$n_{c\ min} = n\left(1 - \frac{h}{K\prime_{min}}\right) \tag{11}$$

Here, K'_{max} is the proportional constant for α_{min}^{sut} and K'_{min} is the proportional constant for α_{max}^{sut}. Therefore, generally, the suitable α can be expressed as

$$\frac{2C_x}{n_{c\max} - 1} < \alpha < \frac{2C_x}{n_{c\min} - 1} \tag{12}$$

By substituting equation (8) into (9),

$$\frac{2Cx}{n\left(1 - \frac{h}{K\prime_{max}}\right) - 1} < \alpha < \frac{2Cx}{n\left(1 - \frac{h}{K\prime_{min}}\right) - 1} \tag{13}$$

Since the values of the proportional constants K and K' are unknown, in this study, the loosening experimental results for M16 in table 2 is used. From $\alpha_{min}^{suit} = 34\ \mu m < \alpha^{suit} < \alpha_{max}^{suit} = 41\ \mu m$ for M16 with$C_x = 72\ \mu\ m$, equation (14) is obtained by using equation (1).

$$4.5 \le n_c \le 5.2 \tag{14}$$

By substituting $n_{c\ max}$ and $n_{c\ min}$ into (8), we have

$$5.6 \le l_r \le 7 \tag{15}$$

Here, For M16 bolt–nut connection, substituting by $l = 16\ mm, h = 5\ mm$(M16), equations (2) and (3).

$$K'_{min} = 11.4 \le K' \le K'_{max} = 14.3 \tag{16}$$

Assumed this K'_{max} and K'_{min} can be for various nominal thread diameter. By using equation (2) and equation (3) into (13)

$$\frac{2.9764d + 82.478}{\mathrm{n}\left(1 - \frac{0.3099\mathrm{d} + 0.4253}{14.3}\right) - 1} < \alpha < \frac{2.9764d + 82.478}{\mathrm{n}\left(1 - \frac{0.3099\mathrm{d} + 0.4253}{11.4}\right) - 1} \tag{17}$$

Table 3. Suitable pitch difference estimated for M12~M16.

Diameter	Pitch p(mm)	Nut length l(mm)	Number of threads n	Nut thickness h(mm)	Clearance C_x(μm)	Suitable Pitch difference range α(μm)
M12	1.75	10	5.7	4.1	59.1	38.6<α<44.8
M14	2	11	5.5	4.8	62.1	46.5<α<56.4
M16	2	13	6.5	5.4	65.1	42.6<α<53.5
	2	16	8	5.4	65.1	32.6<α<40.4

4.2. Validity of equation (17)

Suitable pitch difference can be estimated by using equation (17). Table 3 shows suitable pitch difference obtained from equation (17) for M12~M16 bolt-nut connection. As an example, for M12 the suitable α can be expressed as $38.6\ \mu m < \alpha < 44.8\ \mu m$. By using Junker vibration test, we have confirmed that $\alpha = 40\ \mu m$ have no loosening equation (17) is valid. Therefore, the validity of equation (17) is confirmed for M12.

5. Conclusion

In the previous study, a suitable pitch difference between the bolt-nut was obtained as $\alpha = 33\ \mu m$ for M16 JIS bolt-nut through loosening experiment and FEM simulation for tightening process. In this paper, suitable pitch difference was considered for other nominal thread diameters to realize anti-loosening performance. Suitable pitch difference was determined by assuming that loosening force should be balanced to the anti-loosening force. Then, the suitable pitch difference was expressed as approximate formula. By using Junker vibration test, we have confirmed that the proposed formula is valid for M12 since $\alpha = 40\ \mu m$ have no loosening equation (17) is valid.

References

[1] Hard Lock Kogyo KK 2002 Hard lock nut *Japan Patent* 2002-195236
[2] Izumi S, Yokoyama T, Teraoka T, Iwasaki A, Sakai S, Saito K, Nagawa M and Noda H 2005 Verification of anti-loosening performance of super slit nut by finite element method *Trans Jpn Soc Mech Eng* **71** 380-6
[3] Chen D H, Shimizu E and Masuda K 2012 Relation between thread deformation and anti-loosening effect for nut with circumference slits *Trans Jpn Soc Mech Eng* **78** 390-402
[4] Goodier J N and Sweeney R J 1945 Loosening by vibration of threaded fastenings *Mechanical Engineering* **67** 798-802
[5] Nishida S I 1944 Failure Analysis in Engineering Applications, Butterworth Heinemann, Oxford
[6] Maruyama K 1975 Stress analysis of a bolt-nut Joint by the finite element method and the copper-electroplating method (3nd Report) *Trans Jpn Soc Mech Eng* 2292-302
[7] Noda N A, Chen X and Kuhara M 2008 The reduction of stress connection by tapering threads *Trans Jpn Soc Mech Eng* **74** 1305-13
[8] Xiao Y, Wang Q, Noda N A, Akaishi Y, Takase Y and Nishida S 2011 Stress reduction effect of tapering thread bolts and nuts which have slightly different pitches *Trans Soc Automotive Engineers Jpn* **42** 927-33
[9] Akaishi Y, Chen X, Yu Y, Tamasaki H, Noda N A, Sano Y and Takase Y 2013 Fatigue strength analysis for bolts and nuts which have slightly different pitches considering clearance *Trans Soc Automotive Engineers Jpn* **44** 1111-7
[10] Chen X, Noda N A, Maruyama H, Wang H, Sano Y and Takase Y 2014 Experimental investigation on the suitable pitch difference improving the fatigue strength significantly for high strength bolts and nuts *Proceedings of the 3rd International Conference on Fracture Fatigue and Wear* **2** 195-200
[11] Chen X, Noda N A, Wahab M A, Akaishi Y I, Sano Y, Takase Y and Fekete G 2015 Fatigue failure analysis in bolt-nut connection having slight pitch difference using experiments and finite element method *Acta Polytechnica Hungarica* **12** 61-79
[12] Takase Y, Noda N A, Chen X, Sano Y, Fujisawa R, Maruyama H and Wahab M A 2015 Fatigue strength evaluation for bolt-nut connections having slight pitch difference considering incomplete threads of nut *Proceedings of 4th International Conference on Fracture Fatigue and Wear* **3** 43-9
[13] Noda N A, Sano Y, Takase Y, Chen X, Maruyama H, Wang H and Fujisawa R 2015 Anti-loosening performance of special bolts and nuts having enhanced fatigue life by introducing pitch difference *Transactions of Society of Automotive Engineers of Japan* **46** 121-6

[14] Chen X, Noda N A, Wahab M A, Sano Y, Maruyama H, Wang H, Fujisawa R and Takase Y 2015 Fatigue life improvement by slight pitch difference in bolt-nut connections *Journal of the Chinese Society of Mechanical Engineers* **37** 11-9

[15] Noda N A, Chen X, Sano Y, Wahab M A, Maruyama H, Fujisawa R and Takase Y 2015 Effect of pitch difference between the bolt-nut connections upon the anti-loosening performance and fatigue life *The Journal of Strain Analysis for Engineering Design* **96** 476-89

IOP Conf. Series: Materials Science and Engineering **372** (2018) 012008 doi:10.1088/1757-899X/372/1/012008

Effect of hydrogen on void initiation in tensile test of carbon steel JIS-S25C

S Sugawa[1], N Tsutsumi[2,3] and K Oda[2]

[1]Graduate School of Oita University, 700 Dannoharu, Oita 870-1192, Japan
[2]Division of Mechanical Engineering, Oita University, 700 Dannoharu, Oita 870-1192, Japan

E-mail: tsutsumi-noriko@oita-u.ac.jp

Abstract. In order to investigate the effect of hydrogen on tensile fracture mechanism of a carbon steel, tensile tests were conducted. Pre-strain specimens (0%, 5% and 10%) were used to study the effect of hydrogen content, since saturated hydrogen content in specimens increases in increasing dislocation density. The tensile strength and the yield stress of hydrogen specimens were almost the same as uncharged. In contrast, the reduction of area of hydrogen charged specimens was smaller than that of uncharged. To reveal the reasons of decrease of the reduction of area, the fracture surface and longitudinal cross section near the fracture surface were observed. On the fracture surface of uncharged specimens, only dimples were observed. On the other hand, dimples and flat fracture surface were observed on the fracture surface of hydrogen charged. On the longitudinal cross section of hydrogen charged specimens, many voids were observed compared to uncharged. From these observations, it is showed that hydrogen gives a rise to the increase of voids and the hydrogen charged specimens break without sufficient necking, thus hydrogen makes the reduction of area smaller.

1. Introduction

In recent years, many researchers [1-7] have been studied about the effect of hydrogen on tensile strength properties. It has been reported that tensile strength and yield stress of hydrogen charged specimens were almost the same as those of uncharged. In contrast, the reduction of area of hydrogen charged specimens was smaller than that of uncharged. During tensile tests, voids are initiated in the specimens and lead to failure. However, there are no studies about the effect of hydrogen on tensile fracture mechanism in terms of void initiation and distribution.

In this study, the effect of hydrogen on void initiation and distribution of a carbon steel was studied to reveal the tensile fracture mechanism. Fracture surface and longitudinal cross section near fracture surface in broken specimens were observed. Longitudinal cross section around necking of unbroken specimens was also observed. And voids distribution in specimens before break was compared with those after break.

2. Materials and experimental procedure

2.1. Materials

Tensile tests were conducted using a normalized 0.24% carbon steel, JIS (Japanese Industrial Standards) S25C [8]. Table 1 shows the chemical compositions of the carbon steel JIS-S25C. Figure 1 shows the shape and dimensions of the tensile specimens. The specimen surface was polished with emery paper #800. Pre-strain specimens (pre-strain: ε_{pre} = 0%, 5% and 10%) were used to study the effect of hydrogen content, since saturated hydrogen content in specimens increases in increasing dislocation density. Pre-strain was introduced in specimens at cross-head speed of 1 mm/min.

Table 1. Chemical compositions of JIS-S25C. (mass%)

	C	Si	Mn	P	S	Cu	Ni	Cr	NC
S25C	0.24	0.22	0.51	0.015	0.017	0.08	0.05	0.16	0.21

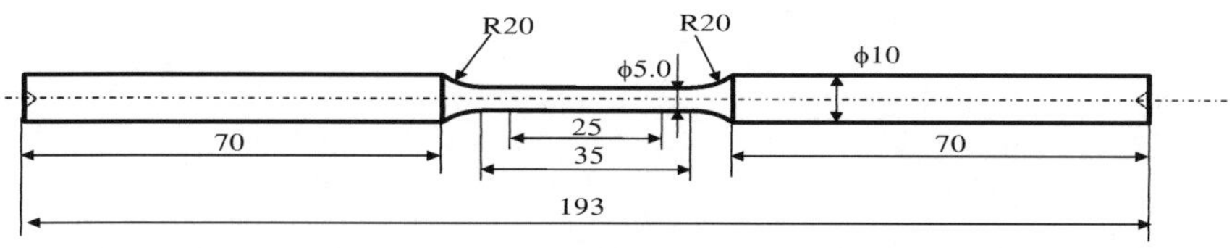

Figure 1. Shape and dimensions of tensile specimens (in mm).

2.2. Experimental procedure
The specimens were charged with hydrogen by immersing in 20 mass% NH_4SCN solution at 313 K for 48 hours. According to the results of the experiment by Kaneko et al. [9], the hydrogen content of uncharged specimens of JIS-S25C was 0.0 ppm. The hydrogen content of 0% pre-strained hydrogen charged specimens was 0.24 ppm, 5% pre-strained hydrogen charged specimens was 0.67 ppm, 10% pre-strained hydrogen charged specimens was 1.20 ppm. Hydrogen content of hydrogen charged specimens increased in increasing pre-strain. Tensile test was carried out with crosshead speed of 1 mm / min at room temperature in air.

3. Results

3.1. Tensile properties
Table 2 shows tensile properties. Figure 2 shows the stress-strain diagram. The tensile strength σ_B and yield stress σ_{UY}, σ_{UL} of hydrogen charged specimens were almost the same as those of uncharged. In contrast, elongation at fracture of hydrogen charged specimens was smaller than that of uncharged.

Figure 3 shows the reduction of area. The reduction of area of hydrogen charged specimens was obviously smaller than uncharged and decrease in increasing pre-stain. Thus, the reduction of area tends to decrease in increasing hydrogen content of specimens.

Table 2. Tensile properties.

S25C	Uncharged			Hydrogen charged		
	Upper yield stress σ_{UY} [MPa]	Lower yield stress σ_{LY} [MPa]	Tensile strength σ_B [MPa]	Upper yield stress σ_{UY} [MPa]	Lower yield stress σ_{LY} [MPa]	Tensile strength σ_B [MPa]
Pre-strain 0%	386	312	499	399	325	494
Pre-strain 5%	382	313	498	375	316	497
Pre-strain 10%	379	311	499	378	309	498

3.2 Fracture surface
Figure 4 shows fracture surface of uncharged and hydrogen charged specimens. On fracture surface of uncharged specimen (Figure 4 (a), (b) and (c)), only typical dimples were observed. On the other hand, on fracture surface of hydrogen charged specimen (Figure 4 (d), (e) and (f)), typical dimples and flat fracture surface were observed. Most of the flat fracture surfaces were formed circularly around inclusions. Figure 5 shows the relationship between pre-strain and area of the flat fracture surface of hydrogen charged specimens. As the pre-strain increased, the area of the flat fracture surface increased. This result shows that the area of flat fracture surface increase in increasing hydrogen content.

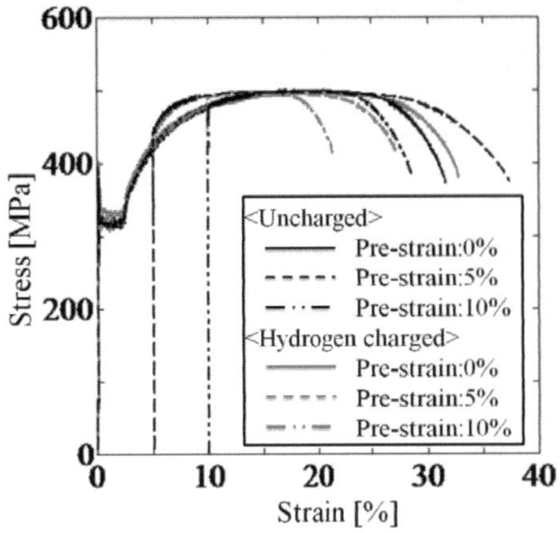

Figure 2. Stress-Strain diagram.

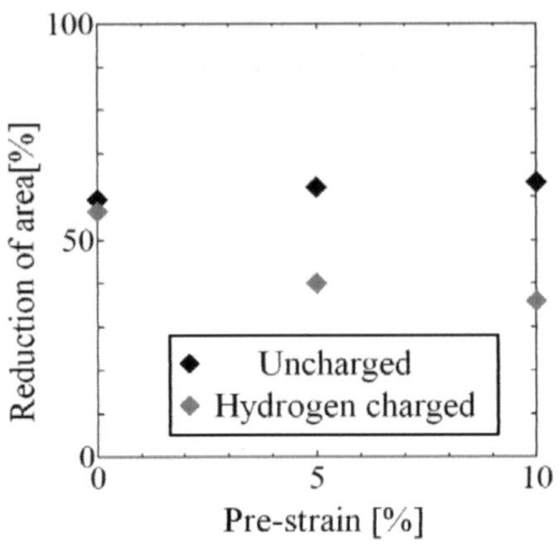

Figure 3. Reduction of area.

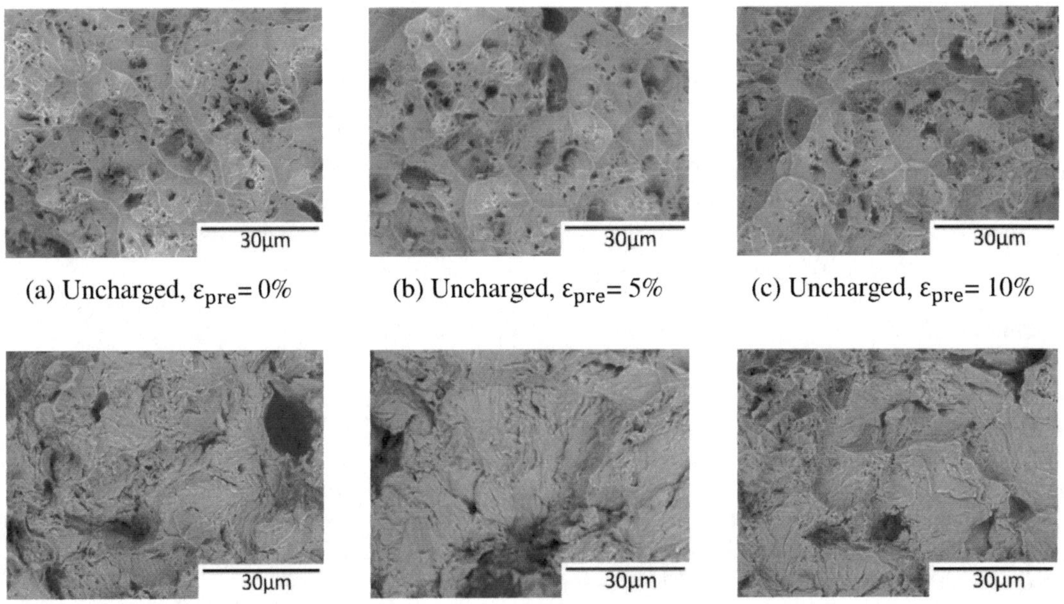

(a) Uncharged, ε_{pre}= 0% (b) Uncharged, ε_{pre}= 5% (c) Uncharged, ε_{pre}= 10%

(d) Hydrogen charged, ε_{pre}= 0% (e) Hydrogen charged, ε_{pre}= 5% (f) Hydrogen charged, ε_{pre}= 10%.

Figure 4. Fracture surface.

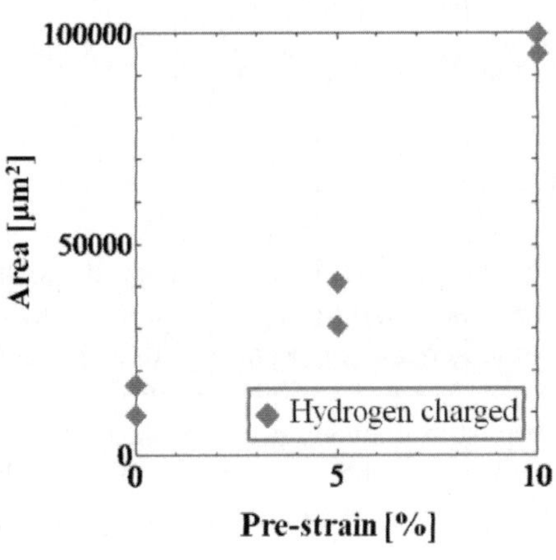

Figure 5. Relationship between pre-strain and area of flat fracture surface.

3.3. Longitudinal section through fracture surface

In order to reveal the tensile fracture mechanism of uncharged and hydrogen charged specimens, longitudinal section of broken and unbroken specimens was observed. Broken specimens were carried out tensile tests to failure. And unbroken specimens were carried out tensile tests up to initiation of necking, just after tensile strength. The longitudinal section of broken specimens was observed near fracture surface and that of unbroken was observed around necking. To study the relationship between voids and metal structure, the cross section was polished by buff and etched by natal (solution of 97% alcohol and 3% nitric acid).

Figure 6 shows the longitudinal section of fracture surface. Almost the voids of the uncharged specimens were formed in longitudinal direction. On the other hand, almost the voids of the hydrogen charged specimens were formed in perpendicular to the longitudinal direction. Almost all voids of hydrogen charged and uncharged specimens were initiated from pearlite structure and inclined to loading direction since voids might occur by slip deformation.

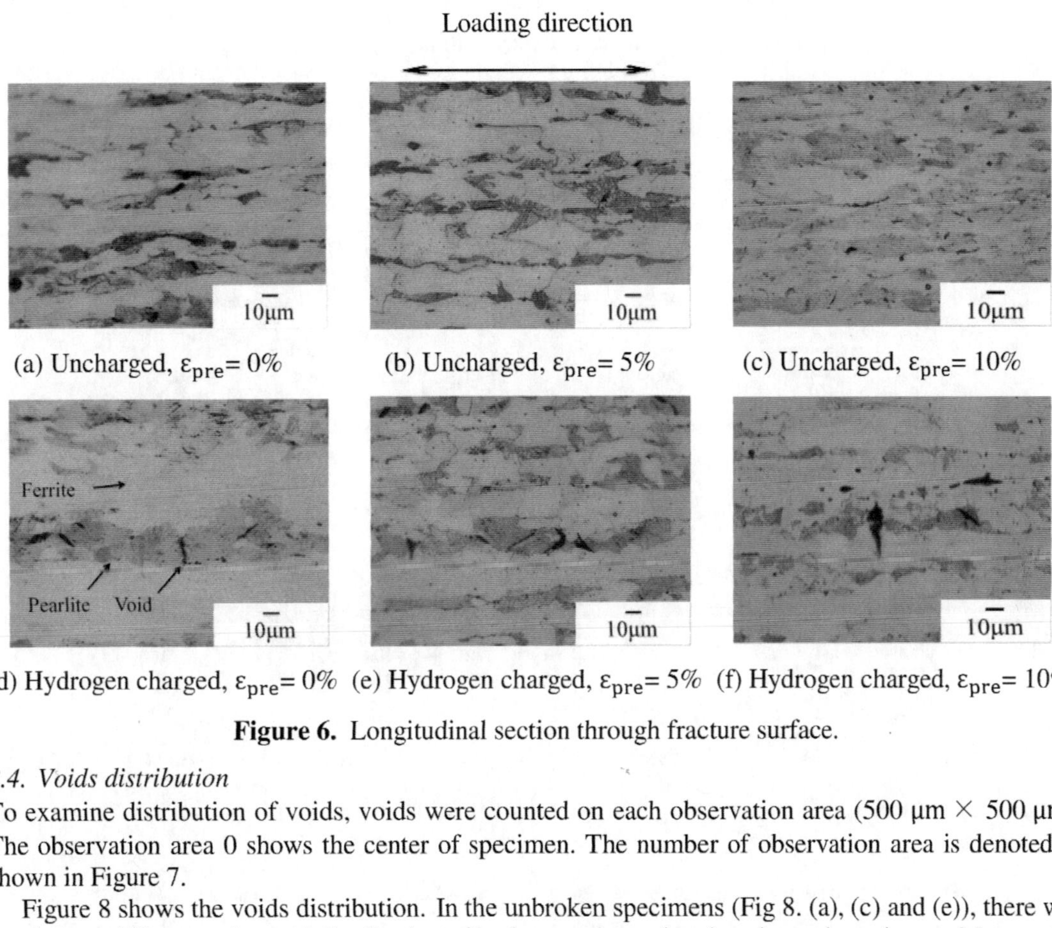

(a) Uncharged, ε_{pre}= 0% (b) Uncharged, ε_{pre}= 5% (c) Uncharged, ε_{pre}= 10%

(d) Hydrogen charged, ε_{pre}= 0% (e) Hydrogen charged, ε_{pre}= 5% (f) Hydrogen charged, ε_{pre}= 10%

Figure 6. Longitudinal section through fracture surface.

3.4. Voids distribution

To examine distribution of voids, voids were counted on each observation area (500 μm × 500 μm). The observation area 0 shows the center of specimen. The number of observation area is denoted as shown in Figure 7.

Figure 8 shows the voids distribution. In the unbroken specimens (Fig 8. (a), (c) and (e)), there was no obvious difference the void distribution of hydrogen charged and uncharged specimens. Many voids on longitudinal cross section of uncharged broken specimen (Fig 8. (b), (d) and (f)) were observed near the center of specimens and a few voids near specimen surface. On the other hand, many voids on longitudinal cross section of hydrogen charged broken specimen (Fig 8. (b), (d) and (f)) were observed over the cross section.

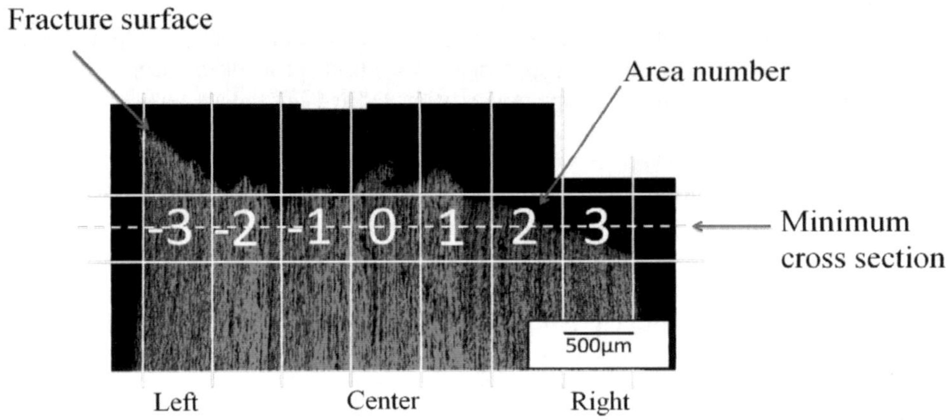

Figure 7. Observation area of longitudinal section.

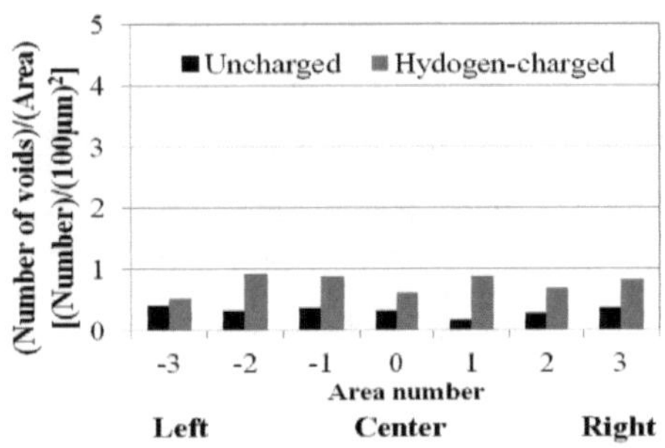

(a) Unbroken specimens, ε_{pre}= 0%

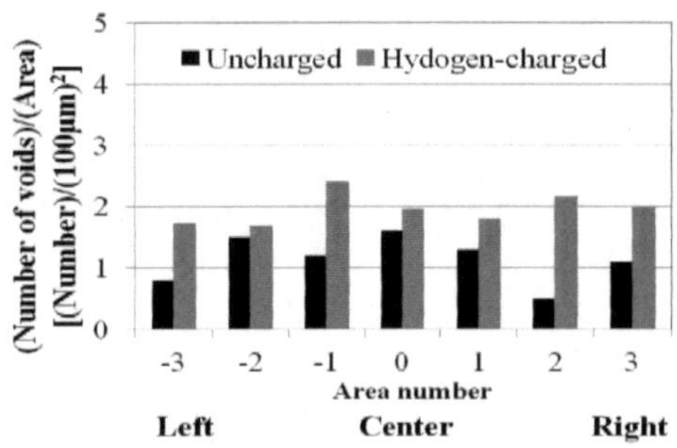

(b) Broken specimens, ε_{pre}= 0%

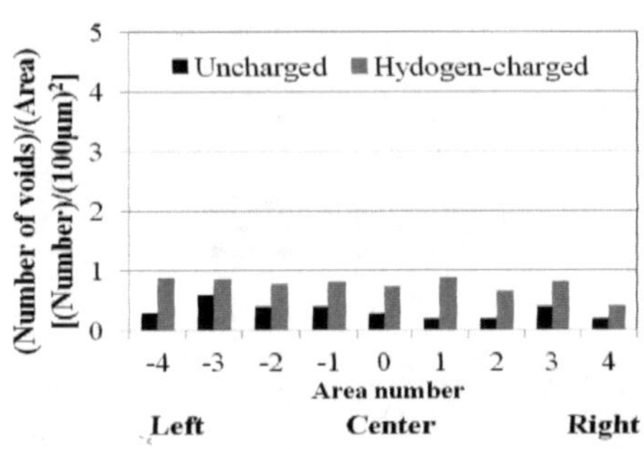

(c) Unbroken specimens, ε_{pre}= 5%

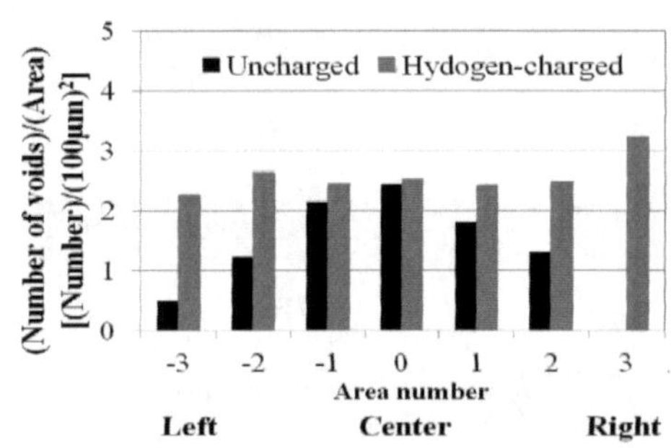

(d) Broken specimens, ε_{pre}= 5%

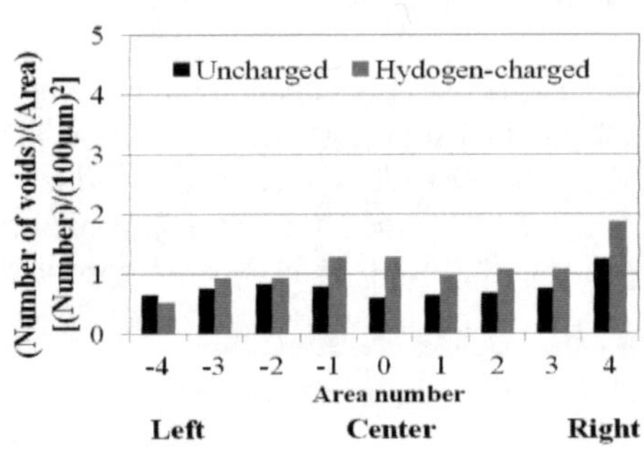

(e) Unbroken specimens, ε_{pre}= 10%

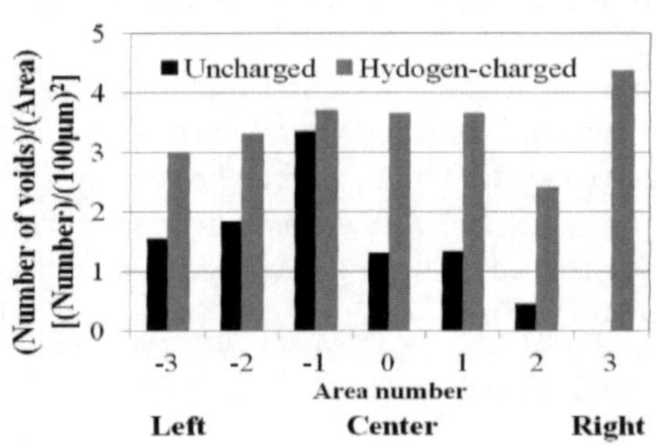

(f) Broken specimens, ε_{pre}= 10%

Figure 8. Voids distribution near fracture surface.

4. Discussion

From the above results and observations, the mechanism of tensile fracture can be explained as follows. Figure 9 shows the mechanism of tensile fracture. In hydrogen charged specimens, many voids initiate over the cross section. And flat fracture surface and voids grow and merge to others, thus hydrogen charged specimens break before necking progress substantially. On the other hand, in uncharged specimen, fewer voids initiate compared with hydrogen charged. Necking progress and the stress distribution at the center of specimens is triaxial stress state. Because stress at the center of specimens is higher than that near surface, more voids initiate in the center of specimens. And the uncharged specimen breaks after the reduction of area becomes larger sufficiently. Hydrogen charged specimens fractured faster than uncharged specimens, because the number of voids is more for hydrogen charged specimens.

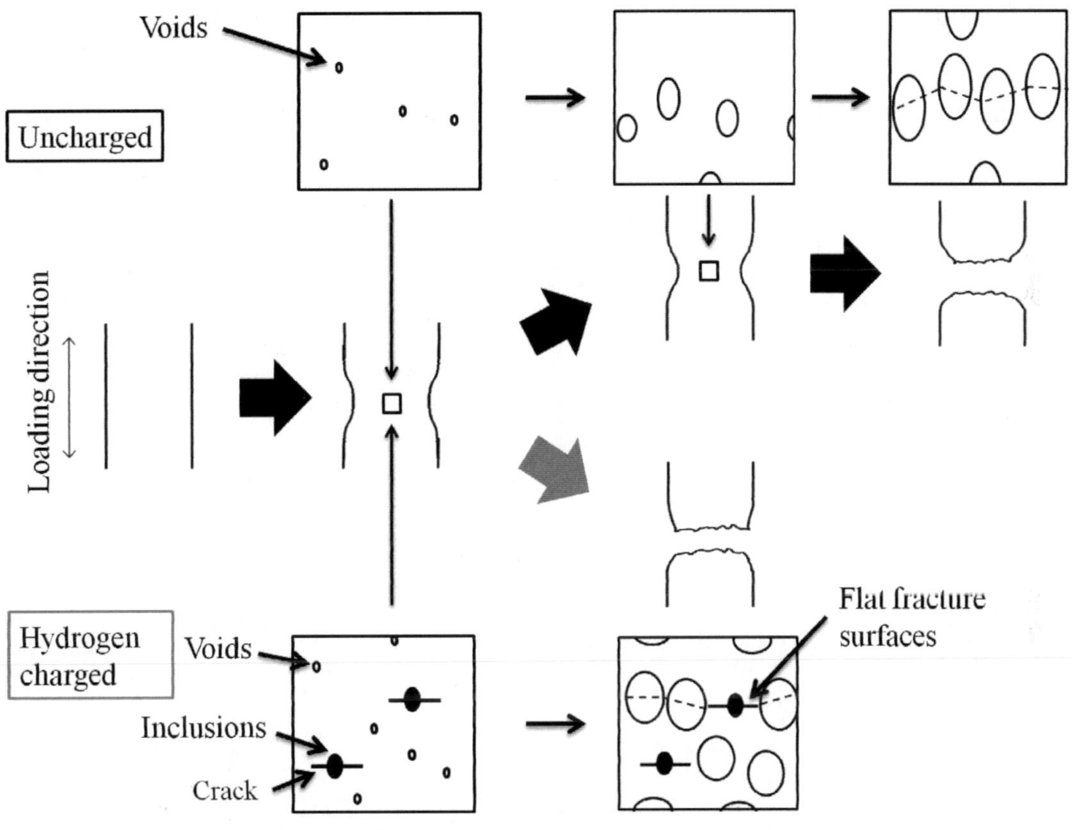

Figure 9. Mechanism of tensile fracture.

5. Conclusions

Tensile tests were conducted using hydrogen charged and uncharged specimens of a 0.24% carbon steel JIS-S25C to reveal the effects of hydrogen on void initiation in tensile tests of a carbon steel. The conclusions can be summarized as follows.

(1) The tensile strength and the yield stress of hydrogen specimens were almost the same as those of uncharged. In contrast, the reduction of area of hydrogen charged specimens was smaller than that of uncharged.

(2) On fracture surface of uncharged specimen, only typical dimples were observed. On the other hand, on fracture surface of hydrogen charged specimen, typical dimples and flat fracture surface were observed. As pre-strain increased, the area of the flat fracture surface on hydrogen charged specimens also increased because of hydrogen content.

IOP Conf. Series: Materials Science and Engineering **372** (2018) 012008 doi:10.1088/1757-899X/372/1/012008

(3) Many voids on longitudinal cross section of uncharged specimen were observed near the center of specimens and a few voids near specimen surface. On the other hand, many voids on longitudinal cross section of hydrogen charged specimen were observed over the cross section.

(4) A flat fracture surface in the hydrogen charged specimens might initiate before the voids initiation.

References

[1] Vesely Jr. E J, Jacobs R K, Watwood M C and McPherson W B 1994 Influence of Strain Rate on Tensile Properties *Hydrogen Effects in Materials* 363–74

[2] NASA 1997 *Safety Standard for Hydrogen and Hydrogen Systems* NSS 1740.16

[3] Matsuoka S, Homma N, Tanaka H, Fukushima Y and Murakami Y 2006 Effects of Hydrogen on the Tensile Properties of 900-MPa-Class JIS-SCM435 Low-Alloy-Steel for Use in Storage Cylinder of Hydrogen Station, *J. Japan Inst. Metals* **70** 1002–11.

[4] Matsuo T, Homma N, Matsuoka S and Murakami Y 2008 Effect of Hydrogen and Prestrain on Tensile Properties of Carbon Steel SPG (0.078 C-0.012 Si-0.35 Mn, mass%) for 0.1 MPa Hydrogen Pipelines *Trans. Japan Soc. Mech. Eng. A Series A* **74** 120–9.

[5] Nishiguchi H, Fukushima Y and Matsuoka S 2008 Effects of Hydrogen and Prestrain on Tensile Properties of Low-Carbon Steel Used in Hydrogen Pipelines *Int. Hydrogen Conf.* 365–72

[6] Nishiguchi H, Fukushima Y, Matsuoka S and Murakami Y 2010 Effect of Hydrogen on Tensile Properties of Ferritic-Pearlitic Carbon Steels *Trans. Japan Soc. Mech. Eng. A* **76** 77–86.

[7] Matsuo T, Yamabe J and Matsuoka S 2014 Effects of hydrogen on tensile properties and fracture surface morphologies of Type 316L stainless steel *Int. J. Hydrogen Energy* **39** 3542–51

[8] Japan Industrial Standards Committee 2009 Carbon steels for machine structural use *Japanese Industrial Standards* JIS G4051

[9] Kaneko S and Tsutsumi N 2011 Effects of Hydrogen on Surface Formation of Carbon Steels in Tensile Tests *Proc. Japan Soc. Mech. Eng. Kyushu Branch* 83–4

Special sliding door with storable handrail to support senior and handicapped persons to walk by themselves

K Saitou[1,4], N-A Noda[2], Y Sano[2], Y Takase[2], K Murai[2], Z F Wang[2], S Q Li[2], X Liu[2], H Tanaka[2] and Y Kubo[3]

[1]Yahata Rolling Mechanical Engineering Department, Engineering Division, Mechanical Engineering Unit, Nippon Steel & Sumikin Texeng. Co. Ltd, Kukioka 2142-1 Maeda Yahata higashi-Ku, Kitakyushu-Shi, Fukuoka, Japan
[2]Mechanical Engineering Department, Kyushu Institute of Technology, Sensui-Cho 1-1 Tobata-Ku, Kitakyushu-Shi, Fukuoka, Japan
[3]Kei products Co. Ltd, Wakahuji-Cho1-26-203 Kokurakita-Ku, Kitakyushu-Shi, Fukuoka, Japan

E-mail: saitou.kinjirou.ks@tex.nssmc.com

Abstract. In this paper, special sliding door is designed in order to support senior and handicapped persons to walk by themselves in hospitals and nursing facilities. This semi-automatic lifting equipment is utilized for the storable handrail to make sure the bad health persons are able to open the door by using a weak force. In this study, to design the equipment of the handrail, the theoretical formula of opening force is derived. Then the simulation is performed by varying geometrical conditions. The simulation results are compared with the experiment results.

1. Introduction

For elderly people needing nursing care, handrails are often installed all over the entire side of the corridor and walking places in their homes and hospitals so that they can grab the handrail and walk by themselves. Keep walking is very important for elderly people to maintain their walking ability as well as to restore their walking function [1-3]. However, Japanese hospitals and homes usually have a lot of sliding doors, which prevent installing the standard handrail. In other words, since Japanese homes and hospitals have a lot of no handrail space at sliding doors, elderly people cannot walk by themselves by using handrail. From this point of view, in this research, we aim to develop a special sliding door with storable handrails [4,5] to support eldering people walking by themselves. Figure 1 shows the prototype special sliding door considered in this study. To obtain the most desirable handrail geometry, theoretical formula will be derived to estimate the opening force of the sliding door. Also experimental measurement will be performed for the prototype sliding door to confirm the validity of the theory.

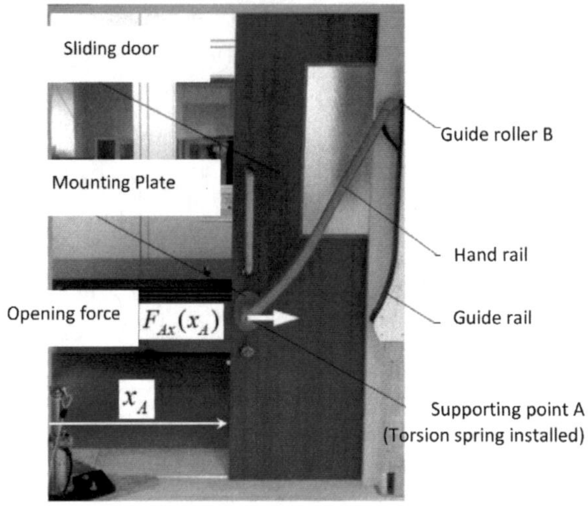

Figure 1. Prototype sliding door with storable handrail.

2. Structure of the sliding door

2.1. Structure of sliding door

Figure 1 shows the prototype handrail storable sliding door considered in this study. This special door consists of a sliding door, a storable handrail, a fulcrum handle, a rotating roller at the end of a storable handrail and a guide rail. The torsion spring is used to reduce the torque due to the handrail weight W for opening the door (see the moment M in figure 2). By using the torsion spring, the opening force can be reduced. The spring constant is chosen to reduce the opening force in the prototype handrail. When the sliding door is completely closed, the storable railing serves as a handrail to support elderly people walking. When the sliding door is being opened by someone, the storable handrail is rotating on the point A in anticlockwise direction and the guide roller B is going upward along the guide rail. When the door is being closed, the storable handrail is rotating on the point A in clockwise direction and the guide roller B is going downward along the guide rail due to the own weight of the handrail.

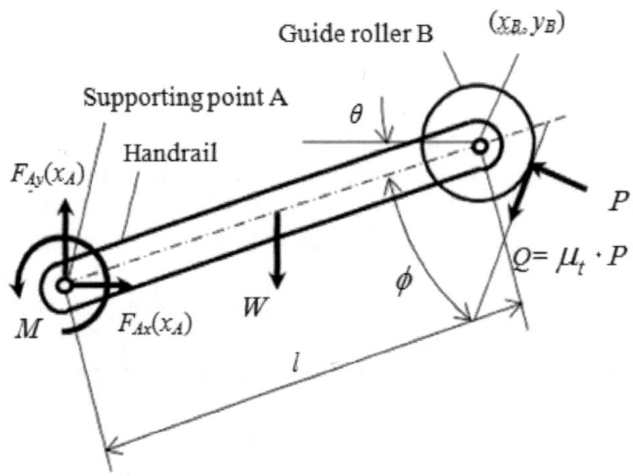

Figure 2. Equilibrium of force with a torsion spring.

2.2. Notations

The following notations will be used in this paper.

$F_{Ax}(x_A)$: Opening force applied at point A in figure 2. The target value $F_{Ax}(x_A)$=19.6 N

$F_{Ay}(x_A)$: Reaction force applied to the handrail at point A $(x_A, 0)$ in the y-direction in figure 2

$(x_A, y_A) = (x_A, 0)$: Coordinate of the supporting point A in figure 3, x_A = Opening distance of sliding door

(x_B, y_B) : Coordinate of point B in figure 3, point B = Center of guide roller

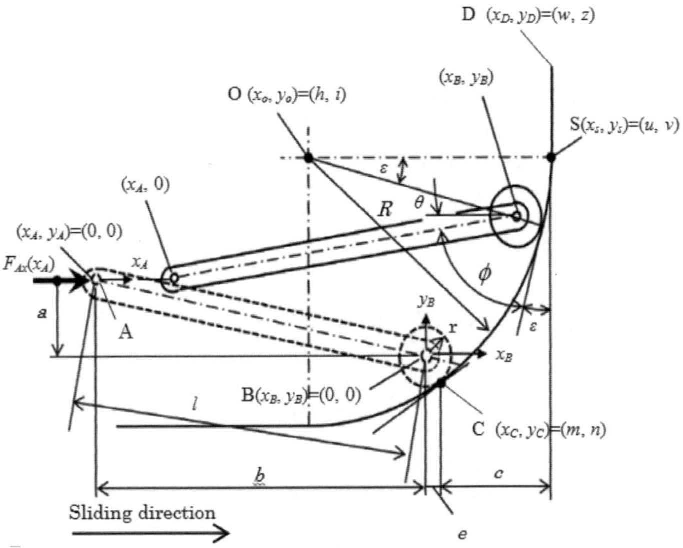

Figure 3. Model of storable handrail.

(x_0, y_0) : Coordinate of point O in figure 3, point O= Center of arc portion of guide rail

Q : $Q = \mu_t \cdot P$, Running resistance (see figure 2)

P : Reaction force to the handrail from the guide rail at point B in figure 2

R : Rolling surface radius of guide rail stand in figure 3 (R = 478 mm for prototype)

θ : Angle between the storable handrail and the horizontal line (see figure 3)

ϕ : Handrail angle between the tangential direction of the guiderail at B and hand rail (see figure 3)

ε : Guide rail angle between the tangential direction of the guiderail and vertical direction (see figure 3)

W : Weight of handrail including guide roller in figure 2 (W = 13.7 N for prototype steel handrail)

M : Moment due to torsion spring in figure 2 ($M = k(0.5\pi - \theta)$)

k : Spring constant (k =2395N・mm/rad)

μ_t : Friction coefficient of bearing 0.03 + friction coefficient of rotating roller 0.05 in figure 2 (μ_t =0.03+0.005=0.035) [6]

a : a= $y_b - y_a$, AB in the y-direction in figure 3 (a = 22.6 mm for prototype)

b : Horizontal difference, b= x_A - x_B , between point A – point B in figure 3 (b = 910.3 mm for prototype)

c : Distance from the guide roller contact point to the rail vertical point in figure 3 (c =58.2 mm for prototype)

r	: Radius of the guide roller B in figure 3 (r =17.5 mm for prototype)
e	: The distance in the x direction from the horizontal distance between center point of the guide roller B and the contact point of the roller and the rail in figure 3
l	: Length of storable handrail rod in figure 3 (l =910.3 mm for prototype)
C	: Contact point of the roller in figure 3
S	: End point of the curved portion of the guide rail in figure 3
D	: Guide rail end in figure 3

3. Analysis of sliding door opening force

3.1. Equilibrium of the handrail to obtain the opening force

Figure 3 shows the handrail model to be considered. Assume the torsion spring is not installed to the rotary fulcrum shaft. Figure 2 shows the free body diagram for the handrail where a torsion spring is installed to the rotation support shaft. Equations (1)-(3) are derived from the equilibrium in figure 2.

$$F_{Ax}(x_A) = Q\cos(\theta+\phi) + P\sin(\theta+\phi) \tag{1}$$

$$F_{Ay}(x_A) + P\cos(\theta+\phi) = W + Q\sin(\theta+\phi) \tag{2}$$

$$M + lP\cos\phi = \frac{1}{2}lW\cos\theta + lQ\sin\phi \tag{3}$$

Therefore, the following equations can be obtained [7].

$$F_{Ax}(x_A) = \mu_t P\cos(\theta+\phi) + P\sin(\theta+\phi) \tag{4}$$

$$F_{Ay}(x_A) = W + \mu_t P\sin(\theta+\phi) - P\cos(\theta+\phi) \tag{5}$$

In figures 2 and 3, two coordinates are used. One has the origin of the center coordinate (x_A, y_A) = (0, 0) at the rotation fulcrum A. The other has the origin (x_B, y_B) = (0, 0) at the center of the radius of curvature R. It is taken as reference coordinates.

Next, the central coordinates (x_B, y_B) of the guide roller B with respect to the moving amount x_A of the rotation fulcrum A, and θ and ϕ determined from the curvature radius R and its center point O (h, i) are obtained by the following equations (figure 3).

$$\theta = \tan^{-1}\frac{y_B - a}{\sqrt{l^2 - (y_B - a)^2}} \tag{6}$$

$$\phi = 90° - (\theta + \varepsilon) \tag{7}$$

$$\varepsilon = \sin^{-1}\frac{i - y_B}{R - r} \tag{8}$$

From equations (1) to (8), $F_{Ax}(x_A)$ can be obtained as equations (4), (5). Note that the effect of inertial force on opening force $F_{Ax}(x_A)$ is small enough and can be negligible.

3.2. Experimental measurement of the opening force

Figure 4 shows the measured opening force for the prototype sliding door as the dashed. In figure 4, the results of the equation (4) are also indicated as the solid lines by varying the curvature radius R of the guide rail. The experimentally obtained opening force was measured by using a spring scale when the door is opening at a constant speed of 15 mm/s. The theoretical value was obtained for five kinds of curvature radii in the range of R = 200 mm to 600 mm. The opening force $F_{Ax}(x_A)$ gradually increases with increasing x_A and taking a peak value at the vicinity of x_A = 79 - 99 mm, and thereafter decreases suddenly. It is found that the maximum value of the opening force is generated when the handrail end B is at the end of curved position. The maximum value of the opening force increases with decreasing the radius of curvature R. The maximum load obtained theoretically when the radius of curvature R = 478 mm is $F_{Ax\max}$ = 19.5 N and the maximum load obtained experimentally $F_{Ax\max}^{Exp}$ = 22.4 N. Those difference 13% may be caused by the error of the analytical and experimental results. The experimental error may be caused by the fabrication and assembly errors of guide rails. The suitable opening force for elderly people can be estimated as $F_{Ax\max}$ =19.6 N. For the elderly people, it is dangerous if the opening force $F_{Ax}(x_A)$ is too small for this sliding door because this door is used as the handrail. In figure 4, this target load can be satisfied theoretically as shown the dashed line in figure 4. However, the experimentally measured opening force for the prototype door is larger than this target value by 2.8 N, which is 14% larger than the target value. Such load difference of 2.8 N can be reduced by improving the manufacturing proficiency for the storable handrail and by adjusting the torsion spring attached to the rotational fulcrum A. The load difference of 2.8 N corresponds to the torsion spring torque error about 400 N·mm. In other words, the target opening force can be realized by the design data shown in Section 2.2.

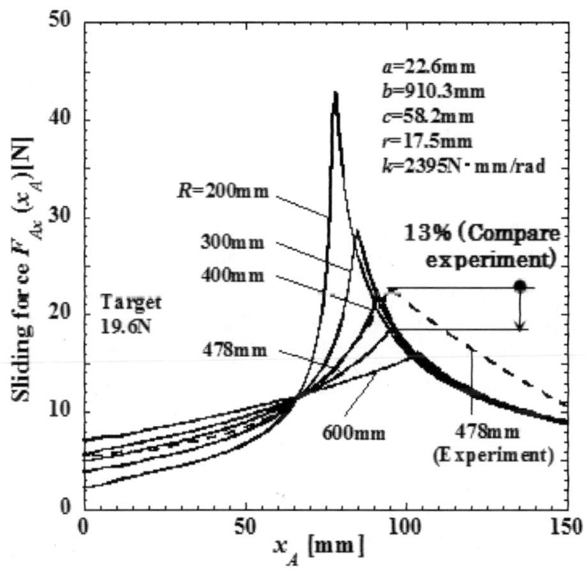

Figure 4. Sliding force $F_{Ax}(x_A)$ calculated by theoretical.

3.3. Consideration of sliding door opening force behavior

In figure 4, there is a peek point where the opening force $F_{Ax}(x_A)$ abruptly increases to the maximum value. It can be seen that this maximum value increases with decreasing the curvature radius R of the guide rail. Figure 4 shows the both results of $F_{Ax}(x_A)$ obtained theoretically and experimentally. Figure 5 compares with increases $F_{Ax}(x_A)$ and Handrail angle ϕ , and takes peak value.

It should be noted that the maximum value of $F_{Ax}(x_A)$ appear at $x_A \cong$ 90mm and the maximum value ϕ also appears at $x_A \cong$ 87 mm.

Note that in figure 5 contact of Point S (x_S, y_S) =(*u*, *v*).

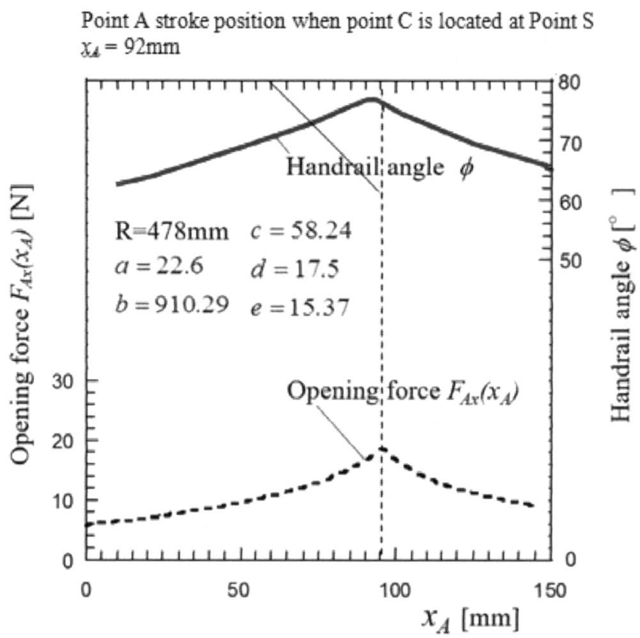

Figure 5. Comparison between the opening force $F_{Ax}(x_A)$ and the angle ϕ.

It should be noted that the position where the maximum opening force appears to the end Point S where the curved guide roller changes to the straight as shown in figure 3.

4. Conclusion

In this study, in the sliding doors with handrails, a theoretical formula for simulating the opening force was obtained, and the force to open the sliding door was analyzed using it.

- The maximum value of the sliding door opening force coincides with the actual measurement result within the error of 13% by using the simulation method of this study.
- By using a torsion spring for the rotation fulcrum A, by appropriately combining the height difference between the rotation fulcrum and the roller and the radius of curvature of the guide rail, the sliding door opening force can be designed to the target 19.6 N or less.

References

[1] Ministry of Health 2006 Research on health promotion effect of walking support program for community elderly people *Labor and Welfare Business report on Health Promotion, etc. for the Elderly* 29
[2] Kim H K, Yoshida H and Suzuki T 2001 *Jounal of Japan Geriatrics Society* **38** 805-11
[3] WHO Global Report on Falls Prevention in Older Age: WHO Library Cataloguing-in-Publication Data ISBN 978 92 4 156353 6 (NLM classification:WA 288) 20-1
[4] Kei・Products Co. Ltd, Handrail for sliding door, sliding door device, Sliding door device, Patent application 2010-102602, Public number 2011-231513
[5] Kei・Products Co. Ltd, Handrail for sliding door, sliding door device, Sliding door device, Patent application 2012-133492, Public number 2013-256808.
[6] Kakuta K 1992 *Machine Design Handbook* 3th Edi. p 652
[7] Andou T 1968 *JSME Mechanical Engineers' Handbook* Rolling friction 5th Edi. **3** p 34

IOP Conf. Series: Materials Science and Engineering **372** (2018) 012010 doi:10.1088/1757-899X/372/1/012010

Rolling contact fatigue strengths of shot-peened and crack-healed ceramics

K Takahashi[1,2] and T Oki[1]

[1]Yokohama National University, 79-5 Tokiwadai, Hodogaya, Yokohama 240-8501, Japan

E-mail: takahashi-koji-ph@ynu.ac.jp

Abstract. The effects of shot-peening (SP) and crack-healing on the rolling contact fatigue (RCF) strengths of Al_2O_3/SiC composite ceramics were investigated. Non-shot-peened, shot-peened, and shot-peened + crack-healed specimens were prepared. SP was performed using ZrO_2 beads. The shot-peened + crack-healed specimen was crack-healed after SP. X-ray diffraction clearly showed that SP induced a compressive residual stress up to 300 MPa at the specimen surfaces. Furthermore, the shot-peened + crack-healed specimen retained a compressive residual stress of 200 MPa. The apparent surface fracture toughness of the shot-peened specimens increased owing to the positive effects of the compressive residual stress. RCF tests were performed using a thrust load-bearing test device. The RCF lives of the shot-peened specimens did not improve compared to that of the non-shot-peened specimen, because the numerous SP-introduced surface cracks could act as crack initiation sites during the RCF tests. However, the RCF life of the shot-peened + crack-healed specimen did improve compared to those of non-shot-peened and shot-peened specimens, implying that combining SP and crack-healing was an effective strategy for improving the RCF lives of Al_2O_3/SiC composite ceramics.

1. Introduction

Alumina ceramics (Al_2O_3) possess superior wear resistance, and electric insulation. Thus, they are widely used in several types of industrial equipment. However, there are few applications of alumina for rolling contact machine elements. In contrast, silicon nitride ceramics (Si_3N_4) are widely applied to contact elements such ball bearings because of their superior rolling contact fatigue (RCF) strengths [1,2]. The fracture toughness of Al_2O_3 is lower than that of Si_3N_4. Thus, surface cracks and defects decrease the structural integrity of Al_2O_3 components. We attempted to overcome the problem by shot-peening (SP) and crack-healing. SP is a surface treatment for increasing the fatigue strength of metallic materials. The propagation of fatigue cracks is retarded by the effects of compressive residual stress. Recent studies have shown that SP could introduce compressive residual stress near the surface of silicon nitride (Si_3N_4) [3,4], Al_2O_3 [4,5], and partially stabilized zirconia (PSZ) [6,7]. If surface cracks are induced during SP, the reliabilities of ceramic components will decrease.

Ceramics containing silicon carbide (SiC) particles and whiskers have crack-healing ability [8,9]. If crack-healing could be combined with SP, the surface strengths and reliabilities of ceramics could be increased. From this perspective, Takahashi *et al* proposed a method of combining SP with crack-healing (SP + crack-healing) to improve the surface strengths of ceramics, which was effective for increasing the bending strength [10] and contact strength [11] of Si_3N_4/SiC. Zhang *et al* reported that combining SP with multiple crack-healing processes was effective for increasing the fracture resistance and bending strength of ZrB_2/SiC [12]. Oki *et al* also reported that the SP + crack-healing process was effective for enhancing the contact strength of Al_2O_3/SiC [13]. However, the effects of SP and crack-healing on the RCF strengths of ceramics have not yet been investigated. Therefore, the objective of this study was to investigate the effects of SP and SP + crack-healing on the RCF strength of Al_2O_3/SiC. To this end, SP and SP + crack-healing were applied to Al_2O_3/SiC, and residual stress distributions and apparent surface fracture toughness were investigated. The RCF lives of the shot-peened and shot-peened + crack-healed specimens were also investigated.

2. Specimens and test procedure

2.1. *Materials and specimens*

Al_2O_3 reinforced by 15 vol.% SiC particles (Al_2O_3/SiC) was selected as the test material. Rectangular plates (90 × 90 × 6 mm) of Al_2O_3/SiC were hot-pressed at 1973 K for 2 h under pressure of 35 MPa in N_2. The processing details have been reported by Ando *et al* [9]. The hot-pressed Al_2O_3/SiC had a relative density of 99 % and an average grain size of approximately 1 μm. Although the SiC particles were uniformly distributed in the grain boundaries, SiC nanoparticles were distributed in the alumina grain, as reported in a previous study [9]. After hot-pressing, the plates were cut into 30 × 30 × 4-mm plate specimens. One of the surface of each specimen was polished to a mirror finish. These specimens are denoted as "non-shot-peened" specimens.

2.2. *SP and crack-healing*

We performed SP on the surfaces of the non-shot-peened specimens using a shot-peening system at a 0.2-MPa shot for 30 s. Commercial 180-μm-diameter ZrO_2 beads (Vickers hardness = 1150 *HV*) were used as the shot material. Subsequently, the shot-peened specimens were heated at 950°C for 100 h to heal any SP-induced surface cracks. The crack-healing conditions were selected because the SP-introduced surface cracks could be healed while maintaining the SP-introduced compressive residual stress. The shot-peened + crack-healed specimen surfaces were repolished to a depth of 1–2 μm to reduce the surface roughness since the friction coefficient was expected to increase owing to the oxide films formed on the sample surfaces.

2.3. *Residual stress and apparent fracture toughness*

The residual stresses of the shot-peened, shot-peened + crack-healed, and non-shot-peened specimens were measured using X-ray diffraction (XRD) with a Cr-Kα beam and the 2θ-$\sin^2\Psi$ relationship. The X-ray beam spot diameter was 2.0 mm. The diffraction plane of Al_2O_3 [146] and the diffraction angle $2\theta = 139°$ were selected to measure the residual stress. The residual stress distributions of the polished specimens were measured. The residual stress was successively measured for each layer. The apparent fracture toughness, K_c, was estimated by indentation-fracture (IF) method. The K_c can be calculated using the following equation [14]:

$$K_C = 0.018(E/HV)^{0.5}(P/c)^{1.5} \tag{1}$$

where HV is Vickers hardness, E is Young's modulus, P is the indent load, and c is half the surface crack length. The material had the following properties: E = 324 GPa and HV = 18.6 GPa (1898 *HV*). Vickers indentations were introduced on the polished surfaces by applying P = 49 N for 20 s.

2.4. *Rolling contact fatigue tests*

The RCF strengths were evaluated using a thrust load-bearing test device. The tests were conducted using six 4.0-mm-diameter bearing-steel balls (JIS-SUJ2) as the rolling element, and the rotation speed was set at n = 500 rpm. A commercial spindle oil was used as the lubricant. The test loads per ball were P = 0.21, 0.33, and 0.46 kN. A vibrometer on the test machine was adjusted to automatically stop in response to test specimen surface failure. The number of contact loadings at the automatic stop was evaluated as the RCF life, N_f.

3. Test results

3.1. Residual stress distribution

Figure 1 shows the distribution of the residual stress induced on each specimen. Compressive residual stresses were induced near the surfaces of the shot-peened and shot-peened + crack-healed specimens. Approximately 300 and 200 MPa of compressive residual stresses were induced on the outermost surfaces of the shot-peened and shot-peened + crack-healed specimens, respectively. Since the compressive residual stresses on the outermost surfaces of the non-shot-peened specimens were approximately 40 MPa, the compressive residual stress measured on the shot-peened specimen surface was induced by SP. The compressive residual stress near the surface of the shot-peened + crack-healed specimen was lower than that near the surface of the shot-peened specimen, indicating that the SP-induced lattice strain was relieved by heating. However, the heating-induced relief in the residual stress was not evident at or deeper than 5 μm. Therefore, the compressive residual stress remained on the shot-peened specimen surface even after heating the specimen in air at 950°C for 100 h.

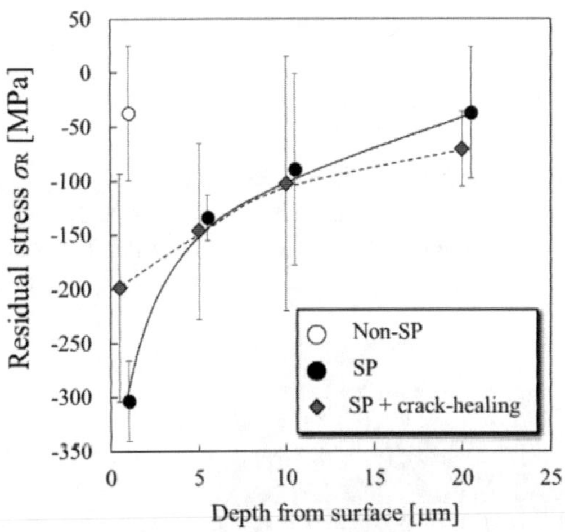

Figure 1. Residual stress distributions of surface-treated Al_2O_3/SiC specimens.

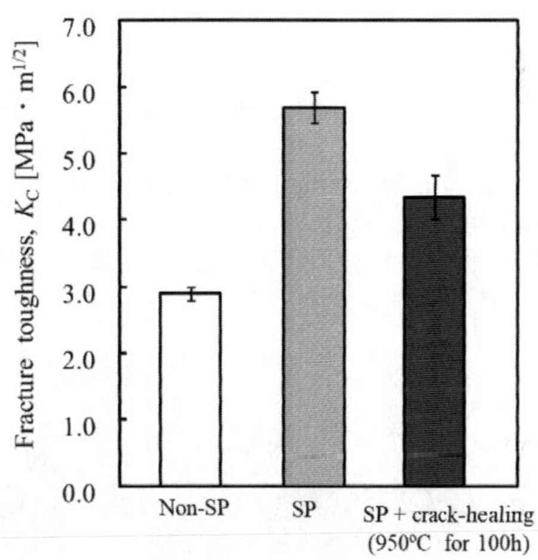

Figure 2. Apparent fracture toughness (K_c) of surface-treated Al_2O_3/SiC specimens.

3.2. Apparent fracture toughness

Figure 2 shows K_c of each specimen. The K_c of the shot-peened and shot-peened + crack-healed specimens improved by 96 and 50%, respectively, compared with that of the non-shot-peened specimen because the compressive residual stress prevented the development of indentation-fracture-induced cracks. The K_c of the shot-peened + crack-healed specimen was lower than that of the shot-peened specimen owing to the heating-induced decrease in the compressive residual stress, as described in section 3.1.

3.3. RCF life

Figure 3 shows the RCF test results. The RCF life of the shot-peened specimen was equivalent to that of the non-shot-peened one. In contrast, the RCF life of the shot-peened + crack-healed specimen was higher than that of the non-shot-peened and shot-peened ones. The crack initiation loads of the monotonic indentation tests [13] are plotted on the left in the figure. Although SP increased the crack initiation loads in the monotonic indentation tests, it did not increase the RCF life in the RCF tests. The reasons for the extended rolling fatigue life achieved by combining SP and crack healing are

IOP Conf. Series: Materials Science and Engineering **372** (2018) 012010 doi:10.1088/1757-899X/372/1/012010

discussed based on the damage behaviors.

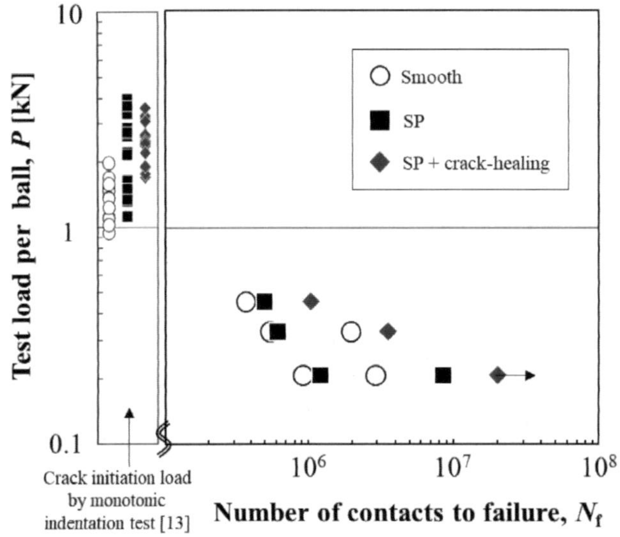

Figure 3. Results of RCF tests for surface-treated Al_2O_3/SiC specimens.

Figures 4-6 show the surface images of the test specimens subjected to a test loading of P = 0.21 kN. The surface images taken at $N = 2.4 \times 10^6$ cycles and at failure (N_f), are shown in figures 4(a)-6(a) and 4(b)-6(b), respectively. Figure 7 shows the schematic images of the RCF mechanisms in each specimen.

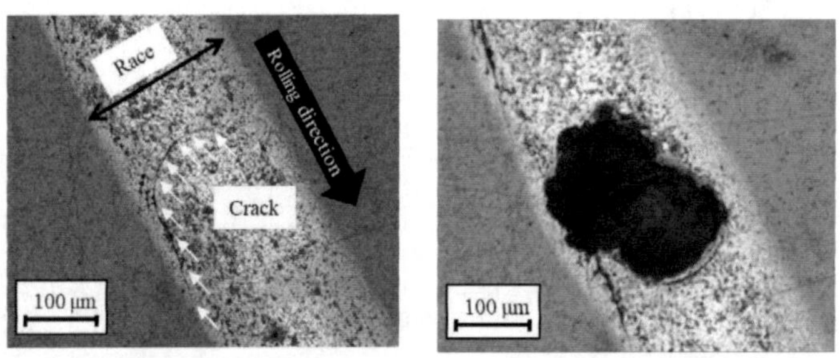

(a) N=2.40 × 10^6 cycles (b) N_f=2.92 × 10^6 cycles

Figure 4. Optical microscopy images of non-shot-peened specimen (P = 0.21 kN).

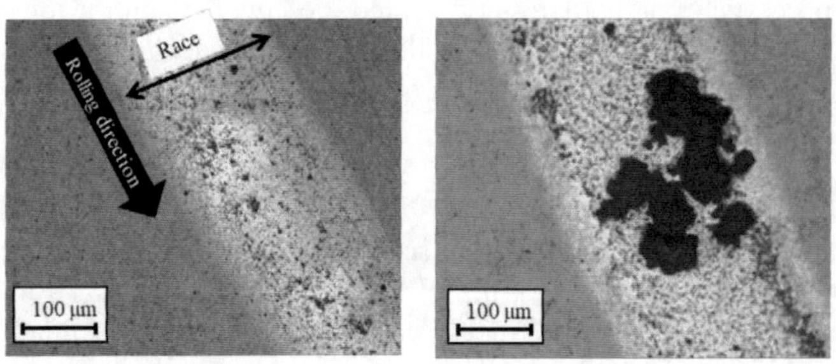

(a) N=2.40 × 10^6 cycles (b) N_f=8.40 × 10^6 cycles

Figure 5. Optical microscopy images of shot-peened specimen (P = 0.21 kN).

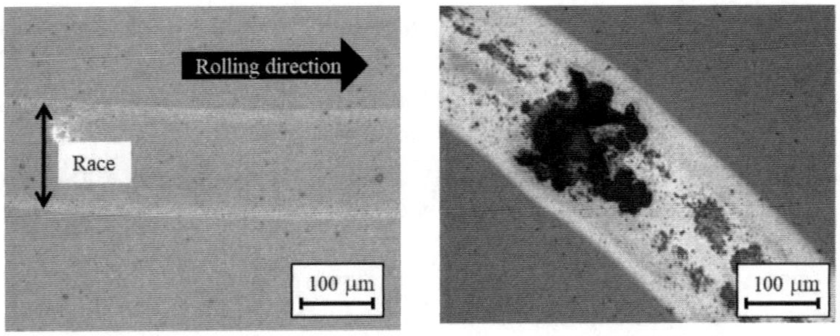

(a) N=2.4 × 10[6] cycles (b) N_f=2.0 × 10[7] cycles

Figure 6. Optical microscopy images of shot-peened + crack-healed specimen (P = 0.21 kN).

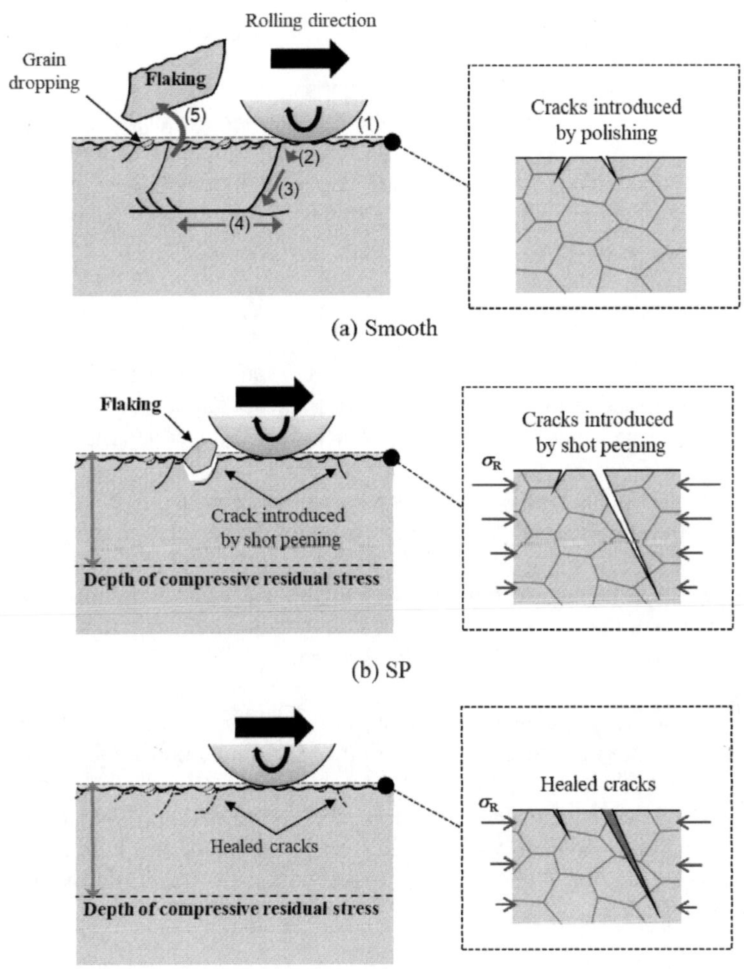

(c) SP + crack-healing

Figure 7. Schematic images of rolling contact fatigues in specimens.

Figure 4 shows optical microscopy images of the non-shot-peened specimen. After a certain number of contact cycles, arc-shaped ring cracks were initiated at the wear race on the specimen surface, as shown in figure 4(a). The cracks led to flaking damage, as shown in figure 4(b). Figure 7(a) shows the schematic images of the RCF mechanisms in the non-shot-peened specimen. The damage mechanism of the non-shot-peened specimen can be divided into five stages: (1) surface layer wear, (2) ring cracks initiating from minute cracks remaining after polishing, (3) ring cracks propagating in the tensile mode (mode I), (4) subsurface main cracks propagating in shear modes (modes II and III) and developing upward and downward branches from their tips, and (5) upward cracks reaching the surface and causing flaking. This flaking process is quite similar to the one for Si_3N_4, as reported by Kida *et al* [2].

Figure 5 shows optical microscopy images of the shot-peened specimen. Although figure 5(a) shows numerous SP-induced minute dents, the wear race of the shot-peened specimen subjected to the same number of contact cycles as the non-shot-peened specimen did not show any arc-shaped ring cracks. Figure 7(b) shows the schematic images of the RCF mechanisms in the shot-peened specimen. The small cracks that initiated from the SP-induced dents presumably led to several small points of flaking damage, which coalesced resulting in the final failure, as shown in figure 5(b). Although the compressive residual stress inhibited the propagation of the ring cracks, the RCF life (N_f) of the shot-peened specimen was approximately the same as that of the non-shot-peened one owing to the negative effects of the SP-induced small dents and cracks.

Figure 6 shows optical microscopy images of the shot-peened + crack-healed specimen. As shown in figure 6(a), the shot-peened + crack-healed specimen did not show any SP-induced dents or any ring cracks. All the SP-induced surface cracks had healed by crack-healing, as shown in figure 7(c). Thus, the RCF life increased owing to the positive effects of the compressive residual stress, implying that applying additional heating to heal SP-induced surface cracks was an effective strategy for improving the RCF lives of Al_2O_3/SiC composites.

4. Conclusion

The effect of combining SP and crack-healing on the RCF strength of an Al_2O_3/SiC composite was investigated. Residual stress distributions and apparent fracture toughness (K_c) were also measured for non-shot-peened, shot-peened, and shot-peened + crack-healed specimens, and the following results were obtained:

- SP induced a compressive residual stress up to 300 MPa at the specimen surfaces. Furthermore, the shot-peened + crack-healed specimen retained a compressive residual stress of 200 MPa.
- The apparent surface fracture toughness of the shot-peened and shot-peened + crack-healed specimens increased by 96 and 50%, respectively, compared with that of the non-shot-peened specimen. The compressive residual stress prevented the development of indentation-fracture-induced cracks.
- The RCF life of the shot-peened specimen did not improve compared with that of the non-shot-peened specimen because the SP-introduced surface cracks could act as crack initiation sites during the RCF tests.
- The RCF life of the shot-peened + crack-healed specimen improved compared with those of the non-shot-peened and shot-peened specimens, implying that applying additional heating to heal SP-induced surface cracks was an effective strategy for improving the RCF lives of Al_2O_3/SiC composites.

Acknowledgments
This work was supported by JSPS KAKENHI Grant Numbers 22686013 and 16H04231.

References

[1] Kanematsu W 2018 A review of rolling contact fatigue behavior of silicon nitride focusing on testing practices and crack propagation analysis *Wear* **400-401** 10–20
[2] Kida K, Saito M and Kitamura K 2005 Flaking failure originating from a single surface crack in silicon nitride under rolling contact fatigue *Fatigue Fract. Eng. Mater. Struct.* **28(12)** 1087–

IOP Conf. Series: Materials Science and Engineering **372** (2018) 012010 doi:10.1088/1757-899X/372/1/012010

97

[3] Tanaka K, Akiniwa Y and Morishita Y 2005 Residual stress distribution in the sub-surface region of shot-peened ceramics *Jpn. Soc. Mech. Eng. A* **71(712)** 1714–21

[4] Pfeiffer W and Frey T 2006 Strengthening of ceramics by shot peening *J. Eur. Ceram. Soc.* **26(13)** 2639–45

[5] Moon W-J, Ito T, Uchimura S and Saka H 2004 Toughening of ceramics by dislocation sub-boundaries *Mater. Sci. Eng. A* **387-389** 837–9

[6] Takahashi K, Iwanaka K, Osada T and Koike H 2015 Increase in strength of partially stabilized zirconia after shot peening *J. Mater. Eng. Perform.* **24(9)** 3573-78

[7] Shukla P P and Lawrence J 2015 Micro-shot peening of zirconia-advanced ceramic: An examination of surface integrity *J. Mater. Sci.* **50(4)** 1728–39

[8] Takahashi K, Yokouchi M, Lee S and Ando K 2003 Crack-healing behavior of Al_2O_3 toughened by SiC whiskers *J. Am. Ceram. Soc.* **86(12)** 2143–7

[9] Ando K, Kim B, Chu M, Saito S and Takahashi K 2004 Crack-healing and mechanical behaviour of Al_2O_3/SiC composites at elevated temperature *Fatigue Fract. Eng. Mater. Struct.* **27(7)** 533–41

[10] Takahashi K, Nishio Y, Kimura Y and Ando K 2010 Improvement of strength and reliability of ceramics by shot peening and crack healing *J. Eur. Ceram. Soc.* **30(15)** 3047–52

[11] Takahashi K and Nishio Y 2012 Improvement of the contact strength of Si_3N_4/SiC by a combination of shot peening and crack-healing *J. Solid Mechanics Mater. Eng.* **6(2)** 144–53

[12] Zhang R, Han B and Wang Y 2015 Strengthening of ultra high-temperature ceramics by shot peening and multiple crack healing *Mater. Lett.* **160** 132–4

[13] Oki T, Yamamoto H, Osada T and Takahashi K 2013 Improvement of the contact strength of Al_2O_3/SiC by a combination of shot peening and crack-healing *J. Powder Technol.* 946984

[14] Japan Standards Association, Testing methods for fracture toughness of fine ceramics at room temperature 2015 *Japan Industrial Standards R1607* (Japan Standards Association, Tokyo)

IOP Conf. Series: Materials Science and Engineering **372** (2018) 012011 doi:10.1088/1757-899X/372/1/012011

Relationship between notch strengthening threshold and mechanical property for ductile cast iron

T Ikeda[1,3], N-A Noda[2], Y Sano[2], T Umetani[1] and N Kai[1]

[1]Research & Development Center, Hinode, Ltd., Azaiwasaki, Miyaki-cho, Miyaki-gun, Saga, 849-0101, Japan
[2]Mechanical Engineering Department, Kyushu Institute of Technology, 1-1 Sensui-cho, Tobata-ku, Kitakyushu-shi, Fukuoka 804-8550, Japan

E-mail: t_ikeda@hinodesuido.co.jp

Abstract. In this study, dynamic tensile tests were conducted at the various strain rates and temperatures for traditional ductile cast iron. Then, the notch strength $\sigma_{\mathrm{B}}^{\mathrm{notch}}$ and the static tensile strength at room temperature $\sigma_{\mathrm{B,\ RT}}^{\mathrm{smooth}}$ were discussed in terms of the strain rate-temperature parameter R, which is known to be useful for evaluating the combined influence of strain rate and temperature. This study focuses on the notch strengthening threshold $R \geqq R_{\mathrm{th}}$ where $\sigma_{\mathrm{B}}^{\mathrm{notch}}$ is larger than $\sigma_{\mathrm{B,\ RT}}^{\mathrm{smooth}}$ and therefore notched components can be used safely. In other words, if $R \geqq R_{\mathrm{th}}$, $\sigma_{\mathrm{B,\ RT}}^{\mathrm{smooth}}$ can be used to evaluate notched components in mechanical design to prevent the instantaneous fracture. In this study, it was found that the R_{th} value can be predicted from the static tensile property and Brinell hardness. Since the traditional ductile cast iron considered in this paper has a broad range of mechanical properties, the present approach and discussion can be applied to evaluate other materials under various temperature and strain rate.

1. Introduction

Mechanical properties of traditional ductile cast iron can be changed widely to meet the demanding requirements [1]. The higher grade of strength can be attained with increasing the pearlitic matrix. For example, the pearlite content 45 ~ 55% has the tensile strength = 500 MPa but the pearlite content 80 ~ 90% in ductile cast iron reaches tensile strength ranging over 700 MPa. The size and weight can be reduced in structures by using the high strength grade material. Meanwhile, Charpy impact absorbed energy decreases. As an example, figure 1 shows that Charpy absorbed energy decreases drastically by changing the material from JIS-FCD500 to JIS-FCD700 [2,3]. This is the main reason why industrial application is still limited for the high strength grade ductile cast iron at the room temperature or lower temperature.

However, Charpy absorbed energy cannot be directly used for the mechanical design to prevent the instantaneous fracture. The tensile strength is used as the most basic value. Therefore, a dynamic tensile test is studied for several materials [4-9]. If the strength of notched specimen is larger than the tensile strength of smooth specimen, the tensile strength of smooth specimen can be safely used to the mechanical design of the real product having notched components under the instantaneous fracture. In this paper, those strengths will be compared in terms of the strain rate-temperature parameter R [7-14]. By using this R parameter, a notch strengthening threshold value of $R \geqq R_{\mathrm{th}}$ will be focused when the strength of notched specimen becomes larger than the tensile strength of smooth specimen. Then, it will be shown that the R_{th} value can be predicted by the tensile strength and Brinell hardness for all types of traditional ductile cast iron.

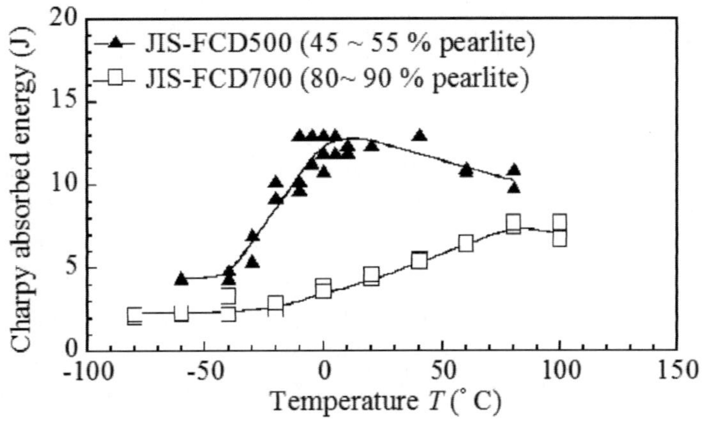

Figure 1. Charpy absorbed energy for traditional ductile cast iron [2].

2. Conditions where notch strength can be larger than smooth tensile strength

Figure 2 shows the smooth specimens and the notched round bar specimens used in the present high speed tensile test [11]. The tensile strength of smooth specimen σ_B^{smooth} and strength of notched specimen σ_B^{notch} are defined in equation (1) [9].

$$\left.\begin{aligned} \sigma_B^{smooth} &= 4\,P_{max} / \pi\, d^2 \\ \sigma_B^{notch} &= 4\,P_{max} / \pi\, d^2 \end{aligned}\right\} \quad (1)$$

Here, P_{max} is maximum load (N), d is diameter of specimen. Strain rate-temperature parameter R is defined in equation (2) [7-14].

$$R = T\ \ln\,(\,A\,/\,\dot{\varepsilon}\,) \quad (2)$$

Here, T is temperature (K), A is 10^8 s^{-1} [10], $\dot{\varepsilon}$ is strain rate (s^{-1}). The strain rate for the smooth specimen is defined $\dot{\varepsilon}^{smooth}$. The strain rate for the notched specimen is defined $\dot{\varepsilon}^{notch}$. Those strain rates are obtained from the tensile speed v as shown in equation (3) [7-9].

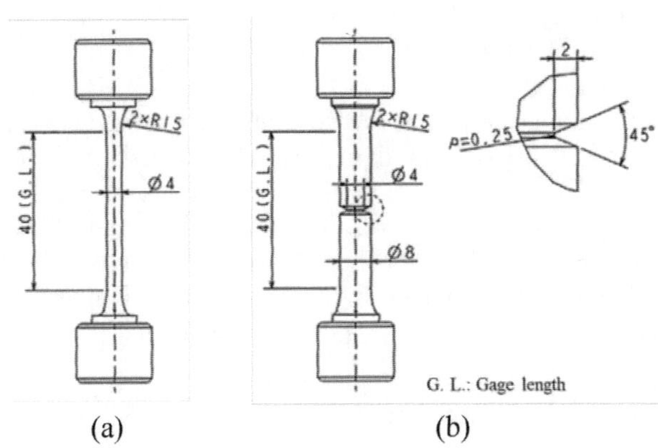

Figure 2. Configuration of high speed tensile test specimen (in mm), (a) smooth specimen, and (b) notched round bar specimen.

$$\left.\begin{aligned}\dot{\varepsilon}^{\text{smooth}} &= v / \ell \\ \dot{\varepsilon}^{\text{notch}} &= K_{t\dot{\varepsilon}}\, v / \ell\end{aligned}\right\} \tag{3}$$

Here, ℓ is gauge length (40 mm). Strain rate concentration factor $K_{t\dot{\varepsilon}}$ [4-9] should be used to obtain $\dot{\varepsilon}^{\text{notch}}$ because it is difficult to measure the strain rate at the notch root. For the specimen of figure 2(b), the value of $K_{t\dot{\varepsilon}} = 9.49$ was accurately obtained in the previous study [7-9].

In the previous research, R parameter is introduced to evaluate σ_B^{notch} as a master curve [7-9]. Figure 3 shows an example of master curve of σ_B^{notch} as a function of R parameter for high silicon ferritic ductile cast iron [9]. This new material is now expected to be used as structural components [15-18]. Figure 3 also shows the static tensile strength $\sigma_{B,\ RT}^{\text{smooth}}$ for smooth specimen at room temperature. The intersection between σ_B^{notch} curve and $\sigma_{B,\ RT}^{\text{smooth}}$ line can be regarded as notch strengthening threshold R_{th}. If $R \geqq R_{th}$, σ_B^{notch} is always larger than $\sigma_{B,\ RT}^{\text{smooth}}$. Therefore, if the real products are used under $R \geqq R_{th}$, $\sigma_{B,\ RT}^{\text{smooth}}$ can be used for the mechanical design of the real product having notched components under the instantaneous fracture. This paper deals with the traditional ductile cast iron whose mechanical properties can be changed widely to meet demanding requirements by focusing on R_{th}.

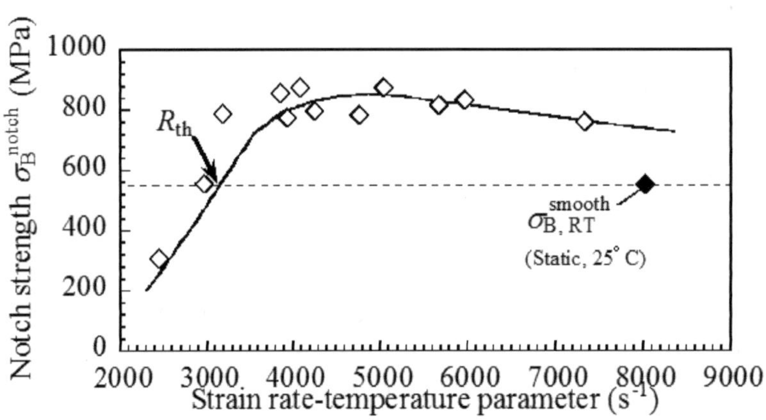

Figure 3. Influence of strain rate- temperature parameter R on σ_B^{notch} for high silicon ferritic ductile cast iron [9].

3. Experimental procedure

3.1. Preparation of specimens

In this study, we consider traditional ferrite-pearlite ductile cast irons (DCI). The tensile test was conducted according to the JIS-Z2241 standard. In addition, Brinell hardness test was conducted according to the JIS-Z2243. The tensile strength of DCI500 is approximately 500 MPa and DCI700 is approximately 700 MPa. In addition, fully pearlitic ductile cast iron (PDI) is also considered. The tensile strength of PDI is approximately 900 MPa. The pearlite ratio of DCI500 is 52.2% and DCI700 has pearlite ratio of 83.6%. PDI has fully pearlitic matrix. PDI is obtained by heat treatment indicated figure 4. The test specimen for the dynamic tensile test is taken from JIS Type II Y-shaped blocks (JIS-G 5502).

3.2. High speed tensile test

Figure 2 shows the configuration of dynamic tensile test specimens [9]. The tests are conducted at strain rate of $2.1 \times 10^{-4} \sim 1.8 \times 10^{1}\ \text{s}^{-1}$, and at temperature of -180 ~ 22°C. The experimental data of σ_B^{notch} is plotted by using the R parameter (equation (2)).

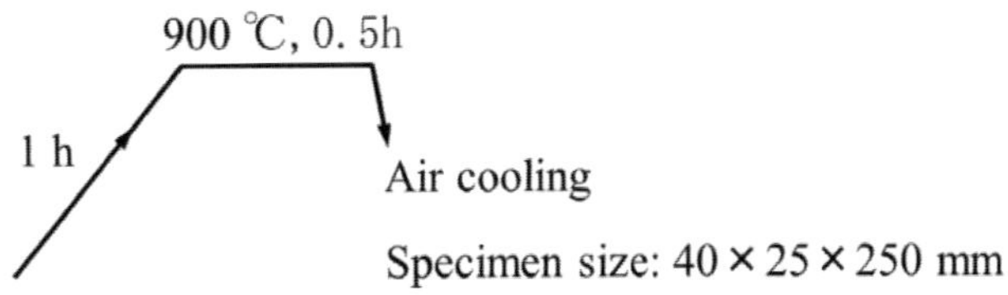

Figure 4. Normalizing condition for PDI.

4. Results and discussion

4.1. Relationship between tensile properties and pearlite area fraction

Figure 5 shows influence of pearlite area fraction on the static tensile property for smooth tensile test specimen (JIS No. 4). The static tensile strength σ_B and the 0.2% proof stress increase by increasing the pearlite area fraction. The Brinell hardness also increases by increasing the pearlite area fraction. The elongation decreases by increasing the pearlite area fraction. These results are in agreement with the other researcher's studies [2,14].

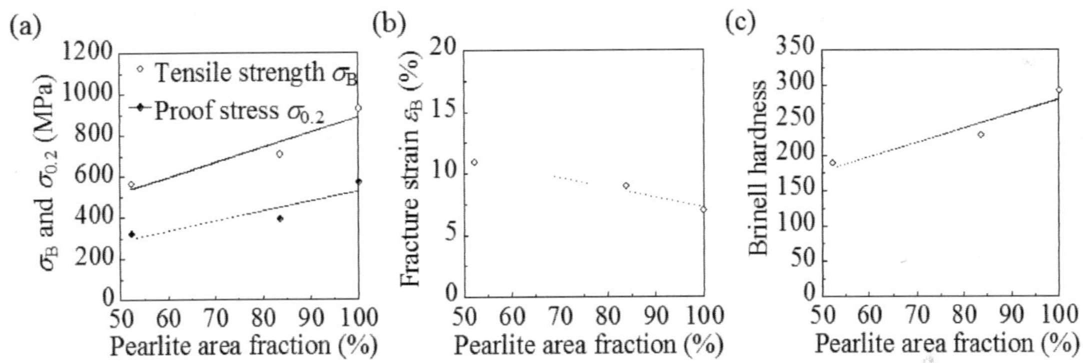

Figure 5. Influence of pearlite area fraction on static tensile properties, (a) σ_B and $\sigma_{0.2}$, (b) ε_B, (c) Brinell hardness.

4.2. Relationship between notch strengthening threshold R_{th} and static tensile property

In the dynamic tensile tests, DCI500, DCI700 and PDI showed the same tendency of the previous research on high silicon ferritic ductile cast iron in figure 3 [9]. The intersection between σ_B^{notch} curve and $\sigma_{B,\ RT}^{smooth}$ line can be regarded as notch strengthening threshold R_{th}. Figure 6 shows the relationships between the R_{th} value and the static tensile properties for the ferrite-pearlite ductile cast iron. There is a good correlation between the R_{th} value and the static tensile properties. The R_{th} value linearly increases with increasing the tensile strength and the Brinell hardness, and the R_{th} value linearly decreases with increasing the elongation. Therefore, it is considered that the R_{th} value is related to the deformation resistance of the materials.

From figure 6, the R_{th} value can be predicted from the static tensile property and Brinell hardness in the traditional ductile cast iron. Figure 7 shows the relationship between the R_{th} value and pearlite area fraction for the traditional ductile cast irons. A good correlation can be seen between the R_{th} value and pearlite area fraction. As stated earlier, in figure 5, pearlite area fraction was also correlated to the static tensile property. Therefore, a good correlation was recognized for traditional ductile cast iron in figure 6. In the traditional ductile cast iron, the increasing of the pearlite area fraction means the increasing of the deformation resistance because the pearlitic matrix doesn't have deformability. The higher grade of strength can be attained with increasing the pearlitic matrix. Therefore, it is considered

IOP Conf. Series: Materials Science and Engineering **372** (2018) 012011 doi:10.1088/1757-899X/372/1/012011

that the dependence of the mechanical property on the R_{th} value differs depending on the strengthening method in the materials.

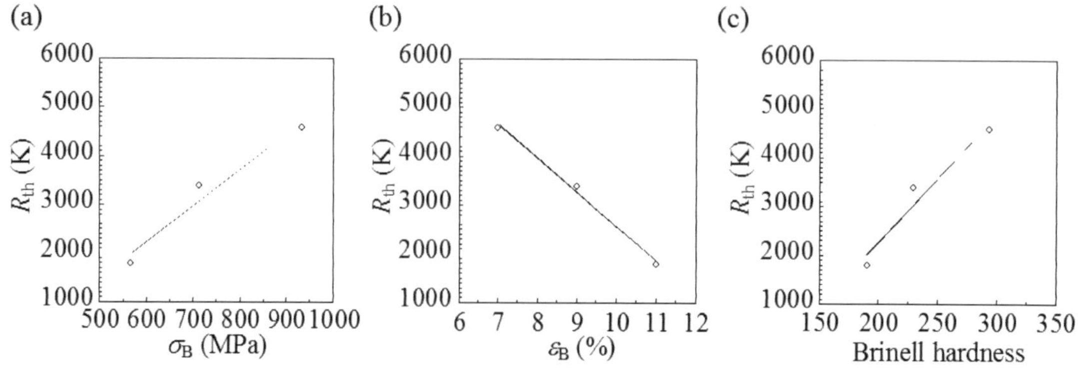

Figure 6. Relationship between R_{th} and static tensile properties, (a) σ_B, (b) ε_B, (c) Brinell hardness.

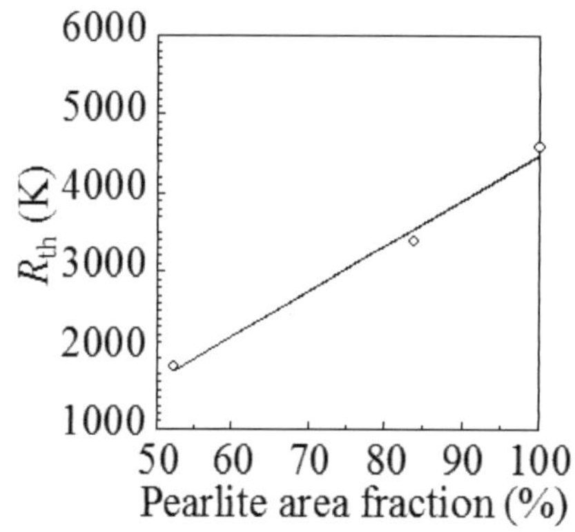

Figure 7. Relationship between R_{th} and pearlite area fraction.

5. Conclusions

In this study, for the traditional ductile cast iron, the strength of the notched specimen σ_B^{notch} was evaluated for the various strain rates and temperatures in comparison with the static tensile strength of the smooth specimen at room temperature $\sigma_{B,\ RT}^{smooth}$. The results can be applied to the mechanical design of the real product having notched components under the instantaneous fracture. From the results of this study, the conclusions can be summarized as follows.

- The intersection between σ_B^{notch} curve and $\sigma_{B,\ RT}^{smooth}$ line can be regarded as notch strengthening threshold R_{th}. If $R \geqq R_{th}$, σ_B^{notch} can be always larger than $\sigma_{B,\ RT}^{smooth}$. Therefore, if the real product having notched components is used under $R \geqq R_{th}$, $\sigma_{B,\ RT}^{smooth}$ can be used in the mechanical design to prevent the instantaneous fracture.
- The notch strengthening threshold R_{th} can be predicted from static tensile property and Brinell hardness in the traditional ductile cast iron (see figure 6).
- Since mechanical properties can be widely changed in the traditional ductile cast iron, the present discussion can be applied to evaluate other materials under various temperature and strain rate.

References

[1] Japanese Standards Association 2012 *JIS G 5502 Spheroidal Graphite Iron Castings* pp 1880-

97

[2] Umetani T, Ikeda T, Sura N, Ashizuka K, Nemoto T, Takada H and Ogi K 2014 Tensile strength, fatigue strength and impact strength of solution strengthend high silicon ferritic ductile cast iron *Journal of Japan Foundry Engineering Society* **86** 36-42

[3] Ikeda T, Umetani T, Kai N, Noda N-A and Sano Y 2016 Strain rate and temperature insensitiveness of notch-bend strength for high Si ductile cast iron *ISIJ International* **56** 868-74

[4] Noda N-A, Ohtuka H, Ando M, Sano Y, Takase Y, Shinozaki T and Guan W 2013 Analysis of dynamic stress concentration and strain rate concentration for notched specimens used for high speed tensile test *Transactions of the Japan Society of Mechanical Engineers Series A* **79** 1182-90

[5] Ando H, Noda N-A, Kuroshima Y, Ishikawa Y and Takeda H 2014 Impact properties of polydimethylsiloxane copolymerized polycarbonate and application of the time-temperature superposition principle *Transactions of the JSME* **80** doi: 10.1299/transjsme. 2014smm0149

[6] Noda N-A, Ohtsuka H, Zheng H, Sano Y, Ando M, Shinozaki T and Guan W 2015 Strain rate concentration and dynamic stress concentration for double-edge-notched specimens subjected to high-speed tensile loads *Fatigue & Fracture of Engineering Materials & Structures* **38** 125-38

[7] Noda N-A, Shen Y, Takaki R, Akagi D, Ikeda T, Sano Y and Takase Y 2017 Relationship between strain rate concentration factor and stress concentration factor *Theoretical and Applied Fracture Mechanics* https://doi.org/10.1016/j.tafmec.2017.05.017

[8] Noda N-A, Akagi D, Shen Y, Takaki R, Ikeda T, Sano Y and Takase Y 2017 Strain rate concentration factor in comparison with stress concentration factor of a circumferential notch in a round bar specimen *Transactions of the JSME* **83** doi: http://doi.org/10.1299/transjsme.17-00034

[9] Ikeda T, Umetani T, Kai N, Ogi K, Akagi D, Noda N-A and Sano Y 2017 Notch strength of high silicon ductile cast iron and wide applicability to structural design *Transactions of the JSME* **83** doi: http://doi.org/10.1299/transjsme.16-00455

[10] Bennett P E and Sinclair G M 1966 Parameter representation of low-temperature yield behavior of body-centered cubic transition metals *Transactions of the ASME* **65** 518-24

[11] Fuji E, Ohkuma Y, Kawaguchi Y and Tsukamoto M 1985 Effects of temperature and strain rate on dynamic fracture toughness of steel *The Society of Naval Architects of Japan* **158** 619-29

[12] Goto K, Hirasawa H and Toyosada M 1994 A simple estimating method of constitutive equation for structural steel as a function of strain rate and temperature *The Society of Naval Architects of Japan* **176** 501-7

[13] Yamamoto H, Kobayashi T and Fujita H 1999 Strain rate-temperature dependency of impact tensile properties and ductile fracture behavior in ductile cast iron *Tetsu-to-Hagané* **85** 765-70

[14] Minami F, Hashida T, Toyoda M, Morikawa J, Ohmura T, Arimochi K and Konda N 1998 *The Society of Naval Architects of Japan* **184** 453-63

[15] British Standards Institution 2011 *EN1563Founding-Spheroidal Graphite Cast Iron* pp 23-6

[16] Larker R 2009 Solution strengthened ferritic ductile iron ISO 1083/JS/500-10 provides superior consistent properties in hydraulic rotators *China Foundry* **6** 343-51

[17] Larker R 2011 Paradigm shift in revised EN 1563 (GJS) enables improved properties and production economy in both as-cast and austempered (ADI) states *Proc. NEWCAST Forum* (Düsseldorf: Bundesverband der Deutschen Gießerei Industrie ed.) pp 29-34

[18] Löblich H and Stets W 2014 Die Einführung von mischkristallverfestigtem Gusseisen mit Kugel-graphite in die Industrie-eine Erfolgsstory *Proc. Deutscher Gießereitag 2013 und 5. NEWCAST Forum* (Fellbach: Verein Deutscher Giessereifachleute ed) pp 14-7

IOP Conf. Series: Materials Science and Engineering **372** (2018) 012012 doi:10.1088/1757-899X/372/1/012012

CFD-PBM coupled simulation of a nanobubble generator with honeycomb structure

F Ren[1], N A Noda[1,3], T Ueda[1], Y Sano[1], Y Takase[1], T Umekage[1], Y Yonezawa[2] and H Tanaka[1]

[1]Mechanical Engineering Department, Kyushu Institute of Technology, 1-1 Sensui-cho, Tobata-ku, Kitakyushu-shi, 804-8550, Japan
[2]Mrufukusuisan Corp., Ltd., 94-22 Nishiminato-machi, Kokurakita-ku, Kitakyushu-shi, 803-0801, Japan

E-mail: noda@mech.kyutech.ac.jp

Abstract. In recent years, nanobubble technologies have drawn great attention due to their wide applications in many fields of science and technology. The nitrogen nanobubble water circulation can be used to slow the progressions of oxidation and spoilage for the seafood long-term storage. From previous studies, a kind of honeycomb structure for high-efficiency nanobubble generation has been proposed. In this paper, the bubbly flow in the honeycomb structure was studied. The numerical simulations of honeycomb structure were performed by using a computational fluid dynamics–population balance model (CFD-PBM) coupled model. The numerical model was based on the Eulerian multiphase model and the population balance model (PBM) was used to calculate the gas bubble size distribution. The bubble coalescence and breakage were included. Considering the effect of bubble diameter on the fluid flow, the phase interactions were coupled with the PBM. The bubble size distributions in the honeycomb structure under different work conditions were predicted. The experimental results were compared with the simulation predictions.

1. Introduction

Recently, a kind of honeycomb structure was designed for high-efficiency nanobubble generation [1]. The aggregation and breakage happen and cause a wide range bubble size distribution when the bubbles flow in the honeycomb structure. Calculation of the bubble size and number density distribution and understanding of the bubble aggregation and breakage in bubbly flow are very important for the evaluation and improvement of the nanobubble generating ability.

Nowadays, many researches on gas-liquid flows have been carried out with computational fluid dynamics (CFD) simulations for engineering purposes [2]. The Euler–Lagrange (E-L) approach [3,4] and the Euler–Euler (E-E) approach [5,6] are primarily used. And the population balance method is a well-known method for tracking the size distribution of the dispersed phase and accounting for the breakage and coalescence effects in bubbly flows [7,8]. This approach is concerned with maintaining a record of the number of bubbles initially and tracking their evolution in space over time.

A progress in the CFD simulation of bubbly flow is the coupling of the population balance model (PBM) into CFD models, namely the CFD-PBM coupled model [2,9-11]. CFD-PBM coupled model combines the PBM into the CFD framework so that bubble breakup and coalescence can be taken into account, beneficial for describing the bubble size distribution and gas holdup in different flow regimes, and it also considers the influence of bubble size on the interphase interaction, which allows it to well predict the local gas-liquid interfacial area and the flow behavior in diverse flow regimes [12].

In this work, the CFD-PBM coupled model along with the RNG k-ε turbulence model was used to simulate the bubble aggregation and breakage and predict the bubble number density distribution for the case of gas-liquid bubbly flow in the honeycomb structure under different work conditions.

2. Model descriptions

2.1. Euler–euler two-fluid model

For the bubbly flow in the honeycomb structure, the liquid is treated as a continuous primary phase and the gas as a dispersed secondary phase. The main continuity and momentum conservation equations for the liquid and gas phases are given as follows,

$$\frac{\partial}{\partial t}\left(\alpha_q \rho_q\right) + \nabla\cdot\left(\alpha_q \rho_q \vec{u}_q\right) = 0 \tag{1}$$

where α_q, ρ_q and $\vec{u}_q$ are the volume fraction, density and velocity of the qth phase respectively. The momentum balance for the qth phase is

$$\frac{\partial}{\partial t}\left(\alpha_q \rho_q \vec{u}_q\right) + \nabla\cdot\left(\alpha_q \rho_q \vec{u}_q \vec{u}_q\right) = -\alpha_q \nabla p + \nabla\cdot\bar{\tau}_q + \alpha_q \rho_q \vec{g} + \vec{F}_{drag,\,q} + \vec{F}_{lift,\,q} + \vec{F}_{VM,\,q} \tag{2}$$

where $\bar{\tau}_q$, p and $\vec{g}$ are the qth phase stress-strain tensor, pressure and gravity acceleration respectively. $\vec{F}_{drag,\,q}$ is the drag force, $\vec{F}_{lift,\,q}$ is the lift force and $\vec{F}_{VM,\,q}$ is the virtual mass force acting on the qth phase.

The RNG-based *k-ε* turbulence model [13] is derived from the instantaneous Navier-Stokes equations, using a mathematical technique called "renormalization group" (RNG) methods. It has greater potential to give accurate predictions for complex flows. The *k* and ε for continuous phase are computed by the follows,

$$\frac{\partial}{\partial t}\left(\alpha_q \rho_q k_q\right) + \nabla\cdot\left(\alpha_q \rho_q \vec{u}_q k_q\right) = \nabla\cdot\left(\alpha_q \left(\mu_q + \frac{\mu_{t,q}}{\sigma_k}\right) \nabla k_q\right) + \alpha_q G_{k,q} - \alpha_q \rho_q \varepsilon_q \tag{3}$$

and

$$\frac{\partial}{\partial t}\left(\alpha_q \rho_q \varepsilon_q\right) + \nabla\cdot\left(\alpha_q \rho_q \vec{u}_q \varepsilon_q\right) = \nabla\cdot\left(\alpha_q \left(\mu_q + \frac{\mu_{t,q}}{\sigma_\varepsilon}\right) \nabla \varepsilon_q\right) + \alpha_q \frac{\varepsilon_q}{k_q}\left(C_{1\varepsilon} G_{k,q} - C_{2\varepsilon} \rho_q \varepsilon_q\right) \tag{4}$$

In these equations, σ_k and σ_ε are the turbulent Prandtl numbers for *k* and ε, respectively. $C_{1\varepsilon}$ and $C_{2\varepsilon}$ are constants. $G_{k,q}$ is the production of turbulent kinetic energy.

2.2. Discrete population balance model

Assuming the gas phase to be made of spherical bubbles of diameter *L*, the dispersed phase volume fraction can be expressed as

$$\alpha_p(\vec{x},t) = \int_0^{+\infty} n(\vec{x},\mathcal{V},t)\frac{\pi}{6}L^3 dL \tag{5}$$

The bubble state vector is characterized by a set of external coordinates ***(x)*** which denote the spatial position of the bubble and $(\mathcal{V})$ is referred to as an internal coordinate which could include particle size, composition, and temperature. $n(\vec{x},\mathcal{V},t)$ is the number density function. Assuming that $\mathcal{V}$ is the bubble volume, the transport equation for $n(\vec{x},\mathcal{V},t)$ can be given as

$$\frac{\partial}{\partial t}[n(V,t)] + \nabla\cdot[\vec{u}n(V,t)] = B_{ag}(V,t) + B_{br}(V,t) - D_{ag}(V,t) - D_{br}(V,t) \tag{6}$$

The terms $B_{ag}(V,t)$, $B_{br}(V,t)$, $D_{ag}(V,t)$, $D_{br}(V,t)$ represent birth due to aggregation, birth due to breakage, death due to aggregation, and death due to breakage, respectively.

3. Nanobubble generating device and experimental details

From previous studies, a kind of honeycomb structure for high-efficiency nanobubble generation was proposed. Figure 1 shows the bubble generating device equipped with several parallel honeycomb structure units. The water and nitrogen gas are pumped into the bubble generator together as is shown in figure 1(a). There are 15 honeycomb structure units placed in parallel inside the bubble generator as is shown in figure 1(b). Figure 2 shows the details of the honeycomb structure unit. One honeycomb unit consists of two honeycomb structure plates stacked so that the complex crisscross flow channels are formed in the honeycomb unit.

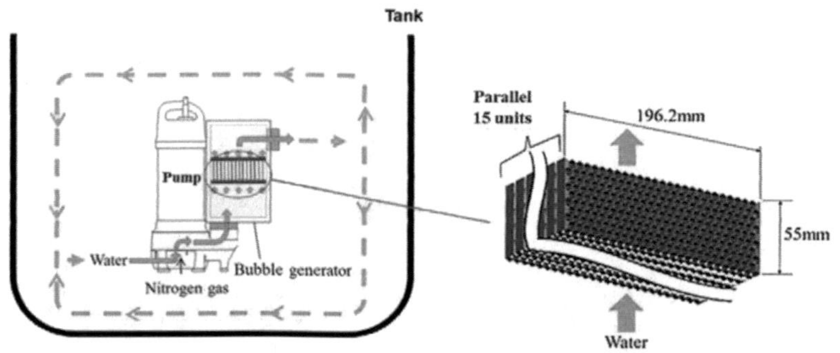

(a) Experimental device (b) Parallel honeycomb units

Figure 1. Nitrogen nanobubble generating device.

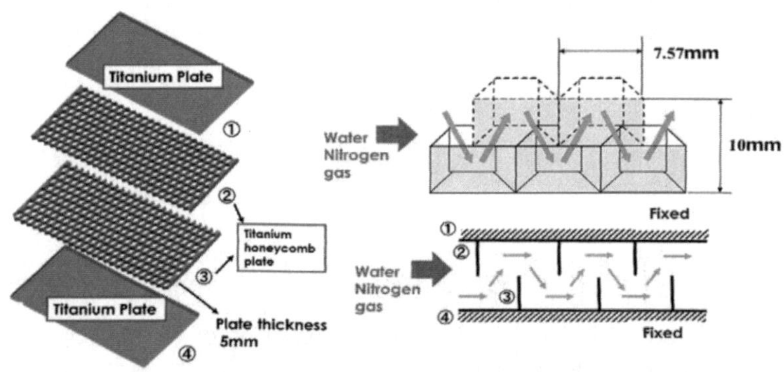

(a) Honeycomb unit composition (b) Honeycomb unit cross section

Figure 2. Honeycomb structure.

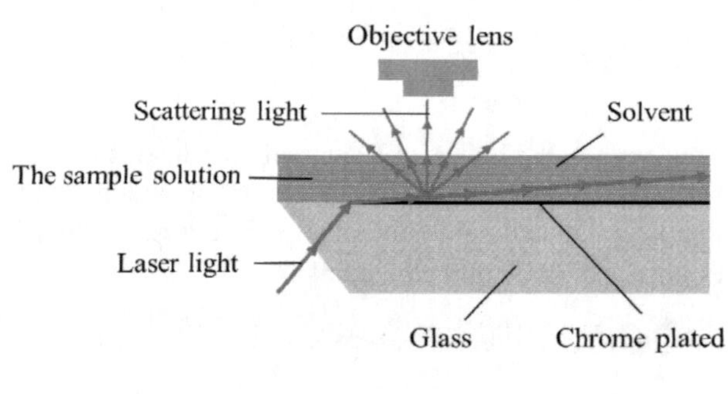

Figure 3. Schematic of Nano Sight LM10-HS.

The nanobubble density in the tank is measured from the sample taken from the tank. The nano-particle analyzer, Nano Sight LM10-HS [14,15], is used to measure the nanobubble density of the sample. Figure 3 shows the principle of Nano Sight LM10-HS. Irradiating a laser beam in the horizontal direction in the sample liquid, the side-scattered light from the nanoparticles is visualized by the objective lens, the movement trajectory of each particle is displayed on the computer. Tracking Brownian motion of all particles recognized on the screen and using the Stokes-Einstein equation, the particle size is obtained from the moving velocity of the particle.

The details of the experimental device are shown in table 1. The main purpose of the bubble generating device is to reduce the dissolved oxygen (DO) in the water by using the nitrogen nanobubbles. The experimental results of different flow rates are shown in table 2. In the experiment, the flow rate of the gas-liquid mixture is controlled by the pump power, and the gas flow rate is a constant 5 L/min. From table 2, it can be seen that the DO decreases and the particle number density increases over the time. It should be noted that not all of the particles are nitrogen nanobubbles because it is hard to distinguish what type of a particle might be. However, it can be inferred that most of the increased particles are the nitrogen bubbles generated by the device, and these bubbles make the DO lower. It can be also found that the device can generate more bubbles with higher flow rate. The bubble size and number density distribution will be analyzed in the next chapter.

Table 1. Details of the experimental device.

Item	Specification
Pump	Submersible pump(80TM23.7)
Output [kW]	3.7, (60Hz)
Tank [mm]	1580×1100×600
Water and amount [kg]	Tap water, 1000
Gas	Nitrogen
Flow rate [L/min]	5.0

Table 2. Experimental results of DO.

Time	Flow rate 700 L/min		Flow rate 550 L/min	
	DO [mg/L]	Particle number density [x10^8/mL]	DO [mg/L]	Particle number density [x10^8/mL]
0min	10.4	0.37	10.14	0.41
10min	4.4	0.82	7.79	0.43
20min	2.5	1.12	6.7	0.88
30min	1.7	1.72	5.5	0.94

4. Numerical details

In this work, all computations were performed by using the commercial CFD software FLUENT 16.2. Because of the symmetry, only a part of one honeycomb structure unit will be modeled to simplify the CFD analysis, as is shown in figure 4. According to the device in the experiment, the length of the side of the hexagon is 3.5 mm and the thickness of the wall between two hexagons is 0.5 mm. Pressure-outlet boundary condition is applied. Along the walls, no-slip boundary conditions are adopted. The symmetry boundary conditions are applied on the two sides of the model. The inlet velocities are 0.456 m/s and 0.358 m/s according to the flow rates 700 L/min and 550 L/min. An assumption is that the gas phase has the same velocity as the liquid phase at the inlet because the gas volume fraction is small and the bubble size is small. The number of bubble bins is specified 10, from 10 nm to 5200 nm. It is assumed that all the bubbles at the inlet have the diameter 1280 nm initially.

Figure 5 shows the bubble number density distributions of different flow rates at the inlet and outlet. Due to the fixed gas flow rate 5 L/min, the gas volume fraction is smaller when the flow rate is high, the bubble number density when the flow rate is 700 L/min is smaller than that when the flow rate is 550 L/min. It can be found that most of the bubbles at the inlet aggregate into larger bubbles after flowing through the honeycomb structure. The bubble breakage in the bubbly flow is not as obvious as the aggregation. Figure 6 shows the number density distribution of the bubbles with size less than 350 nm at the outlet. It can be seen that more small bubbles are generated when the flow rate is 700 L/min.

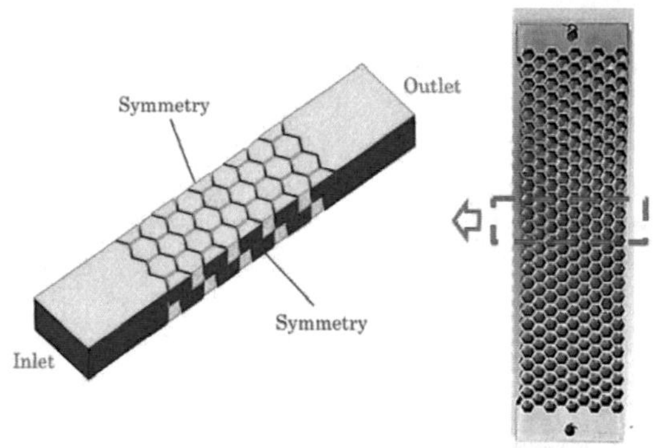

Figure 4. Simplified honeycomb unit model.

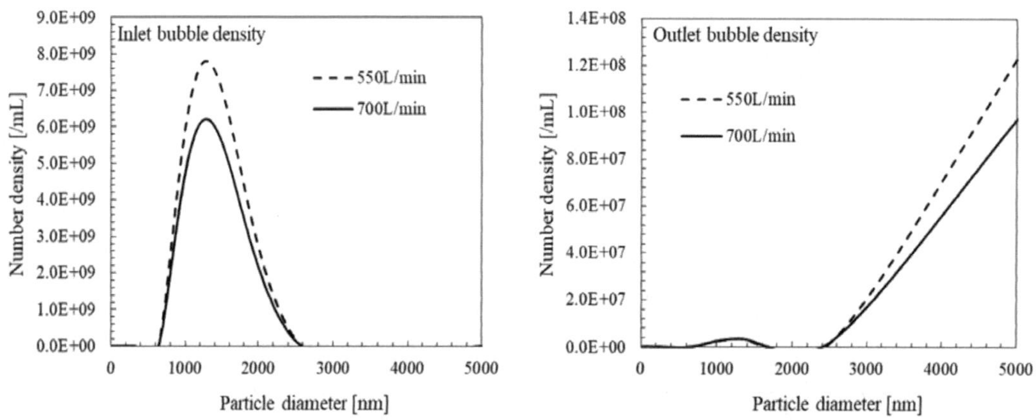

Figure 5. The bubble number density distribution of different flow rates.

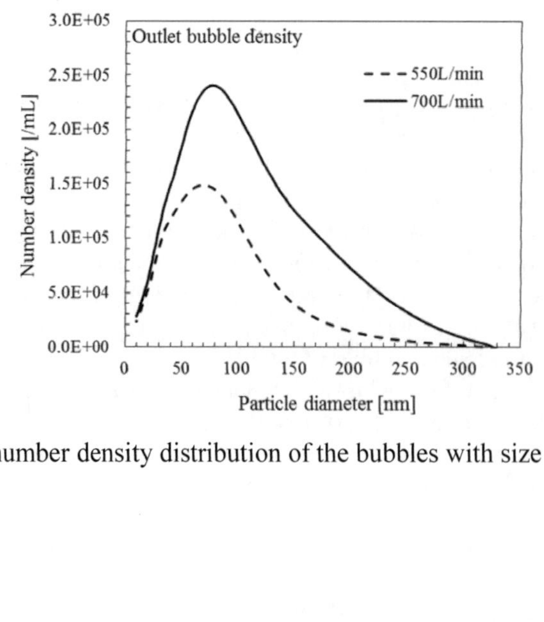

Figure 6. The number density distribution of the bubbles with size less than 350 nm.

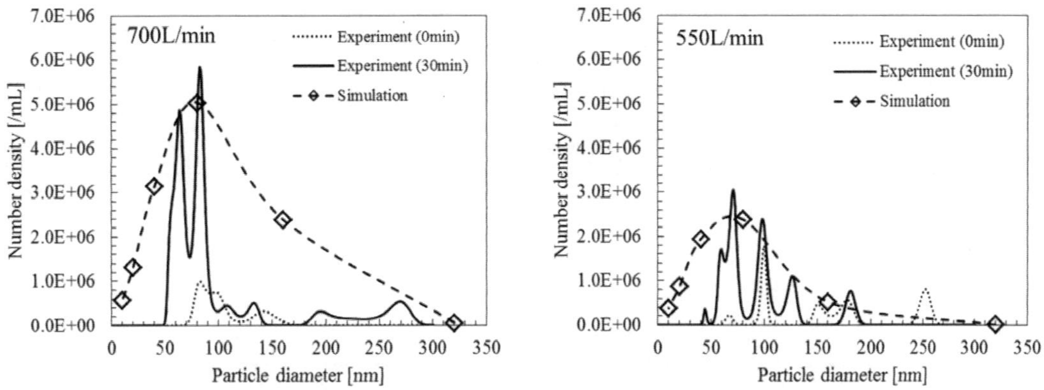

Figure 7. The bubble number density distribution in the tank.

The large bubbles in the tank quickly rise to the surface of water and collapse, only the tiny bubbles remain in the water tank and are detected by Nano Sight LM10-HS. Figure 7 shows the bubble number density distribution in the tank. The simulation result is obtained by

$$N_{pbm}^{tank}=\frac{N_{pbm}^{outlet}\cdot Q_{total}\cdot t}{V_{water}^{tank}} \quad (7)$$

where N_{pbm}^{outlet} is the number density at the outlet as is shown in figure 6, Q_{total} is the flow rate (550L/min or 700 L/min), t is the experiment time (30 min) and V_{water}^{tank} is the volume of water in the tank (1 m^3). N_{pbm}^{tank} denotes the number density of the bubbles generated from the device, the original particles in the water tank when t=0min are not included. The 30-minute experimental result includes the original particles. In figure 7, the simulation result can approximately reflect the trend of number density distribution. In the range of 0nm to 350 nm, only 6 bin sizes are calculated because of the limit of computer power, the more accurate solution can be obtained by calculating more bubble bins using more computing resource. In the experiment, the number of bubbles with diameter of 70 nm is the largest when the flow rate is 550 L/min, the number of bubbles with diameter of 82 nm is the largest when the flow rate is 700 L/min. Table 3 shows the number densities of the bubbles with diameters of 70 nm and 82 nm. Simulation results are obtained by interpolation from the results of 6 bin sizes shown in figure 7. The original particles should not be counted so that the effective number density can be calculated by subtracting the 0-minute experimental result from the 30-minute experimental result. It can be found in Table 3 that the simulation result can accurately predict what bubble size is predominant in the water tank. As shown in figure 7, only 6 bin sizes are calculated in the range of 0 nm to 350 nm, and their number densities are shown in table 4. The results of these 6 bins do not coincide with the experimental results very well, however, in figure 7 it can be seen that the curve of these 6 bins can be used to predict the trend of number density distribution.

Table 3. Number density details of the bubbles with diameters of 70 nm and 82 nm.

		Flow rate 550 L/min Bubble size 70 nm	Flow rate 700 L/min Bubble size 82 nm
Experimental result	0 min	96727/mL	989257/mL
	30 min	3056609/mL	5843111/mL
	30 min-0 min	2959882/mL	4853854/mL
Simulation result		2608186/mL	4999691/mL
Error		12%	2%

Table 4. Number density details of 6 bins.

Bin size [nm]	Flow rate 550 L/min			Flow rate 700 L/min		
	Experimental result (0 min) [/mL]	Experimental result (30 min) [/mL]	Simulation result [/mL]	Experimental result (0 min) [/mL]	Experimental result (30 min) [/mL]	Simulation result [/mL]
320	0	0	1891	0	0	52733
160	204324	186	527698	90351	57	2394528
80	941	213412	2382769	931425	5019031	5036182
40	0	364302	1950508	0	0	3142667
20	0	0	880177	0	0	1317974
10	0	0	376941	0	0	582056

5. Conclusions

In this work, the bubble aggregation and breakage processes in bubbly flow in honeycomb structure were studied. From the results of simulation and experiment, most of the nitrogen bubbles aggregate into the larger bubbles and get out of the water quickly. Few bubbles become the tiny bubbles and remain in the water. The gas utilization rate is not very high, however, it is still a simple and effective way to increase the gas content in water. This kind of honeycombed generator also has the advantages of simple structure, easy assembly and production, low requirements for use and maintenance, and large quantity of bubbles for wide applications.

The bubble aggregation and breakage in turbulence were simulated by using the CFD-PBM coupled model along with the RNG k-ε turbulence model. The bubble number density distribution was predicted. Comparing with the experimental results, the simulation results can approximately reflect the trend of number density distribution and accurately predict the predominant bubble size in the water tank. The more accurate solution of the number density distribution can be obtained by calculating more bubble bins using more computing resource.

References

[1] Kudo Y and Hiraki K 2011 Characteristics comparison of the honeycomb mixing device by considering the mixing energy in the gas-liquid mixing *The Proceedings of the Symposium on Micro-Nano Science and Technology* **3** 26-7

[2] Liang X F, Pan H, Su Y H and Luo Z H 2016 CFD-PBM approach with modified drag model for the gas-liquid flow in a bubble column *Chem. Eng. Res. Des.* **112** 88-102

[3] Delnoij E, Lammers F, Kuipers J and Van Swaaij W 1997 Dynamic simulation of dispersed gas-liquid two-phase flow using a discrete bubble model *Chem. Eng. Sci.***52** 1429–58

[4] Sokolichin A, Eigenberger G, Lapin A and Lübert A 1997 Dynamic numerical simulation of gas-liquid two-phase flows Euler/Euler versus Euler/Lagrange *Chem. Eng. Sci.* **52** 611–26

[5] Drew D 1983 Mathematical modeling of two-phase flow *Annu. Rev. Fluid. Mech.* **15** 261–91

[6] Krishna R, Urseanu M I, van Baten J M and Ellenberger J 1999 Influence of scale on the hydrodynamics of bubble columns operating in the churn-turbulent regime: experiments vs. eulerian simulations *Chem. Eng. Sci.* **54** 4903–11

[7] Wang T F, Wang J F and Jin Y 2005 Population balance model for gas–liquid flows: influence of bubble coalescence and breakup models *Ind. Eng. Chem. Res.* **44** 7540-9

[8] Li L M, Liu Z Q, Li B K, Matsuura H and Tsukihashi F 2015 Water model and CFD-PBM coupled model of gas-liquid-slag three-phase flow in ladle metallurgy *ISIJ. Int.* **55** 1337-4

[9] Duan X Y, Cheung S C P, Yeoh G H, Tu J Y, Krepper E and Lucas D 2011 Gas–liquid flows in medium and large vertical pipes *Chem. Eng. Sci.* **66** 872–83

[10] Wang T and Wang J 2007 Numerical simulations of gas-liquid mass transfer in bubble columns with a CFD–PBM coupled model *Chem. Eng. Sci.* **62** 7107–18

[11] Wang T, Wang J and Jin Y 2006 A CFD-PBM coupled model for gas-liquid flows *Am. Ins.Chem. Eng. J.* **52** 125–40

IOP Conf. Series: Materials Science and Engineering **372** (2018) 012012 doi:10.1088/1757-899X/372/1/012012

[12] Xing C, Wang T and Wang J 2013 Experimental study and numerical simulation with a coupled CFD–PBM model of the effect of liquid viscosity in a bubble column *Chem. Eng. Sci.* **95** 313–22

[13] Yakhot V, Orszag S A, Thangam S, Gatski T B and Speziale C G 1992 Development of turbulence models for shear flows by a double expansion technique *Phys. Fluids. A.* **4** 1510-20

[14] Malloy A and Carr B 2006 Nanoparticle tracking analysis *Part. Part. Syst. Char.* **23** 197-204

[15] Filipe V, Hawe A and Jiskoot W 2010 Critical evaluation of Nanoparticle Tracking Analysis (NTA) by NanoSight for the measurement of nanoparticles and protein aggregates *Pharm. Res.* **27** 796-810

IOP Conf. Series: Materials Science and Engineering **372** (2018) 012013 doi:10.1088/1757-899X/372/1/012013

Fracture mechanics analysis of cracked structures using weight function and neural network method

J G Chen[1,2], F G Zang[1], Y Yang[1], K K Shi[1] and X L Fu[1]

[1]Science and Technology on Reactor System Design Technology Laboratory, Nuclear Power Institute of China, China

E-mail: chenjianguomm@126.com

Abstract. Stress intensity factors(SIFs) due to thermal-mechanical load has been established by using weight function method. Two reference stress states sere used to determine the coefficients in the weight function. Results were evaluated by using data from literature and show a good agreement between them. So, the SIFs can be determined quickly using the weight function obtained when cracks subjected to arbitrary loads, and presented method can be used for probabilistic fracture mechanics analysis. A probabilistic methodology considering Monte-Carlo with neural network (MCNN) has been developed. The results indicate that an accurate probabilistic characteristic of the K_I can be obtained by using the developed method. The probability of failure increases with the increasing of loads, and the relationship between is nonlinear.

1. Introduction

Many parameters in deterministic fracture mechanics evaluation are very conservative, actually many parameters in the evaluation process, such as fracture toughness and nil-ductility transition temperature, have statistical distribution characteristics. For example, a probabilistic model was developed by Rahman [1] for analysis of circumferential through wall cracks in pipes. Other researchers do some work on J-integral estimation [2-5]. Essentially, this presents a rational success for probabilistic analysis of some structures. When come to complex external loads, it becomes a great challenge to obtain the fracture prameters. The pipes often subject to thermal loads as well as mechanical loads, determining thermal-mechanical fracture parameters is difficult because of the nonlinearity of thermal stress. One needs to evaluate thermal-mechanical fracture problem by employing FEM or other numerical methods. To date, reports are found considering both thermal stress and mechanical stress, which results in a nonlinear stress problem.

A rapidly way has been presented to predict the stochastic properties of SIFs and failure probability of cracked structures subject to random thermal-mechanical loads. A general weight function was derived to evaluate the thermal stress intensity factors of a circumferential crack in cylinders. The accuracy of the analysis has been examined using the finite element method results.

IOP Conf. Series: Materials Science and Engineering **372** (2018) 012013 doi:10.1088/1757-899X/372/1/012013

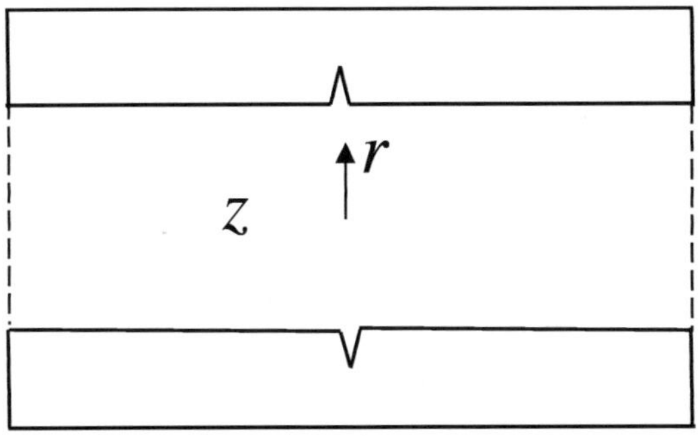

Figure 1. 360-degree internal circumferential flaw.

2. Thermal stress

There is a circumferential crack on the internal surface of an axisymmetric cylinder, as shown in Figure 1. And the inner surface subjected to a sudden cooling, and there is an internal pressure inside.

Governing equation [6] of the temperature distribution along the thickness is as follows.

$$\frac{\partial^2\theta}{\partial r^2}+\frac{1}{r}\frac{\partial\theta}{\partial r}=0 \tag{1}$$

In which, θ is the body temperature in the cylinder. and $\theta = T - T_0$, T_0 is the ambient temperature. Considering the third kind boundary condition, boundary conditions on the outer and inner surfaces are as following.

$$-\lambda\frac{\partial\theta}{\partial n}=\lambda\frac{\partial\theta}{\partial r}=h(\theta-\theta_f) \text{ on } r=R_i$$
$$\theta=0 \text{ on } r=R_o \tag{2}$$

In which, λ is coefficient of heat conductivity, h is convection coefficient. θ_f is the temperature of the fluid in the pipe ($\theta_f = T_f - T_0$).

Solve equation (1) under the conditions of equation (2). The temperature distribution is as following:

$$\theta(r)=\frac{\theta_f \ln(R_o/r)}{1/B_i+\ln(R_o/R_i)} \tag{3}$$

In which, $B_i = ah/\lambda$ is the Biot number.

The mechanical equilibrium equation in terms of displacement u as

$$\frac{d}{dr}\left[\frac{1}{r}\frac{d(ur)}{dr}\right]=\beta\frac{d\theta}{dr} \tag{4}$$

The parameter β is defined as $\beta=\alpha(1+\nu)/(1-\nu)$, in which α is the thermal expansion coefficient and ν is the Poisson's ratio.

Integration of the equilibrium equation yields displacement as follows.

$$u=\frac{\beta}{r}\int_a^r r\theta dr+C_1 r+\frac{C_2}{r} \tag{5}$$

The stress related to displacement and temperature can be obtained as follows.

$$\sigma_{rr} = \frac{2\mu}{1-2\nu}\left[(1-\nu)\frac{du}{dr} + \nu\frac{u}{r} - \alpha(1+\nu)\theta\right]$$

$$\sigma_{\phi\phi} = \frac{2\mu}{1-2\nu}\left[(1-\nu)\frac{u}{r} + \nu\frac{du}{dr} - \alpha(1+\nu)\theta\right] \tag{6}$$

$$\sigma_{zz} = \nu(\sigma_{rr} + \sigma_{\phi\phi}) - 2\mu(1+\nu)\alpha\theta$$

In which, μ is shear modulus of the cylinder.

It is assumed that only the inner surface of the cylinder is subjected to pressure boundary condition. So, stress boundary conditions are described as follows.

$$\sigma_{rr} = -P_i \quad \text{at} \quad r = R_i$$

$$\sigma_{rr} = 0 \quad \text{at} \quad r = R_o \tag{7}$$

The boundary conditions above can be used to determine the unknown parameters C_1 and C_2. Then, the axial stress can be computed as follows.

$$\sigma_{zz} = \frac{2\mu}{1-2\nu}\left[\nu(\frac{du}{dr} + \frac{u}{r}) - \alpha(1+\nu)\theta\right] \tag{8}$$

So, the axial stress, which is also the normal stress on the crack surface, can be rewritten as following.

$$\sigma_{zz} = -4\mu\xi_1 \ln(r) + \phi_1 \tag{9}$$

In which,

$$\xi_1 = -\frac{\beta\theta_f}{2\left(1/B_i + \ln(R_o/R_i)\right)}$$

$$\phi_1 = 4\mu\xi_1\left(\ln(R_o) + \nu\frac{R_i^2\ln\left(R_o/R_i\right)}{R_o^2 - R_i^2} - \frac{\nu}{2}\right) + \frac{2\nu PR_i^2}{R_o^2 - R_i^2} \tag{10}$$

3. Weight function method

Geometry of a cracked cylinder is as shown in figure 2.

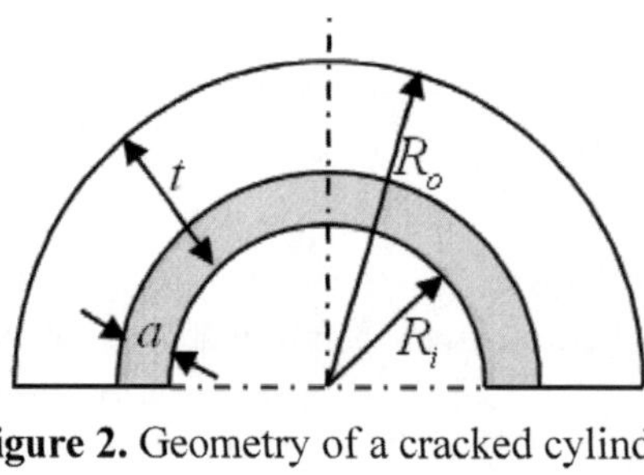

Figure 2. Geometry of a cracked cylinder.

The weight function method [7-9] can be calculated generally as follows.

$$K_I = \int_0^a \sigma(r) m(r,a) dr \tag{11}$$

In which, a is the crack depth, $\sigma(r)$ is the stress distribution across the plane of the crack in the non-cracked body and $m(r,a)$ is the weight function. The general expression of weight function can be written as follows [10].

$$m(r,a) = \sqrt{\frac{2}{\pi}}\sqrt{\frac{1}{R_i + a - r}} + M_1\sqrt{\frac{2}{\pi a}} + M_2\frac{1}{a}\sqrt{\frac{2}{\pi}}\sqrt{R_i + a - r} + M_3\frac{1}{a}\sqrt{\frac{2}{\pi a}}(R_i + a - r) \tag{12}$$

In which M_i , $i = 1,2,3$ are constants. Two reference stress state are used to obtain the constant coefficients [11]. The third condition is that the second derivative of the weight function with respect to a is zero at inner surface $r = R_i$, which results in $M_2 = 3$ [12]. Two reference stress state, which are homogeneous and linear stress distributions, are suggested to derive the constants.

$$\sigma_{1ref}(r) = \sigma_0$$

$$\sigma_{2ref}(r) = \sigma_0\left(\frac{r - R_i}{t}\right) \tag{13}$$

In which, t is the thickness. The SIFs due to the two stress distributions above in cylinders were found as.

$$K_{1ref} = \sigma_0\sqrt{\pi a}S_1$$

$$K_{2ref} = \sigma_0\sqrt{\pi a}S_2 \tag{14}$$

In which, S_1 and S_2 are boundary correction factors. Expressions for S_1 and S_2 are boundary correction factors, which can be expressed as following.

$$S_j = \sum_{i=0}^{4}\left(A_{ji}e^{B_{ji}(R_o/R_i)} + C_{ji}\right)\left(\frac{a}{t}\right)^i, \ j=1,\ 2 \tag{15}$$

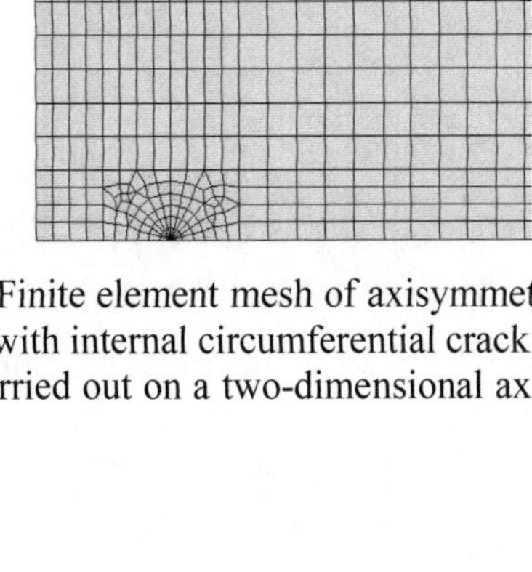

Figure 3. Finite element mesh of axisymmetric cylinder specimen with internal circumferential crack.

Finite-element analyses are carried out on a two-dimensional axisymmetric model. Figure 3 shows the finite element model established by ANSYS 13.0 [13].

Parameters *A*, *B* and *C* in equation (15) are shown in table 1.

Table 1. Coefficients: *A*, *B* and *C*.

	i=0	*i*=1	*i*=2	*i*=3	*i*=4
A_{1i}	0.42536	6.16171	1378.9905	6405.4262	0.93451
B_{1i}	0.04526	-1.39118	-6.5556	-22.36137	0.36609
C_{1i}	0.66445	-1.16945	2.73214	-4.02993	1.38286
A_{2i}	2.50134×10^{-4}	7.66246	395646.6828	826865.3	-50828.7
B_{2i}	2.69514	-2.17119	-1.78362×10^{6}	-12.3363	-9.78166
C_{2i}	0.58398	-0.70693	2.24895	-3.41457	2.48187

Substituting equations (12) and (13) into equation (11), and note that integrate region becomes R_i to R_i+a, constants M_i, i=1,3 can be found as follows.

$$M_1=-\sqrt{2\pi}S_1+3\sqrt{2\pi}S_2 t/a-4.8$$
$$M_3=3\sqrt{2\pi}S_1-6\sqrt{2\pi}S_2 t/a+1.6 \tag{16}$$

4. Determination of the SIFs

Since the continuously distributed stress field in equation (9) and weight function in equation (12) are obtained, SIFs of the crack tip considering both mechanical and thermal loads can be determined as following.

$$K_I(a)=\int_{R_i}^{R_i+a}(\sigma_{zz}(r)+P_i)m(r,a)dr \tag{17}$$

Note that, internal pressure is considered when calculate the SIFs of the crack tip. The SIFs is as following.

$$K_I(a)=(\varphi_1+P_i)\sqrt{\frac{a}{2\pi}}4+2M_1+\frac{4}{3}M_2+M_3\quad 4\mu\xi_1\sqrt{\frac{8a}{\pi}}\ln(R_i)\quad 2\quad\sqrt{\frac{R_i+a}{a}}\ln\frac{\sqrt{R_i+a}\quad\sqrt{a}}{\sqrt{R_i+a}+\sqrt{a}}$$
$$4\mu\xi_1\sqrt{\frac{2}{\pi a}}[(R_i+a)\ln(R_i+a)\quad R_i\ln(R_i)\quad a]M_1 \tag{18}$$
$$4\mu\xi_1\frac{2}{9a}\sqrt{\frac{2}{\pi}}3a\sqrt{a}\ln(R_i)\quad 8a\sqrt{a}\quad 6a\sqrt{a}R_i\quad 3(R_i+a)\sqrt{R_i+a}\ln\frac{\sqrt{R_i+a}\quad\sqrt{a}}{\sqrt{R_i+a}+\sqrt{a}}M_2$$
$$+4\mu\xi_1\sqrt{\frac{1}{8\pi a^3}}[3a^2+2aR_i+2R_i(R_i+2a)\ln(R_i)\quad 2(R_i+a)^2\ln(R_i+a)]M_3$$

The SIFs for homogenous stress state is developed as following.

$$K_I(a)=P_i\sqrt{\frac{a}{2\pi}}\left(4+2M_1+\frac{4}{3}M_2+M_3\right) \tag{19}$$

To verify the accuracy, consider a normal pressure $P_i=10MPa$. The material properties are assumed to as Shear modulus μ =80 Gpa, Possison's ratio ν =0.3. The non-dimensional stress intensity factor is defined as $K_N=K_I/(P_i\sqrt{\pi a})$. The results are indicted in figure 4. Different relative depths (a/t) varied from 0.1 to 0.8 and ratio of outer radius to internal radius of the cylinder varied from 1.1 to 2.0.

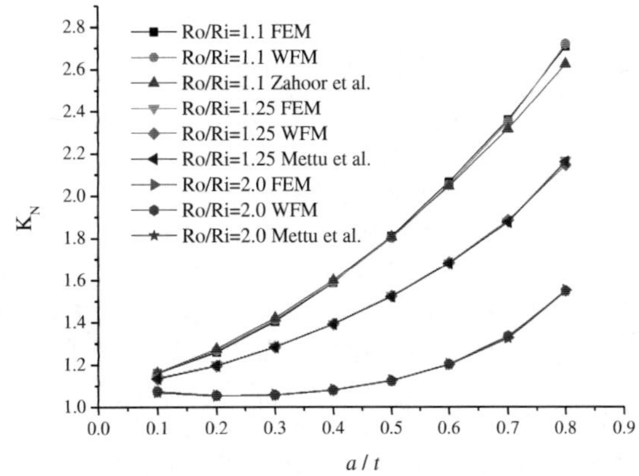

Figure 4. Results of the predicted K_I.

The results of R_o/R_i=1.1 are compared with the literature [14] and the results for R_o/R_i=1.25 and 2.0 are compared with those of Mettu [15]. From figure 4 we can see that the predicted K_I is reasonable at a wide range of radius ratio.

It is assumed that sudden cooling ($B_i = \infty$) happened at internal surface of the cylinder. Thermal expansion coefficient of the cylinder is $\alpha=14.2\times 10^{-6}$. The results of the normalized thermal stress intensity factors are shown in figure 5. The results show an excellent agreement.

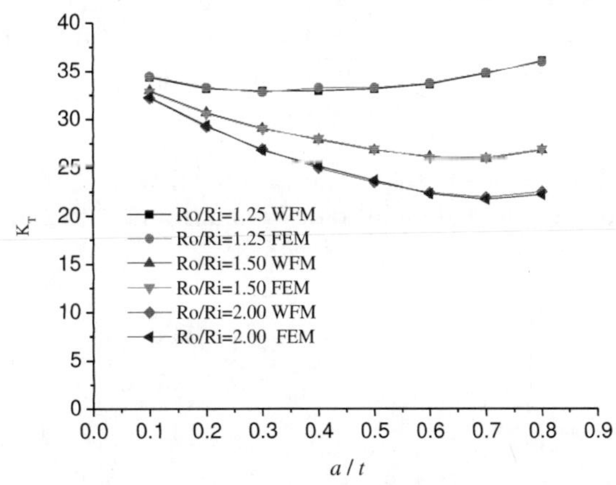

Figure 5. Comparisons of predicted thermal K_I for cylinder specimen with FEM.

5. Reliability analysis

There are several criteria to determine the failure. K_I based criteria is often used in engineering, described as follows.

$$K_I = K_{Ic} \tag{20}$$

Structure fails when the driving force (K_I) is greater than the fracture toughness (K_{Ic}), failure criterion due to crack initiation is widely used [16].

Consider a cracked cylinder subject to random loads and the structure fails when $K_I > K_{Ic}$, in which, K_I depends on input geometric parameters and loads which are random and K_{Ic} itself is a random variable. The probability of failure P_F, can defined as following.

$$P_F = \Pr[g(X) < 0] \tag{21}$$

Where $g(X)$ is as following.

$$g(X) = K_{Ic} - K_I \tag{22}$$

Monte-Carlo with neural network (MCNN) was used to compute these probabilities. Firstly, we produce many SIFs using the presented weight function method. The parameters associated with the weight function method are sampled form its own statistical distribution. Then, a comparison between the random SIF and the critical SIF has been done, and if the random SIF is greater than the critical SIF, make the tag equals 1, otherwise 0. We input all of the random parameters and the result tag in the neural network, and train the neural network until we get a stable neural network. At last, we can use the neural network to get the failure of the structure by input a large number of random parameters and a large number of result tags also can be obtained. The failure probability is obtained by total number of tag which equals 1 divided by total number of simulations. The cylinder with 360-degree circumferential crack subjected to internal pressure with P_i, the inner surface cooling with temperature θ_s . The internal radius is 100mm, Biot number is $B_i = \infty$. The input statistical variables are as shown in Table 2. The probabilistic characteristics of P_i were chosen arbitrarily.

Table 2. statistical properties of random input for cylinder.

Statistical Variable	Mean	COV[a]	Probability distribution
Elastic modulus (E)	208GPa	0.05	Gaussian
Internal pressure (P_i)	10MPa	0.1	Gaussian
Fluid temperature (θ_f)	-100°C	0.1	Gaussian
Crack size (a/t)	Variable[b]	0.2	Gaussian
Initiation fracture toughness (K_{Ic})	10MPa$\sqrt{m}$	0.5	Lognormal

[a] COV=coefficient of variation.
[c] varies from 0.1 to 0.8.

Figure 6 shows P_F vs a/t, as obtained using standard reliability methods. No significant differences are found between them. Outer to internal radius ratio varied from 1.1 to 2.0.

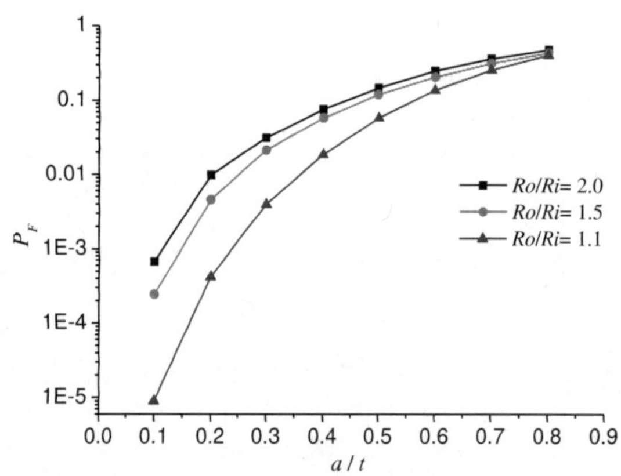

Figure 6. Probability of failure for cylinder specimen.

6. Summary and conclusions

An accurate weight function method to achieve thermal-mechanical SIFs for cracked cylinder has been established. Reference stress state is used to determine the coefficient. Results show a good agreement with the documents. The SIFs can be determined quickly using the weight function obtained when cracks subjected to arbitrary loads, and can be used for probabilistic fracture mechanics analysis.

A probabilistic fracture mechanics analysis methodology has been obtained for cracked cylinder. Computational reliability methods of the K_I based fracture have been presented to analysis the statistical characteristics. The results from the example show that a precise statistical characteristics of the K_I can be determined using the developed methodology. The failure probability of the structure increases with the increasing of loads, and the relationship between is not linear.

References

[1] Rahman S, Ghadiali N, Paul D and wilkowski G 1995 Probabilistic pipe fracture evaluations for leak-rate-detection applications *NUREG/CR-6004* (Washington, D.C.: U.S. Nuclear Regulatory Commission)
[2] Rahman S 1995 A stochastic model for elastic-plastic fracture analysis of circumferential through-wall-cracked pipes subject to bending *Engng Fract Mech* **52(2)** 265-88
[3] Dillstrom P 2000 ProSINTAP-A probabilistic program implementing the SINTAP assessment procedure *Engng Fract Mech* **67** 647-68
[4] Rahman S 2001 Probabilistic fracture mechanics: J-estimation and finite element methods *Engng Fract Mech* **68** 107-25
[5] Rahman S and Kim J S 2001 Probabilistic fracture mechanics for nonlinear structures *Int J Pres Ves Pip* **78** 261-9
[6] Hetnarski R B and Eslami M R 2009 *Thermal Stresses – Advanced Theory and Applications* (Netherlands: Springer)
[7] Bueckner H F 1970 A novel principle for the computation of stress intensity factors *Z Angew Math Me* **50** 129-46
[8] Rice J R 1972 Some remarks on elastic crack-tip stress fields *Int J Solids Struct* **8** 751-8
[9] Nabavi S M and Ghajar R 2010 Analysis of thermal stress intensity factors for cracked cylinders using weight function method *Int J Eng Sci* **48** 1811-23
[10] Glinka G and Shen G 1991 Universal features of weight functions for cracks in mode I *Engng Fract Mech* **40** 1135-46
[11] Shen G and Glinka G 1991 Determination of weight functions from reference stress intensity factors *Theor Appl Fract Mec* **15** 237-45
[12] Fett T, Mattheck C and Munz D 1987 On the calculation of crack opening displacement from the stress intensity factor *Engng Fract Mech* **27** 697-715
[13] ANSYS Release 13.0 2010 (Canonsburg, PA: ANSYS, Inc)
[14] Zahoor A 1985 Closed form expressions for fracture mechanics analysis of cracked pipes *J Press Vess-T ASME* **107** 203-5
[15] Mettu S R and Forman R G 1993 Analysis of circumferential cracks in circular cylinders using the weight-function method *Fracture Mechanics: Twenty-Third Symposium ASTM STP 1180* (Philadelphia) 417-40
[16] ASME Boiler and Pressure Vessel Code, Section XI 2015 Rules for inservice inspection of nuclear power plant components (New York: The American Society of Mechanical Engineers)

Strength evaluation of butt joint by stress intensity factor of small edge crack near interface edge

T Sato[1], K Oda[2,3] and N Tsutsumi[2]

[1]Graduate School of Engineering, Oita University, 700 Dannoharu, Oita 870-1192, Japan
[2]Division of Mechanical Engineering, Oita University, 700 Dannoharu, Oita 870-1192, Japan

E-mail: oda-kazuhiro@oita-u.ac.jp

Abstract. Failure of the bonded dissimilar materials generally initiates near the interface, or just from the interface edge due to the stress singularity at the interface edge. In this study, the stress intensity factor of an edge crack close to the interface between the dissimilar materials is analyzed. The small edge crack is strongly dominated by the singular stress field near the interface edge. The analysis of stress intensity factor of small edge crack near the interface in bi-material and butt joint plates is carried out by changing the length and the location of the crack and the region dominated by the interface edge is examined. It is found that the dimensionless stress intensity factor of small crack, normalized by the singular stress at the crack tip point in the bonded plate without the crack, is equal to 1.12, independent of the material combination and adhesive layer thickness, when the relative crack length with respect to the crack location is less than 0.01. The adhesive strength of the bonded plate with various adhesive layer thicknesses can be expressed as the constant critical stress intensity factor of the small edge crack.

1. Introduction

The development of high-strength adhesives has encouraged the use of bonded joint for various engineering applications. In adhesively bonded joint components under mechanical or thermal loading condition, the singular stresses near the interface free-edge occur and these high stresses may result in the initiation of failure from the interface edge. Many studies on adhesive strength have been carried out analytically and experimentally using the intensity of the singular stress field at the interface corner of bonded heterogeneous material. It is well known that the singular stress distribution in the vicinity of the interface edge can be expressed by two parameters, that is, the index and the intensity of the singular stress. In late years, Kubo-Ohji [1] and Chen-Nisitani [2] have reinvestigated the singularities of stress field in the vicinity of the interface corner by the method of eigenvalue expansion. Munz-Yang [3] have showed the extrapolation technique for single term singularity corner in order to determine the intensity of the singular stress distribution at the edge of interface, and Xu *et al* [4] have proposed the numerical method to calculate the indexes of multiple singular terms and the corresponding stress intensity coefficients. Noda *et al* [5,6] have reported the efficient technique to evaluate the intensity of the singular stress field (ISSF) at the interface corner by using the ratio of the stress value near the interface edge obtained by the finite element method (FEM). Furthermore, Noda *et al* [7-10] have presented that the fictitious interface crack model in bonded materials is useful for evaluating the adhesive strength of the bonded structures.

On the other hand, the problem of cracks induced near the free-edge of interface has also been researched. Fett *et al* [11,12] have examined about the weight functions of the sub-interface cracks. Xu *et al* [13] have presented the approximate analytical solution of the stress intensity factor (SIF) for edge crack close to the interface by using the superposition method.

In this study, a problem of small edge crack close to the interface in a bi-material plate and butt joint plate will be considered. The stress intensity factors (SIFs) for small edge crack will be analysed by changing the crack length and the distance from the interface, and the reasonable expression of solution for the edge crack will be examined. In addition, the fracture criterion of the bonded plate will be discussed by using the small edge crack model.

2. Crack tip stress method

In this study, a small edge crack problem near the interface in the bonded plates is analyzed. As shown in figure 1, '*a*' expresses the crack length and '*c*' stands for the distance from the interface to the crack. When the crack is close to the free-edge of interface, the edge crack is strongly affected by the corner stress singularity. The stress intensity factor of the edge crack is calculated by the finite element analysis and the crack tip stress method [14,15]. In this analysis, the versatile FEM code MSC.Marc/Mentat is employed and the four-node-quadrilateral element is used. The pattern of finite element mesh around the crack tip is systematically and finely created.

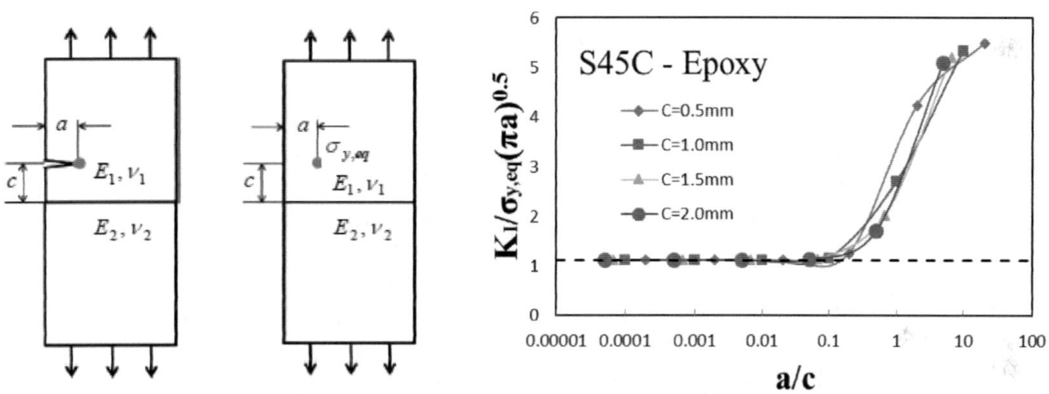

Figure 1. Relation between F_I and t *a*/*c* for bi-material plate (S45C-Epoxy).

The crack tip stress method has been originally proposed by Nisitani [14,15] and a stress intensity factor for two dimensional crack problem can be obtained accurately by using the FEM analysis. According to the principle of the method, the ratio between the stress intensity factors of two different problems is nearly equal to the ratio between the stress values at the crack tip nodes in y-direction obtained by FEM. When the same mesh patterns around the crack tip are used for two different crack problems, the singular stress fields can be represented by the crack tip stress $\sigma_{y0,FEM}$ [14,15]. Based on the above fact, the equation (1) is obtained. In this relation, K_I is the mode I stress intensity factor, $\sigma_{y0,FEM}$ means the stress value in y-direction at the crack-tip-node calculated by FEM and the asterisk * indicates the value of the reference problem.

$$\frac{K_I}{\sigma_{y0,FEM}} \fallingdotseq \frac{K_I^*}{\sigma_{y0,FEM}^*} \tag{1}$$

If we know the K_I^* for the reference problem in advance, the value of K_I for the unknown problem can be evaluated from the equation (1) because the stress values $\sigma_{y0,FEM}^*$ and $\sigma_{y0,FEM}$ for the reference and the unknown problems can be readily computed by FEM. As the reference problem, the closed-form solution of a crack in an infinite homogeneous plane is selected in this calculation.

IOP Conf. Series: Materials Science and Engineering **372** (2018) 012014 doi:10.1088/1757-899X/372/1/012014

3. Numerical results and discussion

3.1. Stress intensity factors of small edge crack close to interface in bi-material plate

The stress intensity factors of small edge crack close to the interface illustrated in figure 1 are analyzed. In figure 1, the plate length L=400 mm, the plate width W=100 mm and the distance from the interface to the crack c=2 mm. As the combination of the materials used for the analysis, the medium carbon steel S45C (E_1 = 206 GPa, ν_1= 0.30 for material '1') and the epoxy resin (E_2=4.93 GPa, ν_2= 0.33 for material '2') are assumed. The singularity index of interface corner is $1-\lambda$=0.2823.

In this study, the normalized stress intensity factors are defined by the singular stress near the free-edge of interface as the next expression [16]:

$$F_I = K_I/\sigma_{y,eq}\sqrt{\pi a}, \quad F_{II} = K_{II}/\sigma_{y,eq}\sqrt{\pi a} \tag{2}$$

As shown in figure 1, $\sigma_{y,eq}$ stands for the y-directional stress value at the crack tip position in the same bi-material plate without the crack. The stress $\sigma_{y,eq}$ close to the edge of interface in material '1' has been given as following equations [13]:

$$\begin{aligned}\sigma_{y,eq} = \frac{\rho(\sin\theta)^{1-\lambda}}{4}\{&[(\lambda+2-\cos\lambda\pi)d + b\sin\lambda\pi]\cos(\lambda-1)\theta \\ &-[(\lambda+2+\cos\lambda\pi)b + d\sin\lambda\pi]\sin(\lambda-1)\theta \\ &+(\lambda-1)[d\cos(\lambda-3)\theta - b\sin(\lambda-3)\theta - b\sin(\lambda-3)\theta]\}\end{aligned} \tag{3}$$

Here,

$$\begin{aligned}&\rho = \frac{\lambda c^{\lambda-1}K_\sigma}{\lambda^2-\sin^2(\lambda\pi/2)}, \quad \theta = \tan^{-1}(c/a), \\ &b = 2F - \tfrac{1}{\lambda}[\sin\lambda\pi - F(1-\cos\lambda\pi)], \quad d = 2 - \tfrac{1}{\lambda}(1-\cos\lambda\pi + F\sin\lambda\pi), \\ &F = \frac{\lambda D + \sin(\lambda\pi/2)\cos(\lambda\pi/2) - D\sin^2(\lambda\pi/2)}{\lambda + \sin^2(\lambda\pi/2) + D\sin(\lambda\pi/2)\cos(\lambda\pi/2)}, \\ &D = \frac{2\beta[\lambda^2 + (2\lambda+1)\sin^2(\lambda\pi/2)] - 2\alpha[\lambda^2 - \lambda\cos(\lambda\pi)]}{\sin(\lambda\pi)[1-2\lambda(\beta-\alpha)]}.\end{aligned} \tag{4}$$

In the equations (3) and (4), K_σ is the fracture mechanics parameter expressing the intensity of the singular stress field in the vicinity of the interface corner, and α, β are Dundurs' composite parameters describing the elastic mismatch.

Figure 1 illustrates the numerical results of the normalized stress intensity factor F_I for various crack lengths and crack locations. As shown in this figure, it is found that the normalized stress intensity factors are converged 1.12 when the relative crack length with respect to the crack distance a/c<0.01, independent of the crack distance from interface. In this state, the mode II factors F_{II} are nearly zero. Therefore, the stress intensity factor of small edge crack close to the interface can be represented by

$$K_I = 1.12\sigma_{y,eq}\sqrt{\pi a}, \; K_{II} = 0. \tag{5}$$

Furthermore, in the case of a/c<0.01, it is shown that the equations (3), (4) and (5) express the linear relationship between the stress intensity factor K_I of the small edge crack and the intensity K_σ of corner singular stress.

3.2. Stress intensity factor of bonded plate with adhesive layer

Next, the edge crack problem in butt joint plate with adhesive layer is considered. As illustrated in figure 2, the S45C (material 1) rectangular plates are adhered the epoxy resin (material 2). The stress intensity factors of the edge crack for various crack lengths and locations are calculated by using the crack tip stress method. Figure 2 shows the relation between the normalized factor F_I and the relative crack length with respect to the bond line thickness a/h when the bond line thickness is h=1 mm. In this case, the stress value $\sigma_{y,eq}$ is calculated by FEM, not the equation (3). As shown in figure 2, the normalized stress intensity factors for the different crack locations are also converged with 1.12 when $a/h<0.01$. Therefore, even in the case of the bonded plate with the adhesive layer, the stress intensity factor of the small edge crack within the corner singularity area can be evaluated by $K_I = 1.12\sigma_{y,eq}\sqrt{\pi a}$.

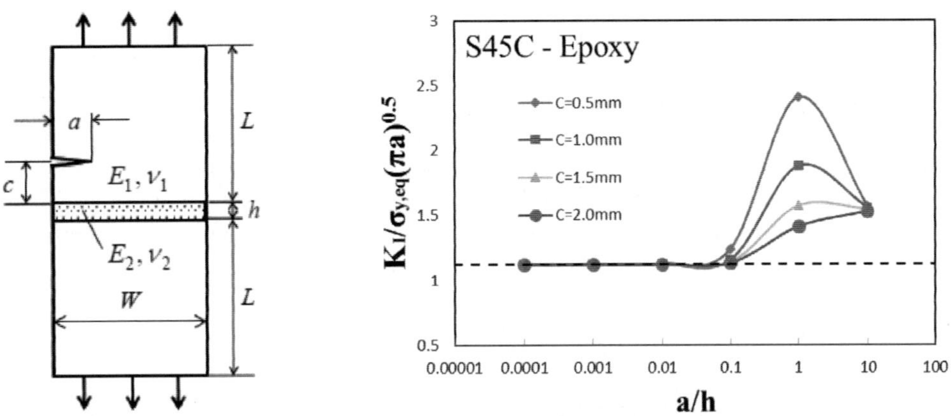

Figure 2. Relation between F_I and a/h for butt joint (S45C-Epoxy) when h=1 mm.

Table 1 shows the comparison of K_σ between the present results obtained from equations (3), (5) and the results analyzed by Noda *et al* [5] when the crack length is fixed and the crack position is systematically changed in the range of $a/c<0.01$ and $a/h<0.01$. Table 2 also presents the present K_σ calculated from equations (3), (5) when the crack position is fixed and the crack length is systematically changed. From tables 1 and 2, even when the crack position is changed within the range of $a/c<0.01$ and $a/h<0.01$, the results of both are in good agreement.

Table 1. Comparison of K_σ. (a=0.0009 mm, h=0.1 mm)

c (mm)	K_σ	
	Present results	Noda *et al* [5]
0.09	70.7	68.9
0.07	70.56	68.9
0.05	70.34	68.9
0.03	70.18	68.9

Table 2. Comparison of K_σ. (c=0.09 mm, h=0.1 mm)

a (mm)	K_σ	
	Present results	Noda *et al* [5]
0.0009	70.76	68.9
0.0008	70.73	68.9
0.0007	70.70	68.9
0.0006	70.66	68.9

Table 3 indicates the stress intensity factors K_I calculated from equation (5) and by the crack tip stress method (CTSM) when c = 0.09 mm and the crack length is changed from a = 0.0006 to 0.0009 mm. As shown in table 3, the difference between both results is small within 0.1% error and equation (5) is useful for evaluating the stress intensity factor of small edge crack close to the interface when a/c<0.01 and a/h<0.01.

Table 3. Comparison of stress intensity factors. (c=0.09 mm, h=0.1 mm).

a (mm)	K_I		Error (%)
	Equation (5)	CTSM	
0.0009	6.015	6.019	0.067
0.0008	5.671	5.670	0.018
0.0007	5.305	5.306	0.005
0.0006	4.912	4.912	0.003

4. Application to adhesion strength evaluation

Next, we examine the strength evaluation of butt joint specimens by using the stress intensity factor of edge crack near the interface. The tensile test for the bonded plates with various adhesive thicknesses has been carried out by Suzuki [17]. In this study, the critical stress intensity factor K_{IC} is calculated by using the experimental fracture stress σ_c obtained by Suzuki [17]. The material properties for two patterns of bonded specimens used by Suzuki are shown in table 4.

Table 4. Material combinations [17].

	E (GPa)	ν
S35C Epikote828	210	0.3
	3.14	0.37
S35C Epikote871	210	0.3
	2.16	0.38

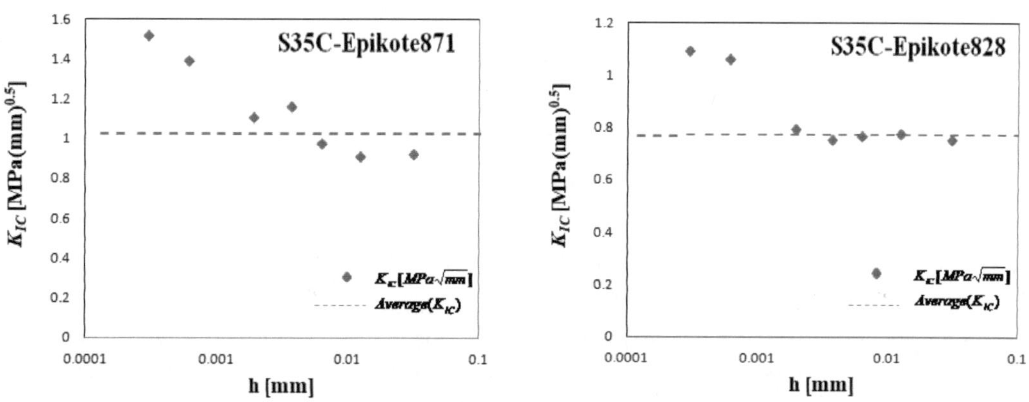

(a) S35C-Epikote828 (b) S35C-Epikote871

Figure 3. Relation between K_{IC} and h when $a = 10^{-5}$mm and $c = 10^{-3}$mm ($a/h \leq 0.1$).

Figure 3 shows the relationship between the critical stress intensity factor K_{IC} and the adhesive layer thickness h when the crack location $c = 10^{-3}$ mm and the crack length $a = 10^{-5}$ mm. The left figure is S35C-Epikote828 and the right is S35C-Epikote871. As indicated in figure 3, when the bond line thickness h is larger than the edge crack distance c from the interface, that is, c/h<1, the values of K_{IC} are almost constant. On the other hand, when c/h>1, the values of K_{IC} deviate because the dominated area of the interface corner singularity is related to the adhesive layer thickness.

If the distance c from the interface is larger than the adhesive thickness h, the edge crack is out of the singularity area, the stress intensity factor K_I does not correspond to the intensity of corner singular stress field K_σ.

Figure 4 also indicates the relationship between the critical stress intensity factor K_{IC} and the adhesive layer thickness h when the location and length of the edge crack are changed, $c = 10^{-4}$ mm and $a = 10^{-6}$ mm. In this case, since the adhesive layer thickness h is always greater than the distance c to the interface in the calculation range, the values of K_{IC} are approximately constant including experimental error, as indicated by the blue dashed line in figure 4. These results are equivalent to the results evaluated by the intensity of the corner singular stress field K_σ [9,10].

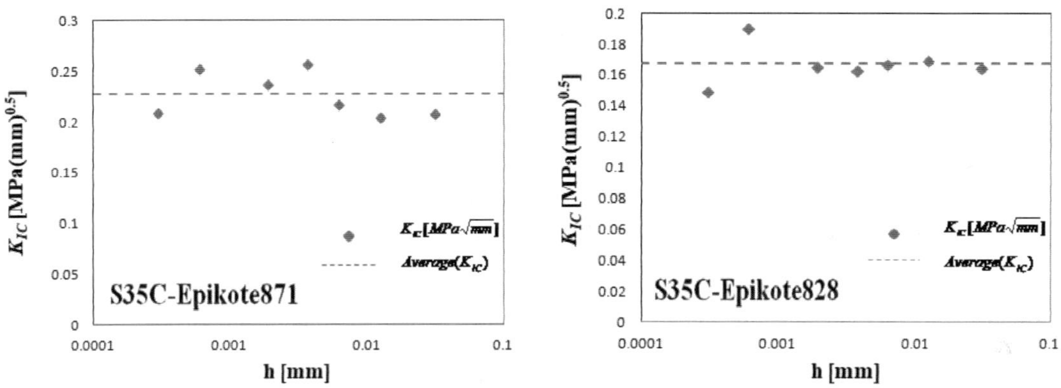

(a) S35C-Epikote828 (b) S35C-Epikote871

Figure 4. Relation between K_{IC} and h when $a = 10^{-6}$ mm and $c = 10^{-4}$ mm ($a/h \leq 0.01$).

5. Conclusions

In this study, the evaluation method of the adhesive strength for butt joint specimens were considered by means of the edge crack model. The stress intensity factors of the small edge crack close to the interface were analyzed by the crack tip stress method and the relation between the stress intensity factor (SIF) of edge crack and the intensity of corner singular stress field (ISSF) were examined. Then, the numerical results of edge crack model were applied to estimation of the fracture strength of butt joint specimens. The conclusions can be summarized as follows:

- The normalized stress intensity factors on the basis of the singular stress $\sigma_{y,eq}$ were converged with the constant value 1.12 when the ratio between the length and the location of the crack a/c<0.01. The stress value $\sigma_{y,eq}$ at the crack tip position in the bonded dissimilar plate without the crack was calculated by the equation (3) or the finite element method.
- The stress intensity factor of the small edge crack close to the free-edge of interface can be shown by $K_I = 1.12\sigma_{y,eq}\sqrt{\pi a}$. Since the singular stress $\sigma_{y,eq}$ is proportional to the intensity K_σ of the singular stress field at the interface corner, K_I and K_σ have a one-to-one relationship.
- The critical stress intensity factors K_{IC} became almost constant irrespective of the bond line thickness h when a/h<0.01 and c/h<1. Therefore, the stress intensity factor of the small edge crack close to the interface can be an alternative parameter of the intensity of the interface edge singular field, and useful for evaluating the adhesive strength of bonded specimen.

References

[1] Kubo S and Ohji K 1991 Geometrical conditions of no free-edge stress singularities in edge-bonded elastic dissimilar wedges *Trans the Japan Society of Mechanical Engineers* **57** 632–6 (in Japanese).
[2] Chen D H and Nisitani H 1992 Singular stress field in two bonded wedges *Trans the Japan Society of Mechanical Engineers* **58** 457–64 (in Japanese)
[3] Munz D and Yang Y Y 1993 Stresses near the edge of bonded dissimilar materials described by

two stress intensity factors *Int J Fract* **60** 169-77

[4] Xu J Q, Liu Y H and Wang X G 1999 Numerical methods for the determination of multiple stress singularities and related stress intensity coefficients *Eng Fract Mech* **63** 775-90

[5] Noda N-A, Zhang Y, Lan X and Takaishi K 2011 Strength analysis for the adhesive layer on the basis of intensity of singular stress *Procedia Engineering* **10** 722-7

[6] Noda N-A, Zhang Y, Takaishi K, Lan X and Oda K 2012 Intensity of singular stress for single-lap joints *Trans the Japan Society of Mechanical Engineers* **78** 651-5 (in Japanese)

[7] Noda N-A, Lan X, Michinaka K, Zhang Y and Oda K 2010 Stress intensity factor of an edge interface crack in a bonded semi-infinite plate *Trans the Japan Society of Mechanical Engineers* **76** 1270-7 (in Japanese)

[8] Oda K, Lan X, Noda N-A and Michinaka K 2012 Effect of arbitrary bi-material combination and bending loading conditions on stress intensity factors of an edge interface crack *Int J Structural Integrity* **3** 457-75

[9] Noda N-A, Miyazaki T, Uchikoba T, Li R, Sano Y and Takase Y 2014 Convenient debonding strength evaluation based on the intensity of singular stress for adhesive joints *J the Japan Institute of Electronics Packaging* **17** 132-42 (in Japanese)

[10] Noda N-A, Misyazaki T, Li R, Uchikoba T, Sano Y and Takase Y 2015 Debonding strength evaluation in terms of the intensity of singular stress at the interface corner with and without fictitious crack *Int J Adhesion & Adhesives* **61** 46-64

[11] Fett T, Tilscher M and Munz D 1997a Weight functions for cracks near the interface of a bimaterial joint and application to thermal stresses *Eng Fract Mech* **56** 87-100

[12] Fett T, Tilscher M and Munz D 1997b Weight functions for sub-interface cracks *Int J Solids and Structures* **34** 393-400

[13] Xu J-Q, Wang X-G and Mutoh Y 2001 Stress intensity factors of a surface crack near an interface end *Int J Fract* **111** 251-64

[14] Nisitani H, Kawamura T, Fujisaki W and Fukuda T 1999 Determination of highly accurate values of stress intensity factor or stress concentration factor of plate specimen by FEM *Trans the Japan Society of Mechanical Engineers* **65A** 26-31 (in Japanese)

[15] Nisitani H and Teranishi T 2004 *KI* of a circumferential crack emanating from an ellipsoidal cavity obtained by the crack tip stress method in FEM *Eng Fract Mech* **71** 579-85

[16] Oda K, Kawakita M and Tsutsumi N 2017 Stress intensity factor for small edge crack in interfacial singularity region *IOP Conf. Series: J. Physics: Conf. Series* **843** 012010

[17] Suzuki Y 1987 Adhesive tensile strengths of scarf and butt joints of steel plates, (Relation between adhesive layer thicknesses and adhesive strengths of joints) *JSME Int. J.* **30** 1042-51

IOP Conf. Series: Materials Science and Engineering **372** (2018) 012015 doi:10.1088/1757-899X/372/1/012015

Evaluation of adhesive scarf joint strength by using singular stress field of small interface crack

S Haraga[1], N Okazaki[1], K Oda[2,3] and N Tsutsumi[2]
[1]Graduate School of Engineering, Oita University, 700 Dannoharu, Oita 870-1192, Japan
[2]Division of Mechanical Engineering, Oita University, 700 Dannoharu, Oita 870-1192, Japan

E-mail: oda-kazuhiro@oita-u.ac.jp

Abstract. In the present study, the adhesive strength of scarf joint is examined by using the stress intensity factor of the fictitious small interface crack. The stress intensity factor of small crack near the interface edge is dominated by the singular stress field of the interface corner. In this study, to evaluate the joint strength, small crack is assumed at the interface corner of the scarf joint. The stress intensity factor of the interfacial crack is calculated by changing the thickness of the adhesive layer and the scarf angle. By using the experimental fracture strength of the scarf adhesive joint specimens under tension, the values of the critical stress intensity factor are calculated. From the analysis result, when the combination of adhesive materials and the scarf angle are fixed, the critical stress intensity factors of the small interface crack are constant value irrespective of the adhesive layer thickness. Therefore, the adhesive joint strength can be evaluated as the constant stress intensity factor of small interface crack. In addition, it is possible to evaluate easily the stress intensity factor by using the dimensionless coefficients depending only on the material combination when the crack length is sufficiently smaller than the thickness of the adhesive layer. The effects of adhesive layer thickness and adhesion angle of scarf joint specimen were discussed and the effectiveness of the proposed method was indicated.

1. Introduction

In recent years, bonded dissimilar materials are used as structural materials in various industrial fields. It is generally known that the adhesive strength of a scarf joint is higher than that of a butt joint structure. In these bonded joint components, due to the mismatch of the material elastic properties, singular stresses occur at the interface corner. These stresses are especially high and may result in the initiation of failure from the interface-edge, which is a problem in terms of adhesive strength. Then, many studies have been conducted to evaluate the adhesive strength focusing on the singular stress field at the interface corner. Recently, in order to estimate the fracture stress of bonded plates, the small crack approach has been proposed [1,2]. In this method, by assuming a small crack at the interface edge, the stress intensity factor of small interface crack is used to evaluate the intensity of corner singular stress and the criterion of interface fracture can be expressed as the constant critical stress intensity factor [1,2]. Oda *et al* [3,4] have analyzed the singular stress of butt joint specimens with small crack in detail, and they have reported that when the interface crack is sufficiently smaller than the adhesive layer thickness, the stress intensity factor of the small interface crack is equivalent to the intensity of the singular stress field at the interface corner and it can be converted to each other.

However, it has not been studied whether the small crack model is appropriate for the scarf joint.

In this study, a scarf joint model assuming a small crack will be analyzed. The effect of the scarf angle and the adhesive layer thickness on the stress intensity factor of small interface crack will be discussed. By using experimental fracture loads [5], the evaluation method of the fracture strength for the scarf adhesive joint will be examined.

2. Analysis method

Figure 1 shows an adhesive joint model under tension analyzed in this study. The scarf joint consists of a thin adhesive layer sandwiched between two different plates. A small crack is assumed at the interface corner where fracture is expected. The subscripts 1, 2 and 3 of E and ν in figure 1 refer to SUS304 stainless steel, Epoxy resin and AL YH5 aluminum alloy, respectively. The material constants used in this analysis are shown in table 1. The detailed material properties are described in [5]. In addition, Dundurs composite parameters α and β defined by equation (1) are shown in table 2. The parameters represent the material combination of the interface.

$$\alpha = \frac{G_n(\kappa_m + 1) - G_m(\kappa_n + 1)}{G_n(\kappa_m + 1) + G_m(\kappa_n + 1)}, \quad \beta = \frac{G_n(\kappa_m - 1) - G_m(\kappa_n - 1)}{G_n(\kappa_m + 1) + G_m(\kappa_n + 1)},$$

$$G_m = \frac{E_m}{2(1 + \nu_m)}, \kappa_m = \begin{cases} \dfrac{3 - \nu_m}{1 + \nu_m} & \text{(Plane stress)} \\ 3 - 4\nu_m & \text{(Plane strain)} \end{cases} \quad (m, n = 1, 2, 3) \tag{1}$$

The joint has the length of the plate L, the plate width W, the plate thickness t, the adhesive layer thickness h and the scarf angle θ. The same joint specimen has been used in the tensile fracture strength test conducted by Afendi *et al* [5]. The scarf angles are set to be θ = 45°, 60° and 75°.

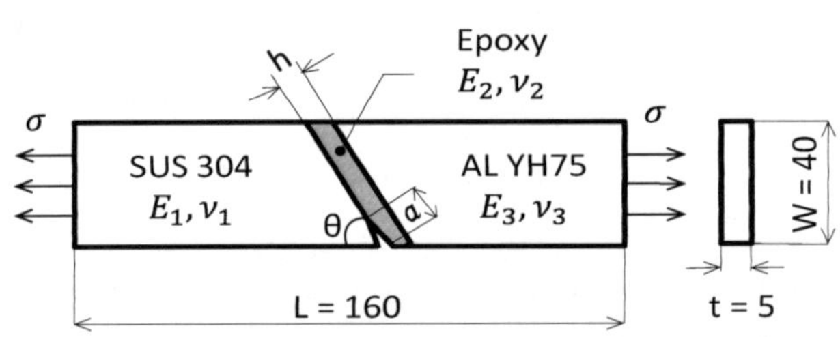

Figure 1. Analysis model.

Table 1. Mechanical properties used in the analysis.

Material	Young's modulus E (GPa)	Poisson's Ratio ν
SUS304	206	0.3
Epoxy	3.4	0.396
AL YH5	71	0.33

Table 2. Dundurs composite parameters α, β.

Material combination	Parameter α	Parameter β
SUS / Epoxy	0.9650	0.1642
Epoxy / YH5	-0.9037	-0.1517

The singular stress occurs at the interface free edge in the adhesive joint without the crack. The order of stress singularity in the vicinity of the interface corner is different from that of crack, and it

IOP Conf. Series: Materials Science and Engineering **372** (2018) 012015 doi:10.1088/1757-899X/372/1/012015

varies depending on the material combination and the edge configuration. The exact solution of 1-λ is obtained by the following eigen-equation [6]:

$$A\beta^2 + 2B\alpha\beta + C\alpha^2 + 2D\beta + 2E\alpha + F = 0 \tag{2}$$

Here,

$$\begin{aligned}
A &= 4K(\lambda,\theta_1)K(\lambda,\theta_2-\theta_1),\\
B &= 2\lambda^2\sin^2\theta_1 K(\lambda,\theta_2-\theta_1) + 2\lambda^2 sin^2(\theta_2-\theta_1)K(\lambda,\theta_1),\\
C &= 4\lambda^2(\lambda^2-1)sin^2\theta_1 sin^2(\theta_2-\theta_1) + K(\lambda,\theta_2-2\theta_1),\\
D &= 2\lambda^2[sin^2(\theta_2-\theta_1)sin^2(\lambda\theta_1) - sin^2\theta_1 sin^2\{\lambda(\theta_2-\theta_1)\}],\\
E &= -D + K(\lambda,\theta_1) - K(\lambda,\theta_2-\theta_1),\\
F &= K(\lambda,\theta_2),\\
K(\lambda,x) &= \sin^2(\lambda x) - \lambda^2\sin^2(x).
\end{aligned} \tag{3}$$

Table 3 shows the values of the order of singularity 1-λ at SUS/Epoxy and Epoxy/AL YH75 interface corners when the scarf angle is 45°, 60° and 75°. Figures 2 and 3 indicate the order of stress singularities 1-λ at the interface corner as a function of scarf angle. From these figures, we assume the small interfacial crack along the SUS/Epoxy interface because the singularity index 1-λ of SUS 304 / Epoxy interface is always larger than that of Epoxy/YH5 interface.

Table 3. The order of singularity 1-λ at interface corner.

Interface	75°	60°	45°
SUS/Epoxy	0.3648	0.3619	0.2796
Epoxy/YH75	0.2369	0.1179	0.0000

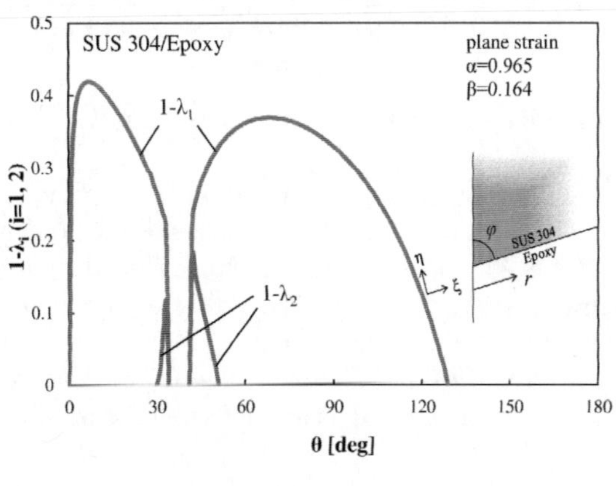

Figure 2. The order of stress singularity λ at SUS/Epoxy vs scarf angle θ.

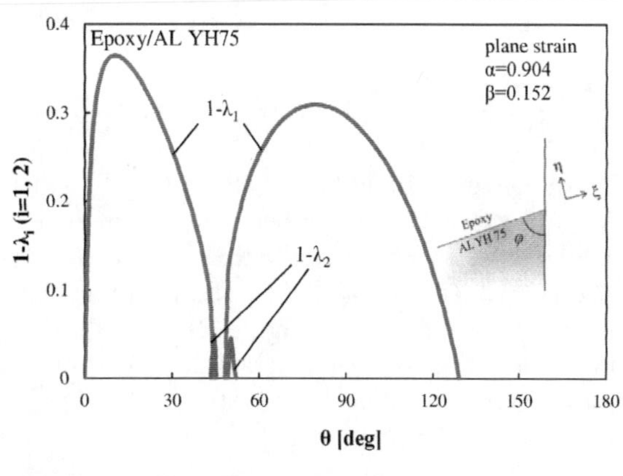

Figure 3. The order of stress singularity λ at Epoxy/Al vs scarf angle θ.

In this analysis, we use the versatile FEM program MSC.Marc/Mentat. The stress intensity factor of the small interface crack is calculated by the crack tip stress method [7,8]. In this method, the stress value at the crack tip node obtained by FEM is compared with the crack tip stress value of a single interface crack in an infinite bonded plate. When two analysis models have the same mesh pattern around the crack tip, the numerical error due to the mesh division can be canceled. By using this method, it is possible to easily determine the stress intensity factor with high accuracy in FEM. In the finite element calculation, the plane strain 8-node quadrilateral element is selected and the minimum element size e around the crack tip is set to be $e / a = 8.14\times10^{-4}$ with respect to the crack length a.

3. Numerical results and discussion

3.1. Effect of crack length on stress intensity factor of small interface crack

We investigate the effect of crack length on the stress intensity factor of the assumed small interfacial crack. The stress intensity factors K_1 and K_2 under the fixed adhesive thickness h=1 mm (h/W=0.025) are calculated by the crack tip stress method when the crack length is changed. The dimensionless stress intensity factors F_1 and F_2 are determined as follows:

$$K_1 + iK_2 = (F_1 + iF_2)\sigma\sqrt{\pi a}(1 + 2i\varepsilon)\ , \qquad \varepsilon = \frac{1}{2\pi}\ln\left(\frac{1-\beta}{1+\beta}\right) \tag{4}$$

Here, a is the crack length, F_1 and F_2 are the dimensionless stress intensity factors, σ is the tensile stress, and ε is the bi-material constant according to the combination of materials. The relations between the relative crack length a/h and the dimensionless stress intensity factor are presented in figures 4 and 5.

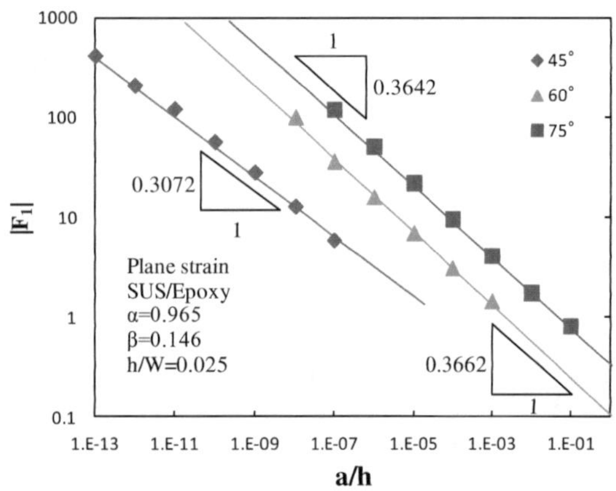

Figure 4. Relationship between F_1 and a/h.

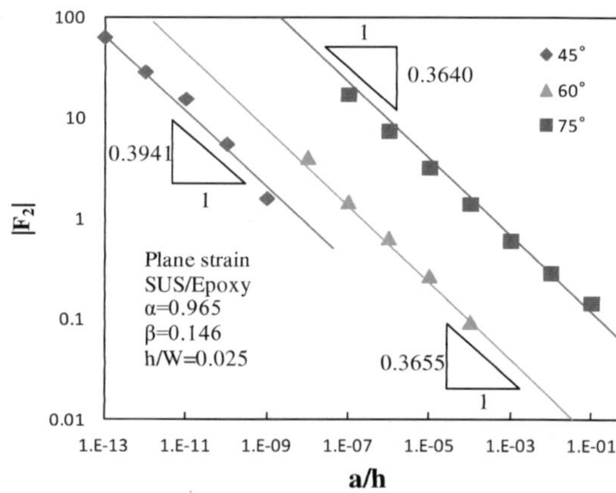

Figure 5. Relationship between F_2 and a/h.

From figures 4 and 5, it is found that F_1 and F_2 show a constant slope on the double logarithmic plot for each scarf angle. When the scarf angle is 75° or 60°, the slope almost coincides with the order of singularity 1-λ in table 3. In the case that the scarf angle is 45°, however, the slops for F_1 and F_2 are different from the value of 1-λ. The reason is that when the scarf angle is 45°, two different orders of singularity are obtained (figure 2), and the singular stress field is not dominated by only one singularity. Therefore, it is found that the dimensionless factors F_1 and F_2 are related to the relative crack length a/h and the singularity index 1-λ and F_1 and F_2 can be defined as a function of the relative crack length a/h as follows [3].

$$F_1 = C_1(h/a)^{1-\lambda}\ , \qquad F_2 = C_2(h/a)^{1-\lambda} \tag{5}$$

In figures 6 and 7, the relations between the coefficients C_1, C_2 obtained from equation (5) and the relative crack length a/h are shown. For θ = 45°, the strong singularity index $1-\lambda_1$ is used for calculation. As shown in figures 6 and 7, it can be seen that the coefficients C_1 and C_2 in equation (5) converge to the constant values when the relative crack length $a/h<10^{-3}$ for θ = 60° and 75°. However, for θ = 45°, we cannot find the converged values of C_1 and C_2 even when $a/h=10^{-12}$.

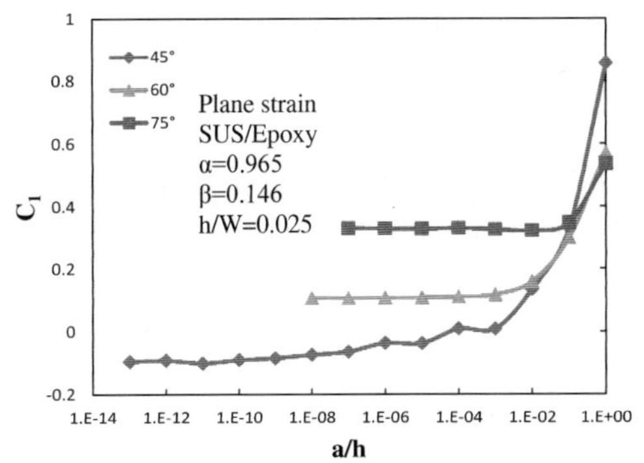

Figure 6. Relationship between C_1 and a/h.

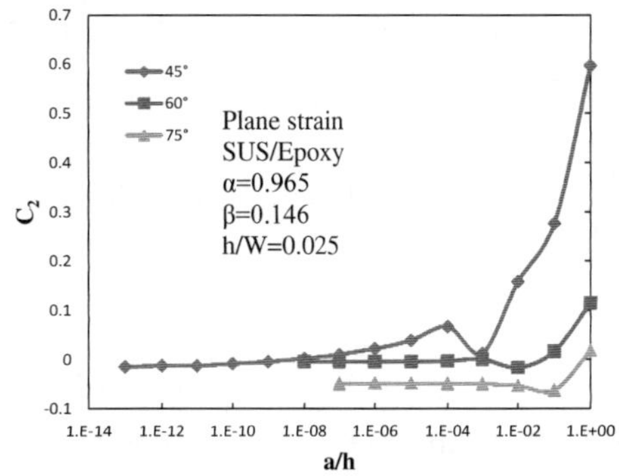

Figure 7. Relationship between C_2 and a/h.

From these results, in the case that the singular stress field near the interface corner is dominated by the single singularity, the values of the coefficients C_1 and C_2 are constant and the dimensionless factors F_1 and F_2 can be represented by equation (5) when $a/h < 10^{-3}$. The coefficients C_1 and C_2 depend only on Dundurs composite parameters.

3.2. Consideration of interface fracture criteria

Afendi *et al* [5] have investigated experimentally the fracture strength by using the scarf joint specimens shown in figure 1. The fracture load of the scarf joint varies by changing the adhesive thickness and the scarf angle. In this study, the stress intensity factors of the small interface crack are calculated by using the fracture loads in order to consider the fracture criteria.

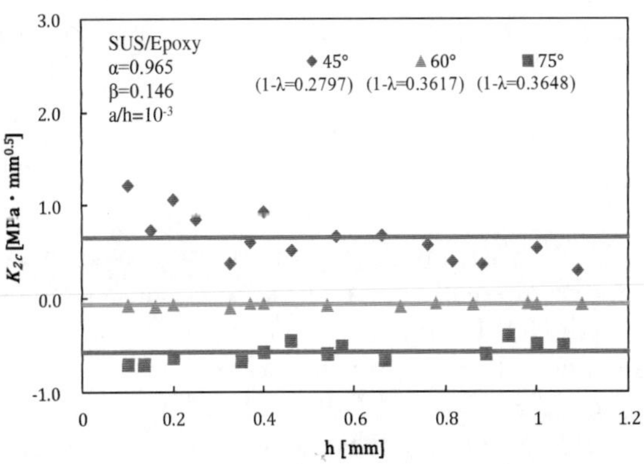

Figure 8. Relationship between K_{1C} and h.

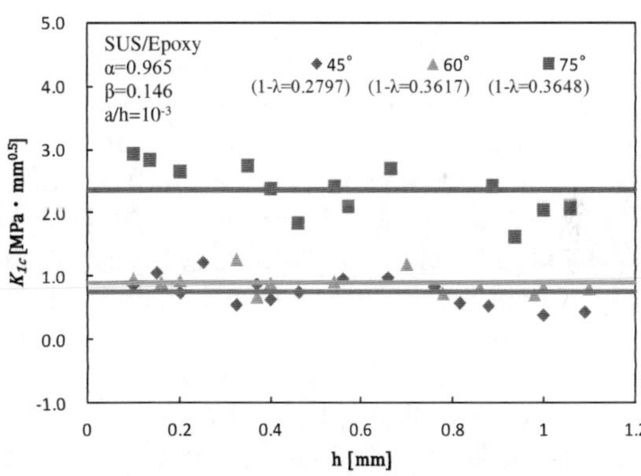

Figure 9. Relationship between K_{2C} and h.

Figures 8 and 9 show the critical stress intensity factors K_{1C} and K_{2C} for each adhesive layer thickness at each scarf angle. From figures 8 and 9, it can be seen that the critical stress intensity factors are almost constant for each scarf angle even when the adhesive layer thickness h changes widely. Here, it is noted that the assumed crack length should be set to less than $a/h=10^{-3}$. The fracture criterion can be represented as K_{1C}=constant by using the small crack model when the combination of adhesive materials and the scarf angle are the same. Consequently, the application of small crack model is useful for evaluating the fracture strength for the scarf joint for any thickness of adhesive layer.

4. Conclusions

In this study, the adhesive joint strength evaluation based on the small interface crack approach was considered. The stress intensity factors of the small interface crack were analyzed by changing the adhesive layer thickness, the scarf angle and the crack length of the scarf joint. The conclusions can be summarized as follows:

- The dimensionless stress intensity factors F_1 and F_2 were calculated by using the crack tip stress method and examined for the relation between the dimensionless factors and the relative crack length a/h. The dimensionless factors can be presented as following expressions when the singular stress filed at the interface corner has a single singularity index 1-λ.

$$K_1 + iK_2 = (F_1 + iF_2)\sigma\sqrt{\pi a}(1 + 2i\varepsilon),$$

$$F_1 = C_1(h/a)^{1-\lambda}, F_2 = C_2(h/a)^{1-\lambda}.$$

- When the relative crack length is sufficiently small, that is $a/h<10^{-3}$, the dimensionless coefficients C_1 and C_2 become constant. The coefficients C_1 and C_2 depend only on the material combination. If the coefficients C_1 and C_2 can be known in advance, the stress intensity factor of the small interface crack can be evaluate easily without numerical analysis.
- The fracture criterion can be represented as K_{1C}=constant by using the small crack model and the experimental fracture loads when the combination of adhesive materials and the scarf angle are the same. By using the K_{1C} criterion, the fracture strength for any thickness of adhesive layer can be evaluated for the scarf joint.

Therefore, it is concluded that the use of the stress intensity factor of the small interface crack is effective for evaluating the bonding strength of the interface.

References

[1] Noda N-A, Miyazaki T, Li R, Uchikoba T, Sano Y and Takase Y 2014 Convenient debonding strength evaluation based on the intensity of singular stress for adhesive joints *J. the Japan Institute of Electronics Packaging* **17** 132-42 (In Japanese)

[2] Noda N-A, Miyazaki T, Li R, Uchikoba T, Sano Y and Takase Y 2015 Debonding strength evaluation in terms of the intensity of singular stress at the interface corner with and without fictitious crack *Int. J. Adhesion and Adhesives* **61** 46-64

[3] Oda K, Takahata Y, Noda N-A and Tsutsumi N 2015 Effect of adhesive thickness on stress intensity factor for edge interface crack in adhesively- bonded joint *Trans. the Japan Society of Mechanical Engineers* **81** (In Japanese)

[4] Oda K, Takahata Y, Noda N-A and Tsutsumi N 2016 Relation between intensity of singular stress at interface edge and stress intensity factor of small edge interface crack in butt joint *Trans. the Japan Society of Mechanical Engineers* **82** (In Japanese)

[5] Afendi M, Abdul M S, Daud R, Abdul A and Teramoto T 2013 Strength and reliability of brittle epoxy adhesively bonded dissimilar joint *Int. J. Adhesion and Adhesives* **45** 21-31

[6] Chen D and Nisitani H 1992 Singular stress field in two bonded wedges *Trans. the Japan Society of Mechanical Engineers* **58** 457-64 (In Japanese)

[7] Oda K, Kamisugi K and Noda N-A 2009 Analysis of stress intensity factor for interface cracks based on proportional method *Trans. of the Japan Society of Mechanical Engineers* **75** 476-82 (In Japanese)

[8] Oda K, Lan X, Noda N-A and Michinaka K 2012 Effect of arbitrary bi-material combination and bending loading conditions on stress intensity factors of an edge interface crack *Int. J. Structural Integrity* **3** 457-75

IOP Conf. Series: Materials Science and Engineering **372** (2018) 012016 doi:10.1088/1757-899X/372/1/012016

Experimental study on the effect of shape of bolt and nut on fatigue strength for bolted joint

T Matsunari[1], K Oda[2,5], N Tsutsumi[2], T Yakushiji[3], N A Noda[4] and Y Sano[4]

[1]Graduate School of Engineering, Oita University, 700 Dannoharu, Oita 870-1192, Japan
[2]Division of Mechanical Engineering, Oita University, 700 Dannoharu, Oita 870-1192, Japan
[3]National Institute of Technology, Oita College, 1666 Maki, Oita 870-0152, Japan
[4]Mechanical Engineering Department, Kyushu Institute of Technology, 1-1 Sensui-cho Tobata-ku, Kitakyushu-shi, 804-8550, Japan

E-mail: oda-kazuhiro@oita-u.ac.jp

Abstract. In this study, the effect of curvature radius of the thread bottom and the pitch difference between of M16 bolt and nut on fatigue strength for bolted joint is considered experimentally. The M16 bolt-nut specimens having the two kinds of thread bottom radii and the pitch differences are prepared. The S-N curves for bolted specimens with different thread shapes are obtained by the stress-controlled fatigue test (stress ratio R>0). The experimental results are compared and discussed in terms of stress analysis. The finite element method is used to make a simulation of the fatigue experiment and the mean stress and stress amplitude at each thread bottom of bolt are analysed. It is found that the initiation and propagation of crack are changed by introducing the pitch difference of α=15 μm, from the crack observation in cross section of the bolt specimens after the experiment. Furthermore, the fatigue life can be extended by increasing curvature radius of thread bottom and introducing the pitch difference.

1. Introduction

The bolted connection is one of the most important mechanical components and is used frequently. For instance, since about 1,000 bolts and nuts joints are used in a single car, high fatigue strength at low cost is required. Damage of the bolt-nut joint is due to high stress concentration at the bolt thread bottoms and its strength is reduced. To ensure the safety and reliability of the bolted joint, researches on anti-loosening performance have been actively conducted [1-3]. However, there are not many studies on fatigue strength improvement of bolted connections. It is not easy to improve the fatigue strength of standard shape of bolt-nut joint because high stress concentration always occurs at the bottom of bolt thread.

Focusing on the stress relaxation at the thread bottom due to the shape change of the bolt [4-7] and the material difference of the fatigue strength [8], several studies have been conducted to improve the fatigue strength of the bolted joint. The effect of the pitch difference between the nut and the bolt on the stress concentration at the screw thread of the bolt has been studied previously and it is reported that the fatigue strength can be improved by changing the contact state between the bolt and the nut [9-14]. In particular, Noda *et al* [9-13] analysed the effects of anti-loosening and stress reduction for the bolt-nut joint with small pitch difference, and showed that the fatigue life of the bolted joint can be improved by using the pitch difference of α=15 μm. However, the studies on the conventional pitch difference were limited to the improvement of fatigue life for the standard shape bolt thread.

In this study, a more detailed fatigue test is systematically carried out for a new bolt specimen with small pitch difference α and large thread bottom radius ρ. Here, the standard bolt-nut joint has α=0 and $\rho=\rho_0$. Then, from the S-N curves obtained for the bolted specimens, the improved fatigue lives are discussed. To clarify the effects of the thread bottom radius and the pitch difference, stress amplitude and average stress at each bolt thread are analysed by the finite element method (FEM). By comparing the experimental results with the results computed using FEM, the mechanism of improvement in fatigue life of bolted specimen is examined.

2. Fatigue experiment to investigate the fatigue life

2.1. Specimens and experimental conditions

In this study, the Japanese Industrial Standard (JIS) M16 bolt-nut joints of strength grade 8.8 are used. The bolt is made from chromium-molybdenum steel SCM435, and the nut is the quenched and tempered medium carbon steel S45C, and their properties are indicated in table 1. Figure 1 shows schematic diagram of bolted joint. Numbers -3, -2 … 7, 8 in figure 1 correspond to the bottom of each thread. The fatigue experiment device assembly drawing is illustrated in figure 2. The Servo Fatigue Testing Machine (392 kN) is used at a cycle frequency of 5 or 10 Hz in this test. The bolt specimen is subjected to a mean tensile force of 30 kN. Since the area A_R of the bolt cross section is 141 mm^2, the stress-controlled fatigue experiment is conducted under the corresponding mean tensile stress σ_m=213 MPa. The stress amplitude in the fatigue test is set in the range of 50 MPa to 160 MPa. The S-N curves are obtained with five stress amplitudes and the fatigue limit of 2×10^6 cycles.

Table 1. Mechanical properties of bolt and nut specimen.

	Young's modulus (GPa)	Poison's ratio	Yield strength (MPa)	Tensile strength (MPa)
SCM435 (Bolt)	206	0.3	800	1200
S45C (Nut)	206	0.3	530	980

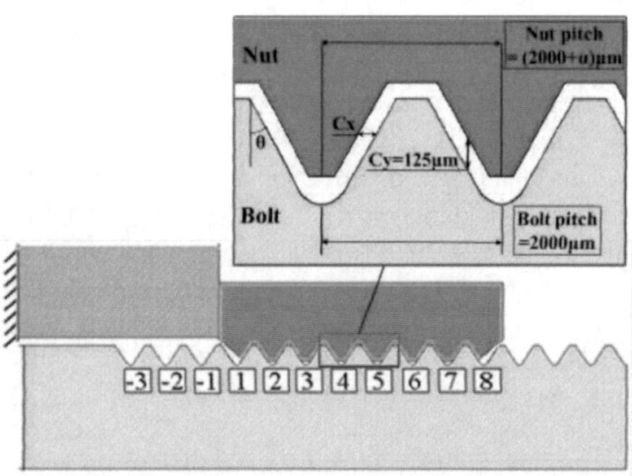

Figure 1. Schematic view of bolted joint.

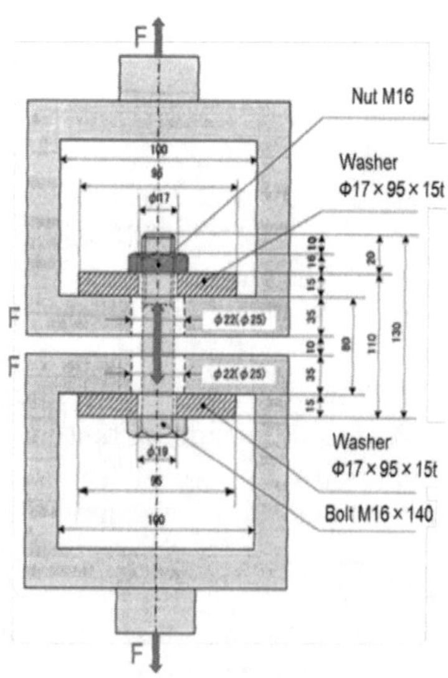

Figure 2. Schematic illustration of fatigue test.

2.2. *Effect of pitch difference of the standard bolt-nut joint*

In the previous study, Noda et al. have reported that the effect of the pitch difference on the fatigue strength in bolted joint [9-13]. The standard M16 bolts and nuts usually have the same pitch dimension of 2000 μm. For different pitch joint, the pitch dimension of the nut is set to be equal or slightly larger than the bolt pitch, that is, 2000 μm+α. Three types of bolted joint specimens of α=0, 15 and 33 μm are investigated. The bolt-nut clearance is assumed to be the standard dimension of 125 μm.

The S-N curves for bolted joints having the pitch difference α are shown in figure 3. It is found that the fatigue lives of bolt-nut joints with the three levels of pitch difference are different clearly. When the stress amplitude is above 80 MPa, the fatigue lives of the bolted joints for α=15 μm and 33 μm are about 1.4 times and 1.2 times larger than that of the standard bolted joint for α=0. However, near the fatigue limit at $N=2\times10^6$ cycles, the fatigue lives of three different pitch specimens are not very different, and the fatigue limit for three levels of pitch difference remains the same value of 60 MPa.

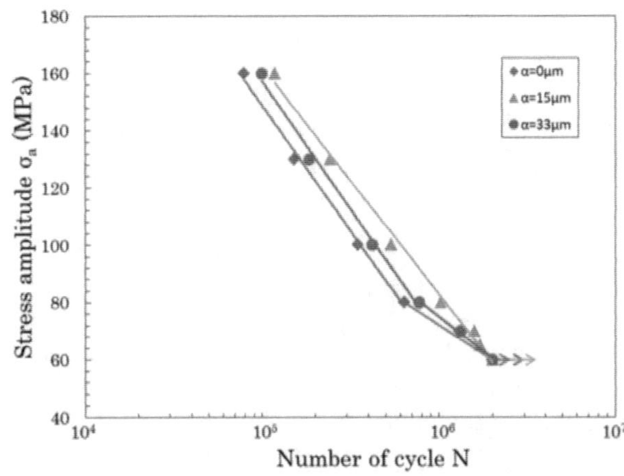

Figure 3. S-N curves for bolt-nut joints having the pitch difference α=0, 15, 33 μm when $\rho=\rho_0$.

2.3. *New bolt-nut joint with larger thread bottom radius and slight pitch difference*

Next, we focus on the effect of the radius of the thread bottom on the fatigue strength of bolted joint. Newly shaped bolt specimens having the thread bottom radius $2\rho_0$ are prepared. Figure 4 shows the shapes and dimensions of the standard and the new bolt-nut joints when the pitch difference is α=0. The stress concentration factors K_t for cylindrical bar having a 60° V-shaped circumferential notch under tension are shown in figure 4. The stress concentration factor decreases considerably from K_t = 4.53 ($\rho=\rho_0$) to K_t = 2.90 ($\rho=2\rho_0$) by doubling the thread bottom radius.

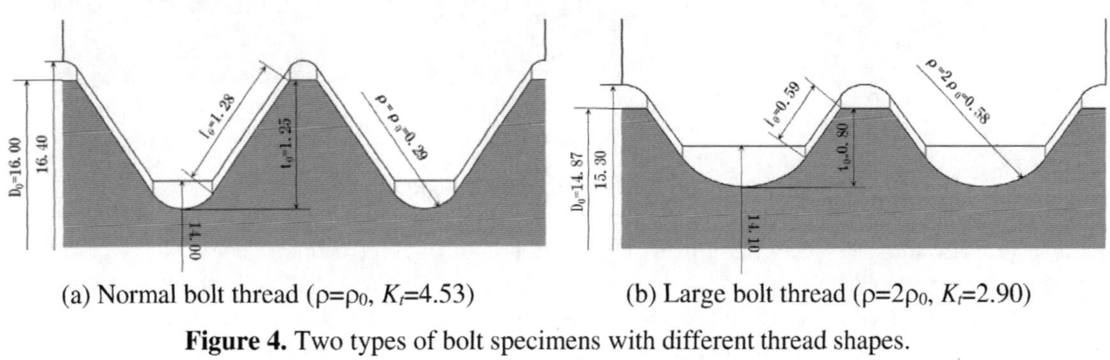

(a) Normal bolt thread ($\rho=\rho_0$, K_t=4.53) (b) Large bolt thread ($\rho=2\rho_0$, K_t=2.90)

Figure 4. Two types of bolt specimens with different thread shapes.

Figure 5 indicates the S-N curves for the two types of bolt specimens, $\rho=\rho_0$ and $2\rho_0$, when the pitch difference is α=0. From figure 5, it is found that the fatigue life for the newly shaped bolt ($\rho=2\rho_0$) is improved to more than 2 times that for the standard bolt specimen ($\rho=\rho_0$). Furthermore, the fatigue limit is about 30% larger than that of the normal bolt-nut joint.

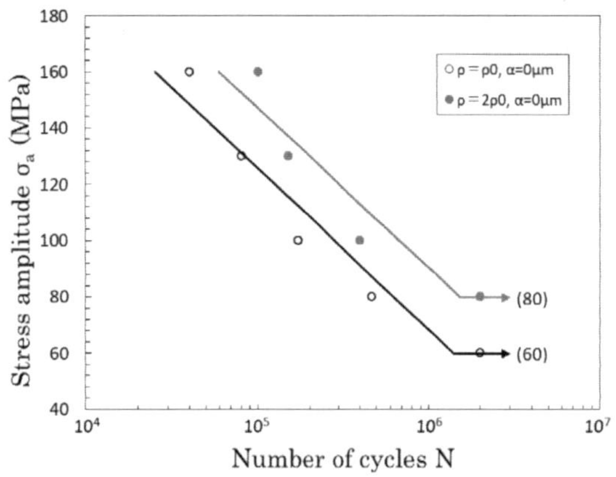

Figure 5. S-N curves for bolted joints with different thread bottom radius $\rho=\rho_0$ and $2\rho_0$ when α=0.

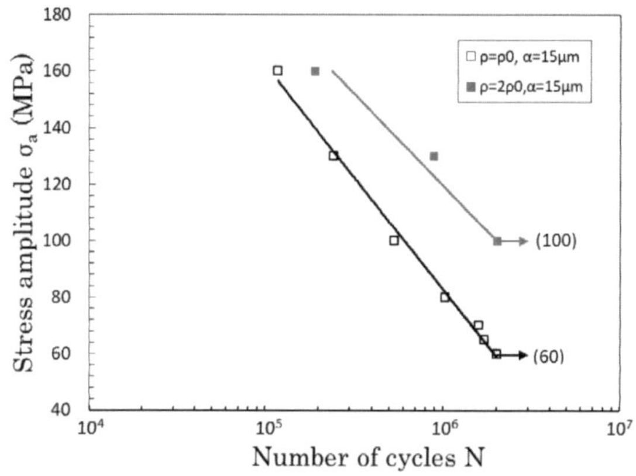

Figure 6. S-N curves for bolted joints with different thread bottom radius $\rho=\rho_0$ and $2\rho_0$ when α=15 μm.

Figure 6 shows the S-N curves for the two types of bolt-nut joints when the pitch difference α=15 μm. In the case that the bolted joints have the pitch difference of α=15 μm, the fatigue life of the new bolted joint with $\rho=2\rho_0$ is over 2 times longer than that of the normal shaped bolt $\rho=\rho_0$. By doubling the thread bottom radius and applying the pitch difference of α=15 μm, the fatigue limit for the bolt-nut joint is improved to 100 MPa.

2.4. Crack observation to investigate the fatigue life improvement mechanism

Figure 7 shows the example of the fractured specimens with the different thread shape and pitch difference when σ_a=100 MPa. In figure 7, N_f means the number of cycles until the specimen breaks.

(a) $\rho=\rho_0$, α=0 μm (N_f=171,790)

(b) $\rho=2\rho_0$, α=0 μm (N_f=404,500)

(c) $\rho=\rho_0$, α=15 μm (N_f=536,690)

Figure 7. Fracture surface of the broken specimens (σ_a=100 MPa).

Figure 8 also indicates the longitudinal cross section photo and the crack trajectories observed from the surface of fractured bolt specimens. For α=0, cracks occur at No. 1 and No. 2 threads as shown in figures 8(a) and 8(b). For α=15 μm, cracks occur between No. 2 and No. 6 threads as illustrated in figure 8(c). It can be seen that for the bolted joints with normal pitch α=0, the crack initiates at No. 1 thread or No. 2 thread causing final fracture. On the other hand, for the specimen of α=15 μm, the cracks initiate at No. 5 thread or No. 6 thread, extending toward No. 1 thread and finally fracture occurs nearby No. 1 thread. From the S-N curves and the crack observation in figures 6 and 8, it is concluded that the fatigue life of the bolt-nut connection can be improved by introducing the suitable pitch difference and thread bottom radius because the initiation and the propagation of cracks may be changed.

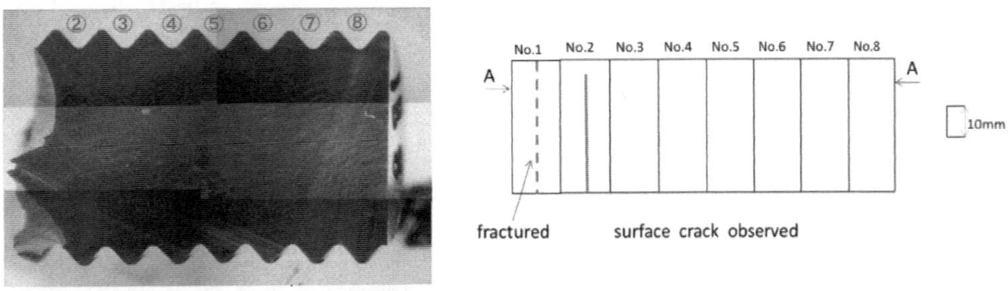

(a) $\rho=\rho_0$, $\alpha=0$ μm

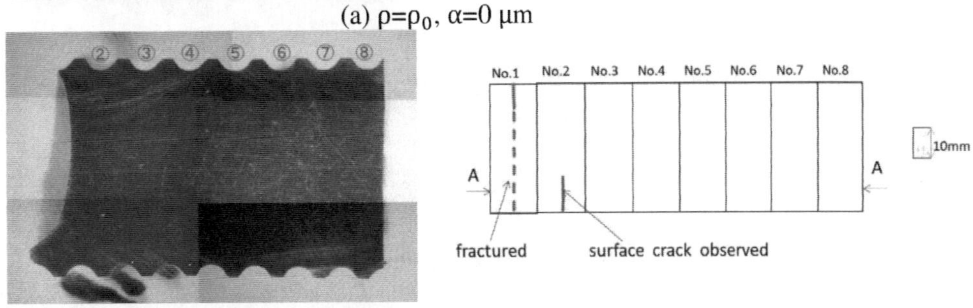

(b) $\rho=2\rho_0$, $\alpha=0$ μm

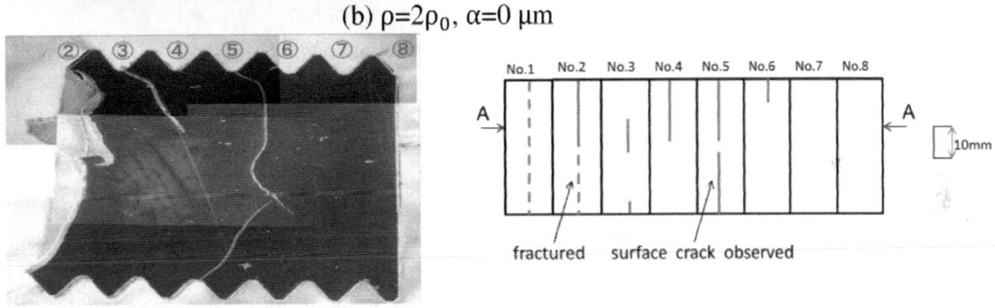

(c) $\rho=\rho_0$, $\alpha=15$ μm

Figure 8. Crack configuration observed from the surface of the fractured specimens (σ_a=100 MPa).

3. Stress state at bolt thread bottom

3.1. Analysis model and conditions

The stress state at each bottom of the bolt thread is investigated by the finite element analysis. The axisymmetric FEM model of bolt-nut joint is illustrated in figure 9. The versatile FEM code MSC.Marc/Mentat 2012 is used for analysis. A four-node axisymmetric solid element is adopted and the minimum element size is 0.015 mm × 0.01 mm near the thread bottom of the bolt. The FEM models have different pitches and thread bottom radii, i.e. α=0, 15 μm and $\rho=\rho_0$, $2\rho_0$ according to the experimental arrangements of the bolt-nut specimens. The material properties of bolt and nut listed in table 1 are applied in the calculation. An elastic-plastic analyses for FEM models are performed under the same loading conditions as experiment. The clamped plate is fixed horizontally and the axial force $F=30\pm14.1$ kN is applied to the bolt head. The force amplitude F_a=14.1 kN corresponds to the nominal stress amplitude σ_a=100 MPa in the minimum section of the bolt.

IOP Conf. Series: Materials Science and Engineering **372** (2018) 012016 doi:10.1088/1757-899X/372/1/012016

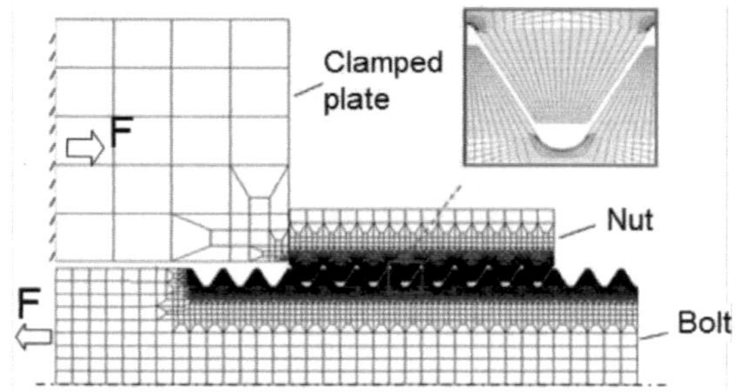

Figure 9. Axisymmetric FEM model.

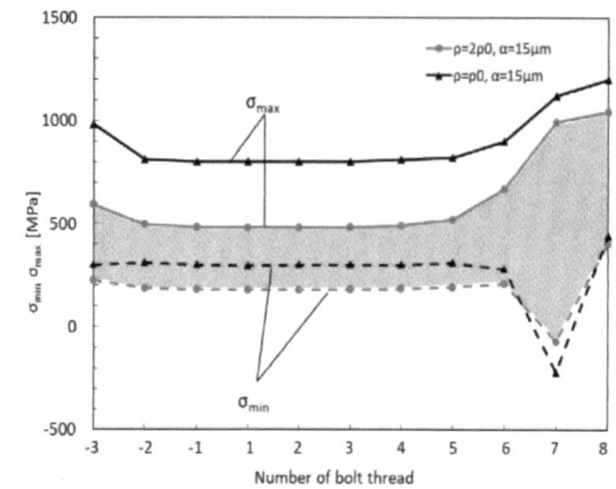

Figure 10. σ_{max} and σ_{min} at each bolt thread.

3.2. Maximum stress amplitude at bottom of bolt thread

Figure 10 shows the maximum and minimum stresses at each thread bottom of the bolt when α=15 μm. In figure 10, σ_{max} is the maximum stress at each thread bottom under the force F=30+14.1 kN and σ_{min} is the minimum stress at each thread bottom under the force F=30-14.1 kN. The black line shows the values for standard bolt ($\rho = \rho_0$) and the red line is the values for newly shaped bolt ($\rho = 2\rho_0$). The mean stress σ_m and the stress amplitude σ_a are obtained by

$$\sigma_m = \frac{\sigma_{max}+\sigma_{min}}{2},\ \sigma_a = \frac{\sigma_{max}-\sigma_{min}}{2} \quad (1)$$

The previous study [13] has reported that the highest stress amplitude appears at the bottom of No. 1 thread when α=0 μm. On the other hand, when the pitch difference of α=15 μm is introduced, the stress amplitude decreases at the bottom of No. 1 and the highest stress amplitude occurs at No. 7 thread. This is roughly in agreement with the crack observation in the fatigue experiment shown in figure 8. As indicated by the red range in figure 10, the stress amplitude σ_a and the mean stress σ_m at the bottom of all threads are reduced by doubling the thread bottom radius of the bolt.

From the figures 8 and 10, by introducing the suitable pitch difference, the load distribution of each thread of bolt is changed, and the fatigue life extends due to the change of crack initiation and propagation. Furthermore, by increasing the thread bottom radius of bolt, the fatigue life and the fatigue limit are greatly improved because the stress amplitude and the mean stress at each bolt thread decrease.

4. Conclusions

In this study, the effect of the pitch difference and large radius of bolt thread bottom on fatigue strength of bolted joint was examined using experimental technique. The fatigue test was conducted by using the bolt-nut specimens having the different pitches and thread bottom radii to investigate the fatigue life and the fatigue limit. The conclusions can be summarized as follows:

- The fatigue life was improved significantly by increasing the radius of bolt thread and introducing the slight pitch difference. As compared with the standard bolt-nut joint, the fatigue limit of newly shaped bolt joint was also improved to about 1.5 times.
- From the crack observation, it was found that the crack propagation path changes by introducing pitch difference. In the case of the pitch difference α=15 μm, cracking started from No. 5 or No. 6 thread and the final fracture occurred at No. 1 or No. 2 because the load distribution of each thread of bolt was changed.
- The FEM analysis showed that the stress amplitude and the mean stress at each bolt thread decrease by increasing the thread bottom radius of bolt when the pitch difference is α=15 μm. Therefore, the present method is effective for improving the fatigue strength of the bolt-nut

joint.

References

[1] Hard Lock Kogyo K K 2002 Hard Lock Nut Japanese Patent 2002-195236 (in Japanese)

[2] Noda N A, Noma S, Saito K, Nagawa M, Yumoto A and Xiao Y 2006 Stress reduction effect and anti-loosening performance of outer cap nut by finite element method *Trans. the Japan Society of Mechanical Engineers* **72** 1318-25 (in Japanese)

[3] Xiao Y, Kuhara M, Noda N A, Saito K, Nagawa M and Yumoto A 2008 Optimum dimensions of thin walled tube on the mechanical performance of super stud bolt *Trans. the Japan Society of Mechanical Engineers* **74** 954-60 (in Japanese)

[4] Nishida S I 1997 A new method for fatigue life improvement of screws *European Structural Integrity Society* **22** 215-25

[5] Hirai K and Uno N 2005 Fatigue strength of super high strength bolt *J. Structural Engineering* **595** 117-22

[6] Noda N A, Xiao Y and Kuhara M 2011 The reduction of stress concentration by tapering threads *J. Solid Mechanics and Materials Engineering* **5** 397-408

[7] Xiao Y, Wan Q, Noda N A, Akaishi Y, Takase Y and Nishida S 2011 Stress reduction effect of tapering thread bolts and nuts which have slightly different pitches *Paper Collection of Automotive Technology Society* **42** 927-33 (in Japanese)

[8] Maruyama K 1976 Stress analysis of a bolt-nut joint by the finite element method and the copper-electroplating method *Trans. the Japan Society of Mechanical Engineers* **19** 360-8 (in Japanese)

[9] Nippon Steel Corporation 2005 Steel Bolt Set Japanese Patent 2005-265150 (in Japanese)

[10] Chen X, Noda N A, Akaishi Y, Sano Y and Takase Y 2013 Effect of pitch difference on anti-loosing performance and fatigue strength for high strength bolts and nuts *13th Int. Conf. on Fracture* 10988-43072

[11] Akanishi Y, Xin C, Yang Y, Tamasaki H, Noda N A, Sano Y and Takase Y 2013 Fatigue strength analysis for bolts and nuts which have slightly different pitches considering clearance *Paper Collection of Automotive Technology Society* **4** 1111-7 (in Japanese)

[12] Noda N A, Sano Y, Takase Y, Xin C, Maruyama H, Huan W and Fujisawa R 2015 Anti-loosing performance of special bolts and nuts having enhanced fatigue life by introducing pitch difference *Paper Collection of Automotive Technology Society* **46** 121-6 (in Japanese)

[13] Chen X, Noda N A, Magd A W, Akaishi Y, Sano Y, Takase Y and Gusztav F 2015 Fatigue failure analysis for bolt-nut connections having slight pitch differences using experimental and finite element methods *Acta Polytechnica Hungarica* **12** 61-79

[14] Hamanaka Nut Corporation 2006 Fastening Part and Fastening Method Japanese Patent 2006-118691

IOP Conf. Series: Materials Science and Engineering **372** (2018) 012017 doi:10.1088/1757-899X/372/1/012017

Scaling analysis and SE simulation of the tilted cylinder-interface capillary interaction

S Q Gao[1,2,3], X Y Zhang[2] and Y H Zhou[2]
[1]National Key Laboratory of Science and Technology on Reactor System Design Technology, Nuclear Power Institute of China, Chengdu 610041, China
[2]Department of Mechanics and Engineering Sciences, College of Civil Engineering and Mechanics, Lanzhou University, Lanzhou 730000, China

E-mail: gaoshq07@163.com

Abstract. The capillary interaction induced by a tilted cylinder and interface is the basic configuration of many complex systems, such as micro-pillar arrays clustering, super-hydrophobicity of hairy surface, water-walking insects, and fiber aggregation. We systematically analyzed the scaling laws of tilt angle, contact angle, and cylinder radius on the contact line shape by SE simulation and experiment. The following in-depth analysis of the characteristic parameters (shift, stretch and distortion) of the deformed contact lines reveals the self-similar shape of contact line. Then a general capillary force scaling law is proposed to incredibly grasp all the simulated and experimental data by a quite straightforward ellipse approximation approach.

1. Introduction

Capillary interaction arises whenever an initial flat interface between two immiscible fluids (or fluid-air) is perturbed by other object (e.g., sphere, vertical cylinder, or other floating anisotropic particles). Deformation of the interface is the result of gravity (including buoyancy force), asymmetric hydrostatic pressure, and wettability around the object surface. Overlap of the menisci surrounding these objects leads to a lateral capillary force between them, which can be either attractive for similar particles or repulsive for dissimilar ones [1,2]. This capillary interaction has been intensely studied theoretically and experimentally (e.g. [3-12], for more details see reviews [1,2,13]), for the insight it provides into the fundamental of statistical physics with tunable many-body system [14] and for its potential applications in areas such as colloidal self-assembly [15,16], stabilization of foams and emulsions [17], and pollination, etc (see [18] for details).

Fully analysis of this inter-particle interaction involves solving the nonlinear Laplace-Yong equation, which is difficult to solve analytically, especially for the complex boundary conditions (e.g., anisotropic floating particles). Hence, plenty of the aforementioned earlier works are conducted mostly based on the linearized Laplace equation and the Nicolson's super-position approximation [19] for some limited symmetry cases. Instead of using the superposition approximation, Kralchewsky *et al* [2,3], as a new moment, directly solved the linearized Laplace equation in bipolar coordinates using the matched asymptotic expansions method. Validity of the aforementioned superposition approximation was analytically proven, since it is the asymptotic case of the general expression. However, two particularly challenges for these approaches are highlighted due to the use of linearized Laplace equation [20]: (i) the great discrepancy between model and experiment results at shorter inter-particle separation or with large meniscus slope; (ii) the disability to reproduce the experimental fact that the interaction between two dissimilar cylinders becomes short-range attraction [4]. Recently, some numerical solutions are devoted to break down these restrictions and to extend to more complex boundary conditions. Some typical works involve using the conformal mapping [21,22], mesh-free finite difference method [20] and the FLEXPDE program [23] to solve the nonlinear Laplace-Yong equation, or using Surface Evolver (SE) to directly minimize the total energy for given configuration [11,12,24-26]. In order not to sink into excess numerical simulation for every configuration, here we present a general capillary force model with clear physical meaning, as an alternative, based on the essential contact line characters revealed by the in-depth scaling analysis of simulated data and experiment. In this paper, we focus on the capillary interaction induced by a single floating tilted cylinder [24,25], varying contact angle θ, tilt angle ϕ and cylinder radius R. Analyzing this single tilted cylinder-interface interaction is not only a first step towards understanding the more complex systems (e.g., micro-pillar arrays clustering [27,28] and fiber aggregation [29]), but also of interest for its own right due to its close connection to water-walking insects (e.g., water strider legs [30,31]), super-hydrophobicity of hairy surface [32,33] and design of water strider robot [34], etc.

This study aims to provide insight into the characters of the contact line shape and the corresponding cylinder-interface interaction induced by the floating tilted cylinder for various contact angle θ, tilt angle ϕ and cylinder radius R. We use Surface Evolver (SE) [35] to perform plenty of simulations to seek for the equilibrium interface shape and position of the contact line, for various cylinder-interface configuration. The excellent accordance of the meniscus profile obtained by our experiment and simulation confirms the effectiveness of the simulation. The following in-depth analysis of the characteristic parameters (shift, stretch and distortion) of the deformed contact lines reveals the self-similar shape of contact line. Both experiment and SE simulation indicate an invariant structure of the contact line (independent of contact angle and radius). Based on these observations, we then propose a general capillary force scaling law to incredibly grasp all the simulated results by the ellipse approximation.

2. Materials and methods

Our system, depicted in figure 1, consists of a long cylinder of radius R piercing an otherwise infinite planar gas-liquid interface (dashed line; figure 1) with tilt angle ϕ. The previous planar interface will deform to fit for the constant-contact-angle condition along the three-phase contact line [2,20]. Shown in figures 2(a) and 2(b) are morphology and isoheight map of one representative simulated interface for specific parameters. The deformation around the tilted cylinder is asymmetrical near the roughly tilted ellipse contact line, but symmetrical at a distance, as a comparison to the quadrupolar structure induced by an anisotropic particle (lying cylinder or ellipsoid) floating at planar interface [13,26], or by a spherical particle floating at curved interface [36].

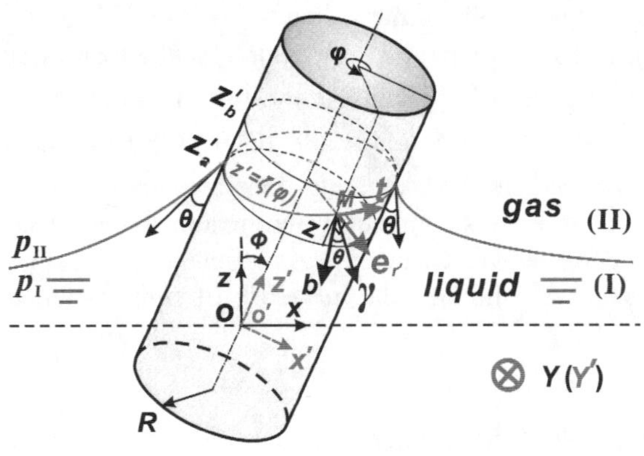

Figure 1. Schematic sketch of a tilted cylinder piercing the gas-liquid interface.

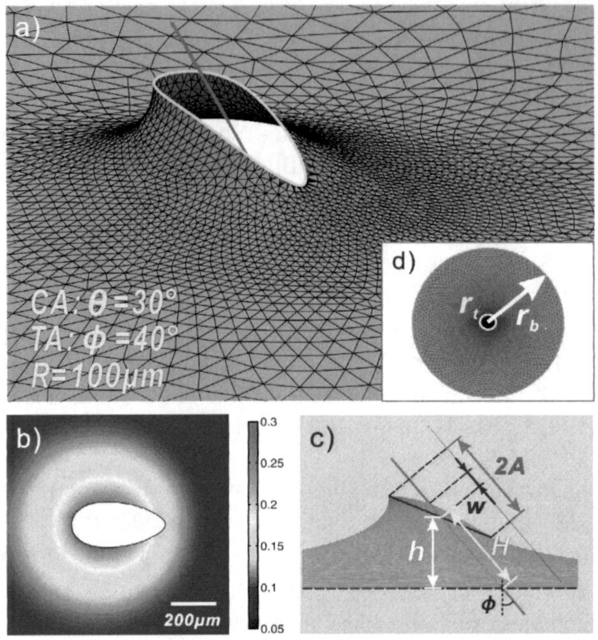

Figure 2. Typical morphology (a) and isoheight map (b) of the simulated interface perturbed by a tilted cylinder of radius $R = 100\ \mu m$, contact angle θ=30°, tilt angle ϕ=40°, for $\gamma = 36.8$ mN/m, $\Delta\rho = 1000$ kg/m^3. (c) Definition of the three parameters to characterize the contact line shape with $\zeta(\varphi) = H + A\cos(\varphi + n\sin\varphi)$: H (shift), A (stretch) and w (distortion). (d) Sketch of the whole calculated and center refined areas for the simulation, with radius $r_b = 250R$ and $r_t = 10R$. Note that (a) is close-up of the center refined area.

2.1. SE simulation

We performed plenty of simulations to characterize the shape of contact line as a function of contact angle θ, tilt angle ϕ, and cylinder radius R. For every given cylinder-interface configuration (θ, ϕ, R), we use Surface Evolver [35] to minimize the gravitational and interfacial energy, seeking for the equilibrium interface shape that satisfies the constant-contact-angle boundary condition, as described previously [24,25]. Note that following the force approach [2,3], only the location of the points along the contact line needed to be recorded, which is sufficient to reveal the scaling laws of the contact line and the corresponding capillary interaction.

We first created a large fixed circular frame of radius r_b and a small transition region (with mesh refinement) of radius r_t (see figure 2(d)). To ensure the robustness of the simulation, we have tested three groups of parameter pairs (r_b/R, r_t/R): a) (250, 10), b) (250, 20) and c) (500, 20), all with two levels of refinement: (i) 12288 triangles and 96 points along the contact line and (ii) 49152 triangles and 192 points. It is found that these 6 cases yield equivalent results about the contact line shape (data not shown). Considering the tradeoff between accuracy and efficiency, we employ the parameter pair a) in our simulation below. Hence, we minimize the influence of finite boundary without adding excess computation.

2.2. Experiment

Four kinds of cylinder fibers with different materials and radius are performed in our experiment, as summarized with table 1. To remove contaminants that may influence the wettability, the fiber surfaces were first cleaned by alcohol and then washed by distilled water. Next the cylinder pierced the otherwise planar distilled water interface with tilt angle ϕ. The image acquisition system, similar as the contact angle measurement set-up, is consists of a high-speed-camera (Memrecam HX-3, NAC. Inc), a zoom lens (AF-S VR Micro-Nikkor 105 mm f/2.8 G, Nikon. Inc), a LED lamp and image analyzer (computer). The LED lamp is carefully adjusted to satisfy the rigorous lighting condition, which is crucial for capturing high-quality images. The observation was carried out by means of the horizontal camera equipped with micro-lens.

Table 1. Radius and corresponding Bond number for the cylinder adopted in experiment and SE simulation. The cylinder radius is measured by the measuring microscope.

Parameters	Material	Radius (μm)	Bond Number
Experiment	Plastic fiber (thick)	464	0.0293
	Soldering tin	503	0.0344
	Copper wire	272	0.0101
	Plastic fiber (thin)	340	0.0157
SE Simulation	N/A	50	0.0007
	N/A	100	0.0027
	N/A	200	0.0107

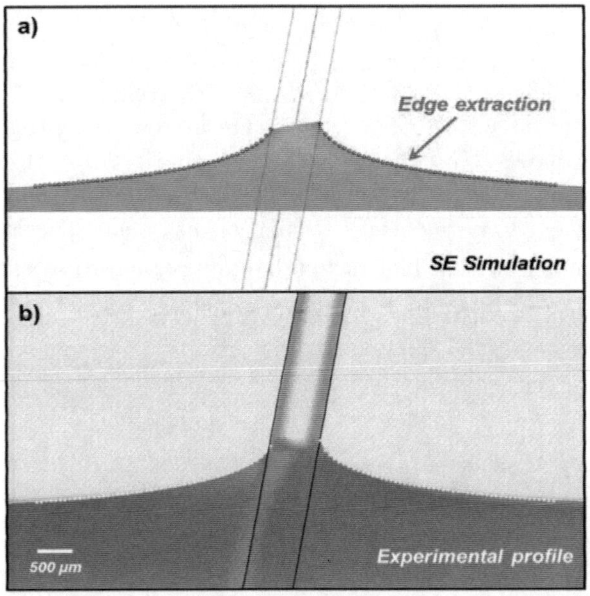

Figure 3. (a) SE simulated and (b) experimental side-view profiles of the meniscus induced by the tilted floating cylinder. The coordinates of selected points on the left- and right-side edges of the SE simulated profile (a) have been digitized and compared with the experimental one in (b). The single-pixel white solid line in (b) corresponds to the experimental profile extracted by the Canny edge operator. Scale bar, 500 μm.

We captured the side-view profile of meniscus at several tilt angles ($\phi \approx 0$ □ $30°$) for each kind of cylinder fibers. Every image is tailored into a standard size of 1201×651 pixels with 256 gray levels for each pixel, where 0 represents black and 255 represents white. Shown in figure 3(b) is a typical photograph of deformation of meniscus. Due to influence of the fiber, we observe the former flat interface deforms obviously. Using a home-developed contact angle measurement program based on automatic ellipse-fitting, the original image is then processed with edge detection, coordinates digitization, and automatic ellipse-arc fitting sequentially. The tilt angle and contact angle at both sides are automatically measured, with tilt angle $\phi = 10.1°$ and contact angle θ_L=20.1°, θ_R=14.6° . The difference of contact angle at the two sides can be attributed to the contact angle hysteresis.

The cylinder radius is measured by the measuring microscope.

To compare the profile obtained by simulation and experiment (using the thin Plastic fiber), we used SE to simulate the equilibrium profile (figure 3(a)) with the former measured parameters f = 10.1鞍q_{Ave}=17.35 . We then digitalized the coordinates of a series of points located on the left- and right-hand-side meniscus profiles and replotted them in figure 3(b) with yellow dots. It shows excellent agreement between those two independent approaches. Hence, we will use them to systemically analyze the characters of the contact line shape.

3. Results and discussion

3.1. Self-similar shape of the contact line

To quantify the contact line, shown in figure 1, we denote the reference frame of the initial flat interface and the cylinder by $\Re(Oxyz)$ and Âï(*Ox y*ïȩ) , with Oz and Oz' along vertical and center line of the cylinder, respectively. Hence, the contact line is clearly given by $z' = \zeta(\varphi)$ in the frame $\Re'$, with the azimuthal angle $\varphi \in [-\pi, \pi]$. If the contact line shape is strictly ellipse, then $\zeta(\varphi) = h_0 + a_0 \cos(\varphi)$, where $h_0 = (z_b' + z_a')/2$, $a_0 = (z_b' - z_a')/2$ and z_b', z_a' correspond to the highest and lowest point of the contact line at frame $\Re'$ (figure 1). However the contact line, in reality, is shifted, stretched and distorted to satisfy the constant-contact-angle condition (figure 2(a)), as also observed in [24].

To characterize deformation, we define three characteristic parameters (figure 2(c)) for each of the deformation modes respectively: H (*Shift*) is the displacement from the virtual ellipse center to the undisturbed flat interface; $2A$ (*stretch*) the projection of the distance between z_b' and z_a' to the z'-axis in frame $\Re'(Ox'y'z')$; and w (*distortion*) the deflection of the contact line at $\varphi = \pi/2$ or $3\pi/2$. In analogy with the aforementioned cosine function for the ellipse case, we use a modified cosine function:

$$\zeta(\varphi) = H + A\cos(\varphi + n\sin\varphi), \qquad (1)$$

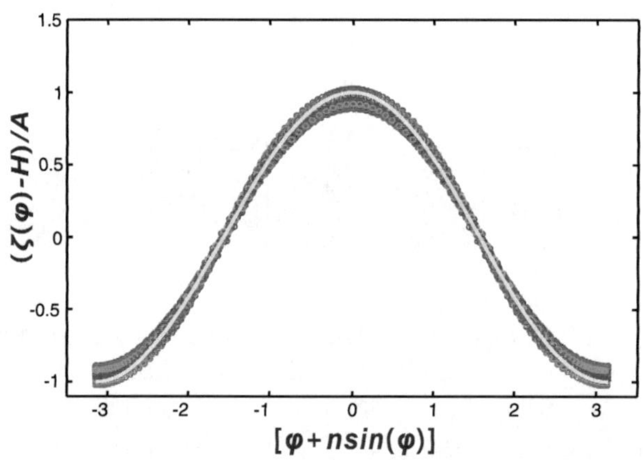

Figure 4. Self-similar shape of the simulated contact lines represented in the normalized coordinates $(\zeta(\varphi) - H)/A$ vs. $\varphi + n\sin\varphi$, for contact angle $\theta = 30 \sim 150°$ (step: 20°), tilt angle $\phi = 0$, 20, 40°, and cylinder radiuses $R = 50$, 100, 200 μm . The data all collapse onto the standard cosine function (solid line), with the normalization using equations (3) and (4).

to fit all the data of the contact lines. Here n controls the shape of curve $\zeta(\varphi)$ (fatter or thinner). Considering the geometry relation, we have $w=-A\sin n$. Plotting profiles of these contact lines at various configuration (θ, ϕ, R) in stretched coordinates, with j and $\zeta(\varphi)$ respectively normalized by φ+nsinφ and $(\zeta(\varphi)-H)/A$, reveals a self-similar shape of the contact line (figure 4). Here, parameters H, A, n are functions of the variables θ, ϕ, R.

It should be noted that we have adopted these relations in the above normalization. All the data collapses onto the standard cosine function (solid line; figure 4) indicates that equation (1) and its accessories can well describe the contact line shape for every configuration. Next, the dependence of the three characteristic parameters on the variables θ, ϕ, R will be presented systematically.

3.2. Detailed characters of the contact line shape

For the Shift mode, the normalized shift displacement H/R is plotted in figure 5(a) (top inset) as a function of the contact angle θ with $R = 200$ μm for various tilt angles $\phi = 0 \sim 50°$. It is no surprise that meniscus around the cylinder will upraised for $\theta < 90°$ and vice versa due to wettability. The Shift displacement H increases with tilt angle ϕ, as the arrow indicates (top inset; figure 5). The solution of the meniscus shape around single vertical cylinder is $\zeta(r) = A_0 K_0(r/l_c)$ [37], with K_i the i th-order modified Bessel function of the second kind [38] and $l_c = \sqrt{\gamma/\Delta\rho g}$ the capillary length.

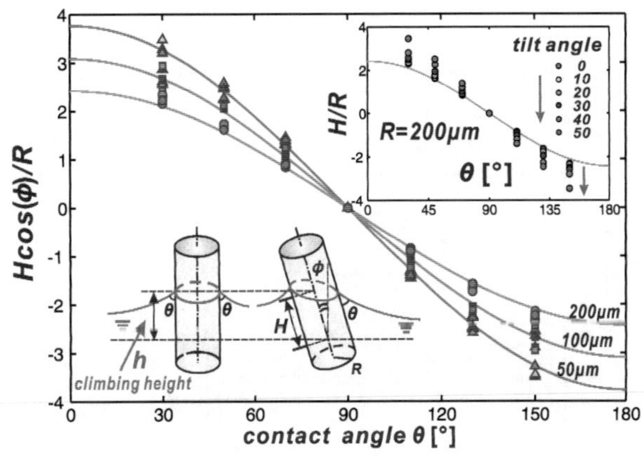

Figure 5. Projection (to vertical; see bottom inset) of the normalized shift displacement $H/R\cos\phi$ as a function of contact angle θ, for various tilt angles $\phi = 0 \sim 50°$ and radiuses $R = 50$ (triangles), 100 (squares), and 200 μm (circles). Solid line: theoretical results (equation (2)). Top inset: H/R as a function of θ with $R = 200$ μm, for various tilt angles ϕ. Solid line: theoretical predicted climbing height h.

Then we determine the constant A_0 by balancing weight of the lifted liquid and surface tension [20]. Finally the normalized climbing height is $h/R = K_0(x)\cos(\theta)/[xK_1(x)]$ (solid line; top inset of figure 5), with $x = R/l_c$ for the case without tilt. It shows that this formula agrees well with the data of $\phi = 0$, but not with the tilted cases. Then simply considering the relation $h = H\cos(\phi)$, the shift displacement for the tilted case can be predicted by (solid lines; figure 5)

$$H\cos(\phi)/R = \frac{K_0(x)}{xK_1(x)}\cos(\theta) , \tag{2}$$

which is validated by plotting $H\cos(\phi)/R$ vs. θ for various ϕ and R. It shows that the former dispersive data (top inset) all collapse onto equation (2), for $R = 200\ \mu\text{m}$.

For the stretch mode, all the normalized stretch A/R is plotted in figure 6(a) as a function of ϕ. It shows contact angle θ and radius R hardly have any effect on A/R, similar as found previously [24,25]. As a first step, we have $A'/R = -\tan\phi$ by ignoring the meniscus deformation. As shown in Fig. 6a, disparity of this law (dashed line) increases greatly with tilt angle ϕ. Considering the independence of θ on the relation $A/R\sim\phi$, we can simply analysis the peculiar case of θ=90° without losing generality. Intuitively, we find that the structure of the deformed contact line shape have the character: $\beta \approx \phi$, as shown in the top inset of figure 6(a).

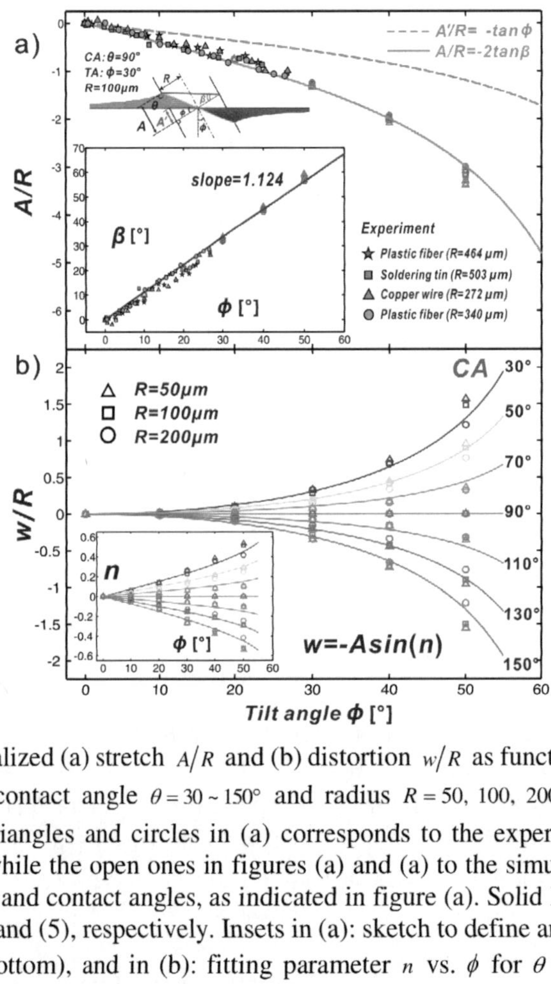

Figure 6. Normalized (a) stretch A/R and (b) distortion w/R as functions of tilt angle ϕ, for various contact angle $\theta = 30\sim150°$ and radius $R = 50,\ 100,\ 200\ \mu\text{m}$. The solid stars, squares, triangles and circles in (a) corresponds to the experiment results of various fibers, while the open ones in figures (a) and (a) to the simulation results of various radiuses and contact angles, as indicated in figure (a). Solid lines correspond to equations (3) and (5), respectively. Insets in (a): sketch to define angle β (top) and angle β vs. ϕ (bottom), and in (b): fitting parameter n vs. ϕ for $\theta = 30\sim150°$, with the relevant fits (equation (4)).

To our surprise, quantitative analysis of this relation reveals a simple linear dependence, that incredibly holds for all the simulated and experimental data of various θ and R. It indicates that an invariant structure, which is independent of contact angle and radius, exists in this tilted case. Combining the linear dependence, the normalized stretch A/R can fully be captured by (solid line; figure 6(a)).

$$A/R = -2\tan(kf),\ \text{with}\ k = 1.124. \tag{3}$$

For the distortion mode, we plotted, in figure 6(b), the fitting coefficient n (see inset) and normalized distortion w/R as functions of ϕ, for various contact angle and radius R. For $\theta = 90°$, the distortion of contact line is very small (e.g., see top inset of figure 6(a)), while as the contact angle extending to the perfect wetting and non-wetting cases, the distortion of contact line is greatly enhanced. As shown in inset of figure 6(b), we found the dependence of coefficient n on ϕ and θ can well be fitted by (solid lines)

$$n = k_0(\theta - \pi/2)\arcsin\phi,\text{with}\ k_0 = -0.4069. \tag{4}$$

Recalling the relation $w = -A\sin n$ and equation (3), the normalized distortion w/R can be predicted by (see solid lines; figure 6(b))

$$w/R = 2\tan(k\phi)\sin[k_0(\theta - \pi/2)\arcsin\phi]. \tag{5}$$

Consequently, shape of the contact line can fully be captured by equation (1) and its accessories (see figure 4). It is worth to mention that only equations (3) and (4) are employed in the normalization since the shift displacement H has no influence on the real contact line shape, and hence on the following capillary force.

3.3. The force scaling law using ellipse approximation

Following the force approach [2,3], the total capillary force exerted on the tilted cylinder can be obtained by adding the components due to surface tension and asymmetric hydrostatic pressure (i.e., variation around the cylinder [20]) together. For our case (figure 1), these two force components are finally given by (see Appendix)

$$\begin{aligned} \boldsymbol{F}^{\gamma} &= \int_{-\pi}^{\pi} \gamma[(\zeta'(\varphi)\boldsymbol{e}_{\varphi} - R\boldsymbol{e}_{z'})\cos\theta + \chi\boldsymbol{e}_{r'}\sin\theta]d\varphi \\ \boldsymbol{F}^{p} &= \frac{1}{2}\Delta\rho g R\cos\phi\int_{-\pi}^{\pi}\zeta^2(\varphi)\boldsymbol{e}_{r'}d\varphi \end{aligned} \tag{6}$$

Here $\zeta'(\varphi)$ is the derivative of equation (1), $\chi = \sqrt{R^2 + \zeta'^2(\varphi)}$. $\Delta\rho$ is the density difference of the two phases (figure 1), and $\boldsymbol{e}_{\varphi}$, $\boldsymbol{e}_{z'}$, $\boldsymbol{e}_{r'}$ are the unit vectors along the corresponding axes in the local cylindrical coordinate. Transforming the local cylindrical frame to the reference one $\Re(Oxyz)$, the inter-action between the tilted cylinder and interface is then obtained numerically either directly with the simulated contact line data or with our approximate scaling law of the contact line (equation (1) and its accessories), as represented by scatters and lines respectively in figure 7(a).

These two approaches show excellent agreements, thus confirming validity of our scaling law of the contact line. Our calculation results indicate that the total force is always vertical (also observed in Ref. [24]), with the normalized x-component $F_x/2\pi\gamma R$ of the order 10^{-2} and y-component $F_y/2\pi\gamma R = 0$. To compare the force induced by surface tension and hydro-static pressure, we plot the two z-component $F_z^{\gamma}/2\pi\gamma R$ and $F_z^{p}/2\pi\gamma R$ separately in figures 7(a) and 7(b) as a function of ϕ, for various contact angle θ and radius R. One sees that $F_z^{\gamma}/2\pi\gamma R$ is independent of the cylinder size, but this is not the case for the component $F_z^{p}/2\pi\gamma R$.

Shown in figure 7(b), we find the force induced by hydrostatic pressure increases with the cylinder radius R. It indicates that the influence of hydrostatic pressure cannot be neglected for large size (approximate millimeter size). But in our case we find F_z^{γ} is much larger than F_z^{p}, since Bond number $Bo = \Delta\rho g R^2/\gamma$ in our parameter range is small enough (see table 1). Hence, we will ignore the force induced by hydrostatic pressure in the following theoretical force scaling analysis.

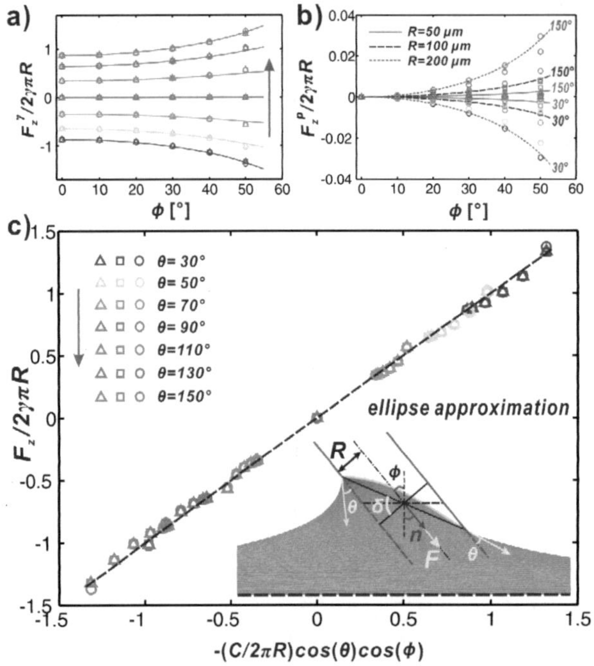

Figure 7. Dimensionless vertical force induced by (a) surface tension and (b) hydrostatic pressure as functions of tilt angle ϕ, for different contact angles $\theta = 30 \sim 150°$ and radiuses $R = 50$ (triangles), 100 (squares), and 200 μm (circles). While the lines in (a) are theoretical results. (c) Data in (a) is replotted with $F_z/2\gamma\pi R$ as a function of variable $x = -(C/2\pi R)\cos(\theta)\cos(\phi)$, where C is perimeter of the ellipse. The data all collapse onto the ellipse approximation model. Inset: sketch for derivation of the model. The color coding (see arrows) represents $\theta = 30°$ (purple), 50° (green), 70° (orange), 90° (grey), 110° (blue), 130° (red), and 150° (pink).

As mentioned before, the contact line shape is roughly a tilted ellipse (figure 2). Hence, the force $\boldsymbol{F}$ can be obtained by integrating the surface tension vector $\boldsymbol{\gamma}$ along this tilted ellipse (figure 1): $\boldsymbol{F} = \gamma C\cos(\theta)\boldsymbol{n}$ ($\boldsymbol{n}$ is the vector along the cylinder center line), as shown in inset of figure 7(c). Recalling the invariant structure predicted in the *stretch* mode, this ellipse have semi-axes of length R and $R/\cos\delta$, where $\delta = \arctan(A/R)$ is the incline angle of the deformed contact line plane. Considering the geometry relation (figure 7(c)), the z-component of the force is consequently derived as

$$F_z = -\gamma C\cos(\theta)\cos(\phi), \tag{6}$$

Where the ellipse perimeter $C = 2\pi R + 4R(1/\cos\delta - 1)$. To evaluate this force scaling, we replotted the data (of figure 7(a)) in figure 7(c) with $F_z/2\gamma\pi R$ as a function of the normalized variable $x = -(C/2\pi R)\cos(\theta)\cos(\phi)$, combining with the equation (3). It is remarkable that all the simulated data of various θ and R indeed collapse onto a single master curve without any fitting parameters, which shows the generality of our force scaling law.

4. Conclusion

In summary, we have investigated the contact line shape and the capillary interaction induced by the floating tilted cylinder, varying the tilt angle, contact angle, and cylinder radius. To gain insight into this problem, we first performed plenty of SE simulations to seek for the equilibrium interface shape and

IOP Conf. Series: Materials Science and Engineering **372** (2018) 012017 doi:10.1088/1757-899X/372/1/012017

position of the contact line, for every cylinder-interface configuration. The subsequent in-depth analysis of the three characteristic parameters (shift, stretch and distortion) of the simulated deformed contact lines reveals the self-similar shape and invariant structure of contact line that is independent of contact angle and radius. Based on these observation of the characteristic parameters, we then propose a simple ellipse force scaling law to incredibly grasp all the simulated results. We believe these results should be useful for better understanding the water-walking insects and for guidance of the water strider robot design. Our future work will focus on the more complex interaction among multiple-tilted-cylinder, for their potential applications as micro-pillar arrays clustering, fiber aggregation and self-assembly.

Acknowledgments
We thank Dr. Hua-Dong Yong for the valuable discussions and proof reading, and Dr. Cong Liu for help in experiment operation.

Appendix
Following the force approach [3,24], the force due to surface tension is given by $F^{\gamma} = \oint \gamma dl$, where the infinitesimal arc length $dl = \chi d\varphi$ with $\chi = \sqrt{R^2 + \zeta'^2(\varphi)}$. Re-calling the constant-contact-angle condition (figure 1), the vector γ is then decomposed as

$$\gamma = \gamma(\boldsymbol{b}\cos\theta + \boldsymbol{e}_{r'}\sin\theta), \tag{8}$$

where the bi-normal vector $\boldsymbol{b} = \boldsymbol{t} \times \boldsymbol{e}_{r'}$, running tangent vector $\boldsymbol{t} = [R\boldsymbol{e}_{\varphi} + \zeta'(\varphi)\boldsymbol{e}_{z'}]/\chi$ and $\mathbf{e}_j$, $\mathbf{e}_{z\text{ii}}$, $\mathbf{e}_r$ are the unit vectors along the respective axes of the local cylindrical coordinate, as shown in figure 1. Combining these relations, the force components due to surface tension is finally given by

$$\mathbf{F}^{g} = \grave{\text{o}}_{-p}^{p} g[(z ¢ j)\mathbf{e}_j - R\mathbf{e}_{z\text{ii}})\cos q + c\,\mathbf{e}_r \sin q]dj \tag{9}$$

For the force due to hydrostatic pressure, we know this force is due to the variations of hydrostatic pressure around the cylinder. The hydrostatic pressure in the two neighboring phases (I and II; figure 1) is given by

$$p_{\text{I,II}} = p_0 - \rho_{\text{I,II}} g z' \cos\phi, \tag{10}$$

where $p_0 = \text{const}$ corresponds the pressure at level $z = 0$, ρ_i the density of the i th phase. Then considering the geometry (figure 1), the force due to hydrostatic pressure can be given by

$$\boldsymbol{F}^{p} = -\int_{-\pi}^{\pi} R\boldsymbol{e}_{r'} d\varphi \left[\int_{z'_a}^{\zeta(\varphi)} p_{\text{I}} dz' + \int_{\zeta(\varphi)}^{z'_b} p_{\text{II}} dz'\right], \tag{11}$$

where z'_b, z'_a is the highest and lowest points of the contact line. Substituting equation (10) into (11), one obtains

$$\boldsymbol{F}^{p} = \frac{1}{2}\Delta\rho g R\cos\phi \int_{-\pi}^{\pi} \zeta^2(\varphi)\boldsymbol{e}_{r'} d\varphi, \tag{12}$$

where $\Delta\rho = \rho_{\text{I}} - \rho_{\text{II}}$ is the density difference of the two phases.

References
[1] Kralchevsky P A and Nagayama K 2000 Capillary interactions between particles bound to interfaces, liquid films and biomembranes *Advances in Colloid and Interface Science* **85** 145-92
[2] Kralchevsky P and Nagayama K 2001 *Particles at Fluid Interfaces and Membranes: Attachment of Colloid Particles and Proteins to Interfaces and Formation of Two-Dimensional Arrays* **10**

(Elsevier Science)
[3] Kralchevsky P, *et al* 1993 Energetical and force approaches to the capillary interactions between particles attached to a liquid-fluid interface *J Colloid Interf Sci* **155** 420
[4] Velev O D, *et al* 1993 Direct measurement of lateral capillary forces *Langmuir* **9** 3702-9
[5] Dushkin C, *et al* 1995 Lateral capillary forces measured by torsion microbalance *Phys Rev Lett* **75** 3454-7
[6] Dushkin C D, *et al* 1996 Torsion balance for measurement of capillary immersion forces *Langmuir* **12** 641-51
[7] Loudet J-C, *et al* 2005 Capillary interactions between anisotropic colloidal particles *Phys Rev Lett* **94** 018301
[8] Müller M M, Deserno M and Guven J 2005 Interface-mediated interactions between particles: a geometrical approach *Phys Rev E* **72** 061407
[9] Di Leonardo R, Saglimbeni F and Ruocco G 2008 Very-long-range nature of capillary interactions in liquid films *Phys Rev Lett* **100** 106103
[10] Lewandowski E P, *et al* 2010 Orientation and self-assembly of cylindrical particles by anisotropic capillary interactions *Langmuir* **26** 15142-54
[11] Botto L, *et al* 2012 Capillary bond between rod-like particles and the micromechanics of particle-laden interfaces *Soft Matter* **8** 4971-9
[12] Rezvantalab H and Shojaei-Zadeh S 2013 Capillary interactions between spherical Janus particles at liquid-fluid interfaces *Soft Matter* **9** 3640-50
[13] Botto L, *et al* 2012 Capillary interactions between anisotropic particles *Soft Matter* **8** 9957-71
[14] Oettel M and Dietrich S 2008 Colloidal interactions at fluid interfaces *Langmuir* **24** 1425-41
[15] Furst E M 2011 Directing colloidal assembly at fluid interfaces *PNAS* **108** 20853-4
[16] Aubry N, *et al* 2008 Micro-and nanoparticles self-assembly for virtually defect-free, adjustable monolayers *PNAS* **105** 3711-4
[17] Dickinson E 2010 Food emulsions and foams: Stabilization by particles *Current Opinion in Colloid & Interface Science* **15** 40-9
[18] Singh P, Joseph D D and Aubry N 2010 Dispersion and attraction of particles floating on fluid–liquid surfaces *Soft Matter* **6** 4310-25
[19] Nicolson M 1949 The interaction between floating particles *Proc. Cambridge Philos. Soc.* **45** 288-95
[20] Cooray H, Cicuta P and Vella D 2012 The capillary interaction between two vertical cylinders *J Phys: Condens Matter* **24** 284104
[21] Hill A and Pozrikidis C 2011 On the shape of a hydrostatic meniscus attached to a corrugated plate or wavy cylinder *J Colloid Interf Sci* **356** 763-74
[22] Pozrikidis C 2010 Computation of three-dimensional hydrostatic menisci *IMA Journal of Applied Mathematics* **75** 418-38
[23] Emami B, *et al* 2011 Predicting shape and stability of air–water interface on superhydrophobic surfaces with randomly distributed, dissimilar posts *Appl Phys Lett* **98** 203106-3
[24] Raufaste C and Cox S 2013 Deformation of a free interface pierced by a tilted cylinder: variation of the contact angle *Colloids and Surfaces A: Physicochemical and Engineering Aspects* **438** 126-31
[25] Raufaste C, *et al* 2012 Deformation of a free interface pierced by a tilted cylinder *Europhysics Letters* **99** 24001
[26] Cavallaro M, *et al* 2011 Curvature-driven capillary migration and assembly of rod-like particles *PNAS* **108** 20923-8
[27] Chandra D and Yang S 2010 Stability of high-aspect-ratio micropillar arrays against adhesive and capillary forces *Accounts Chem Res* **43** 1080-91
[28] Pokroy B, *et al* 2009 Self-organization of a mesoscale bristle into ordered, hierarchical helical assemblies *Science* **323** 237-40
[29] Py C, *et al* 2007 3D aggregation of wet fibers *Europhysics Letters* **77** 44005

[30] Bush J W, Hu D L and Prakash M 2007 The integument of water-walking arthropods: form and function *Adv Insect Physiol* **34** 117-92
[31] Ji X-Y, Wang J-W and Feng X-Q 2012 Role of flexibility in the water repellency of water strider legs: Theory and experiment *Phys Rev E* **85** 021607
[32] Yu Y, Zhao Z-H and Zheng Q-S 2007 Mechanical and superhydrophobic stabilities of two-scale surfacial structure of lotus leaves *Langmuir* **23** 8212-6
[33] Bernardino N R, Blickle V and Dietrich S 2010 Wetting of surfaces covered by elastic hairs *Langmuir* **26** 7233-41
[34] Song Y S and Sitti M 2007 Surface-tension-driven biologically inspired water strider robots: theory and experiments *IEEE Transactions on Robotics* **23** 578-89
[35] Brakke K A 1992 The surface evolver *Experimental Mathematics* **1** 141-65
[36] Zeng C, *et al* 2012 Capillary interactions among spherical particles at curved liquid interfaces *Soft Matter.* **8** 8582-94
[37] Lo L L 1983 The meniscus on a needle–a lesson in matching *J. Fluid Mech* **132** 65-78
[38] Abramowitz M E 1964 *Handbook of Mathematical Functions: With Formulas, Graphs, and Mathematical Tables* **55** (Courier Dover Publications)

Design and performance of honeycomb structure for nanobubbles generating apparatus having different cell dimensions

T Ueda[1], H F Zhai[1], F Ren[1], N-A Noda[1,3], Y Sano[1], Yasushi Takase[1], Y Yonezawa[2] and H Tanaka[1]

[1]Mechanical Engineering Department, Kyushu Institute of Technology, 1-1 Sensui-cho Tobata-ku, Kitakyushu-shi, 804-8550, Japan
[2]Mrufukusuisan Corp., Ltd., 94-22 Nishiminato-machi Kokurakita-ku, Kitakyusyu-shi, 803-0801, Japan

E-mail: noda@mech.kyutech.ac.jp

Abstract. In recent years, nanobubble technology has drawn great attention due to their wide applications in various fields of science and technology, such as water treatment, biomedical engineering, and nanomaterials. This study focuses on the application to seafood long term storage. The nitrogen nanobubble water circulation may reduce the oxygen in water and slow the progressions of oxidation and spoilage. Our previous study showed the pressure reduction and shear stress are involved in nanobubble generation apparatus with honeycomb cells. In this work, the nanobubble generating performance is studied experimentally for honeycomb structures by varying the cell size and the flow velocity. Computational Fluid Dynamics analysis is also performed to simulate the experiment and find out the efficient nanobubble generation.

1. Introduction

In recent years, nanobubble technology has drawn great attention due to their wide applications in wide fields of science and technology, such as water treatment, medical engineering, and chemical and food industry. Among them, this study focuses on the application to seafood long term storage. The nitrogen nanobubble water circulation may reduce the oxygen in water and slow the progressions of oxidation and spoilage [1,2]. In this study, we mainly focus on the technique of honeycomb structure which have excellent efficiency of nanobubble generation [2-4].

Our previous study showed the pressure reduction and shear stress are involved in nanobubble generation when a large apparatus in figure 1(a) is used for 1000 liters of water [3-5]. In this study, a small apparatus in figure 1(b) will be used for 100 liters of water in comparison with the large apparatus under different cell size and fluid velocity. Experimental studies will be conducted to investigate the nanobubble generating performance in honeycomb structures by varying the cell size and the flow velocity. The design and performance of the honeycomb structures will be discussed to generate nanobubble efficiently.

2. Nanobubble generation experiments

Figure 1 shows the experimental apparatus with parallel honeycomb units immersed in water in a container and table 1 shows the details of the experimental apparatus. The water and the nitrogen gas are pumped into this apparatus together flowing into the honeycomb units. Figure 2 shows the model of the honeycomb structure consisting of a pair of honeycomb plates and upper and lower plates for sealing both ends. Through the complex flow channels composed by the overlapped honeycomb plates, the turbulence for the gas-liquid mixture may generate nanobubles. During many times of the gas-liquid mixture flow from a wide area to a narrow area, the shear field and the pressure reduction are formed many times, which contribute to the nitrogen bubbles finer. This cycle has been repeated, and miniaturization effect continues. The experimental conditions are shown in table 2.

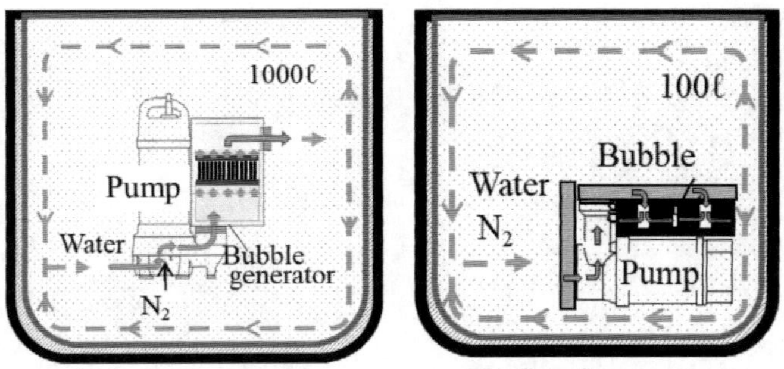

(a) Large apparatus (b) Small apparatus

Figure 1. Nitrogen nanobubbles generating apparatus.

Table 1. Details of the experimental apparatus.

Item	Large apparatus	Small apparatus
Pump	Submersible pump (80TM23.7)	Submersible pump (50TMLS2.4S)
Output [kW]	3.7, (60 Hz)	0.4, (60 Hz)
Flow rate [ℓ/min]	400	270
Container [mm]	1580×1100×600	840×600×500
Water amount [ℓ]	Tap water, 1000	Tap water, 100
Gas	Nitrogen	Nitrogen
Flow rate [ℓ/min]	5.0	1.0
Pressure [MPa]	0.30	0.20

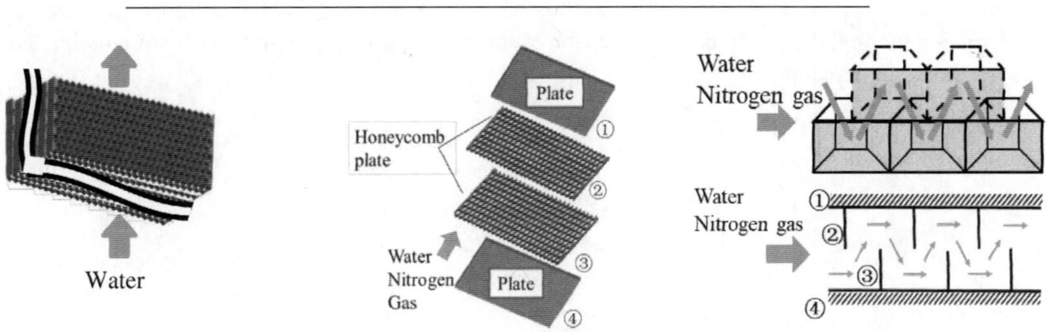

(a) Honeycomb unit (b) 1-unit composition (c) Cross section 1-unit honeycomb structure

Figure 2. Honeycomb structure model of generating apparatus.

IOP Conf. Series: Materials Science and Engineering **372** (2018) 012018 doi:10.1088/1757-899X/372/1/012018

Table 2. Experimental condition.

Experimental conditions	Power [kW]	Wall Thickness [mm]	Inlet Pressure [MPa]	Flow rate [ℓ/min]
Large apparatus	3.7	0.5	0.35	550
Small apparatus	0.4	1	0.06	103

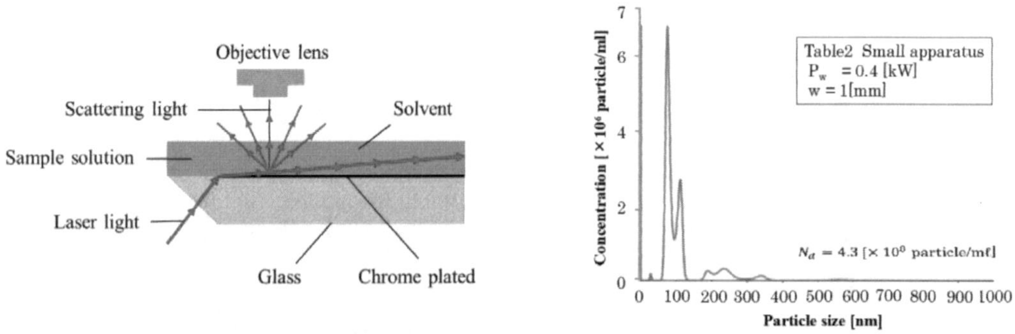

(a) Visualization principle for nanobubbles (b) Measured example for nanobubble density

Figure 3. Visualization principle and measured example for nano sight.

The nano-particle analyzer, Nano Sight LM10-HS [6], is used to measure the nanobubble number density. Figure 3(a) shows the principle of Nano Sight LM10-HS. Irradiating a laser beam in the horizontal direction in the sample liquid, the side-scattered light from the nanoparticles are visualized by the objective lens, the movement trajectory of each particle is displayed on computer. Tracking Brownian motion of all particles recognized on the screen and using the Stokes-Einstein equation, the particle size is obtained from the moving velocity of the particles. Figure 3(b) shows an example of the nanobubble number density distribution measured at 60 minutes after the start of the experiment. After collecting the sample solution, measure the nanobubble number density within the range of 30 minutes to 1 hour at which micro bubbles of 1 μm or more disappear and become stable state.

Figure 4 shows the time variation of nanobubble number density for large and small apparatus. The target value was set to a nanobubble number density of 200 million / mℓ whose effect was confirmed by sensory test using fresh fish. It can be seen that the small apparatus achieves the target value after 10 minutes and then the nanobubble number density further increases. Compared to the experimental results of large apparatus obtained in previous studies, better nanobubble number densities can be obtained for the small apparatus, although the water amount, pump output, etc. are smaller than the ones of the large apparatus.

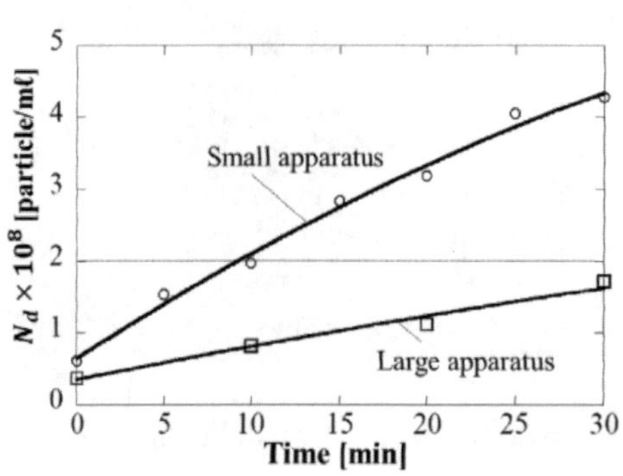

Figure 4. Time variation of nanobubble number density.

3. CFD analysis

Figure 5 shows (a) flow direction, (b) contour plot of velocity and (c) contour plot of shear stress in a honeycomb cell of the small apparatus. Considered flow Path 1-Path 6 are also indicated. The results in figure 5 are obtained under the inlet pressure shown in table 2 with the cell dimensions in table 3.

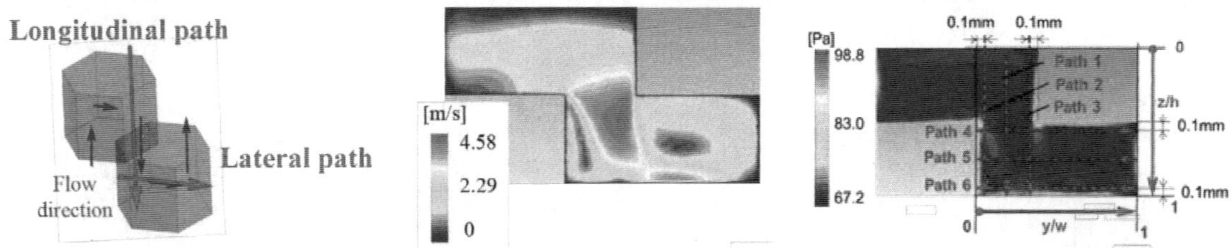

(a) Flow direction (b) Velocity distribution (c) Shear stress with Path 1-6

Figure 5. Schematic illustrations of flow direction, velocity distribution and shear stress with Path 1-6.

Table 3. Experimental results.

	Item	Large apparatus	Small apparatus
Conditions	Cell width [mm]	6.1	3.9
	Cell height [mm]	10	4.0
	Flow rate [ℓ/min]	550	103
	Water amount [ℓ]	1000	100
	Number of cells	3510	1840
Performance	Nanobubble number density after 30 min [particle/mℓ]	1.7×10^{8}	4.3×10^{8}
	Nanobubble number density per cell per cycle [particle/mℓ]	3.13×10^{3}	7.61×10^{3}

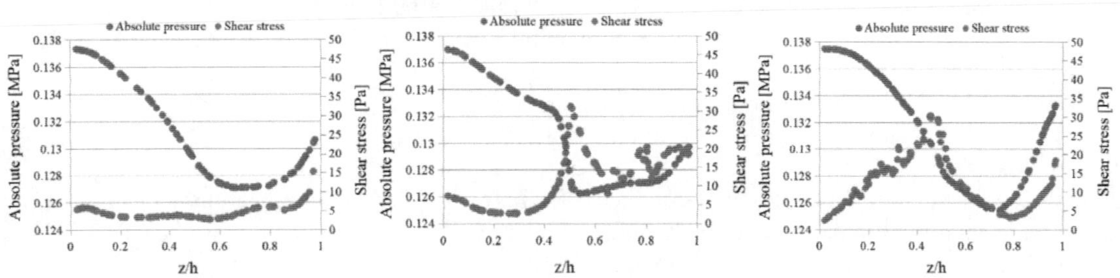

(a) Path 1 (b) Path 2 (c) Path 3

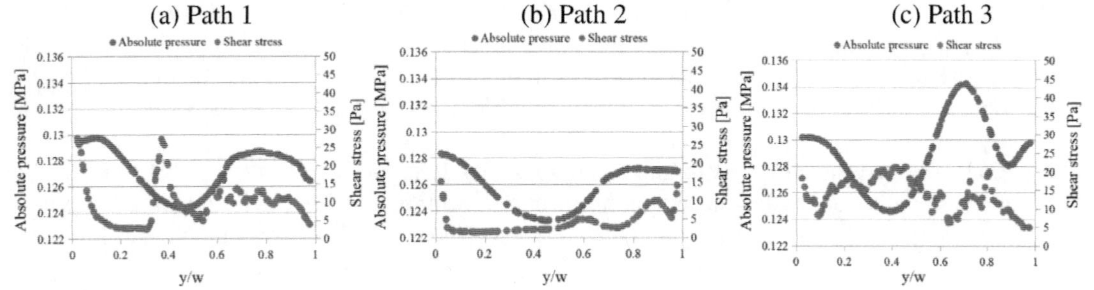

(d) Path 4 (e) Path 5 (f) Path 6

Figure 6. Absolute pressure and shear stress distribution.

As shown in figure 5, in the analysis model, Path 1 and 5 are set as the center lines of the flow path, and Paths 2, 3, 4, 6 are set along the lines 0.1 mm away from the wall. Figure 6 shows the pressure distributions and the shear stress distribution along Path 1-6. The maximum shear stress appears along Path 3. The shear stress was calculated as the product of strain rate and viscosity [3]. Along Paths 1, 2, 3, the pressure decreases with increasing z/h initially and finally increases. The total pressure reduction is about 0.01 in Paths 1, 2, 3, and the maximum shear stress appears as τ = 30 Pa near the corner (z/h = 0.5) in Path 3.

Regarding the lateral Paths 4-6, the pressure decreases with increasing y/w initially and next increases similar to the longitudinal Path 1-3, and the maximum shear stress appears near the corner. Similar results were obtained for the large apparatus. The large apparatus has larger pressure reduction but smaller maximum shear stress. The pressure reduction and the shear stress in figure 6(c) are indicated table 4.

Table 4. Analytical condition and results along the Path 3.

	Item	Large apparatus	Small apparatus	
Analysis condition	Inlet absolute pressure [MPa]	0.45	0.16	0.45
Analysis results	Average velocity [m/s]	0.71	0.40	0.95
	Decrease in pressure [MPa]	0.03	0.01	0.07
	Shear stress [Pa]	24	30	56

4. Results and discussion

Table 3 shows the experimental results of the large apparatus and the small apparatus. The nanobubble number density per cell per cycle after 30 minutes is calculated from the nanobubble number density, the number of cycles and the number of cells. In table 3, the nanobubble number density after 30 min in the small apparatus is 2.5 times larger than the one in the large apparatus. The nanobubble number density per cell per cycle in the small apparatus is 2.4 times larger than the one in the large apparatus. Table 4 shows the CFD results for both apparatuses along the Path 3. The small apparatus has smaller average velocity and smaller pressure reduction but larger maximum shear stress.

According to the experimental results, the nanobubble generation ability of the small apparatus is 2.4 times larger than that of the large apparatus, and the CFD analysis shows that the shear stress is 1.3 times (2.3 times if the inlet pressure is set to the same value to the large apparatus) larger than that of the large apparatus, which means that the experimental results ware confirmed by the CFD analysis. These results show that the shear stress is greatly related to the nanobubble generation ability, and the nanobubble generation ability is better in the small apparatus.

5. Conclusions

In this study, the experiment study was conducted for both large and small apparatuses of honeycomb structural generator with the nitrogen-water mixture. Comparing the experiment with the CFD analysis, the conclusions can be summarized as follows.

- The experiment showed that the nanobubble number density after 30 minutes in the small apparatus is larger than the one in the large apparatus. In addition, the nanobubble number density per cell per cycle in the small apparatus is 2.4 times larger than the one in the large apparatus. Those results indicated the nanobubble generation efficiency is better in the small apparatus.
- The CFD analysis showed that the small apparatus has smaller average velocity and smaller pressure reduction but larger maximum shear stress compared to the large apparatus. It may be concluded that the shear stress has a great effect on the nanobubble generation efficiency.

References

[1] Hiraki K 2014, Efficient long term storage technology of seafood using nitrogen nanobubbles

The Latest Technology on Microbubbles and Nanobubbles **2** 177-83
[2] Kudo Y and Hiraki K 2010 MNM-P4-6 Visualization of mixing process in a conduit with honeycomb structure *The Proceedings of the Symposium on Micro-Nano Science and Technology* **2** 95-6
[3] Noda N-A, Ren F, Yamamoto W, Ueda T, Sano Y, Chen D-H, Takase Y and Yonezawa Y Design and performance of honeycomb structure for nanobubbles generating apparatus *Journal of Japan Society for Design Engineering* (in Japanese) (in press)
[4] Noda N-A, Ren F, Yamamoto W, Sano Y, Chen D-H and Yonezawa Y 2016 Design and performance of honeycomb structure for nanobubbles generating apparatus *International Journal of Fracture Fatigue and Wear* **4** 109-13
[5] Aoki K, Kato K, Okutsu T and Shinohara N 2017 Basic principle of fine bubble generation and characteristics of generator *Journal of Japan Society for Design Engineering* **52** 275-85
[6] Vasco F and Andrea H 2010 Win Jiskoot: Critical evaluation of nanoparticle tracking analysis (NTA) by nanosight for the measurement of nanoparticles and protein aggregates *Pharmaceutical Research* **27** 796-810

Effect of tempering treatment upon the residual stress of bimetallic roll

Y Sano[1], N-A Noda[1,3], Y Takase[1], R Torigoe[1], K Tsuboi[1], M R Aridi[1], Y Sanada[1], and L Y Lan[2]

[1]Department of Mechanical Engineering Kyushu Institute of Technology, 1-1 Sensui-cho Tobata-ku, Kitakyushu-shi, 804-8550, Japan
[2]School of Mechanical Engineering and Automation, Northeastern University, No. 3-11, Wenhua Road, Shenyang, 110819, China

E-mail: noda@mech.kyutech.ac.jp

Abstract. Bimetallic rolls are widely used in steel rolling industries because of the excellent hardness, wear resistance, and high temperature properties. However, thermal stress is produced by heating-cooling thermal cycles, which is a great challenge for their practical application. Indeed, if severe thermal tensile stress is introduced into these rolls, it can assist the thermal cracks to propagate, even lead to the overall failure of rolls. In this paper, we investigated the effect of tempering treatment on the residual stress after the bimetallic rolls were subjected to quenching. Compared with the non-uniform heating-quenching process, the tempering process makes the maximum stress at the core decreased by 15% (from 275 MPa to 234 MPa) with considering martensite transformation but decreased by 26% (from 275 MPa to 201 MPa) without considering martensite transformation. For tempering process after uniform heating quenching, the maximum stress at the core decreases by 24% from 357 MPa to 273 MPa with considering martensite transformation but decreases by 30% from 357 MPa to 246 MPa without considering martensite transformation. And compared with the non-uniform heating-quenching process, the double tempering process makes the maximum stress at the core decreased by 8% (from 275 MPa to 253 MPa) with considering martensite transformation but decreased by 27% (from 275 MPa to 200 MPa) without considering martensite transformation.

1. Introduction

Work rolls are used in the roughing stands of hot strip mill to reduce the steel thickness. They have to meet the requirements of hardness, wear resistance at the surface, and toughness at the center [1-3]. Traditional single material roll cannot satisfy these conflicting properties at the same time. Many studies were done to improve roll performance in the past decades. The bimetallic roll is one of the most important development to resolve this problem [4,5]. High speed steel (HSS) is widely used in the bimetallic roll as shell material. The HSS roll is characterized by excellent hardness, good wear resistance as the shell material and significant toughness of core material [6-9]. As shown in figure 1, the bimetallic roll is manufactured by centrifugal casting method, using HSS as the shell material and the ductile casting iron (DCI) as the core material. During heat treatment, residual stress is inevitably introduced due to temperature gradient and phase transformation [4,10,11]. The residual stress is self-equilibrating within the roll, independent of the any external loads. In addition, thermal stress is produced by heating-cooling thermal cycles during subsequent hot rolling process [12,13]. If severe thermal tensile stress is added under the rolling trouble, the thermal crack starts to propagate. Therefore, suitable compressive stresses are necessary for preventing the thermal crack extension.

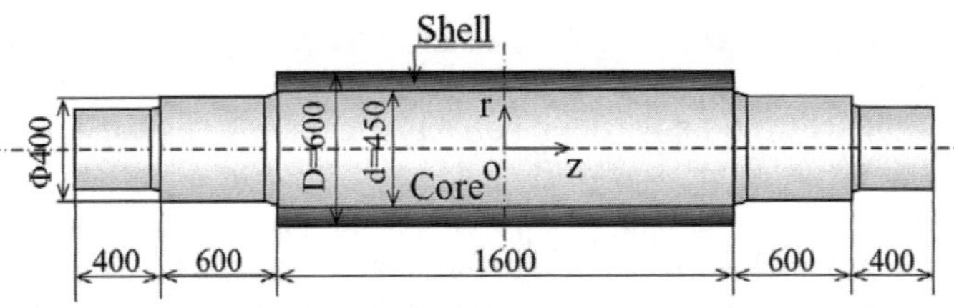

Figure 1. Schematic diagram of the HSS bimetallic roll (mm).

One of the most used numerical modelling techniques is finite element method, which can be used for many engineering applications conveniently [14-19]. In the previous studies [14-16], different quenching methods were discussed through FEM simulations to produce suitable surface compressive residual stresses and to reduce the center tensile residual stress. However, since the tempering process after quenching was not been considered yet, the tempering effect will be discussed after uniform heating quenching and after non-uniform heating quenching on the basis of FEM simulations.

2. Tempering effect to reduce residual stress

2.1. Quenching and tempering heat treatment

Figure 2(a) illustrates the non-uniform heating quenching process in comparison with figure 2(b), which illustrates the uniform heating quenching process. In the uniform heating process, the whole roll is heated up to the higher temperature equaling to T_{Start} before the quenching process. In the non-uniform heating process, the whole roll is heated up to the uniform lower temperature of T_{Heat} and kept at T_{Heat} for some hours, then rapidly heated up to T_{Start} as the non-uniform heating before quenching. This rapid heating provides temperature difference between the roll surface and roll center. The quenching processes after non-uniform heating and uniform heating are similar, but the keeping temperature T_{Keep1} in figure 2(a) > T_{Keep2} in figure 2(b). The quenching process after the non-uniform heating quenching can be described in the following way: the roll is put out from the heating furnace and cooled down rapidly from T_{Start} by using the spray cooling. After rapid cooling, the roll is maintained for several hours when the surface temperature drops to T_{Keep1}. Here, keeping T_{Keep1} is beneficial to relax the excessive thermal stresses caused by rapid surface cooling. After keeping T_{Keep1}, the roll is put out from the furnace and slowly cooled down in air until to T_{Finish}. In this paper, one or two tempering processes will be assumed to clarify the effect of transformation after uniform heating and non-uniform heating.

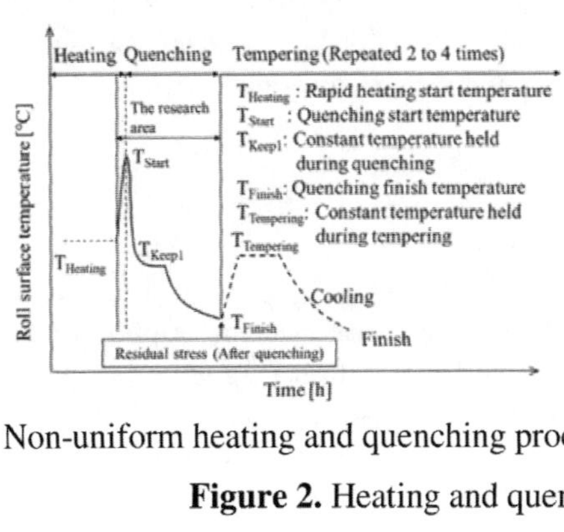

(a) Non-uniform heating and quenching process

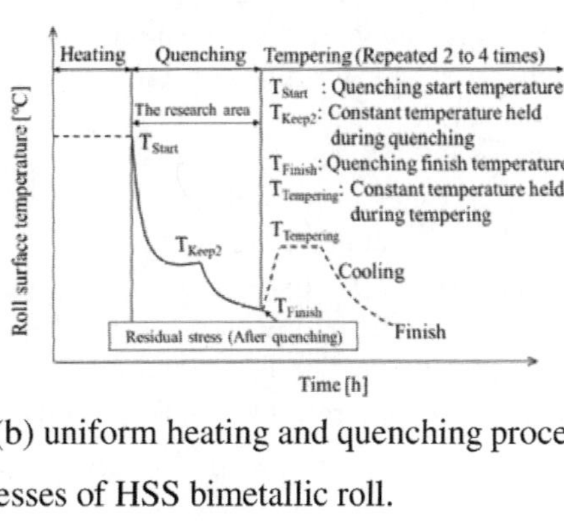

(b) uniform heating and quenching process

Figure 2. Heating and quenching processes of HSS bimetallic roll.

Figure 3 shows the thermal expansion rate used in tempering analysis. The temperature is normalized by the quenching temperature T_{Start}°C. After quenching, the temperature is raised from T_{Finish}°C and held at tempering temperature $T_{Tempering}$, and cooled down. During the cooling down process, martensitic transformation occurs from retained austenite. Although stress reduction may appear by keeping tempering temperature $T_{Tempering}$, martensitic expansion may elliminate this effect. Figure 3 is obtained by using the specimen with dimensions of 8 mm width × 17 mm length × 8 mm thickness, which was cut out from HSS roll. The results obtained from the solid line in figure 3 is denoted as martensite transformation 100%. If the compressive residual stress of 300 to 500 MPa existing in the outer layer of the actual roll is considered, the martensitic transformation may be suppressed. By considering this surface residual effect, in this study, an extreme case will be considered for martensite transformation 0% obtained from the dotted line data shown in figure 3.

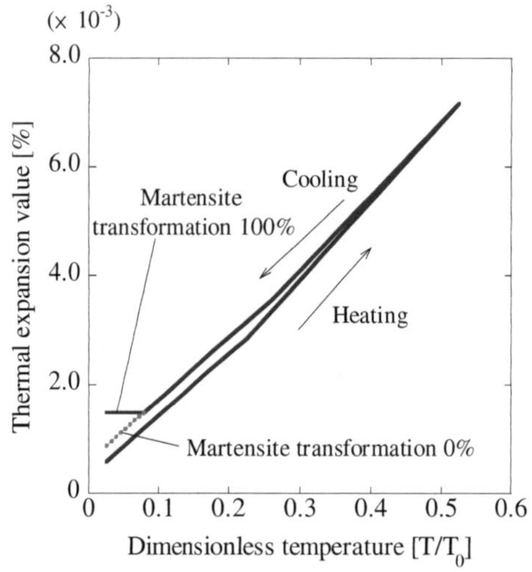

Figure 3. Dilatometric curves during tempering process ($T_0 = T_{Start}$).

2.2. Creep analysis

In the creep analysis, the transient creep strain should be considered as well as the steady creep strain. Among several equations available for creep analysis, the time hardening law, sometimes called power law, is used to express the core material which has low strength under high temperature. The core creep can be given by equation (1).

$$\varepsilon^c = A\sigma^m t^n \tag{1}$$

Where ε^c is the transient strain, σ is stress, t is time, A, m, and n are temperature dependent material constants. This time hardening formulation is used to predict the creep behavior under a variable stress history. In order to determine constants A, m, and n, the creep tests are performed.

2.3. Creep test

The creep testing was conducted by using a miniature creep rupture testing machine based on JISZ2271. The specimens were prepared from core material as shown in figure 4(a). Those specimens were, respectively, heated up to the testing temperatures, T_{Keep1} and T_{Keep2}, and kept at theses temperatures during the testing process. Then, the creep tests were carried out by applying constant loads 100 and 130 MPa. The strain changes were recorded with time. From the strain–time curves obtained, the creep equations can be written as shown in equations (2) and (3).

$$\varepsilon^c = 2.25\times10^{-13}\sigma^{3.44}t^{0.672}(T_{keep1}) \tag{2}$$

$$\varepsilon^c = 8.38\times10^{-19}\sigma^{5.71}t^{0.514}(T_{keep2}) \tag{3}$$

(a) Specimen of the creep test (mm) (b) specimen of the stress relaxation test (mm)

Figure 4. Specimens of the creep test and stress relaxation test.

2.4. Stress relaxation test

To confirm the validity of equations (2) and (3), the following stress relaxation testing was conducted. The specimens were prepared from the core material as shown in figure 4(b). Those specimens were, respectively, heated up to the testing temperature, T_{Keep1} and T_{Keep2}, and kept at theses temperatures during the testing process. Then, the stress relaxation tests were carried out under the constant strain when the primary stress is 130 MPa. The time and stress changes were recorded as shown in figure 5.

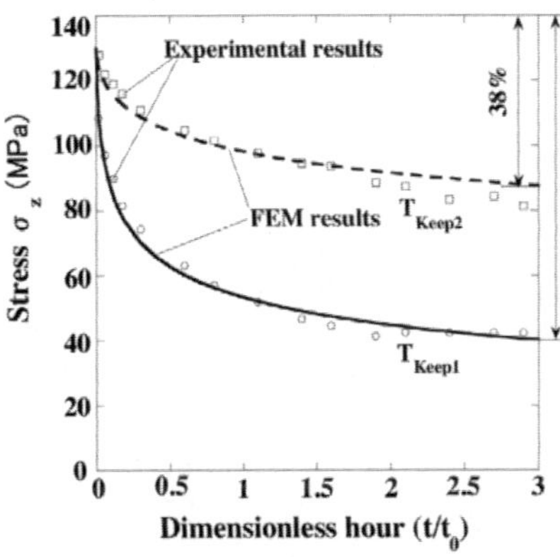

Figure 5. Comparison between FEM simulation results and experimental results for stress relaxation at T_{Keep1} and T_{Keep2}.

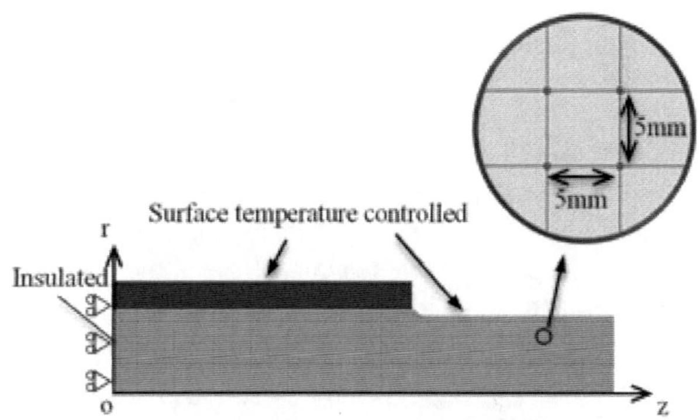

Figure 6. Analytical model and boundary condition for bimetallic roll whose mesh size is 5x5 mm for both inner and core material.

2.5. Results and discussion for creep

The model used for the FEM simulations is shown in figure 6, and the mechanical properties as shown in table 1.

Table 1. Mechanical properties of shell and core at room temperature.

Property	Shell	Core
0.2% proof stress (MPa)	(1270)*1	410
Young's modulus (GPa)	228	168
Poisson's ratio	0.3	0.28
Density (kg/m^3)	7600	7300
Thermal expansion coefficient (/K)	12.6×10^{-6}	13.0×10^{-6}
Thermal conductivity [W/(m·K)]	20.2	23.4
Specific heat [J/(kg·K)]	0.46	0.42
Shore hardness, Hs	85	50

IOP Conf. Series: Materials Science and Engineering **372** (2018) 012019 doi:10.1088/1757-899X/372/1/012019

Table 2. Comparison of keeping process between non-uniformheating quenching and uniform heating quenching.

Heat treatment	Non-uniform	Uniform
Stress relaxation ratio [%]	69	38
Keeping temperature [ºC]	$T_{keep1} > T_{keep2}$	
σ_{eq} at keeping temperature [MPa]	136	103
Keeping time [h]	5.3	6

FEM simulations of stress relaxation were performed to verify equations (2) and (3) as shown in figure 5 in comparison with the experimental result at T_{Keep1} and T_{Keep2}. The results show that the stress decreases by 69% at T_{Keep1} and by 38% at T_{Keep2} as shown in table 2. It can be found that the stress relaxation ratio is larger at high temperature (T_{Keep1}) than relatively low temperature (T_{Keep2}). The FEM results are in good agreement with the experiment results. It is confirmed that equations (2) and (3) are useful for predicting the creep effect on the residual stress.

3. Residual stress after tempering

3.1. Effect of tempering process on residual stress

Figure 7 shows stress distribution σ_z along the central cross section where $z = 0$ after tempering process for non-uniform heating quenching and uniform heating quenching. It is shown that the tempering process effectively reduces the maximum tensile residual stress at the roll core and keeps large enough compressive residual stress at the roll surface. In addition, the stresses become uniformly distributed at the core after uniform heating quenching and also after non-uniform heating quenching.

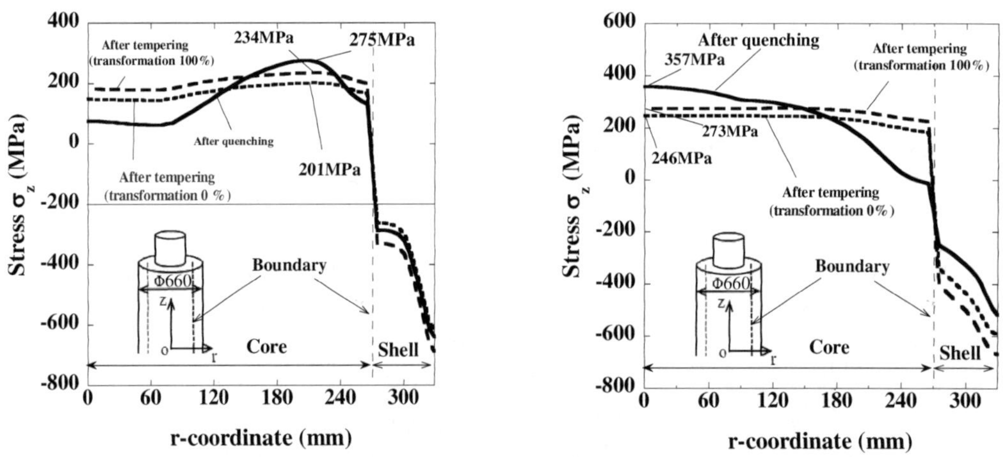

(a) Tempering process after non-uniform heating quenching process

(b) Tempering after uniform heating quenching process

Figure 7. Effect of tempering process on the residual stress.

For tempering process after non-uniform heating quenching, the maximum stress at the core decreases by 15% from 275 MPa to 234 MPa for martensite transformation 100%, and decreases by 26% from 275 MPa to 201 MPa for martensite transformation 0%. For tempering process after uniform heating quenching, the maximum stress at the core decreases by 24% from 357 MPa to 273 MPa for martensite transformation 100%, and decreases by 30% from 357 MPa to 246 MPa for martensite transformation 0%.

4. Conclusion

In this paper, the numerical analysis method which can predict the residual stress reduction considering tempering process was established. The results of the current study can be summarized as follows:

- FEM simulations of stress relaxation were performed. As a result, it can be found that the stress relaxation ratio is larger at a high temperature than at a relatively low temperature.
- Compared with the non-uniform heating-quenching process, the tempering process makes the maximum stress at the core decreased by 15% (from 275 MPa to 234 MPa) with considering martensite transformation but decreased by 26% (from 275 MPa to 201 MPa) without considering martensite transformation.
- Compared with the uniform heating-quenching process, the tempering process makes the maximum stress at the core decreased by 24% (from 357 MPa to 273 MPa) with considering martensite transformation but decreased by 30% (from 357 MPa to 246 MPa) without considering martensite transformation.

References

[1] Choi W J and Kim D 1999 *ISIJ Int.* **39** 823-8
[2] Noguchi H and Watanabe Y 1987 *Kawasaki Steel Tech.* **19** 195-201
[3] Nilsson M and Olsson M 2013 *Wear* **307** 209-17
[4] Ichino K, Kataoka Y and Koseki T 1997 *Kawasaki Steel Tech.* **37** 13-18
[5] Shimizu M, Shitamura O, Matsuo S, Kamata T and Kondo Y 1992 *ISIJ International* **323** 1244-9
[6] Fu H, Xiao Q and Xing J D 2008 *Materials Science and Engineering: A* **474** 82-7
[7] Takigawa H, Tanaka T and Ohtomo S 1997 *Nippon Steel Tech. Rep.* **74** 77-83
[8] Fu G H, Zhao J H, Du Z Z, Feng J Z, Lei P Y, Zhang Y, Li W M, Jiang H Y, Zhou R and Guo X H 2011 *Ironmaking Steelmaking* **38** 338-45
[9] Boccalini M Jr and Sinatora A 2002 *Proc. of 6th Int. Tooling Conference* (Karlstad: Karlstads University) 509-24
[10] Vro L D V Z and Valjanje E 2014 *J. Mater. Technol.* **48** 983-90
[11] Podgornik B, Milanovic S and Vizintin J 2010 *J. Mater. Proc. Technol.* **210** 1083-8
[12] Onisa F C and Farrugia J C D 2008 *Int. J. Mater. Form.* **1** 363-6
[13] Chang F D 1999 *J. Mater. Proc. Technol.* **94** 45-51
[14] Noda N A, Hu K, Sano Y, Hosokawa Y and Wang X 2016 *ISIJ Int.* **57** 1433-40
[15] Noda N A, Hu K, Sano Y, Ono K and Hosokawa Y 2016 *Steel Research Int.* **3** 1448-78
[16] Noda N A, Hu K, Sano Y, Ono K and Hosokawa Y 2017 *Steel Research Int.* **88** 160-15
[17] Noda N A, Chen X, Sano Y, Wahab M A, Maruyama H and Fujisawa R 2016 *Mater. Des.* **96** 476-89
[18] Miyazaki T, Noda N A, Ren F, Wang Z, Sano Y and Iida K 2017 *Int. J. Adhes. Adhes.* **77** 118-37
[19] Noda N A, Suryadi D, Kumasaki S, Sano Y and Takase Y 2015 *Eng. Fail. Anal.* **57** 219-35

IOP Conf. Series: Materials Science and Engineering **372** (2018) 012020 doi:10.1088/1757-899X/372/1/012020

Study on coming out of the shaft from ceramic sleeve in terms of the residual displacement

G W Zhang[1,2], N-A Noda[1], Y Sano[1] and H Sakai[1]

[1]Department of Mechanical and Control Engineering, Kyushu Institute of Technology Sensui-Cho 1-1 Tobata-Ku, Kitakyushu-Shi, Fukuoka, Japan

E-mail: q595105c@mail.kyutech.jp

Abstract. Ceramic roller can be used in the heating furnace conveniently because of its high temperature resistance. However, the coming out of the shaft may often happen from the ceramic sleeve under repeated load. In this paper, a two-dimensional shrink fitted structure is considered by replacing the shaft with the inner plate and by replacing the sleeve with the outer plate. Based on the model with stopper, the FEM simulation is performed under alternate loading with certain intervals newly added. The analysis results show that the coming out failure can be explained from the residual displacement accumulation during these intervals.

1. Introduction

In order to produce high-quality steel plates for automobiles as is shown in figure 1, steel conveying rollers with high temperature resistance working in furnace have been developed. Usually, the conventional roller material is composed by steel and ceramic spray coating on the outside of sleeve. However the thermal expansion mismatch may exceed the adhesion strength of the ceramic layer, which causes failure on the roller surface such as crack, peeling, and wearing [1].

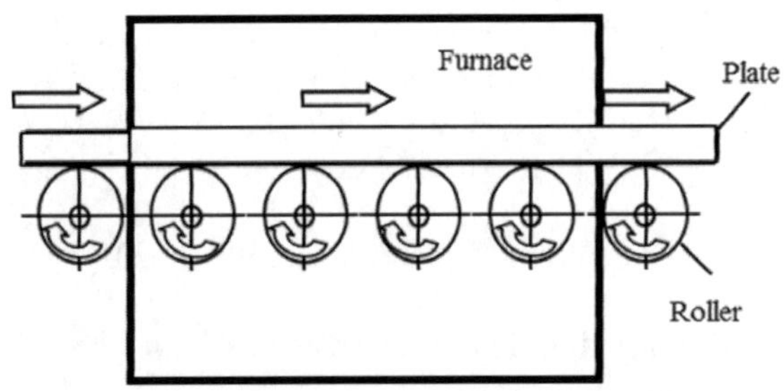

Figure 1. Layout of rollers in heating furnace.

Figure 2 shows a new ceramic roller consisting of steel shaft at both ends and ceramic sleeve with high corrosion resistance and high wear resistance [2,3]. The inside of the roller is usually cooled by water. The method of joining the steel with the ceramic sleeve is an important issue. The previous studies suggested that the shrink fitting system may be the most suitable joining method for cylindrical ceramic structures, which can reduce the maintenance cost and the shaft replacement time [4-10]. However, the coming out of the shaft from the sleeve happens during operation since only a small shrink fitting ratio can be used because of the ceramic brittleness [11-17]. Previously the coming out simulations are performed by using FEM, which can be applied many industrial problems conveniently [4-17]. Suryadi [4] *et al* built 3D FEM model to simulate the coming out but only until 5 loading circles because of the large calculation time. Therefore, Xu [5] *et al* developed 2D model as shown in figure 3 (2D roller old model) to reduce the calculation time and obtained the results until 40 circles or more. As a further study, authors designed a new simplified 2D model with stopper to prevent the coming out behavior as shown in figure 3 [6]. Those studies [4-6] have proved that the coming out behavior can be realized in the 2D and 3D simulation and it can be prevented by the stopper. This paper considers 2D coming out problem under alternate loading with a certain interval, by taking account of the influence of intervals in alternate loading. And try to explain the coming out mechanism in detail with the help of the stopper. Then the coming out failure can be considered from the residual stress displacement during these intervals. Here, the inner plate and outer plate will be considered as the shaft and sleeve as shown in figure 3. The finite element method is applied to simulating the prevention of the coming out behavior in terms of the residual displacement.

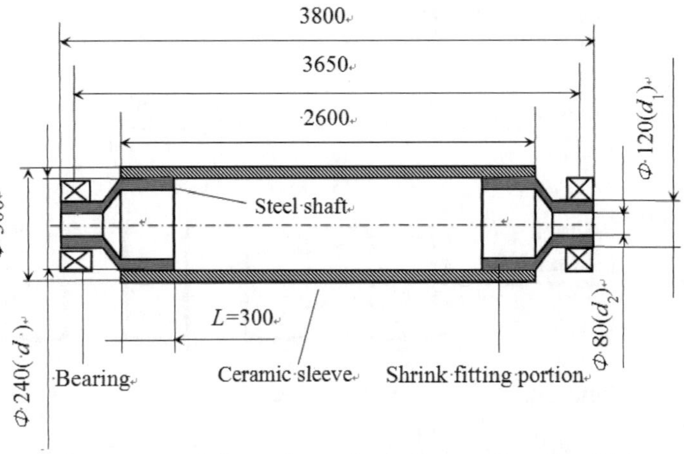

Figure 2. Structure and dimensions of a real ceramic roller (Unit: mm).

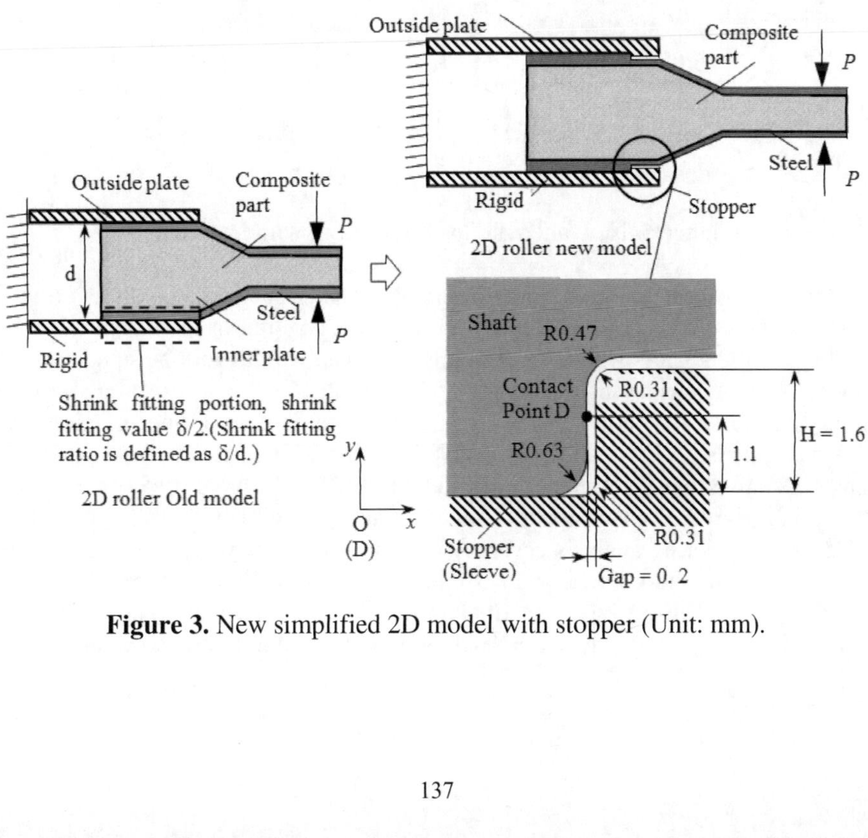

Figure 3. New simplified 2D model with stopper (Unit: mm).

2. Analysis method

Figure 3 illustrates the structure of the newly designed stopper installed on the outer plate and the boundary condition. Figure 4 shows the previously used loading pattern which does not contain intervals between the loading [4-6]. Figure 5 shows the new loading pattern including intervals denoted by Step II and step IV between the loading periods as Step I and step III, where the temporary no-load condition during roller conveyance of the iron plate has been taken account into practically. Table 1 shows the material properties of the models. In this 2D model, the 2D solid inner plate and the 2D rigid outer plate are considered to simplify the model and to reduce the calculation time as described in the literature [5,6].

Table 1. Material properties.

Model	2D−alt.		
	Sleeve	Shaft	
	Rigid	Steel	Composite part
Young's modulus [GPa]	∞	210	55
Poisson's ratio	—	0.3	0.3
Tensile strength [MPa]	∞	600	—
Mass density [kg/m^3]	0	7800	7800

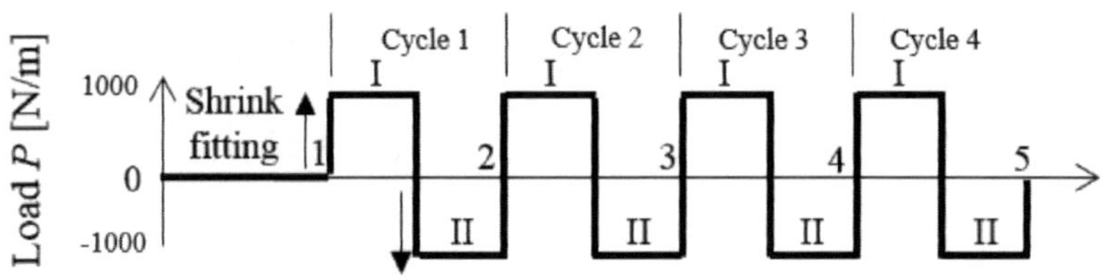

Figure 4. Alternate loading pattern without intervals.

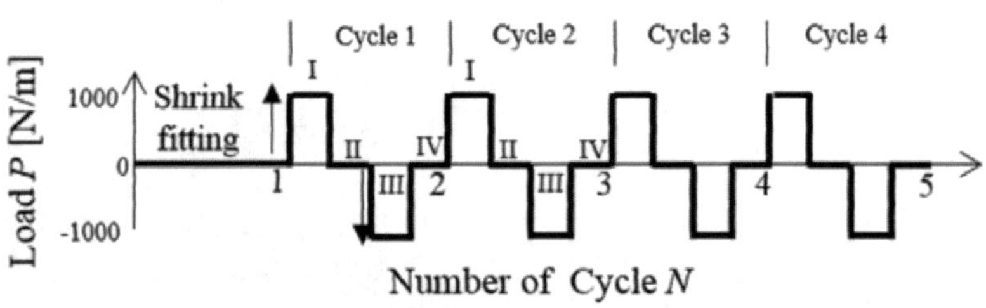

Figure 5. New alternate loading pattern with intervals.

To express the roll rotation, the point load is shifted in the circumferential direction in the 3D roll simulation [4]. This load shifting is replaced by the alternate loading in the 2D simulation [5]. Here, the 2D model has a unit thickness which corresponds to 120 mm width and the standard load 18 KN is represented by 150 N/mm considering the diameter of the shaft.

In this simulation, room temperature is assumed where the coming out occurs more easily. Then, static structural analysis is performed to the models by using MSC. Marc/Mental 2012 with the Full Newton-Raphson iterative sparse solver of multifrontal method [18]. A half model is considered due to the symmetry. Quadrangular contact element under axial symmetry has been used, the number of elements is 24546. Considering analysis accuracy, a minimum element size is 0.15625 mm×0.15625 mm around the part comes into contact with stopper. The shrink fitting ratio δ/d is defined as the height difference δ divided by the inner plate height d = 240 mm. Then, the shrink fitting ratio δ/d = 0.2×10^{-3}, friction coefficient μ = 0.3, load P = 1000 N/mm are used as reference values.

3. Residual displacement of the inner plate

In this section, Point D in figure 2 is focused on representing the contact part. To find out the contact condition between the inner plate and the stopper, the coming out is represented by the displacement u_{xD} in the x-direction at Point D under the alternate loading in figure 6(a).

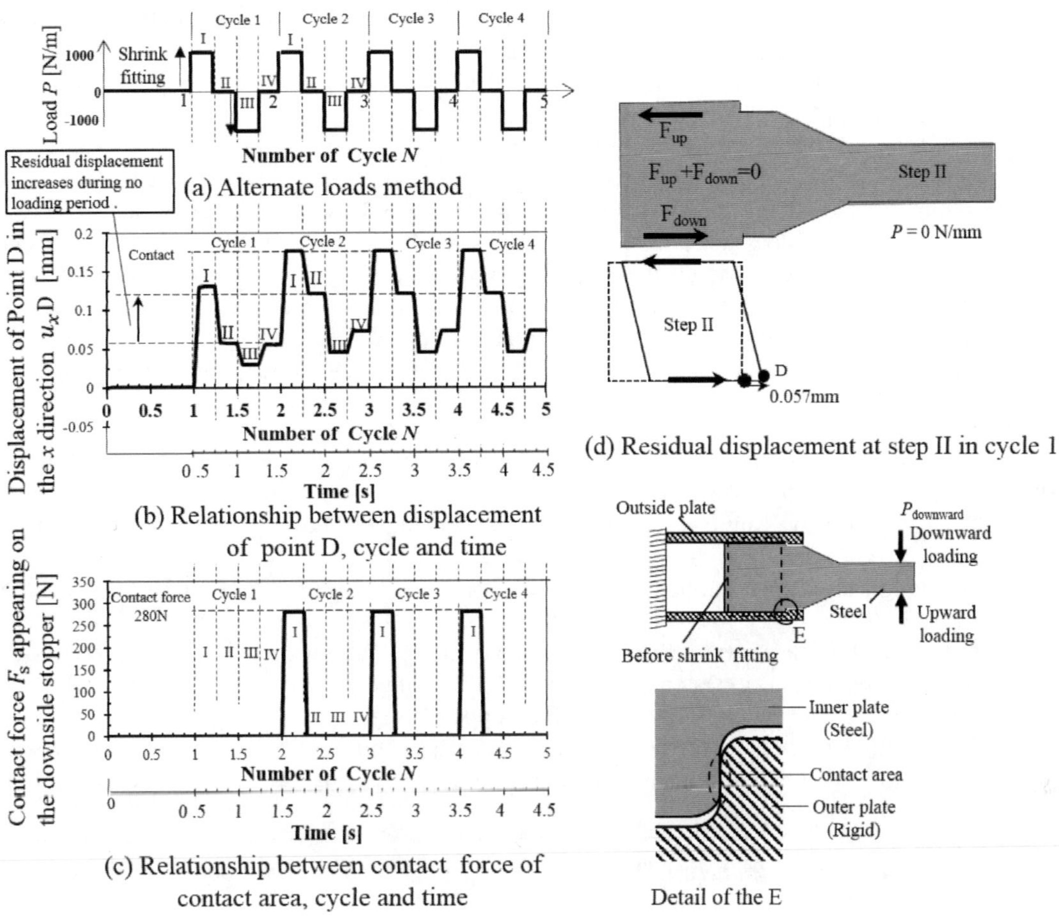

Figure 6. Displacement of Point D and contact force of the downside stopper vs loading time when load P = 1000 N/mm under new alternate loading pattern with intervals.

Figure 6(b) shows the displacement u_{xD} in the x-direction at Point D when load P = 1000 N/mm is repeated 5 times. From the 2nd cycle I, the displacement u_{xD} becomes stable independent of the cycles with the maximum displacement u_{xD} = 0.176 mm at Step I. Note that u_{xD} = 0.176 mm is smaller than the original gap 0.2 mm. In the case of dealing with the contact problemp [19] as in this research, it is necessary to take as much as 1/2 or more of the fitting value as the contact determination amount for calculation, $\delta/d = 0.2 \times 10^{-3}$, the inner diameter d = 240 mm, since $\delta = 0.2 \times 10^{-3} \times 240 = 0.048$, $0.048 \times 1/2 = 0.024$ mm is taken as the contact determination amount. Therefore, in this simulation, u_{xD} = 0.176 mm (0.2 mm - 0.024 mm) can be regarded as the contact distance between the inner plate contact with the stopper.

Figure 6(c) shows the contact force appearing at the stopper under the alternate loading. It can be seen that the inner plate end (part E in figure 6) is in contact with the stopper from the 2nd cycle I. Then, the resultant contact force becomes stable as shown in figure 6(c) independent of the cycles with

IOP Conf. Series: Materials Science and Engineering **372** (2018) 012020 doi:10.1088/1757-899X/372/1/012020

the value of 280 N. If there is no stopper, the shaft will gradually come out as discussed in the previous papers [4,5]. After the coming out is prevented by the stopper, a constant amplitude of the displacement with the contact force can be obtained as shown in figure 6.

During every step II and step IV, the contact force is zero but the displacement is neither zero nor the maximum. During step II, no load is applied after the upward load is removed and also during step IV, no load is applied after the downward load is removed. The displacement during those periods can be regarded as a residual displacement. Therefore, figure 6(b) indicates that the residual displacement increases from u_{xD} = 0.057 mm in cycle 1 to u_{xD} = 0.121 mm in cycle 2, then becomes stable from cycle 2 after the inner plate is in contact with the stopper.

Figure 6(d) shows the detail of the residual displacement at step II in cycle 1. During this no loading step in cycle 1, it should be noted that shear forces F_{up} and F_{down} are balanced. Here, F_{up} and F_{down} represent the shear forces applied along the upside and downside of the inner plate. Therefore, although the coming out does not progress during this no-loading step II, the residual displacement appears due to this balanced frictional forces. During step II in cycle 2, the residual displacement accumulates and increases until the inner plate contact with the stopper. The accumulation of the residual displacement can be regarded as a coming out process. In other words, the coming out phenomenon can be explained from the residual displacement accumulation during the intervals.

4. Conclusions

In this study, the finite element method was applied to simulating the coming out behavior of the shaft which was connected to the sleeve by shrink fitting. To clarify the coming out mechanism, a new alternate loading pattern was applied where intervals was included. The conclusion can be summarized in the following way.

- The design of stopper and the newly loading pattern is helpful to understand the coming out problem.
- The residual displacement can be seen in the intervals that were newly considered during the alternate loading. The residual displacement accumulates and increases by repeating the loading cycle until the shaft contacts to the stopper.
- The accumulation of the residual displacement causes the shaft coming out. In other words, the coming out phenomenon can be explained from the residual displacement accumulation.

References

[1] Fujii M, Yoshida A, Ishimaru J, Shigemura S and Tani K 2006 *Trans. Jpn. Soc. Mech. Eng.* C. **72** 1354-60
[2] Liour C, Mori T, Kobayashi H and Mitamura T 1990 *JCS-Japan.* **98** 348-54
[3] Ono T 1983 *J. Jpn. Soc. Mech. Eng.* **86** 470-5
[4] Noda N, Suryadi D, Kumasaki S, Sano Y and Takase Y 2015 *Eng. Fail. Anal.* **57** 219-35
[5] Noda N, Xu Y, Sano Y and Takase Y 2016 *ISIJ Int.* **56** 303-10
[6] Zhang G, Noda N, Sano Y and Sakai H 2017 *12th Int.* Conf. *on Damage Assessment of Structures (Kitakyushu)* **842** (Bristol/IOP Publishing Ltd) 012051 p1
[7] Li W, Noda N, Sakai H and Takase Y2015 *J. Solid Mech. Mater. Eng.* **5** 14-24
[8] Li W, Noda N, Sakai and Takase H 2011 *Key Eng. Mater.* 452-453, 241-4
[9] Rusin A, Nowak G and Piecha W 2013 *Eng.Fail. Anal.* **34** 217-27
[10] Noda N, Chen X, Sano, Y, Wahab M, Maruyama H and Fujisawa R 2016 *Mater. & Des.* **96** 476-89
[11] Ogawa E, Shimizu K, Hamayoshi S, Kumagai N, Ohtsubo Y, Noda N, Takase Y, Kishi K, Shobu K, Tabaru T, Maeda E, Koga S and Matsuda T 2012 *Hitachi Metals Tech. Rev* **28** 50-6
[12] Tsuyunaru M, Noda N, Hendra and Takase Y 2008 *Trans. Jpn. Soc. Mech. Eng. A* **74** 919-25
[13] Noda N, Hendra, Takase Y and Tsuyunaru M 2008 *J. Solid Mech. Mater Eng.* **2** 1410-9
[14] Noda N, Yamada M, Sano Y, Sugiyama S and Kobayashi S 2008 *Eng. Fail. Anal.* **15** 261-74

[15] Noda N, Hendra, Oosato M, Suzumoto K, Takase Y and Li W 2011 *Key Eng. Material.* **462** 1140-5
[16] Matsuda S, Suryadi D, Noda N, Sano Y, Takase Y and Harada S 2013 *Trans. Jpn. Soc. Mech. Eng.* A **79** 989-99
[17] Noda N, Suryadi D, Matsuda S, Sano Y and Takase Y 2015 *ISIJ Int.* **55** 2416-25
[18] Marc Mentat team 2008 *Theory and User Information* Tokyo MSC.Software A 713-4
[19] Marc Mentat team 2008 *Theory and User Information* Tokyo MSC.Software A 572-3

IOP Conf. Series: Materials Science and Engineering **372** (2018) 012021 doi:10.1088/1757-899X/372/1/012021

Structural analysis and design for the development of floating photovoltaic energy generation system

S J Yoon[1], H J Joo[2] and S H Kim[1,3]

[1]Department of Civil Engineering, Hongik University, 94 Wausan-ro, Mapo-gu, Seoul 04066, Korea
[2]ISIS E & C, 750 Gukhoe-daero, Yeongdeungpo-gu, Seoul 07236, Korea

E-mail: sunnys82@hanmail.net

Abstract. In this paper, we discussed the structural analysis and design for the development of floating photovoltaic energy generation system. Series of research conducted to develop the system from the analysis and design of the structural system to the installation of the system discussed. In the structural system supporting solar panels PFRP materials and SMC FRP materials used. A unit module structure is fabricated and then the unit module structures are connected each other to assemble whole PV energy generation complex. This system connected directly to the power grid system. In addition, extensive monitoring for the efficiency of electricity generation and the soundness of the structural system is in progress for the further system enhancement.

1. Introduction

Fossil fuel is one of the major energy resources used around the world nowadays. However, over consumption of fossil fuel leads to environmental issues. In 2012, Korea adopted the renewables portfolio standard (RPS) to promote the development, use, and diffusion of new and renewable energy. The RPS requires the production of energy from renewable energy resources such as wind, solar, biomass, geothermal, etc [1].

According to the literature review conducted in the study, various types of floating photovoltaic (PV) energy generation systems installed on the idle water surface of dam, reservoir, and sea in USA, Europe, and Asia [2]. For the analysis and design of the system, materials to use, methods to design, design requirements, structural configurations, mooring system including buoy, etc should be predetermined. For the structural system, the glass fiber reinforced polymer plastic (FRP) members manufactured by the pultrusion process and specially produced L-shaped member manufactured by the sheet molding compound (SMC) process are used, respectively [3]. Because of the durability and the construction cost due to lightweight, glass fiber reinforced polyester composite (FRP) structural members having high corrosion resistance and high specific strength and stiffness selected. In general, the pultruded FRP material is considered as an orthotropic material (transversely isotropic more strictly) in the structural analysis and design.

2. Mechanical properties of FRP members

For the FRP material, various manufacturing techniques are developed. In the construction industries, pultrusion process is one of the most cost effective manufacturing processes because of its mass production capability. In general, the mechanical properties of FRP materials are closely related to the material production processes. Prior to analysis and design of structural system, manufacturing process of the FRP member and its mechanical properties need to determine. Figures 1 and 2 show some of manufacturing processes.

Figure 1. Pultrusion process.

Figure 2. Continuous filament (tape) winding process.

2.1. Manufacture of pultruded structural members
The FRP members to fabricate solar panels supporting structural system produced by the pultrusion process. For fabricating the structure to support the solar panels, L-shaped member produced by the SMC process also used.

2.2. Mechanical property of FRP
To determine the mechanical properties of the FRP structural I-shape members and SMC L-shape members, coupon specimens have taken along the reinforcing fiber direction and along the transverse direction, respectively. Uniaxial tension test on the coupon specimens is conducted and the load is applied up to the point of specimen failure. Average test results are given in tables 1 and 2, respectively [3].

Table 1. Material property of pultruded FRP (PFRP) [3].

Location	Tensile strength [MPa]	Modulus of elasticity [GPa]	Poisson's ratio [mm/mm]
Flange	385.01	34.23	0.34
Web	451.52	31.04	0.28

Table 2. Material property of SMC (FRP) [3].

Tensile strength [MPa]	Modulus of elasticity [GPa]	Poisson's ratio [mm/mm]
80.00	14.33	0.25

3. Analysis and design

3.1. Structural analysis
In the analysis of structural system supporting the solar panels and resting on the buoys, the finite element method (FEM) is used. FEM software, MIDAS CIVIL 2012 [4], is used for the structural analysis. In addition, GTSTRUDL version 29 [5] is also used for the analysis. Because the structural system with buoys rests continuously on the water surface with hogging and sagging movement, the fluid-structure interaction (FSI) analysis also performed by using ADINA system finite element

software [6]. From the structural analysis by the finite element analysis (FEA), the maximum stress and the location where the maximum stress occurs in the member found. Loads for the analysis is consisted of the wind load, snow load, load induced by the mooring system, and self-weight of the structural system including weight of workers during construction and maintenance of the system in service.

3.2. Structural design

For the design of floating photovoltaic energy generation structural system relevant specification need to be developed. In USA pre-standard specification is developed and submitted to American Composite Manufacturers Association in 2010. This design method is based on the load and resistance factor design method (LRFD) which may be considered as one of the limit state design methods (LSD) (refer to figure 3). Eurocomp design code and handbook published in 1996 is also available at present (refer to figure 4). In addition, FRP composite manufacturers also developed and continuously revised the design manual, mostly based on the allowable design method (ASD), for their own pultruded FRP (PFRP) materials as shown in figures 5 and 6, respectively.

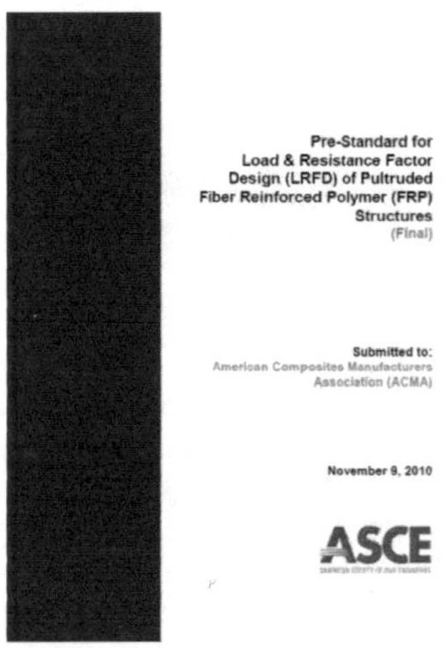

Figure 3. ASCE pre-standard (LRFD).

Figure 4. Eurocomp design code and handbook.

Figure 5. Extren design manual (Strongwell).

Figure 6. Pultex pultrusion design manual (Creative pultrusions).

IOP Conf. Series: Materials Science and Engineering **372** (2018) 012021 doi:10.1088/1757-899X/372/1/012021

4. Connection system

For the pultruded structural members, connection of members for fabricating the structural system is important. Unlike the connection of steel or aluminium members, because of welding of FRP member is not possible, bolting, gluing, gluing with bolting may only be applicable. In the study, bolted connection of PFRP members for the fabrication of structural system adopted based on the test results conducted by Lee *et al* [7-9] and Woo *et al* [10].

4.1. Member connection

In the fabrication of structural system supporting the solar panels (unit module), bolted connection between the FRP members with gusset plates used as shown in figure 7. Stainless steel bolts, nuts, and washers also used as shown in figure 7. To reduce the number of connection points, L-shaped member produced by the SMC process also used as shown in figure 8.

Figure 7. Gusset plate (PFRP angle) [11].

Figure 8. L-shape member produced by the SMC process [3].

4.2. Module connection

In the design of floating PV energy generation structural system, a unit module structure is designed, and then the unit modules are connected each other by C-shape connection devices to assemble the floating PV generation complex (refer to figures 9 and 10 [2]). By this way, we can reduce the stress in structural member in the unit module. In addition, it leads to easy of handling during fabricating, launching, and towing at the job site. Moreover, it is also possible to reduce the construction cost in terms of PFRP material, equipment, and labor cost.

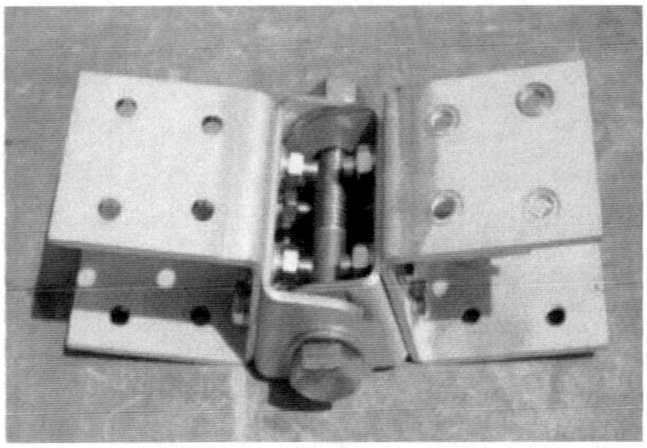

Figure 9. C-shape device [2].

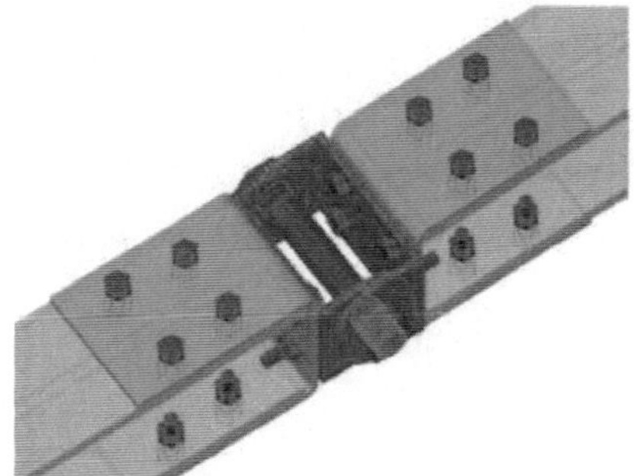

Figure 10. Module connection with C-shape device [2].

IOP Conf. Series: Materials Science and Engineering **372** (2018) 012021 doi:10.1088/1757-899X/372/1/012021

5. Buoy and mooring system

5.1. Buoy

In the development of the system, two types of buoy manufactured considering the cost and durability. Expanded polystyrene (EPS) is filled inside the polyethylene (PE) drum as shown in figure 11. This type of buoy is cheaper but not durable. Filament winding FRP composite drum (FFRP), made by using E-glass fiber and polyester resin, filled inside the drum with expanded polystyrene also produced as shown in figure 12. This buoy is highly durable but expensive compared with buoy made by PE drum.

Figure 11. Buoy (PE).

Figure 12. Buoy (FFRP).

5.2. Mooring system

Planning and designing of mooring system are important because it directly related to the construction cost. In the design, it is necessary to consider the change of water level, the environmental condition of location to install, etc. In addition, the loads induced by the mooring system need to consider in the design of structural system as mentioned earlier. Figure 13 shows one of mooring systems used in the study.

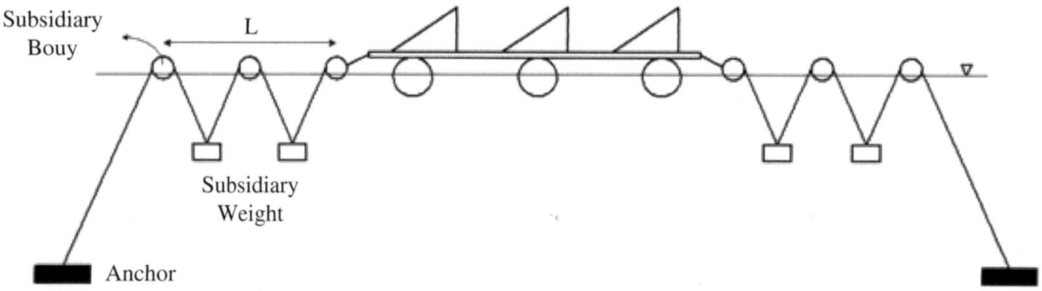

Figure 13. Mooring system.

6. Efficiency of energy generation

Comparison of yearly energy generation between land-fixed type and tracking type floating photovoltaic energy generation system conducted. Summary of comparison result is in table 3.

In the development and installation of the system, energy generation efficiency evaluated by comparing the system installed on the ground as given in table 4. The system installed on the water is slightly more efficient than the system installed on the ground.

Table 3. Comparison of energy generation efficiency [3].

Installation site	Type of PV energy generation	Energy generation capacity [kW]	Yearly energy generation [MWh]	Operation ratio [%]
Samcheon-po, Korea	Land fixed	100.00	113.03	12.90
Hapcheon, Korea	Floating with tracking	100.00	125.00	14.27

Table 4. Comparison of monthly energy production between the floating type and land type system [9].

Month	Floating (0.93 kW) ①	Land (20 kW)②	(20kW/0.93kW)③ (②/①)	①/③
June (21-30)	20.4	313.0	15.34	1.33
July	52.6	931.0	17.70	2.97
August	55.8	1042.0	18.67	2.99
September	46.5	1124.0	24.17	1.92
October (1-10)	12.2	397.0	32.54	0.37

7. Installation site

Figure 14 shows a unit module of floating type photovoltaic energy generation structure. The structure installed at the Tongyeong bay located in the south coast of Korea [1]. A tracking-type floating photovoltaic energy system with 100kW generation capacity installed as shown in figure 15. The structure installed at the Hapcheon dam in Korea. It composed of four unit modules with 24.8 kW generation capacity per each unit module [1]. Figure 16 shows 1MW class floating photovoltaic energy generation structure at cooling water intake channel of power plant in Dangjin-si, Chungcheongnam-do, Korea [1]. The floating photovoltaic energy generation structural system is installed at Otac reservoir, Sangju-si, Gyeongsangbuk-do, Korea (refer to figure 17) [1].

The generation systems installed at different locations showed here for illustration purposes. However, many of similar structural system installed at different sites and they are in service. The electricity generated by the system supplied directly to the power grid system of Korea [1].

Figure 14. Tongyeong bay.

Figure 15. Hapcheon dam.

Figure 16. Dangjin power plant.

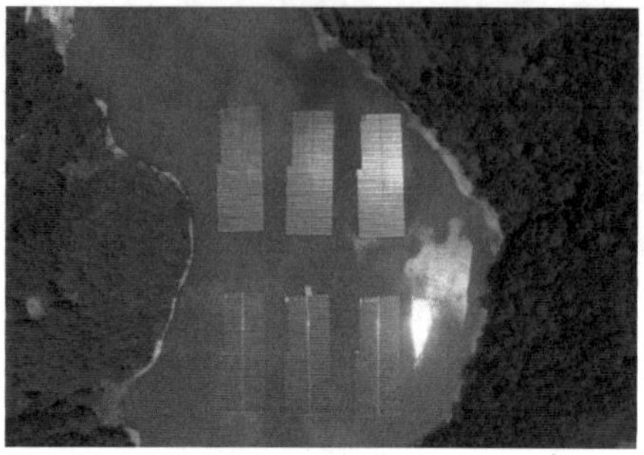

Figure 17. Otae reservoir.

8. Conclusion

In this paper, series of research conducted to develop the floating PV generation system discussed from the analysis and design to the installation. Materials used to fabricate the structural system supporting the solar panels are PFRP profile and SMC FRP members. Mechanical properties of the members determined by the coupon specimen tests. In the fabrication of the structuring system to support the solar panels, a unit module structure designed and fabricated, and then the unit module structures are connected each other to assemble the whole PV generation structural complex.

The floating PV generation system directly connected to the power grid system so that the electricity generated by the system can supply. Extensive monitoring for the efficiency of electricity generation and the behavior of structural system is in progress for further development and structural enhancement.

References

[1] Kim S H, Yoon S J, Choi W C and Choi K B 2016 Application of floating photovoltaic energy generation systems in South Korea *Sustainability* (Basel: MDPI) p 8

[2] Kim S H, Yoon S J and Choi W C 2017 Design and construction of 1 MW class floating PV generation structural sysem using FRP members *Energies* (Basel: MDPI) p 10

[3] Yoo J H, Kim S H, An D J, Choi W C and Yoon S J 2017 Generation efficiency of tracking type floating PV energy generation structure using fiber reinforced polymer plastic (FRP) members *Key Engineering Materials* (Zurich: Trans Tech Publications) **730** pp 212-7

[4] MIDAS/Civil 2012 Analysis Reference MIDAS Information Technoloty Co., Ltd

[5] GTSTRUDL 2009 GTSTRUDL User's Manual, Latest Revision, K, May, GTICS System Laboratory, Georgia Institute of Technology, Atlanta, Georgia

[6] ADINA R&D, Inc. 2009 ADINA Theory and Modeling Guide Volume III, ADINA CFD & FSI. Watertown, Massachusetts

[7] Lee Y G, Lee S, Nam J H and Kim H T 2010 Strength evaluation of bolted connection in PFRP structural member *Materials Science Forum* (Zurich: Trans Tech Publications) **654-656** pp 2640-3

[8] Lee Y G, Kim S H, Won Y S, Cheon J U, Shin K Y and Yoon S J 2014 Strength evaluation of bolt arrangement in PFRP bolted connection with 2 bolts *J Korean Soc Adv Comp Struc* (Seoul: KOSACS) **5** 17-22

[9] Lee Y G, Choi E S and Yoon S J 2015 Effect of geometric parameters on the mechanical behavior of PFRP single bolted connection *Composites Part B* (Amsterdam: Elsevier) **75** pp 1-10

[10] Woo S P, Kim S H, Yoon S J and Choi W C 2017 Effect of bolt-hole clearance on bolted connection behavior for pultruded fiber-reinforced polymer structural plastic members *Int J of Polymer Science* (London: Hindawi) p 12

[11] Lee Y G, Joo H J and Yoon S J 2014 Design and installation of floating type photovoltaic energy generation system using FRP members *Solar Energy* (Amsterdam: Elsevier) pp 13-27

IOP Conf. Series: Materials Science and Engineering **372** (2018) 012022 doi:10.1088/1757-899X/372/1/012022

Delamination of plasters applied to historical masonry walls: analysis by acoustic emission technique and numerical model

A Grazzini[1,2], G Lacidogna[1], S Valente[1] and F Accornero[1]

[1]Department of Structural, Geotechnical and Building Engineering, Politecnico di Torino, Corso Duca degli Abruzzi 24, Turin, Italy

E-mail: alessandro.grazzini@polito.it

Abstract. Masonry walls of historical buildings are subject to rising damp effects due to capillary or rain infiltrations, which in the time produce decay and delamination of historical plasters. In the restoration of masonry buildings, the plaster detachment frequently occurs because of mechanical incompatibility in repair mortar. An innovative laboratory procedure is described for test mechanical adhesion of new repair mortars. Compression static tests were carried out on composite specimens stone block-repair mortar, which specific geometry can test the de-bonding process of mortar in adherence with a stone masonry structure. The acoustic emission (AE) technique was employed for estimating the amount of energy released from fracture propagation in adherence surface between mortar and stone. A numerical simulation was elaborated based on the cohesive crack model. The evolution of detachment process of mortar in a coupled stone brick–mortar system was analysed by triangulation of AE signals, which can improve the numerical model and predict the type of failure in the adhesion surface of repair plaster. Through the cohesive crack model, it was possible to interpret theoretically the de-bonding phenomena occurring at the interface between stone block and mortar. Therefore, the mechanical behaviour of the interface is characterized.

1. Introduction

The historical masonry walls have original plasters often in phase of separation because of climatic conditions as rain infiltrations and damp capillary action. The restoration work needs the application of new dehumidified mortars. The great variety of historical masonry textures requires preliminary tests to assess the structural compatibility of the dehumidified mortars suitable for brick surfaces [1].

The authors have developed a new experimental methodology to assess the mechanical adhesion of new dehumidified mortars applied to historical masonry surfaces. This experimental methodology was carried out at the Politecnico Laboratory, and tested at the Sacro Monte di Varallo, one of the most famous UNESCO heritage site in Italy [2]. This historical site, whose construction has been extended from the 15th to the 18th century, contains 45 chapels that show the Christ's life by particular frescoes and sculptures (figure 1). The main problem of historical plasters' preservation is linked to the rain infiltrations and to the freezing-thawing cycles that compromise their adhesion to stone masonries. Through compression static tests on stone-mortar composite specimens, it was possible to prequalify the dehumidified mortar with greater durability guarantees. Sometime the long-term detachment determined the failure of the restoration work. The test validates the mechanical compatibility between the repair mortars and historical masonries stone, as a guarantee also for durability over time against atmospheric agents. The Acoustic Emission (AE) monitoring technique was employed during static tests for assessing damage evolution inside the adherence surface [3,4].

Figure 1. Chapel of the Christ at the court of Herod.

2. Materials and experimental setup

For carrying out tests, a particular geometry of the composite specimen has been proposed. Four mixed specimens were made, applying a 40 mm mortar layer to both shorter faces of the stone brick suitably cut with the dimensions shown in figure 2(a). The mechanical characteristics of the repair mortar and the stone brick were chosen equivalent to those of the stone masonry at the Sacro Monte di Varallo. The hydraulic lime mortar was a pre-blended transpiring product, for dehumidifying repair work on historical masonry damaged by dampness. For this mortar the Young's modulus was 7000 MPa, the compressive strength was 33.8 MPa.

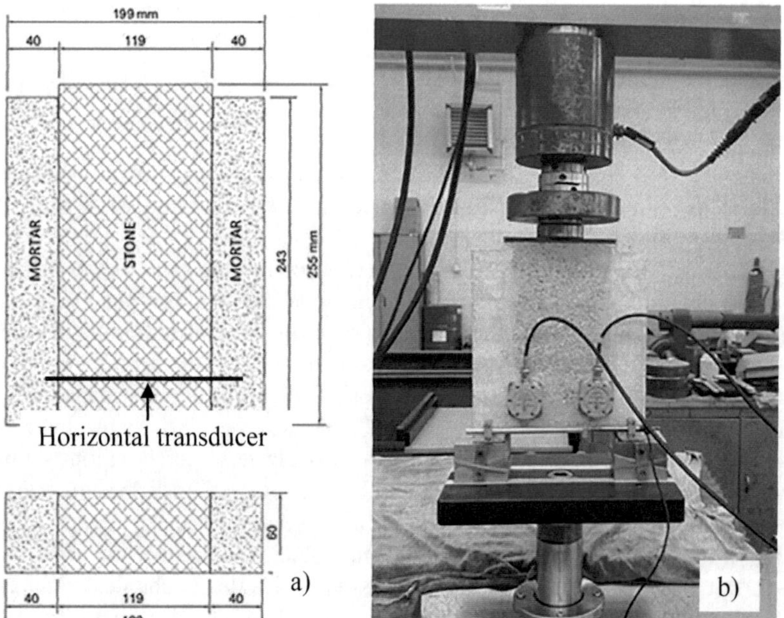

Figure 2. Geometry of composite specimen (a) and test setup (b).

The surface of the stone block has been treated specifically by means of a drill to facilitate the mortar's adhesion. This treatment aimed to simulate the real discontinuities on the wall surface that favor the adhesion of the plaster. The application of the dehumidified mortar took place leaving specific and surface symmetrical discontinuities at the top and bottom of the specimen, as showed in figure 2(a). These notches favored the trigger and propagation of multiple cracks, in order to test the adhesion of two linked materials. An inductive horizontal displacement transducer was applied at the bottom of the specimen for the bulging displacements (figure 2(b)). The vertical displacements were recorded by the piston's translation of the 250 kN servo controlled test machine. The composite specimens were tested by monotonous compression load by horizontal controlled opening. Static tests have been performed after 28 days of maturation. The composite specimen rested, through the side layers mortar, on a double system of steel wedges (figure 3). The wedges were coupled by a 1 mm thick Teflon layer for reducing the horizontal friction during the plaster's expansion. The "SM" (Stone brick-Mortar) label has been associated with each specimen with its sequence number.

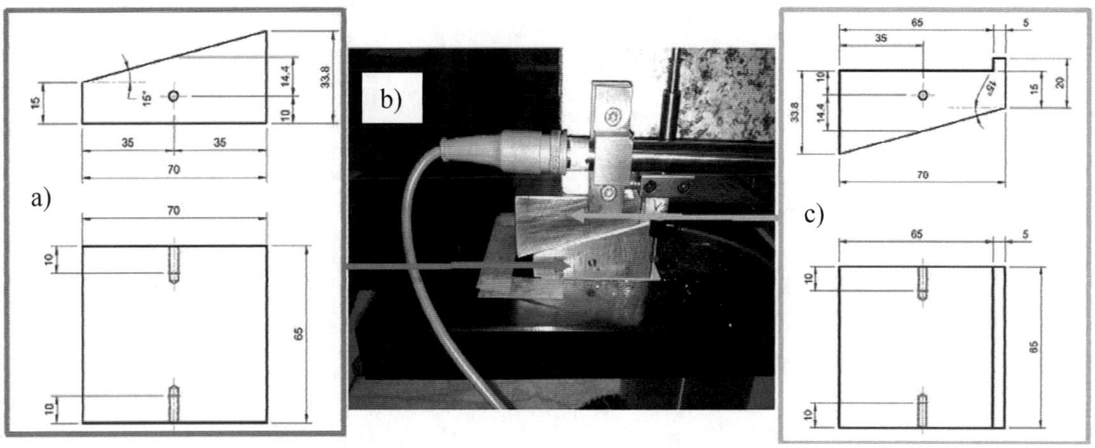

Figure 3. The wedges' geometry (a – c) and test setup (b).

3. Acoustic emission monitoring

As regards the AE monitoring, the microcracks propagation generated AE signals that were detected by dedicated sensors applied to the external specimens surface. The AE waves were amplified with a gain of 60 dB before they have been analyzed, fixing the threshold level detection up to 2 mV. The AE sensors were attached to the specimens by means of a silicon glue. In this way it was possible to guarantee a good contact between the sensor and the specimen, also during the final crack phases.

The AE measurement system was able to recorder each signal waveforms. Rise time and peak amplitude were the AE parameters used to classify active cracks. The shape of the AE waveforms distinguishes the fracture mode: typical shear events are usually characterized by long rise times and high amplitudes, instead low rise time are distinctive of tensile crack propagations. In particular the rise angle (RA), that shows these conditions, was calculated as the ratio of the rise time to the peak amplitude. The first is expressed in ms, the second one in V [3].

Moreover, in order to distinguish the cracking mode, the Average Frequency (AF) expressed in kHz was also used. The AE ringdown count is the number of crossings threshold along the duration time. The ratio between the AE ringdown count and the duration time of each signal gives the AF values. The change of the cracking mode from tensile to shear is followed by the shift from higher to lower values of AF [3]. Nevertheless, when large cracks (Mode I) are forming, the frequency attenuation could be a function of this discontinuity. The wavelength of the AE signals should be larger of these discontinuities to make the waves pass through, therefore the frequencies' shift from higher to lower values could show also a dominant tensile cracking mode.

For the sake of synthesis, the results of the AE monitoring were presented only for specimen SM4.

IOP Conf. Series: Materials Science and Engineering **372** (2018) 012022 doi:10.1088/1757-899X/372/1/012022

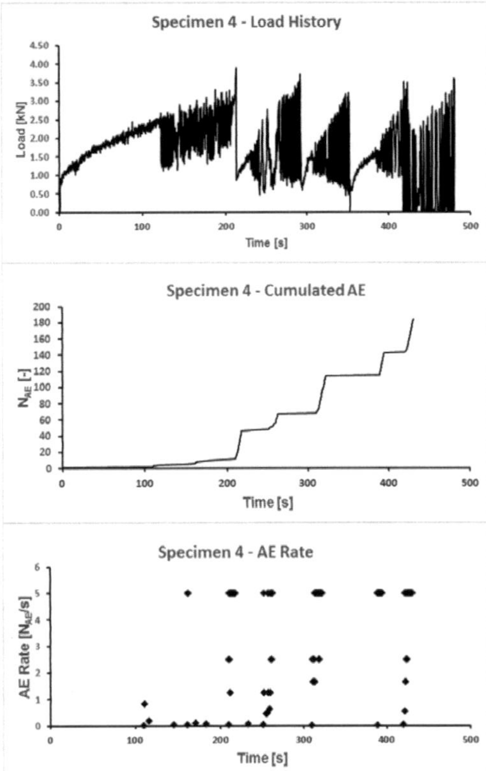

Figure 4. Specimen SM4: Load vs time diagram, Cumulated AE, and AE rate.

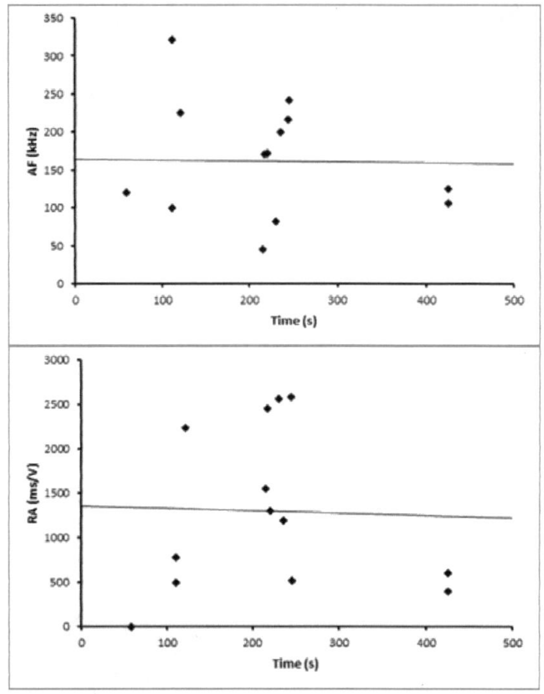

Figure 5. Specimen SM4: Average Frequency (AF) vs time, and Rise Angle (RA) vs time.

The load vs. time diagram of the specimen SM4 is shown in figure 4, together with AE cumulated curve, and AE rate. During the first part of the test, when the load increases proportionally over time, there are few AE signals. However, a clear growth in the AE hits is obtained in the correspondence of each sharply decrease of the load vs. time curve. This supports the fact that the AE signals are mainly associated with the energy emitted by the specimen during the delamination of the mortar from the stone block (snap-back instabilities) [5].

Using the ordinary least squares method, the linear regression of the signal frequencies (AF) and the signals rise angles (RA) were traced during the whole test. Considerable variations from the mean trend were observed in particular immediately after 100 seconds from the beginning of the test, around 200 seconds, and around 420 seconds towards the end of the test (figure 5). In the first phase, from 0 to 120 seconds, rise angles with low values prevail, while the frequencies are oscillating with the higher values more distant from the average trend. This indicates a prevalent Mode I in the delamination process in the initial loading phase. In the second phase, the highest values of the RAs during the whole test were obtained, while the frequencies continue to oscillate with the lowest values more distant from the average. This behavior shows how the delamination process that leads to collapse develops with sudden stress drops and is mainly accompanied by the sliding of the mortar with respect to the stone block (Mode II). In the final phase, over 400 seconds, the frequencies are still lowered below the average line, while the RAs once again become low, so it can be said that there is no prevailing fracture mode before the specimen collapses definitively.

4. Numerical simulation

The Barenblatt – Dugdale - Hillerborg model, also known as the cohesive crack model, is a good numerical approach to simulate the behaviour of quasi-brittle materials, as the stone-mortar delamination. The crack initiation criterion is assumed as:

$$\left(\frac{\sigma_0}{f_t}\right)^2 + \left(\frac{\tau_0}{f_s}\right)^2 = 1 \tag{1}$$

where σ_0 and τ_0 are the stresses respectively orthogonal and tangential directions, f_t and f_s are the respective strengths. The fictional crack tip is the point where equation (1) is proved. The cohesive stresses present on the non-linear fracture process zone (FPZ) are decreasing functions of the effective value of the displacement discontinuity w_{eff} [6-11]. It was assumed:

$$w_{\text{eff}} = \sqrt{\left(\frac{w_n}{w_{\text{nc}}}\right)^2 + \left(\frac{w_t}{w_{\text{tc}}}\right)^2} \tag{2}$$

where w_n is the first component of the mutual-orthogonal displacement, and w_t is the second one (tangential to the interface); the related critical values are w_{nc} and w_{tc}. If $w_{eff} > 1$ there is no stress transfer and the crack is stress free; otherwise, the stresses are decreasing functions of w_{eff} that follow a pre-defined softening law. The above mentioned law is assumed [12]:

$$\frac{\sigma}{\sigma_0} = \frac{\tau}{\tau_0} = \left[1 - \frac{1-\exp(-\alpha \cdot w_{\text{eff}})}{1-\exp(-\alpha)}\right] \tag{3}$$

where $\alpha = 5$ is assumed. Real crack tip is called the point where $w_{eff} = 1$. Outside the FPZ the material had a linear-elastic behavior. The fracture process starts symmetrically, but loses this property afterwards [13-15] because of the propagation of the round-off errors. In this case the numerical simulation was controlled according a uniform downward velocity to the upper edge on the rock. This choice was able to prevent the growth of the round-off errors so that, from an engineering point of view, the propagation of the two cohesive cracks occurs symmetrically. The interface strength properties were assumed as shown in tables 1 and 2. The numerical analyses were performed using the ABAQUS [16] code. Figure 6 shows the deformed mesh of finite element model.

Table 1. The interface strength properties.

f_t (N/mm^2)	f_s (N/mm^2)	w_{nc} (mm)	w_{tc} (mm)
0.8	1.0	0,015	0,015

Table 2. Mechanical characteristics of materials.

	Young's modulus (N/mm^2)	Poisson ratio
Mortar	7000	0.15
Rock	35000	0.20

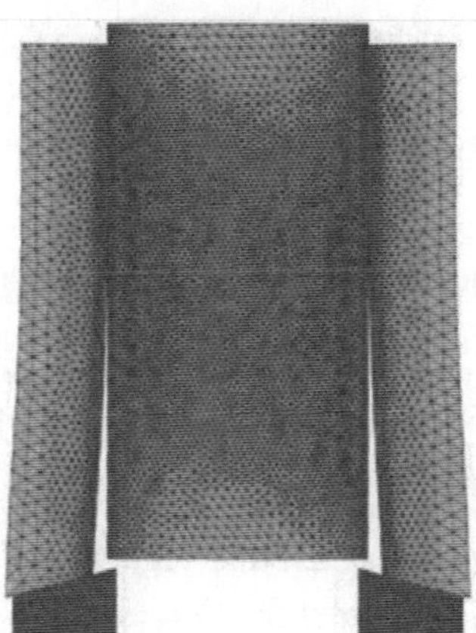

Figure 6. Finite element deformation during max displacement.

5. Numerical and experimental outcomes

The adherence between the repair plaster and masonry stone was tested by the notch tips. The experimental procedure simulated the fatigue loads that can compromise the service-life of a dehumidified plaster. The experimental results shows four stress singularity points in the stone-mortar specimens: two notch tips at the top, and two at the bottom (see figure 2). These are the weakest points involved in the singular stress fields. During the tests, it was possible to observe that the cracks beginning from the bottom were faster than the cracks from the top, because of the wedges. The friction, reduced by the teflon sheet, is not taken into account in the numerical simulation and the materials are considered homogeneous. This is the reason why the response curve is continuous (see figure 7). On the contrary, the properties of the rock specimen can change from point to point. This is the reason why the experimental curve shows many discontinuities (see figure 7). The numerical model represents an ideal behaviour which, in particular in the early stages of loading, fit the average values of the experimental curves, but it is not able to identify the snap-back instabilities that experimentally occur during the delamination process.

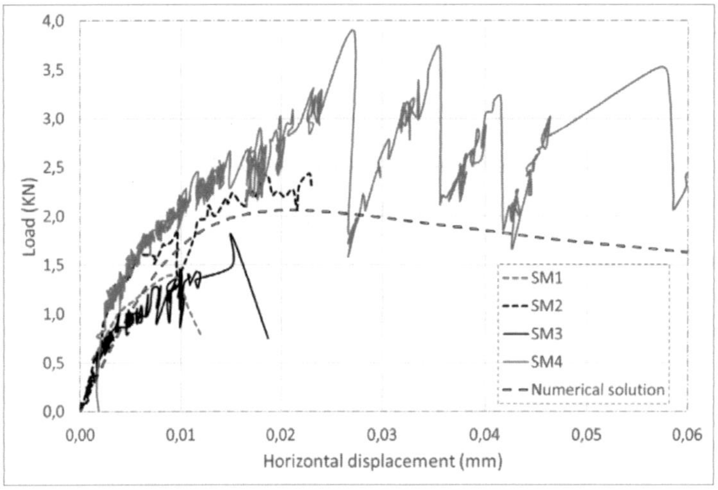

Figure 7. Load – horizontal displacement curves of the specimens vs. the numerical solution.

6. Conclusions

- The aim of this test was to pre-qualify the most durable dehumidified mortar which mechanical compatibility to the historical masonry can avoid delamination problems by dampness and hygrometric effects.
- The experimental procedure, performed by mixed specimens, allowed testing the adherence between the dehumidified repair plaster and the historical masonry stone. This procedure was able to produce in static tests the same interface stresses that caused by freeze and thaw actions.
- The AE signals were associated with the energy emitted by the specimens in the delamination phase between the mortar and the stone block. The variation of the AE parameters (AF and RA) depends strictly on the type of specimen damage.
- From the obtained results it was possible to identify a prevalent Mode I in the delamination process during the initial loading phases. As the test goes on, the delamination process develops with sudden load drops, mainly accompanied by the sliding of the mortar with respect to the stone block (Mode II). While, in the final loading phase, no prevalent fracture mode has emerged before the specimen definitively collapses.
- The non-linear phenomena that occur at the interface between mortar and stone during the de-bonding process were theoretically interpret by means of the cohesive crack model.

IOP Conf. Series: Materials Science and Engineering **372** (2018) 012022 doi:10.1088/1757-899X/372/1/012022

References

[1] Bocca P and Grazzini A 2013 *J Mat Civ Eng* **25** 274-80
[2] Bocca P, Valente S, Grazzini A and Alberto A 2014 *Int J Arch Her* **8** 336-48
[3] Carpinteri A, Lacidogna G, Accornero F, Mpalaskas F, Matikas A and Aggelis D 2013 *Cemente Concr Comp* **44** 9-16
[4] Carpinteri A, Grazzini A, Lacidogna G and Manuello A 2014 *Str Contr Health Monit* **21** 950-61
[5] Carpinteri A and Accornero F 2017 *Eng Fract Mech* doi: 10.1016/j.engfracmech.2017.11.034
[6] Cervenka J, Kishen J M C and Saouma V E 1998 *Eng Fract Mech* **60** 95-107
[7] Barpi F and Valente S 2010 *Eng Fract Mech* **77** 2182-91
[8] Barpi F and Valente S 2002 *Int J Num Anal Meth Geomech* **26** 1005-27
[9] Barpi F and Valente S 2004 *Int J Sol Str* **41** 2607-21
[10] Barpi F and Valente S 2005 *Eng FractMech* **72** 2427-43
[11] Barpi F and Valente S 2008 *Eng Fract Mech* **75** 629-42
[12] Cornelissen H A W, Hordijk D A and Reinhardt H W 1986 *Heron* **31** 45-56
[13] Barpi F and Valente S 1998 *Int J Sol Str* **35** 1851-61
[14] Alberto A, Antonaci P and Valente S *Proc. 11th Int. Conf. Mech Beh Mat* 1151-6
[15] Bocca P, Grazzini A, Masera D, Alberto A and Valente S 2011 *J Phys Conf Series* **305**
[16] Dassault System Simulia Corp., Providence, RI, ABAQUS, release 6.10, 2010

IOP Conf. Series: Materials Science and Engineering **372** (2018) 012023 doi:10.1088/1757-899X/372/1/012023

Finite element analysis of the axial stiffness of a ball screw

L-X Zhou[1] and P-Y Li[1,2]

[1]Department of Mechanical Engineering and Automation, Xi'an University of Technology, Xi'an, 710048, PR China

E-mail: lipengyang@xaut.edu.cn

Abstract. The ball screw was developed for high speed and high precision operation; therefore, increasingly greater demands have been placed on the stiffness of the ball screw. Firstly, ANSYS software was used to compare the axial stiffness of a single-nut and single-arc ball screw and a single-nut and double-arc ball screw when the spiral angle is not considered. On this basis, the model of a single-nut ball screw was established taking into consideration the spiral lead angle, and then the variations in displacement and stiffness when the ball screw pair was subjected to an axial force were determined. The axial contact stiffness of the double-nut ball screw pair, subject to a pre-tightening force, was analyzed, according to the above-mentioned steps. The simulation results demonstrated that under the same working conditions, the stiffness of the double-arc ball screw was larger by between 5～100 N/um than that of the single-arc ball screw. The spiral lead angle increased the axial stiffness of the ball screw pair, and the axial stiffness of the double-nut ball screw pair subject to a pre-tightening force was larger by between 790～1360 N/um than that of the axial stiffness of the single-nut ball screw pair.

1. Introduction

The ball screw pair is a critical component in the machine tool feeding system and other areas, due to its high precision and high stiffness. Therefore, it is of great theoretical and practical significance to study the stiffness characteristics between each joint surface in order to improve the static and dynamic performance of the ball screw pair and therefore the machining precision of the machine tool. Kamalzadeh A and Feng G-H *et al* researched the effect of the pre-stressing force and the gap between the contact pair on the overall precision of the ball screw using the experimental method [1,2]. However, the experimental process they used was complex and not conducive to the process. C-G Zhou and H-T Feng et al. analyzed the working condition of the ball screw when a pre-stressing force was applied [3], which was much closer to the real working conditions of the ball screw.

Considering the whole machine tool, B Fernandez-Gauna and T Liang analyzed the factors that influenced the vibration of the ball screw [4,5]. Sobolewski J Z and Garinei A *et al* analyzed the vibration of the ball screw, and carried out flaw detection on the ball screw through the vibration processes [6,7]. However, the displacement variation of the ball screw during the working process could not be quantitatively deduced. The point of least stiffness in the ball screw, i.e. the joint surface among the nut, lead screw, and the balls, was chosen to be studied and this paper has established three-dimensional models with and without the spiral lead angle, starting from the contact surfaces composed of a single ball and nut or lead screw in the joint part of a single-nut ball screw pair. ANSYS finite element software was used to analyze both models and the axial stiffness characteristics of the ball screw were studied. On this basis, taking into consideration the influence of the pre-stressing force on the stiffness of the ball screw, this paper analyzed the variations of the displacement and the stiffness of the double-nut ball screw pair, subject to a pre-stressing force, when the nut was subjected to an axial force. The principle diagram of the ball screw pair has been shown in figure 1.

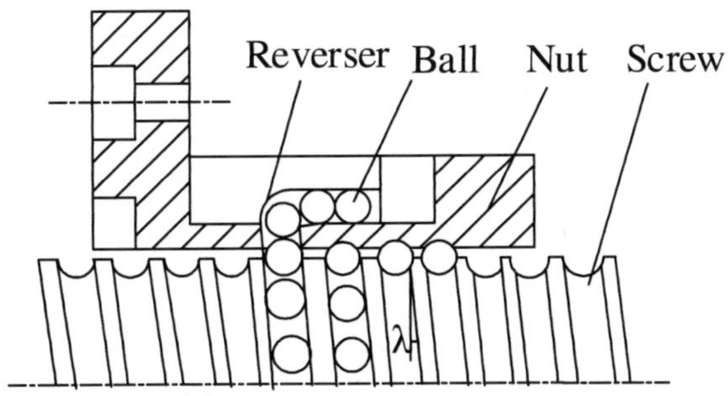

Figure 1. The principle diagram of the ball screw pair.

2. Analysis of axial stiffness without the spiral lead angle

2.1. Establishment of the three-dimensional model

The structural diagram of a nut can be depicted as in figure 2. Since the ball screw is an axisymmetric component, the stress condition of one ball in the ball screw can be taken separately for the analysis; the joint surface structure of a single ball has been shown in figure 3.

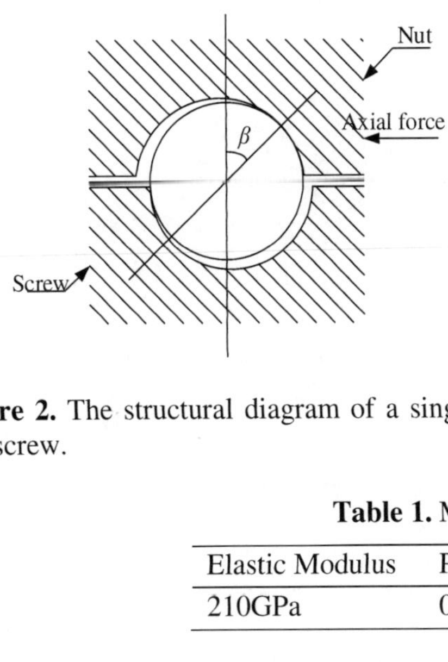

Figure 2. The structural diagram of a single-nut ball screw.

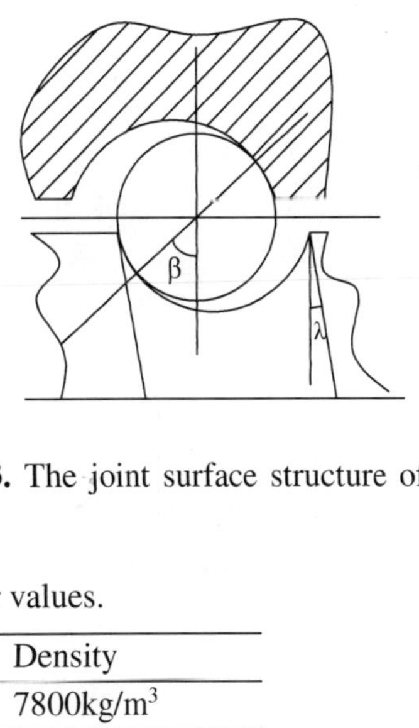

Figure 3. The joint surface structure of a single ball.

Table 1. Material parameter values.

Elastic Modulus	Poisson's Ratio	Density
210GPa	0.3	7800kg/m^3

Table 2. The dimension parameters of the ball screw.

Name	Value	Name	Value
Nominal Diameter of the Ball Screw d_0(mm)	40	Diameter of the Ball d_b (mm)	3.175
Lead Screw P_h(mm)	6	Curvature Ratio of the Raceway	1.04
Overall Length of the Ball Screw(mm)	480	Outer Diameter of the Nut d_n (mm)	63

The material parameters and dimension parameters of the ball screw have been shown in tables 1 and 2, respectively.

The structural diagram of single nut ball screw has been shown in figure 4.

The structure was analyzed without the spiral lead angle λ, and then the three-dimensional model was established, using the SOLIDWORKS software, as shown in figure 5.

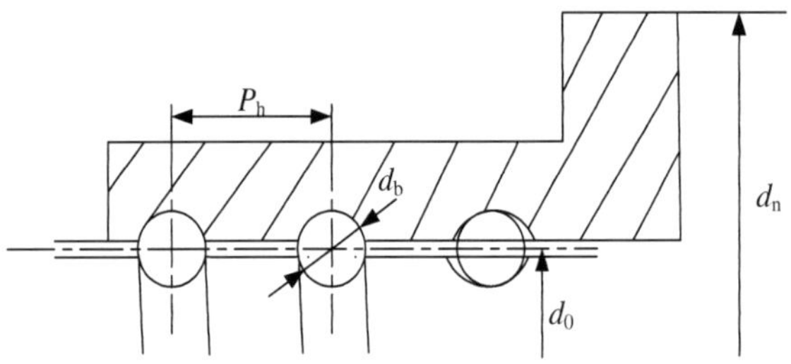

Figure 4. Structural diagram of a single nut ball screw.

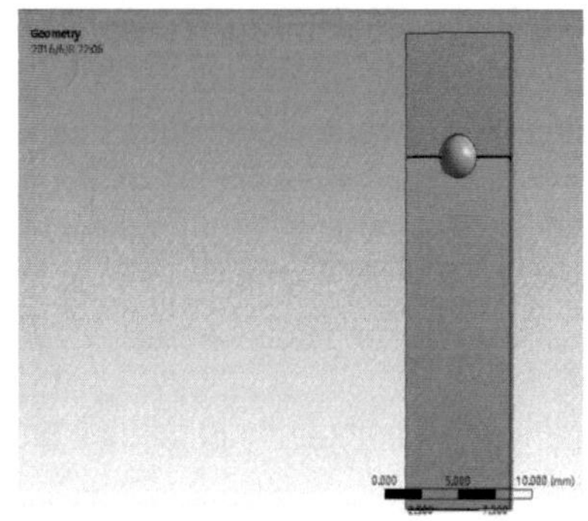

Figure 5. The structural diagram without the spiral lead angle.

2.2. ANSYS analysis process

In ANSYS, the two contact pairs of the ball-lead screw and the ball-nut were set up, respectively, as shown in figure 6. During the simulation process, the lead screw was fully constrained, the displacements for the nut in the Y and Z directions were constrained, and the displacement of the ball in the Z direction was constrained. A 500 N axial force was applied on the right side of the nut in the ball screw model, as shown in figure 7.

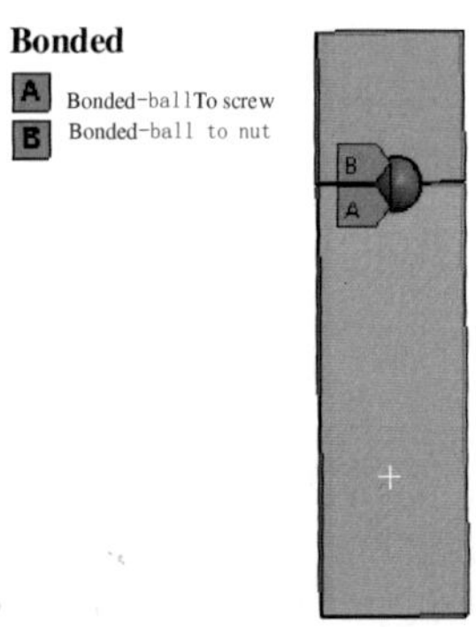

Figure 6. The establishment of the contact pair.

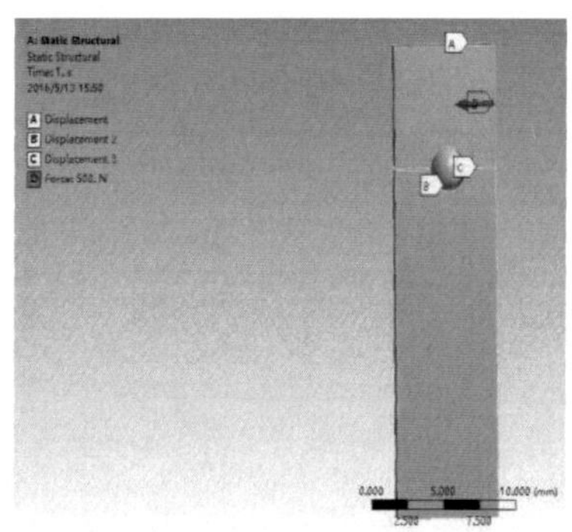

Figure 7. The addition of constraint conditions.

The variations in the axial displacement and the stress distribution of the ball screw have been shown in figures 8 and 9. The maximum stress of the ball screw pair appeared at approximately 45° from the horizontal direction, as can be seen in figure 9.

In this study an applied axial load of 500 N to 6000 N was selected, according to the Hertz point contact theory, the axial contact deformation expression for a single-nut ball screw without the spiral lead angle is:

$$\sigma a = 2.6992 \times 10^{-8} F^{2/3} (m) \quad (1)$$

$$K = \frac{F^{1/3}}{2.6992}(N/m) \tag{2}$$

where σ_a (m) is the axial contact deformation, *F*(N) is the axial force, and *K*(N/m) is the axial contact stiffness.

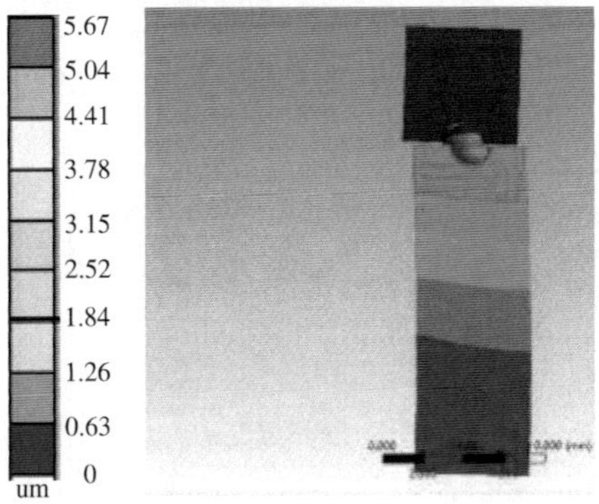

Figure 8. The variation of the axial displacement.

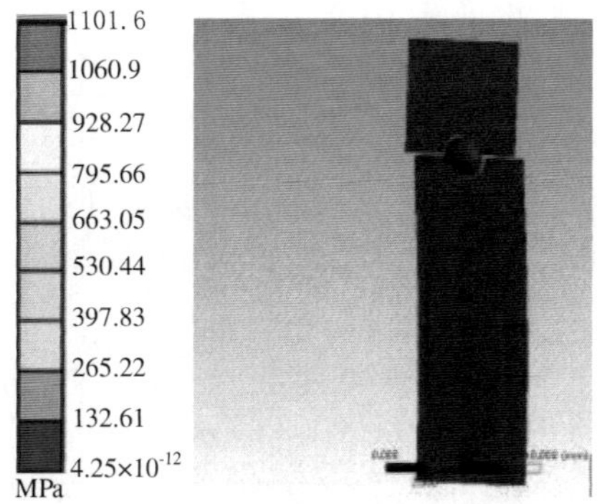

Figure 9. The stress distribution.

The corresponding displacement of the nut was calculated according to this equation. The comparisons of the FEM analysis and the solution of the theoretical model have been shown in figures 10 and 11.

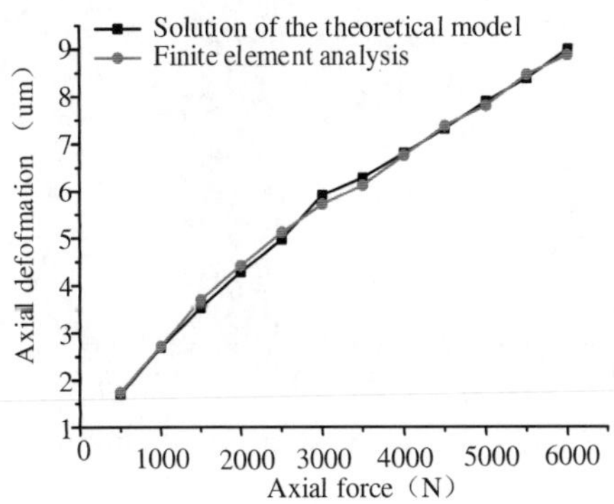

Figure 10. The axial deformation of the ball screw pair.

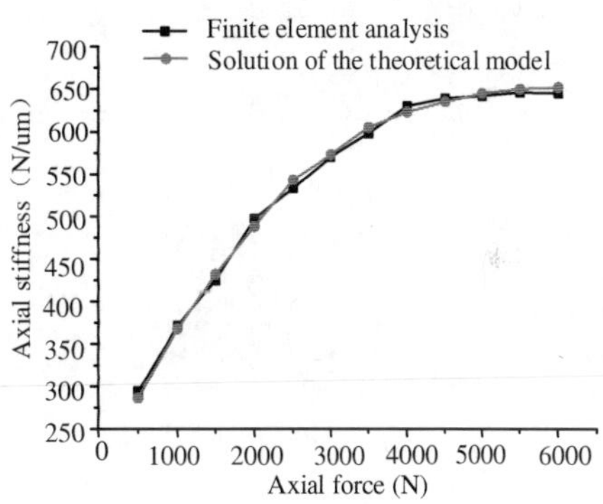

Figure 11. The axial stiffness of the ball screw pair.

It can be seen from figure 10 that the simulation values were basically in accordance with the calculated values from the theoretical model.

The stiffness values have been shown in figure 11, and from this it can be seen that the stiffness of the ball screw pair increased with the increase of the axial load, and the stiffness was relatively low at the start, and it trended to be stable as the load increased. The errors can be analyzed to show that the meshing at non-contact areas was sparse; therefore, the accuracy of the finite element solution was reduced to some extent.

3. Axial stiffness analysis of the single-nut and double-arc ball screw

3.1. Establishment of the three-dimensional model

The structural diagram of the ball screw pair with a double-arc raceway has been shown in figure 12.

The three-dimensional diagram of the ball screw pair with a double-arc raceway was constructed using the SOLIDWORKS software package, as shown in figure 13.

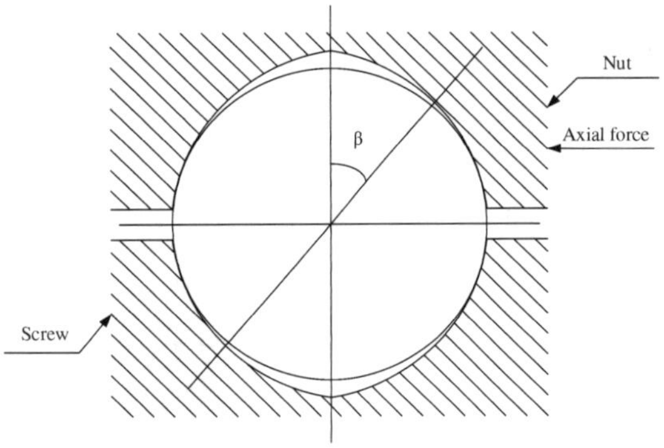

Figure 12. The structural diagram of the ball screw with a double-arc raceway.

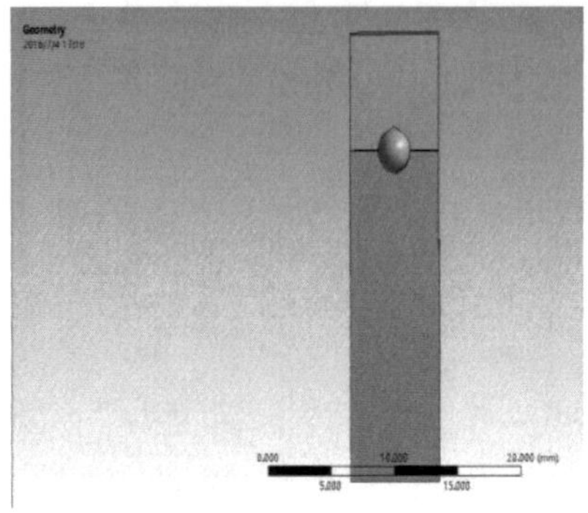

Figure 13. The diagram of the ball screw with a double-arc raceway.

3.2. ANSYS analysis process

The simulation process was the same as mentioned above. Then, the axial displacement variations and the stress distributions under a load of the 500 N axial force were obtained, as shown in figures 14 and 15.

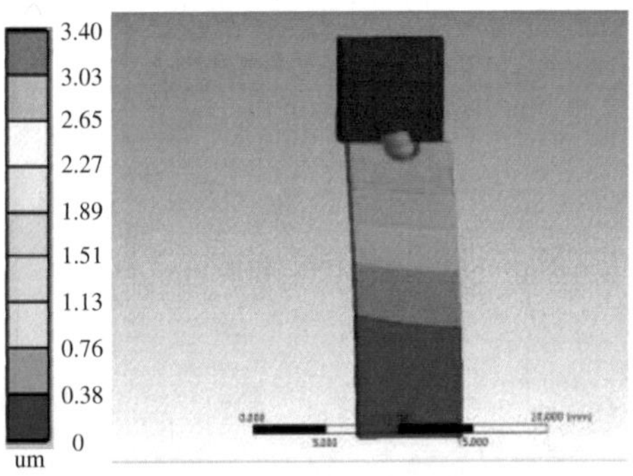

Figure 14. The variation of axial displacement.

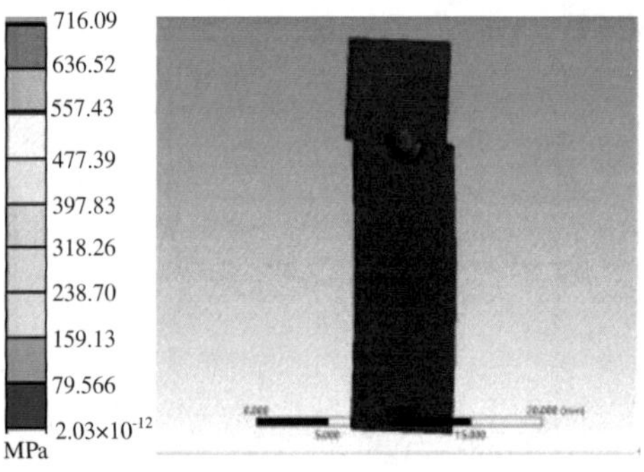

Figure 15. The stress distribution.

Accordingly, the variations in the displacement under the axial load were within the range of 500 N and 6000 N.

3.3. Axial displacement comparison of the single-arc ball screw and the double-nut ball screw

The variations in the axial deformation of the single-arc and double-arc raceway ball screws with the variation of axial load, according to the simulation results, have been shown in figure 16.

It can be seen that under the same axial load, the stiffness of the double-arc ball screw was bigger than that of the single-arc ball screw.

4. Analysis of axial stiffness with the spiral lead angle

The three-dimensional coordinate system of the ball screw pair can be analyzed, as shown in figure 17.

The displacement deformation and the stress distribution under an axial load of 500N have been shown in figures 18 and 19, respectively.

It can be seen that the maximum stress appeared at the locations where contact was made between the ball, the raceways of the lead screw, and the nut, and appeared at approximately 45° from the horizontal direction. Meanwhile, the maximum variation in the axial displacement occurred on the lead screw, which grew larger toward the center, namely, the contact stiffness of the nut's raceway surface was larger than the contact stiffness of the lead screw's raceway surface. Therefore, when checking the strength and fatigue wear of the ball screw pair, it should be based on the deformation at the contact point between the ball and the lead screw. According to the FEM solving steps, the displacement variations for axial loads of between 500 N-6000 N were obtained in turn.

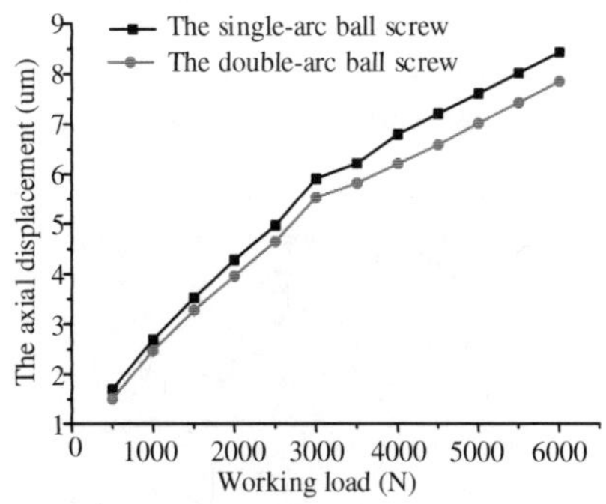

Figure 16. The axial deformation of the ball screw pair.

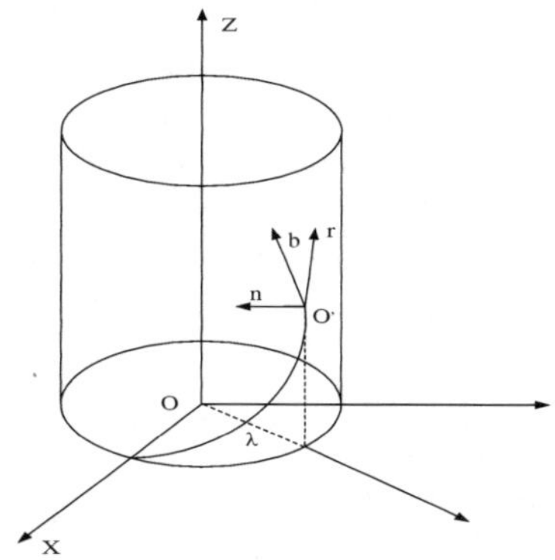

Figure 17. The coordinate system of the ball screw pair.

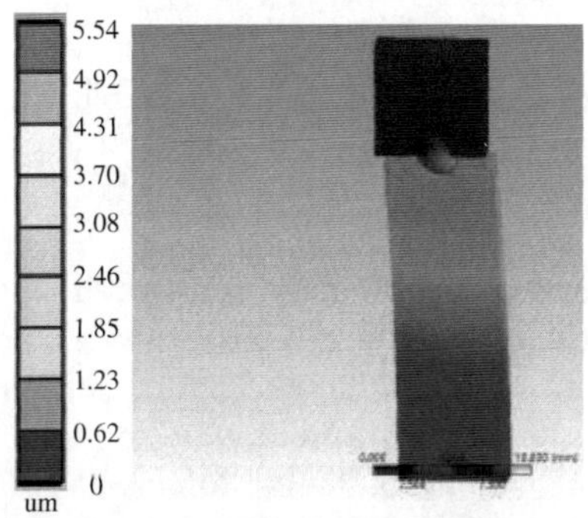

Figure 18. The variation of axial displacement.

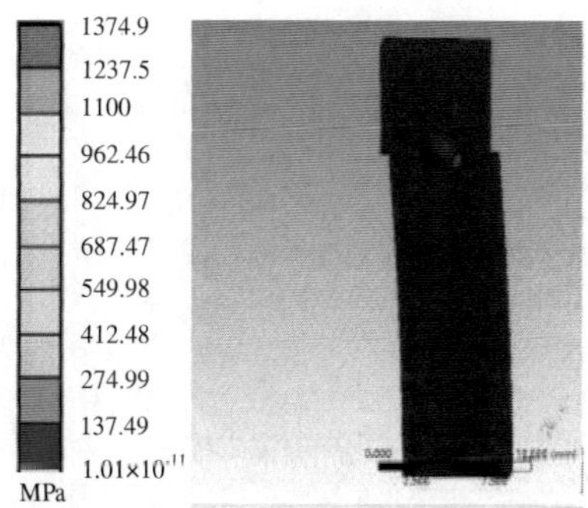

Figure 19. The stress distribution.

5. Comparisons of displacement variations and stiffness for the two models

The displacement variations and stiffness of the nuts in the ball screw pairs, with and without the spiral lead angle, were compared, as shown in figures 20 and 21.

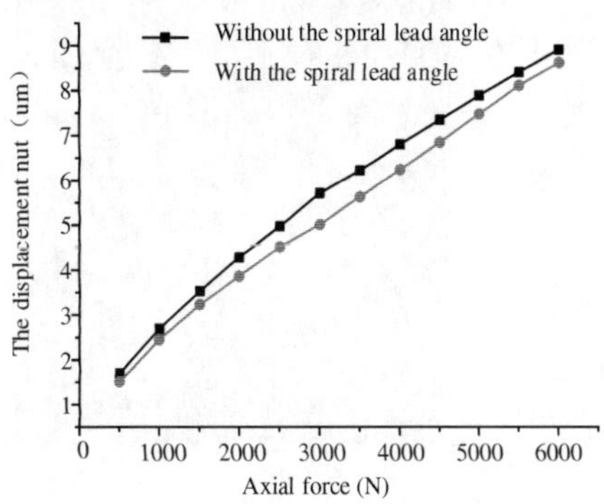

Figure 20. Comparison of the axial deformations.

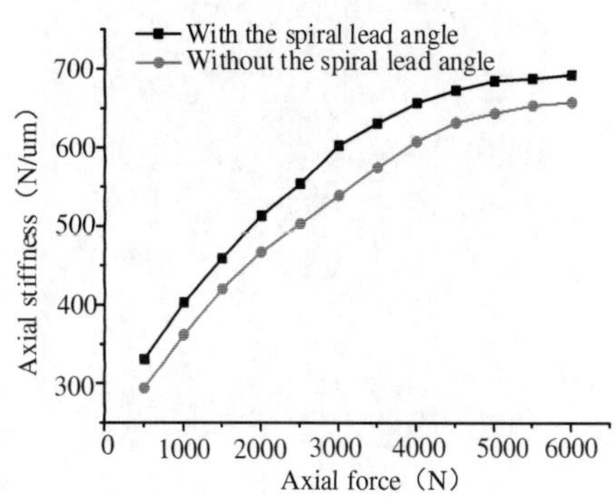

Figure 21. Comparison of the axial stiffness.

It can be seen from figure 21 that the stiffness of the ball screw pair model with the spiral lead angle is greater than that of the ball screw pair without the spiral lead angle. The working accuracy of the ball screw pair model with the spiral lead angle is higher, and the stiffness gradually trends to a stable value with the increase in load.

6. Pre-stressed double-nut ball screw

6.1. Establishment of the three-dimensional model

The structural diagram of the double-nut ball screw subjected to an axial force under working conditions can be obtained, as shown in figure 22, and the three-dimensional model was established, as shown in figure 23.

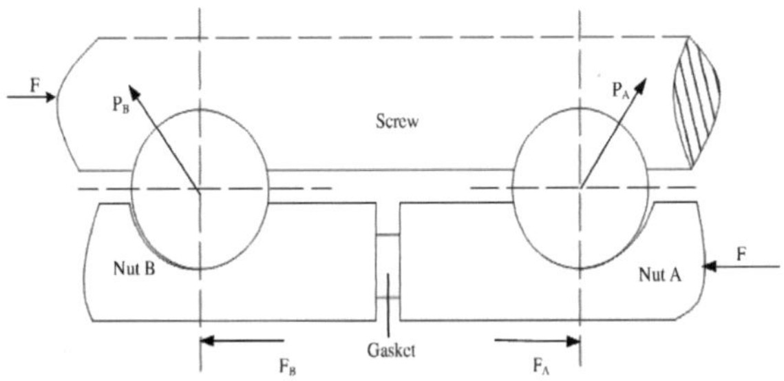

Figure 22. The structural diagram of the double-nut ball screw.

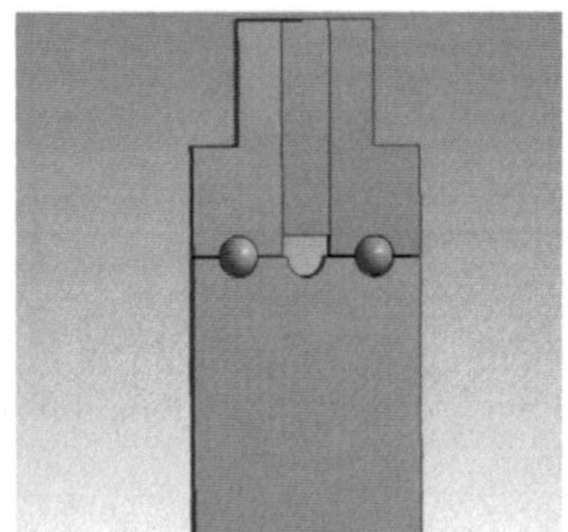

Figure 23. Model of the double-nut ball screw.

6.2. ANSYS simulation process

The simulation process was consistent with the process mentioned above. Preload is a force that is pre loaded on the ball screw to enhance the reliability and tightness of the connections before a working load is applied, in order to prevent gaps or relative slippage between the connected parts after loading. Moreover, the pre-tightening force was approximately one third of the maximum working load F_{max}, i.e. $F_P \approx F_{max}$. Therefore, the pre-tightening force $F_P = 2450$ N. The pre-tightening force was added to the middle of the pre-tightening gasket, based on the pre-tightening unit PRETS179 that was used; meanwhile, the load was applied on the right side of the nut in the ball screw model. The axial displacement variation and the stress distribution under the action of a 500N axial force were obtained, as shown in figures 24 and 25.

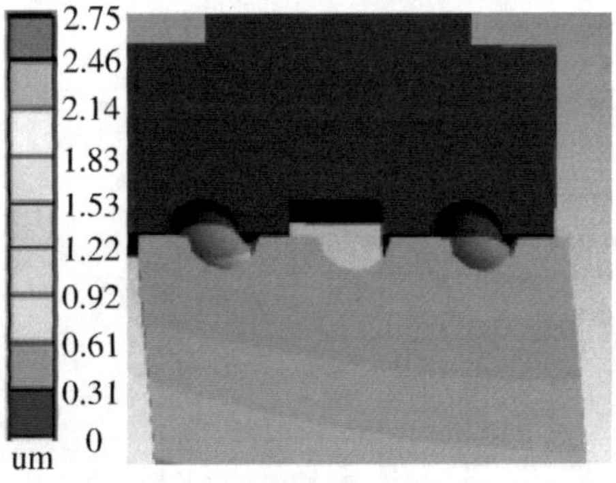

Figure 24. The axial displacement variation.

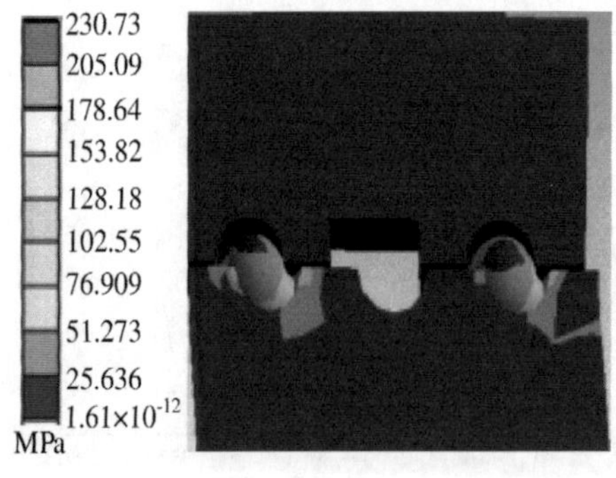

Figure 25. The stress distribution.

The unique variation of the displacement of the nut, based on the sorted simulation data, when an axial force of between 500 N-6000 N was applied in sequence to the right side of the nut, has been shown in figure 26.

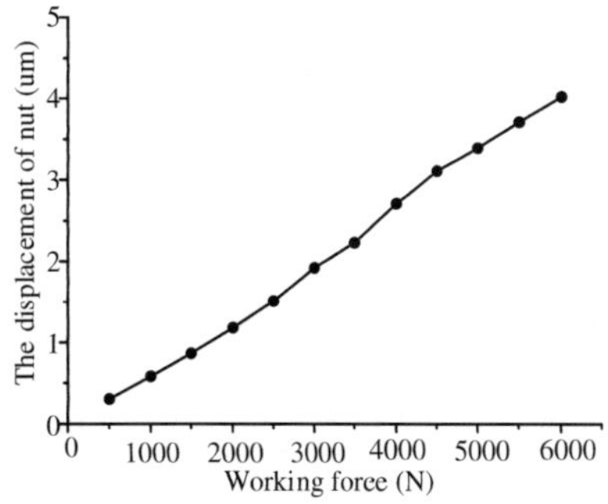

Figure 26. The variation of the displacement of the double-nut ball screw.

It can be clearly seen that the nut displacement is relatively small for the pre-stressed double-nut ball screw under the effect of an axial force, and the corresponding stiffness was larger and the stiffness variation was more moderate, which reflected its structural superiority.

7. Summary

- The location of maximum stress appeared at approximately 45° from the horizontal direction, i.e. the location of the contact angle. The axial stiffness increased with the increase of axial load. Under the same axial load, the stiffness of the double-arc ball screw was larger than that of the single-arc ball screw.
- The contact stiffness of the nut's raceway surface was larger than the contact stiffness of the lead screw's raceway surface. Therefore, when checking the strength and fatigue wear of the ball screw pair, it should be mainly focused on the deformation of the contact point between the ball and the lead screw.
- The stiffness of the ball screw pair model with the spiral lead angle was greater than that of the ball screw pair without the spiral lead angle. The displacement variation of the double-nut ball screw with the spiral lead angle was smaller than that of the ball screw without the spiral lead angle, hence the higher accuracy.
- The working precision of the double-nut ball screw was higher and the stiffness was larger than those of the single-nut ball screw. This reflects the influence of the pre-tightening force, and the nut type of the ball screw on the stiffness and accuracy of the ball screw pair.

Acknowledgments
The authors would like to thank support from the National Science Foundation of China (51675422, 51375381), Shaanxi province Key research and development foundation of China (2017GY-028).

References

[1] Kamalzadeh A, Gordon D J and Erkorkmaz K 2012 Robust compensation of elastic deformations in ball screw drives *International J Mach Tool Manu* **50** 559-74
[2] Feng G-H and Pan Y-L 2012 Investigation of ball screw preload variation based on dynamic modeling of a preload adjustable feed-drive system and spectrum analysis of ball-nuts sensed vibration signals *Int J Mach Tool Manu* **52** 85-96
[3] Zhou C-G, Feng H-T, Chen Z-T and Qu Yi 2016 Correlation between preload and no-load drag torque of ball screws *International J Mach Tool Manu* **102** 35-40
[4] Fernandez-Gauna B, Ansoategui I, Etxeberria-Agiriano I and Graña M 2014 Reinforcement learning of ball screw feed drive controllers *Eng Appl Artif Intel* **30** 107-17

[5] Liang T, Lu D, Yang X-J, Zhang J, Ma X-B and Zhao X-B 2016 Feed fluctuation of ball screw feed systems and its effects on part surface quality *International J Mach Tool Manu* **101** 1-9
[6] Sobolewski J Z 2012 Vibration of the ball screw drive *Eng Fail Anal* **24** 1-8
[7] Garinei A and Marsili R 2012 A new diagnostic technique for ball screw actuators *Measurement* **45** 819-28

IOP Conf. Series: Materials Science and Engineering **372** (2018) 012024 doi:10.1088/1757-899X/372/1/012024

Research and application of key technology of electric submersible plunger pump

K Qian[1,2,4], Y N Sun[2], S Zheng[3], W S Du[2], J N Li[2], G Z Pei[2], Y Gao[2] and N Wu[2]

[1]School of Petroleum Engineering, Northeast Petroleum University, Daqing, 163318, China
[2]Petrochina Daqing Oil Field Production Engineering Research Institute, Daqing, 163712, China
[3]Petrochina Daqing Oil Field Third Oil Production Plant, Daqing, 163113, China

E-mail: qk13836956215@163.com

Abstract. Electric submersible plunger pump is a new generation of rodless oil production equipment, whose improvements and upgrades of key technologies are conducive to its large-scale application and reduce the cost and improve the efficiency. In this paper, the operating mechanism of the unit in-depth study, aimed at the problems existing in oilfield production, to propose an optimization method creatively, including the optimal design of a linear motor for submersible oil, development of new double-acting load-relief pump, embedded flexible closed-loop control technology, research and development of low-cost power cables. 90 oil wells were used on field application, the average pump inspection cycle is 608 days, the longest pump check cycle has exceeded 1037 days, the average power saving rate is 45.6%. Application results show that the new technology of optimization and upgrading can further improve the reliability and adaptability of electric submersible plunger pump, reduce the cost of investment.

1. Introduction

With the development of oilfields, the artificial lift is mainly based on the two rod-lifting methods of pumping units and screw pumps, which account for about 90% of the total number of wells. The technology is very mature, however, a series of problems have exposed in the application of low-permeability reservoirs and heavy-oil reservoirs: big one-time investment, high useless energy consumption; sucker rod and tubing eccentric wear is very serious; ground equipment there is a security risk; In the flood season, low-lying areas affect normal production; There is a lot of work to manage and maintain [1,2].

In China, cylindrical permanent magnet synchronous linear motor used in underground as pumping drive has formed a small-scale field trials [3,4]. In response to the above problems Daqing peripheral oil field low yield and permeability zone block, through the development of electric submersible plunger pump and its supporting technology research, has initially formed a new energy-saving, efficient rodless oil recovery system, with good adaptability. However, there is a general lack of hoisting force in the oil well, vibration larger, which is caused by the design of the submersible linear motor [5,6]; Open-loop control leads to higher energy consumption, system efficiency is low, overload frequently [7]; Power cable is easy to breakdown, high cost caused by checking the pump cycle is shorter, big one-time investment and other issues [8,9]. Therefore, there is still a gap between the technology and scale application in terms of technology, economy and adaptability, so it is urgent to conduct in-depth research and optimization of key technologies.

Based on this, this paper carry out an in-depth study of key technology of electric submersible plunger pump. First, deeply study the operating principle of the unit, then proposed the optimization method of the unit creatively in response to the above shortcomings in the field test, including the optimal design of a linear motor for submersible oil, development of new double-acting load-relief pump, embedded flexible closed-loop control technology, carry out cable design of special materials. Then carried out a large number of oilfield application tests, objective evaluation the application effect, and find a breakthrough technology in the application of large-scale.

2. Operation mechanism and key technology optimization

Electric submersible plunger pump is a new generation of rodless oil production equipment [10], it has gone through three stages: experiencing technological exploration and manufacturing, technical optimization and on-site testing, improving of the supporting technology, now it has entered the stage of further research of key technology and application expanding . Shown in figure 1, the process consists of submersible linear motor, plunger pump, ground control system, dedicated cable and other components, making linear motor as the power-driving device, integrate the linear motor with the tubing pump, dived into the underground oil reservoir. In this process: the linear motor is at the bottom, the tubing pump is at the top; The stator of the motor is connected with the oil pipe and the motor mover is connected with the plunger of the oil pump; The control system transfers the AC power to the downhole motor after the conversion of AC; The use of motor mover reciprocating motion consistent with the characteristics of the plunger to drive the plunger reciprocating cyclically to achieve crude oil lift.

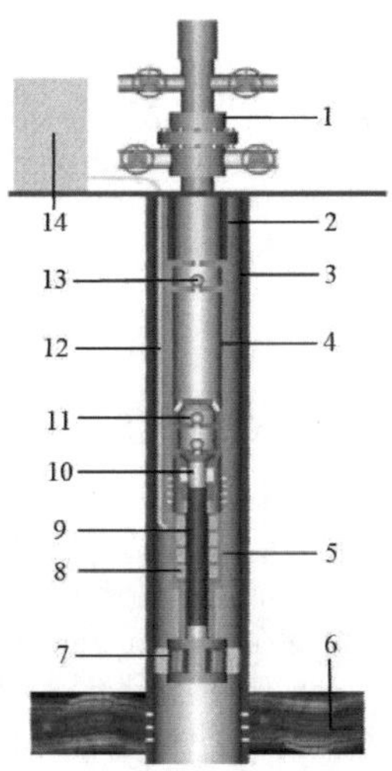

1.Wellhead;
2.Tubing;
3.Casing;
4.Plunger pump;
5.Submersible linear motor;
6.Oil layer;
7.Centralizer;
8.Stator;
9.Mover;
10.Plunger;
11.Swimming valve;
12.Cable connector;
13.Fixed valve;
14.Control system

Figure 1. Electric submersible plunger pump lifting process schematic.

2.1. Optimization design of submersible linear motor

The submersible linear motor is a kind of electromagnetic device that can use the electric energy directly to drive the load to do the linear motion without the intermediate conversion mechanism [11,12]. It belongs to a cylindrical permanent magnet synchronous linear motor. It is a structural evolution of a permanent magnet rotating electrical machine, it mainly consists of a stator, a mover and a motor leading wire. According to actual needs and downhole conditions, it is designed as a short junior long secondary, shown in figure 2, the motor stator is N and S interactive, consisting of the outer shell, the winding iron core and the inner cylinder, where the motor mover consists of a circular permanent magnet, a ring-iron core and a core shaft. The winding is in a radial twisting and axial distribution structure, after the assembly of the winding iron core, it is placed in the oil-filled sealed chamber consisting of the inner tube and the shell. By controlling the frequency and direction of the power supply, the traveling wave magnetic field with periodic alternation of the stator interacts with the excitation magnetic field generated by the permanent magnet of the mover to generate the electromagnetic thrust and achieve the linear reciprocating motion of the mover.

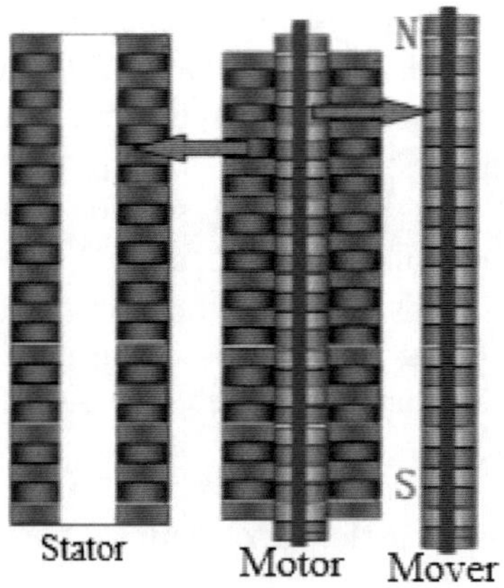

Figure 2. Submersible linear motor structure.

In practical application process, the motor contacts corrosive oil underground, the mover permanent magnet is gradually eroded, and the remanence of the cobalt rare earth permanent magnet material is too low, which seriously affects the back electromotive force constant of the motor and indirectly affects the carrying capacity of the motor; the resistance function of the stator is poor, the magnetoresistance on the yoke is high, the magnetic flux leakage seriously, the magnetic field utilization is low, thrust density is low; When the motor is running, the mover magnet and the spacer ring are easy to rotate, the concentricity with respect to the center connecting rod is poor, causing the inner cylinder of the stator to be worn. Based on this, according to the existing measured data of the motor and the actual demand of field application, this paper optimizes the motor on the basis of stable and mature mechanical structure so as to improve the continuous working performance of the motor in complex environment downhole, reduce vibrations during work.

Linear motor seal design: the stator inner cylinder closed welding, the inner cylinder spray welding nickel alloy powder, a thickness of 0.1 ~ 0.2 mm; Chrome plating on the surface of the mover enhances abrasion resistance and studies on the overall closed process of the mover permanent magnet to ensure complete separation of the mover and the stator from the oil; improve the processing technology, reduce the stator and the rotor axial dimension error, reduce unit vibration at the same time. For 5.5-inch casing design motor diameter 114 mm, 7-inch casing design motor diameter 143 mm, the stator magnetic field strength increased by 20%.

Movable permanent magnet anti-rotation design: movable strut material selection 1Cr18Ni9Ti, after the ring and the permanent magnet with the laser welding as a whole, to ensure that the external body and the strut at the same time, increasing the diameter of the strut, the central spar milling keyway, corresponds to the permanent magnet turning keyway, after assembly to enhance positioning, to prevent radial rotation. At the same time, under the premise of ensuring lifting force, according to the requirements of air gap magnetic field, the air gap is optimized from the original 0.05 mm to 0.3 mm. Avoid motor mover stuck phenomenon.

Thrust output capability improved: movable structure selection axial magnetic circuit structure, optimizing the axial length of the permanent magnet and pole pitch, magnet manufacturing simple magnetization; In order to avoid the influence of temperature and electromagnetic performance on the insulation of the motor, NdFeB permanent magnet with high coercivity is selected as the mover permanent magnet to improve the motor air gap magnetic field index; Under the same thrust index, the winding coil circular cross-section into a rectangular cross-section, in order to increase the coil full slot rate; Meanwhile, on the premise of ensuring the current passing rate, the outer diameter of the winding coil is optimized and the number of turns of the winding is increased, thereby remarkably reducing the thermal load index of the motor and enhancing the reliability of the motor operation.

For the submersible linear motor with rated voltage 1140 v, the motor diameter 114 mm, supporting Φ32 mm pumping pump test indoor evaluation. The results show: the optimized submersible linear motor is tested in an oil-water medium at 30 mpa for 24 hours without deformation and leakage; Winding in the hot state of ambient temperature 200℃, the application of AC 50 hz, 3300 v voltage, one minute after the test dielectric strength, no flashover and breakdown; apply a megohmmeter with a range of 2500 volts to test the insulation resistance, the insulation resistance of no less than 1000 megohms; Winding sintering temperature of 380℃, the motor temperature is not higher than 150℃ in the media run, to ensure that the rated thrust output. The optimized motor mover does not deform under the action of 4.5 t radial force, the maximum lift reaches 3220 m, which enlarges the applicable range of the technology. The comparison of motor performance before and after the optimization is shown in table 1.

Table 1. Submersible linear motor performance optimization before and after comparison.

Motor diameter	Frequency (Hz)	14	12	10	9	8
Before optimization	Lifting force (t)	1.9	2.1	2.4	2.6	2.7
Φ110 mm	Head (m)	1560	1710	2180	2290	2430
	Safety factor	1.0	1.1	1.4	1.6	1.7
After optimization	Lifting force (t)	2.0	2.5	2.9	3.3	4.5
Φ114 mm	Head (m)	2120	2560	2870	2950	3220
	Safety factor	1.2	1.4	1.7	1.8	2.0

2.2. Development of new double-acting load relief pump

Electric submersible plunger pump used in deep wells or supporting the pump diameter is large, it is easy to cause the plunger rod instability, resulting in plunger overload operation or card pump, for this reason we designed a new double-acting load relief pump. The upper and lower stroke of the pump all discharge oil, and the unidirectional load is reduced; The up plunger cross-sectional area of about twice the down cross-sectional area of the plunger, the plunger load up and down roughly the same load, the mover force reduced by about half; For example, a double acting pump with a diameter of 57/38 mm exerts the same force as a conventional piston pump with a diameter of 38 mm. The theoretical displacement corresponds to a conventional piston pump with a diameter of 57 mm; Upper and lower cylinder designed centralizer, to ensure the proper coaxiality, while reducing motor vibration. In order to avoid loosening of motor mover by large load pulling force and design of overcurrent passage in motor influence casing space and motor performance, will double-acting pump fixed valve design at the top, oil from the bottom of the sieve tube into the pump cylinder, traveling valve in internal plunger, all valve balls open in the same direction. Double-acting pump can effectively reduce the operating load, reduce motor power and investment costs, as shown in figure 3, when the plunger at the top dead center or bottom dead center, the leak-proof valve 2 is closed, effectively prevent pump loss, to ensure that the pump is full.

In order to adapt to the lifting of directional well and horizontal well better and to improve the pump efficiency of the new double-acting load relief pump, a spring return pilot valve, as shown in figure 4, in the fixed valve cover and travel valve cover is provided with four-in-one reinforcement bar, where the valve ball is only one degree of freedom, the valve ball can only move along the axis of the mask, the liquid flows around the four reinforcing bars. The structure has a larger flow area and a smaller opening degree, which can reduce the suction resistance and valve ball set lag time.

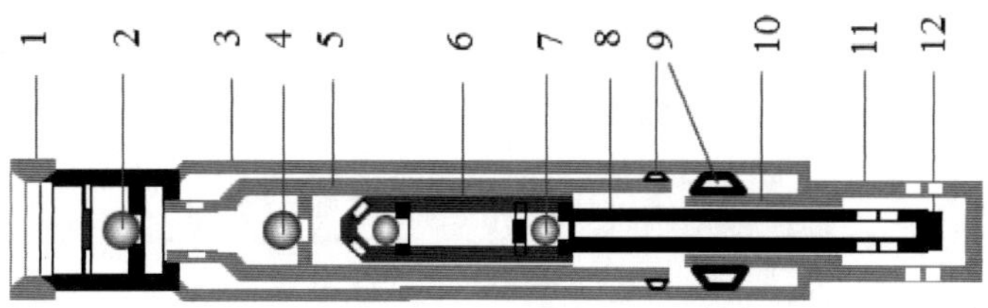

1. Coupling; 2. Leak-proof valve assembly; 3. Outer tube; 4. Fixed valve assembly; 5. Up Pump barrel; 6. Up plunger; 7. Swimming valve assembly; 8. Down plunger; 9. Centralizer; 10. Down Pump barrel; 11. Screening; 12. Pump down connector

Figure 3. New double-acting load relief pump.

Figure 4. Spring return pilot valve.

2.3. Embedded flexible closed-loop control technology

Through the optimization design of submersible linear motor and development of new double-acting load relief pump, unit vibration is reduced, the basic performance has improved. However, the original control system is open-loop control and belongs to a constant rigid drive. In overload operation, the motor mover will not keep up with the speed of the synchronous driving magnetic field and can not make a normal response to the load change, hard start, severe commutation and other issues, affecting the life of units, cables and downhole tools. For the above, the research on flexible closed-loop control of submersible oil linear motor based on the original frequency control technology is carried out. Its core is the long distance sensorless vector control technology, based on the vector control, the induction control is pulled in, and the flexible control algorithm is embedded in the sensorless vector control program, through the advanced intelligent adjustment of PID: automatic generation of full-stroke within the best operating speed distribution and output power distribution programs, and according to changes in operating conditions automatically optimize the operation of the program; Throughout the best speed distribution and power output distribution plan to do continuous variable speed operation, the electrical machinery and the controller completely realizes the closed-loop control by the software method.

The control system is composed of intelligent controller, driver and so on, and possesses functions of remote data transmission. It uses filtering and actuarial elimination of various disturbances to ensure high-speed acquisition and real-time complex operation of multiple operating parameters. And using 32-bit computing processor, through the acquisition and analysis of motor power, adjust the current and voltage, calculate the speed and thrust of the mover, and then automatically analyze the appropriate drive current, to achieve the minimum current efficient operation. It can be flexibly driven in the 0.1-10 specific water-cycle range, with the downhole load changes to adjust the motor output power, reduce starting load and system vibration, achieve flexible operation, flexible start, flexible stop. At the same time, based on the combination of submersible linear motor adaptive method and rotational speed observer method, timely reference downhole pressure sensor feedback, improve the motor out of step monitoring results, improve the accuracy of closed-loop control and adaptive tuning ability. Carry out indoor tests, under the same working conditions, respectively, the use of conventional control systems and flexible closed-loop control system to control the same submersible linear motor, the vibration intensity comparison curve shown in figure 5, the reference data shown in table 2.

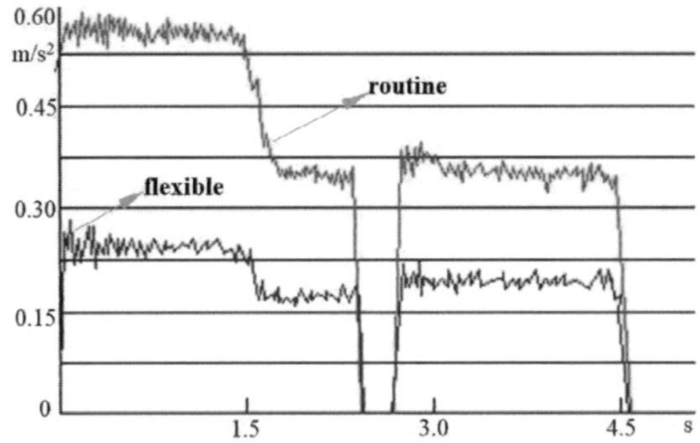

Figure 5. Vibration intensity comparison curve of routine and flexible control.

Table 2. Routine and flexible control of indoor test data comparison.

Control method	Stroke (m)	Stroke times (min^{-1})	Motor thrust (t)	Maximum current (A)	Average current (A)	Maximum power (kw)	Average power (kw)
Routine	1.23	8	1.22	58.71	35.17	51.57	29.74
Flexible				27.04	19.44	23.02	16.30
Reduce the rate (%)				53.94	44.73	55.36	45.19

The test results show that the application of the closed-loop control method, start-up current and operating current significantly reduced, the reduction of 53.94% and 44.73% respectively. At the same time, the system vibration intensity is also greatly reduced, the maximum value of the vibration intensity curve of conventional control method and flexible closed-loop control method are 0.577 m/s^2 and 0.287 m/s^2, reduce the rate of 50.26%.

2.4. Development of new aluminum cables

Electric submersible plunger pump used centrifugal submersible pumps dedicated copper wire core cable, cable costs account for an investment of 1/3. Due to the design of the technology is aimed at low permeability reservoirs, requiring a one-time investment is low, in order to be able to more adapt to the scene needs, conducive to expanding the scale of on-site applications.

In this paper, the characteristics of the operation of the integrated process, for the optimization of the submersible linear motor proposed selection of rare earth aluminum alloy wire instead of copper core, at the same time optimize the cable insulation and sheath material, design active filtering system, filter out the reflected voltage. From the cost input, size standards, material properties and other aspects can meet the site requirements, compared with the conventional copper cable, the cost reduction of about 20%. The new aluminum alloy cable indoor testing evaluation, as shown in table 3, the main indicators meet the relevant performance requirements.

Table 3. Conventional copper cable and aluminum cable performance test data.

Cable type	Width (mm)	Thickness (mm)	Diameter (mm)	Insulation resistance (MΩ,1000m)	Three-phase imbalance rate (%)	Pressure resistance (kV)	DC Leakage (μA)
Copper core	37.56	15.21	4.42	5714	0.048	22	4.7
Aluminum alloy	39.15	14.98	5.59	4762	0.055	22	7.5

Standard value	≤40	≤17	4.4 (copper)	≥2000	≤2	22	≤15

3. Field application

In the low permeability zone of a certain oilfield, a field test was conducted on the optimized electric submersible plunger pump to evaluate its application effect in production. On-site optimization of the submersible linear motor total of 90 wells, among them, 4 wells are used in combination with the new double-acting load relief pump, 12 wells are used in combination with the flexible closed-loop control technology, 1 wells are used in combination with the new aluminum cables.

Up to now the overall application effect is as follows: the average pump hanging 1199 m, the deepest 1865 m; The average liquid level 1032 m, nissan liquid 2.7 t, pump efficiency 55.4%; average pump inspection period 608 days, the longest 1037 days; the average free repair period 438 days, among them, 22 oil wells which were put into trial operation in the early stage have an average maintenance-free period of more than 1000 days and a maximum of 1485 days, respectively, compared with the same displacement rod pumping method, the average saving rate reached 45.6%. Among them, 1 oil well with new cable has been running smoothly for 578 days, after 5 times of normal well washing, the cable did not break down, which initially showed that the performance indexes of the new cable meet the field application requirements.

Field application shows that compared to the optimization before, the average pump inspection cycle increased from 452 days to 608 days, the average system efficiency increased by 6.9%, enhances the reliability of the operation of the process, significantly improves the energy-saving effect and lift capacity, enlarges the applicable range of the process, and achieves cost reduction and efficiency increase.

On-site application of new double-acting load-relief pump, motor mover current value is basically balanced, compared with the conventional φ57 mm plunger pump, nissan liquid volume is basically the same. Head raised 217 m, up current down 27.9%; Compared with the conventional φ38 mm pump, nissan liquid volume increased nearly 1 times. Upstream current is basically the same. The test results show that the pump can effectively reduce the unit operating load, to reduce the motor power and investment costs to create the conditions for extending the life of the unit, the scene test comparison data is shown in table 4.

Table 4. Double-acting pump and conventional piston pump field test comparison data.

Hashtag	Pump type	Pump type (φ,mm)	Liquid level (m)	Stroke times (min^{-1})	Yield (m^3/d)	Uplink current (A)	Downlink current (A)
1	Conventional	57	453	3	11.0	43	5
2	Conventional	38	768	4	6.0	29	11
3	Double-acting	57/38	670	3	11.5	31	31

Table 5. Comparison of routine and flexible control field application data.

Control method	Working Frequency (HZ)		Uplink current (A)	Average input power (kw)	Daily power consumption (kwh)	System efficiency (%)
	Uplink	Downlink				
Flexible	8	15	4.49	3.953	94.90	27.90
Routine			7.55	6.755	162.14	16.33
Flexible	10	20	2.96	2.552	61.20	43.25
Routine			4.50	3.927	94.32	28.07

For the same well, under the condition of the same well conditions and working parameters, flexible closed-loop control system and conventional control system are respectively used for testing and evaluation. Under the two kinds of control modes, the field application effect comparison is shown in table 5. One of the conventional control mode operation, occured two overload shutdown phenomenon, flexible closed-loop control mode of operation, stable and reliable.

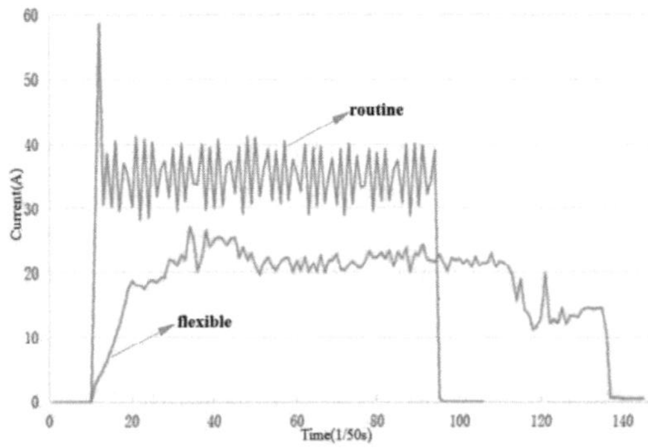

Figure 6. Motor mover uplink real-time current curve comparison.

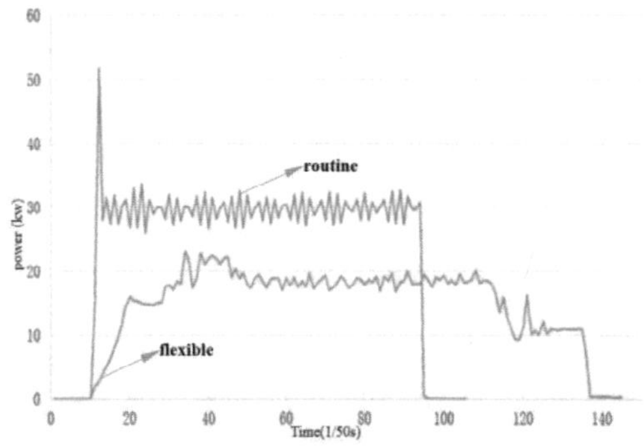

Figure 7. Motor mover uplink real-time power curve comparison.

Field application results show: flexible closed-loop control system can well avoid overload shutdown, energy-saving effect is obvious, the unit system efficiency significantly improved. When the submarine linear motor mover uplink frequency 8 hz, downlink frequency 15 hz, under the two control modes, the uplink real-time current and power curve is as shown in figure 6 and figure 7, it can be seen that the flexible closed-loop control mode effectively reduces the starting impact and running load, and the operation of the unit is stable and reliable.

4. Conclusion

Electric submersible plunger pump is a new generation of rodless oil extraction equipment, the deeply study of the key technology and the large-scale application is the inevitable trend of its development. This paper proposes an optimization method for this production technology creatively: optimization design of submersible linear motor, development of new double-acting load-relief pump, embedded flexible closed-loop control technology, research and development of low-cost power cables. The field application of 90 wells, the average pump inspection cycle 608 days, the longest pump inspection cycle exceeded 1037 days, the average saving rate reached 45.6%. The application results show that the new process of optimization and upgrading can further improve the reliability of the lifting process of the electric submersible plunger pump, reduce the one-time investment cost and broaden the technical and economic adaptability. For the process to digital management, energy efficient, safety and environmental protection in the direction of development, and reasonable control of oil field development costs, improve development effectiveness provided technical support; This technology laid the foundation for the exploration and research of deep well lift.

References

[1] Wang S Z 2007 The technique research of reciprocal pump system working as electric submersible pump (China University of Petroleum)
[2] Zhang F 2015 The design of submersible reciprocating pump lifting process (Northeast Petroleum University)
[3] Zheng G and Wang S W 2013 Application of rodless pump in gproduction of downhole linear motor and matching techniques in ultra-low permeability reservoirs of Changing oil field *Journal of Oil and Gas Technology* **35** 158-160+178
[4] Liu C 2016 The dynamic simulation of submersible linear motor lifting system (Yanshan

University)
[5] Liu Y H and Yu Z W 2016 LLE for submersible plunger pump fault diagnosis via joint wavelet and SVD approach *Neurocomputing* **185** 202-11
[6] Shao K Y and Qian K 2014 Electric submersible plunger pump fault diagnosis research and application based on indirect indicator diagram measurement *Control and Instruments in Chemical Industry* **41** 1023-7
[7] Yu D L and Li Y M 2017 The MMAGA-RBF fault diagnosis method for submersible plunger pump *Journal of Harbin Institute of Technology* **49** 159-65
[8] Liu Y H 2016 Fault diagnosis of submersible reciprocating pump based on locally linear embedding (Harbin Institute of Technology)
[9] Yu D L and Qi W G 2011 Optimization frequency of stroke based on submergence depth SVM forecasting of submersible pump *Proceedings of the CSEE* **31** 138-44
[10] Fu G T and Zhang B 2006 Lifting technology for reciprocating magnetic drive plunger pump *Acta Petrolei Sinica* **27** 122-9
[11] Zhou F and Yang L Y 2015 Overview of linear motor key technology research of underground linear lifting device *Micromotors* **48** 100-8
[12] Cao H 2008 The design and analysis of motor in submersible linear motor pumping unit (Harbin University of Science and Technology)

IOP Conf. Series: Materials Science and Engineering **372** (2018) 012025 doi:10.1088/1757-899X/372/1/012025

Thermal stress analysis of a damaged waste heat boiler

Y-H Wu[1,3], C-C Hsiao[1] and C-H Chen[2]

[1]MCL, Industrial Technology Research Institute, Chutung, Hsinchu, Taiwan 31040, China
[2]Energy Research Center, Taiwan University, Chutung, Hsinchu, Taiwan 31040, China

E-mail: yhwu01@itri.org.tw

Abstract. Waste heat boilers are widely used in the petroleum refining industry to improve energy efficiency. However, improper design, operation or maintenance of these boilers may result in costly failures. Cracking was found repeatedly at the same location on a circumferential weld of a certain waste heat boiler in CPC refinery. Hence a special designed fixture was used to avoid crack extension. Root cause mainly due to improper arrangement of internal tubes and external piping was then mainly found by thermal stress analysis. Recommendations was proposed to shut down the boiler immediately and to rearrange the tubes and piping without changing basic design of the waste heat boiler.

1. Introduction

A typical convective type of waste heat boiler (WHB) used as part of reformer system is shown in figure 1. The hot hydrogen gases from reactor are carried through a number of parallel small diameter tubes, while the incoming water to be heated enters a shell surrounding the tubes and passes directly over the hot tubes in the direction normal to their axes. During operation, cracking was found on a circumferential weld repeatedly at the same location after repairs or a new replacement and, a special designed fixture was used to prevent from crack propagation as shown in figure 2 [1,2]. Since the WHB and surrounding piping system are operating at high temperatures, one of two supports under the heat exchanger is fixed to the base and the other is allowed to slide horizontally to accommodate periodic temperature change. During site survey, the circumferential weld with cracks was found close to the side with sliding support. In addition, the high temperature intake manifold from reactor was parallel to the WHB and connected to the sliding side also. To find the root cause, an analysis model was formulated considering the thermal effect in the heat exchanger and as well from the intake manifold structure under the design and operating conditions as shown in table 1.

Table 1. Design and operating parameters of waste heat boiler.

	Contents	Design temperature (°C)	Operating temperature (°C)	Design pressure (MPa)	Operating pressure (MPa)
Tube side	Hydrogen gas	470/300	396.6	2.45	1.98
Shell side	Boiling water	275	223.88	3.92	3.11

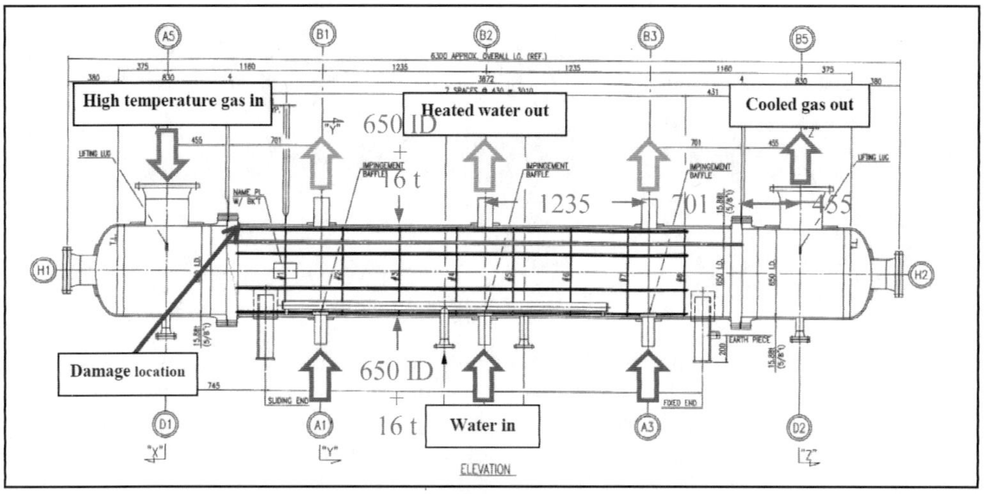

Figure 1. A typical convective type of waste heat boiler (WHB) used as part of reformer system.

(a) (b)

Figure 2. (a) Damaged heat exchanger containing cracks on a circumferential weld, (b) WHB and Fixture covering the flange and welded with shell was used to prevent from crack extension.

2. Modeling and analysis

Based on the structure of WHB, the high temperature gas inlet is defined as tube side and the heat transfer section as shell side. Considering the symmetrical structure of WHB and tube arrangement, a half model to cover half of both tube and shell sides and as well half cross section of tube sheet is used as shown in figure 3. The outer surface temperature of WHB and piping are assumed constant since the whole heat exchanger and piping are fully insulated during operation. In this study, steady state heat transfer in isotropic material and Newton cooling was assumed. The governing equation [3] of thermos-elasticity of an isotropic solid is shown as in equation (1). By ANSYS, temperature distribution of the total model was first calculated. Furthermore, stresses of the structure were then analyzed considering operating pressure and temperature as internal force.

$$(\lambda + \mu) u_{j,ij} + \mu u_{i,jj} - \frac{E\alpha}{1-2\nu} T_{,i} = 0 \tag{1}$$

Where, $\lambda = \frac{\nu E}{(1+\nu)(1-2\nu)}$ and E, μ, ν, α, u, T, the Young's modulus, shear modulus, Poisson's ratio, thermal expansion coefficient, displacement and temperature.

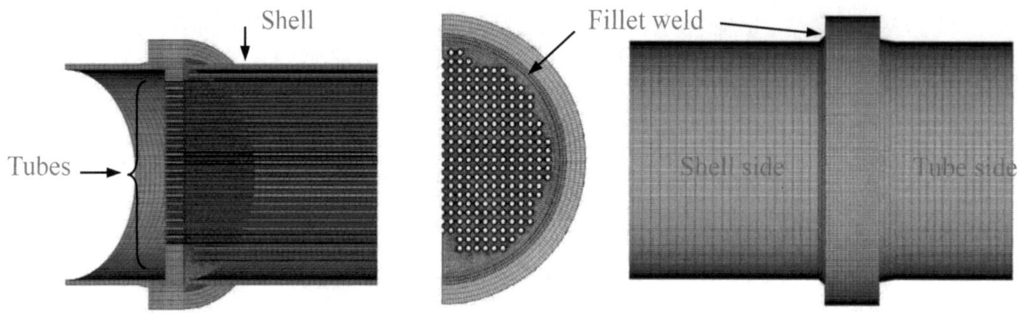

Figure 3. 3D brick element for numerical analysis with circumferential fillet weld in orange color.

To realize the failure mode, three different stresses of principle (S_1), equivalent (S_{eqv}) and stress intensity (S_{int}) are compared as indicated by equations (2), (4) and (5) respectively.

$$S_1 = Max(\eta_1, \eta_2, \eta_3) \quad (2)$$

Where, η_i $(i = 1,2,3)$ is the root of equation (3) and the principle stresses of three principle axis.

$$\det(\sigma_{ij} - \eta\delta_{ij}) = 0 \quad (3)$$

$$S_{eqv} = \left[\sigma_{kk}^2 - \frac{3}{2}(\sigma_{ii}\sigma_{jj} - \sigma_{ij}^2)\right]^{1/2} \quad (4)$$

$$S_{int} = Max[(S_1 - S_2), (S_2 - S_3), (S_1 - S_3)] = 2\tau_{max} \quad (5)$$

For numerical analysis, 3D brick element of 20 nodes and a total number of 513,838 elements are used to cover the shell and tubes as shown in figure 3. Material parameters and boundary conditions are listed in tables 2, 3 and figure 4. The outer surfaces of model are insulated boundaries as indicated earlier and the surface temperature (*T in*) of internal gas medium sides is set to 669.6 (*K*). And, the internal surface temperature (*T_out*) of shell sides which are in contacted with water steam is set to 506.88 (*K*). Thermal transfer with or without convection are both considered with convection coefficients of gas side and steam side assumed as (*h_in*) = 20 (W/m^2K) and (*h_out*) = 30 (W/m^2K) respectively [4]. After the temperature field was acquired, the stress fields are computed based on the conditions as shown in figure 5. The internal pressure (*P_in*) of gas side is set to 1.98 (MPa) and (*P_out*) of steam side to 3.17 (MPa) which is the saturated steam pressure at 506.88 (*K*). In addition, a bending moment was added considering the thermal expansion difference between high temperature manifold and the heat exchanger as shown in figure 6.

Table 2. Material parameters.

	k (W/mK)	E (GPa)	υ	α ($10^{-5}/K$)	ρ (kg/m^3)
Tube side and shell side	40	200	0.28	1.3	7860
Weld	32	160	0.3	1.0	7860

Table 3. Boundary conditions.

	h (W/m^2K)	T (K)	P (MPa)
Inner surface of tube side and heat exchanger tubes	20	669.6	1.98
Inner surface of shell side, Outer surface of heat exchanger tubes within the shell side	30	506.88	3.11

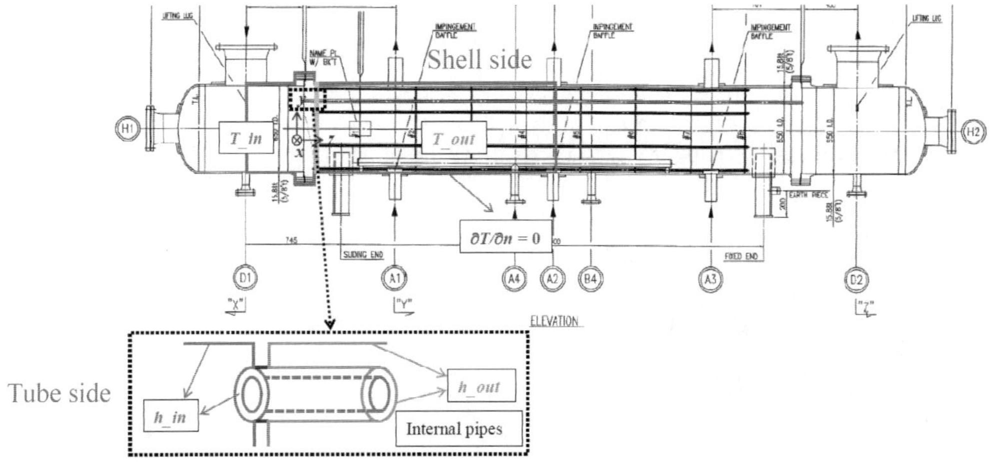

Figure 4. Boundary conditions of thermal analysis of heat exchanger.

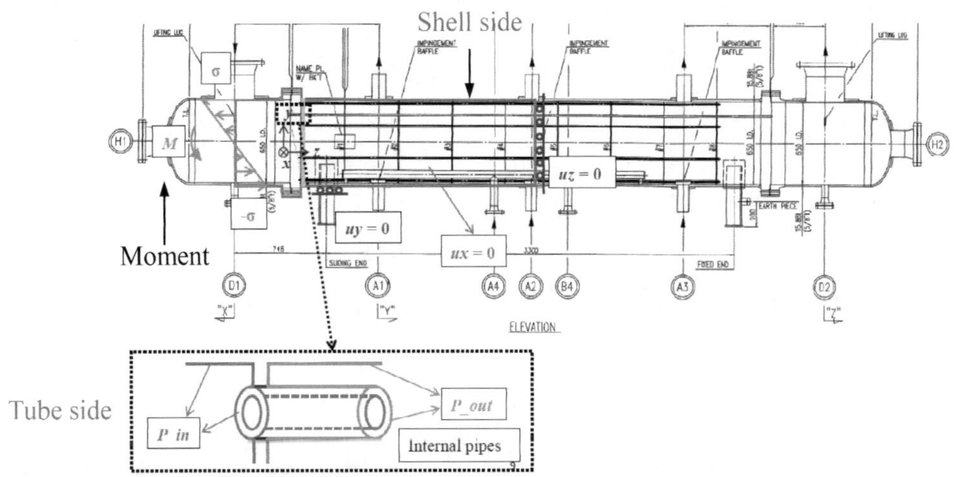

Figure 5. Boundary conditions of stress analysis of heat exchanger.

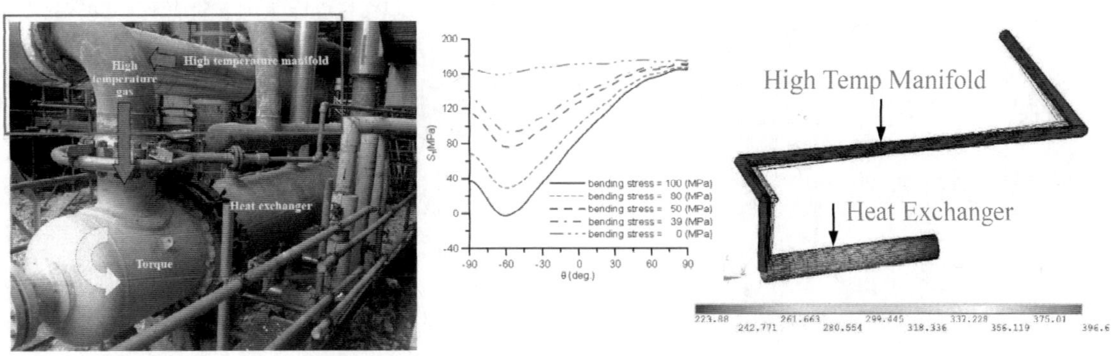

(a) (b)

Figure 6. (a) Bending moment assumed to act on heat exchanger due to thermal expansion of gas intake manifold, (b) Temperature(℃) distribution and deformation of heat exchanger and manifold.

3. Results and discussions

The temperature distribution of heat exchanger based on the assumption with thermal convection

IOP Conf. Series: Materials Science and Engineering **372** (2018) 012025 doi:10.1088/1757-899X/372/1/012025

effect and (h_in) of gas and steam equal to 20 (W/m^{2}K) and 30 (W/m^{2}K) respectively as shown in figure 7. The highest temperature of 382°C (655 K) is found at the inlet of tube side and the lowest temperature of 224°C (497 K) at the outlet of shell side. As a whole, the temperature distribution at the upper part of heat exchanger is higher than that at the lower part and, such a difference is well related to the tube arrangement of heat exchanger which upper tubes are closer to the shell as in figure 8.

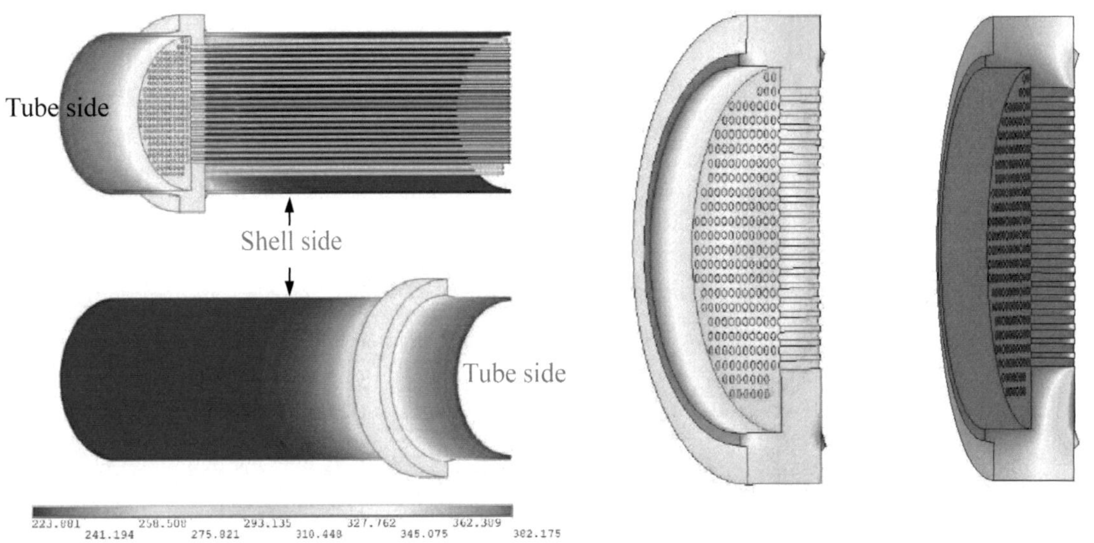

(a) (b) (c)

Figure 7. (a) Temperature distribution of heat exchanger and flange portion (b) with convection and (c) without convection effect.

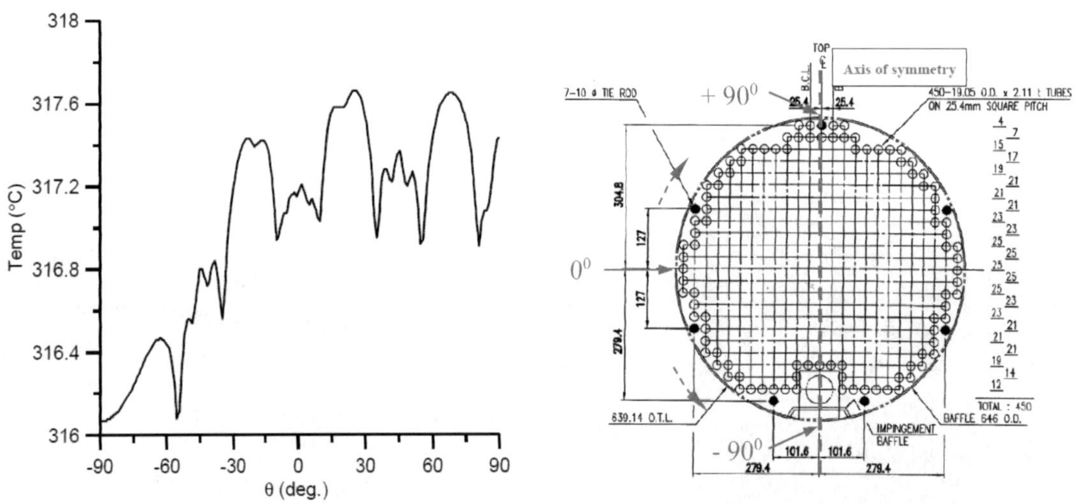

(a) (b)

Figure 8. Temperature distribution at the junction of weld and shell side shows (*a*) The temperature increases with from the lower region(θ = -90°) to upper region (θ = 90°) of heat exchanger and is well related to, (b) the tube arrangement of heat exchanger which upper tubes are closer to the shell.

While the thermal effects of internal convection and external expansion moment by gas intake manifold are both considered, the results of maximum principal stress S_1 show that most area is in tension except the corner of tube sheet on tube side and internal tubes on shell side as in figure 9. Wherein, the tension stress is the highest at the weld region connecting flange and shell. In this case along the circumferential weld of concerned, all three stresses increase gradually with orientation and reach the highest at (θ = 90°) which located at the upper point and outer surface of the shell as in figure 10. In contrast, the stress on the inner surface of weld varied in oscillation along the circumferential direction. Definitely, the stress oscillation is closely related with the temperature variation on the inner surface of weld.

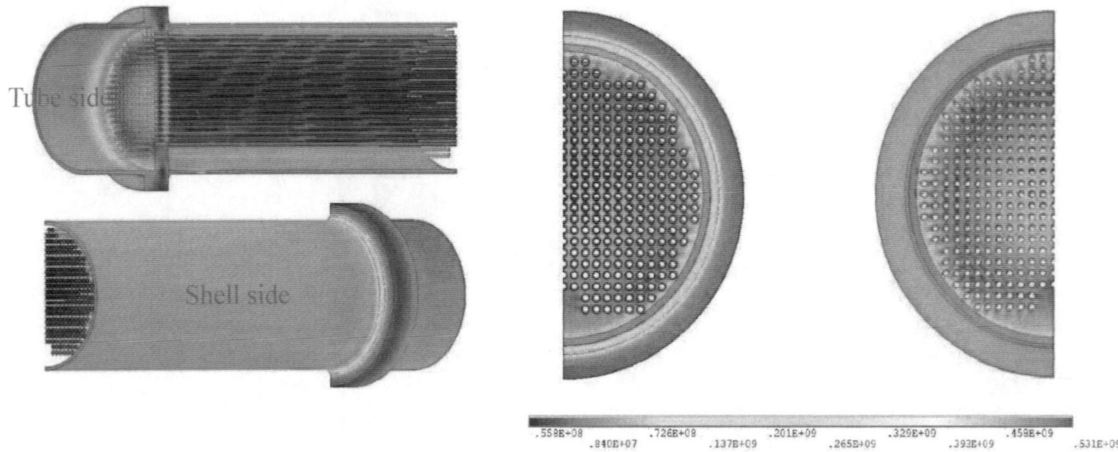

Figure 9. Distribution of maximum principle stress S_1 of heat exchanger in which most area is in tension.

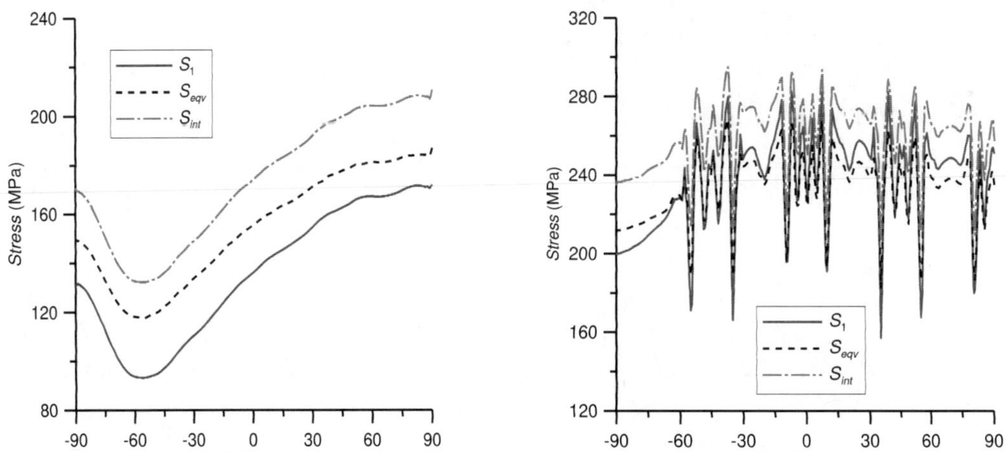

Figure 10. Stress distribution at the junction of weld and shell at (a) outer surface, (b) inner surface. (θ = -90°and θ = 90°are the bottom and top of heat exchanger respectively).

Further referring to the crack configuration appeared on outer surface of weld, it is believed that circumferential crack preceded those two short branch cracks based on the failure principle of crack branching as described in figure 11. Accordingly, it is assumed that circumferential crack initiated from inner surface mainly due to relative high thermal stress. Axial branch cracks followed as circumferential crack propagated further in depth. To verify the crack initiation, distributions of temperature and stress at five different depths (0t, 1/4t, 1/2t, 3/4t and 1t) of weld was investigated as shown in figure 12. The results show that all stresses at inner surface are of the highest in which S_1 is of most concern in terms of crack orientations. Actually, the stress due to thermal effect in this case is much greater than that by internal pressure solely.

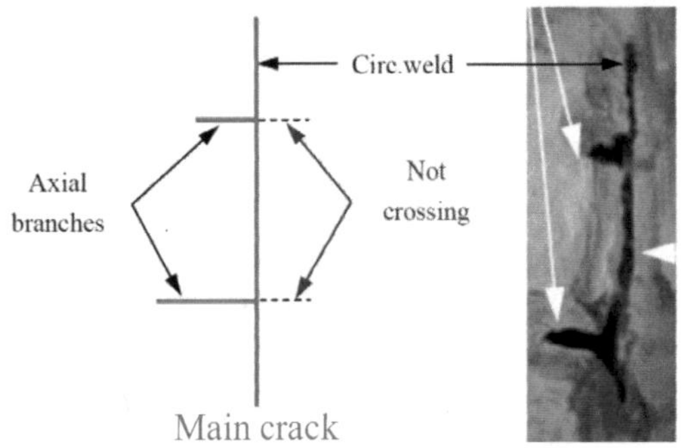

Figure 11. Failure principle of crack branching.

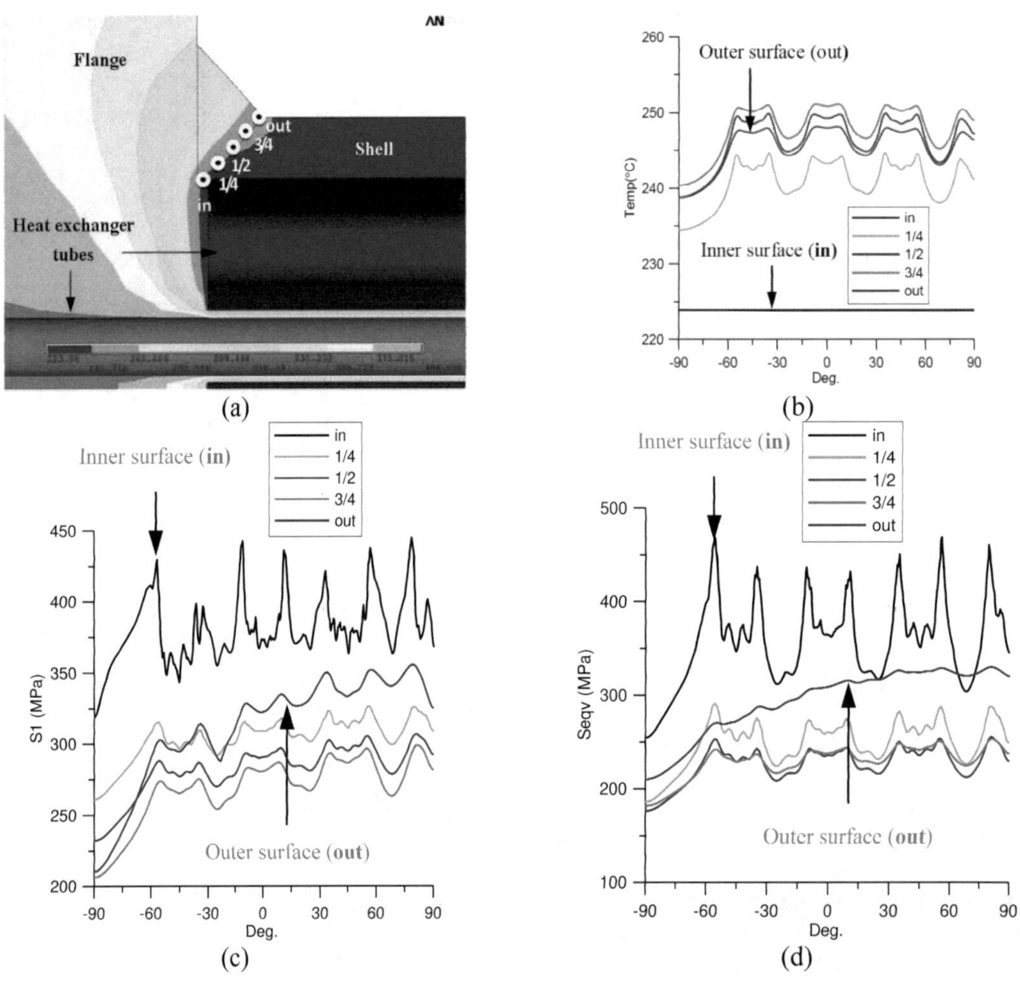

Figure 12. (a) Temperature distribution of weld (a) across the thickness, (b) along the weld in circumferential direction, (c) S_1 stresses at different depth, (d) S_{eqv} stresses at different depth.

4. Conclusions

Repeated cracking at the weld concerned is confirmed by the analysis of temperature, stress and failure principle of crack branching. It is found that internal tubes close to the inner surface at upper part of shell results in higher temperature and stress oscillation. Further study of the temperature and stress at different depths confirm that stress of weld is apparently of the highest at the inner surface which disclosed the evidence of crack initiation from circumferential direction and crack branching to axial ones. Based on these results, it is found that the existing fixture is not proper to avoid the propagation of circumferential crack which could be the major concern of heat exchanger rupture. Hence, proposal was given to shut down WHB immediately and to review the design of internal tubes and the external arrangement of high temperature gas intake manifold before replacing with a new one.

References

[1] Wu Y-H 2011 NDT of crucial structures experienced in refineries in Taiwan GCFNDT (Xiamen and Kinmen)

[2] Wu Y-H and Hsiao C-C 2012 Fitness for service of critical components in refiney - final report (CPC/ITRI, Taiwan)

[3] Boresi A P and Chong K P 2000 *Elasticity in Engineering Mechanics* (New York: John Wiley & Sons)

[4] http://www.engineeringtoolbox.com/overall-heat-transfer-coefficient-d_434.html

IOP Conf. Series: Materials Science and Engineering **372** (2018) 012026 doi:10.1088/1757-899X/372/1/012026

Simulations of interfacial creep generation for shrink-fitted bimetallic work roll

H Sakai[1,2], N-A Noda[1], Y Sano[1], Y Takase[1] and G W Zhang[1]

[1]Mechanical Engineering Department, Kyushu Institute of Technology Sensui-Cho 1-1 Tobata-Ku, Kitakyushu-Shi, Fukuoka, Japan

E-mail: q595103h@mail.kyutech.jp

Abstract. The bimetallic work rolls are widely used in the roughing stands of hot rolling stand mills. The rolls are classified into two types: one is a single-solid type, and the other is a shrink-fitted construction type consisting of a sleeve and a shaft. Regarding the assembled rolls consisting of a sleeve and a shaft, the interfacial creep phenomenon can be seen between the shaft and the shrink-fitted sleeve. This interfacial creep phenomenon causes the relative displacement on the interface between the sleeve and the shaft. Although to clarify this creep mechanism is important issues, experimental simulation is very difficult to be conducted. In this paper, the interfacial creep phenomenon is realized by using the elastic finite element method (FEM) analysis. It is found that the interface creep can be regarded as the accumulation of the relative circumferential displacement on the interface of the sleeve and the shaft.

1. Introduction

The bimetallic roll is used at the roughing stands of hot rolling stand mills as shown in figure 1. The bimetallic roll is usually manufactured by using centrifugal casting method. However, so-called assembled rolls are sometimes used for large diameter roll having more than 1000 mm [1] and H-shaped steel rolling rolls [2,3] by connecting a hollow sleeve and shafts due to shrink-fitting. Although the wear and surface roughness soon appear on the roll body, the shaft can be reused in this assembled structure. Furthermore, it is easy to provide high wear resistance property to the sleeve. However, the shrink-fitted structure has some inherent problems such as roll residual bending deformation [4-6], shaft breakage due to fretting cracks and interfacial creep. This creep phenomenon often causes damage to the roll, although few studies are available.

It is known that this creep phenomenon can be regarded as relative slippage as a common phenomenon in rolling bearing [7]. Mainly, two types of slippage is known as the creep phenomenon in the bearing. One is the relative slippage appearing in the same circumferential direction as the bearing rotating direction [8]. The other is the relative slippage appearing in the opposite circumferential direction to the bearing rotating direction [9]. It is also known that one of the relative slippage appearing in the opposite direction is caused by the accumulation of elastic deformation. However, few study is available to realize the creep phenomenon in the numerical simulation that is useful for investigating the detail.

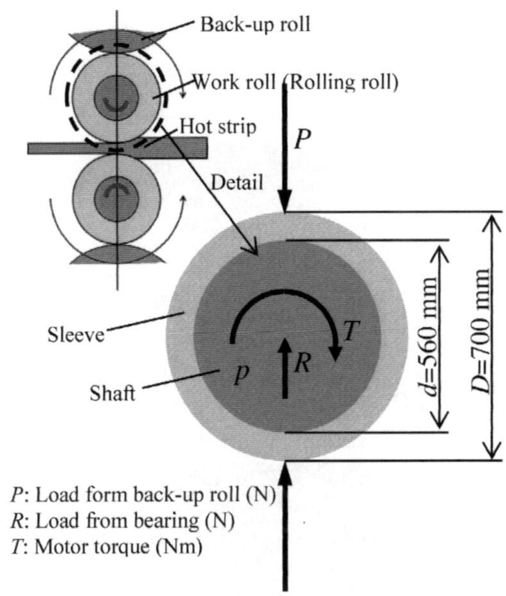

Figure 1. Real rolling roll dimensions with load conditions.

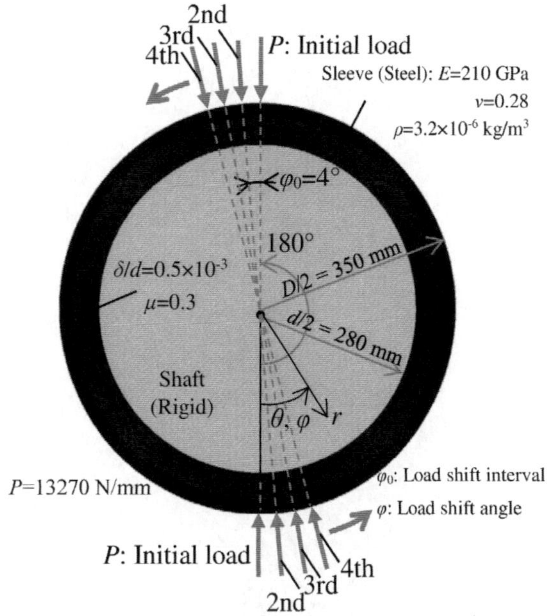

Figure 2. 2D FEM model and boundary conditions.

In this study, therefore, the interfacial creep phenomenon will be realized by using the elastic finite element method (FEM) analysis. Here, elastic deformations will be considered as the most fundamental factors causing the interfacial creep in the rolling roll.

2. FEM modelling to realize the interfacial creep in work roll

In this research, to realize the interfacial creep, a 2D model as shown in figure 2 will be considered. Due to the effect of the elastic deformation on the interfacial creep, the elastic body and rigid body are used in the 2D model. As shown in figure 2, the load from the support roll and the load from the hot strip are applied to this model. The concentrated load from the support roll and the hot strip are balanced, therefore, $R = 0$ in figure 1. The rolling torque from the motor is not considered, therefore, $T = 0$ in figure 1. Additionally, the influence of the thermal expansion of the sleeve on the interfacial creep can be neglected. Therefore we didn't consider it in this paper.

In this study, the interfacial creep will be studied by applying the finite element method (FEM). The FEM is one of the most used numerical modeling techniques, which can be used for many engineering applications conveniently [10-15]. The number of elements of the model used in this study is 46080. Static structural analysis is performed to the roller by using MSC Marc Mentat 2012 with full Newton–Raphson iterative sparse solver of multifrontal method. In this study, a two-dimensional elastic FEM analysis can be applied because the loading condition does not exceed the yielding stress for the steel sleeve.

As shown in figure 2, dimensions of the roller is considered whose outer diameter $D = 700$ mm. Here, the roller consists of steel sleeve and rigid shaft connected by shrink fitting. The shrink fitting ratio is defined as δ/d, where δ is the diameter difference and d is the inner diameter of the sleeve $d = 560$ mm.

As shown in figure 2, the sleeve material is steel whose Young's modulus $E = 210$ GPa, Poisson's ratio $\nu = 0.28$, density $\rho = 3.2\times10^{-6}$ kg/m^3. Also, the friction coefficient between the interface of the sleeve and the shaft is $\mu = 0.3$.

3. Analysis result of the interfacial creep in work roll

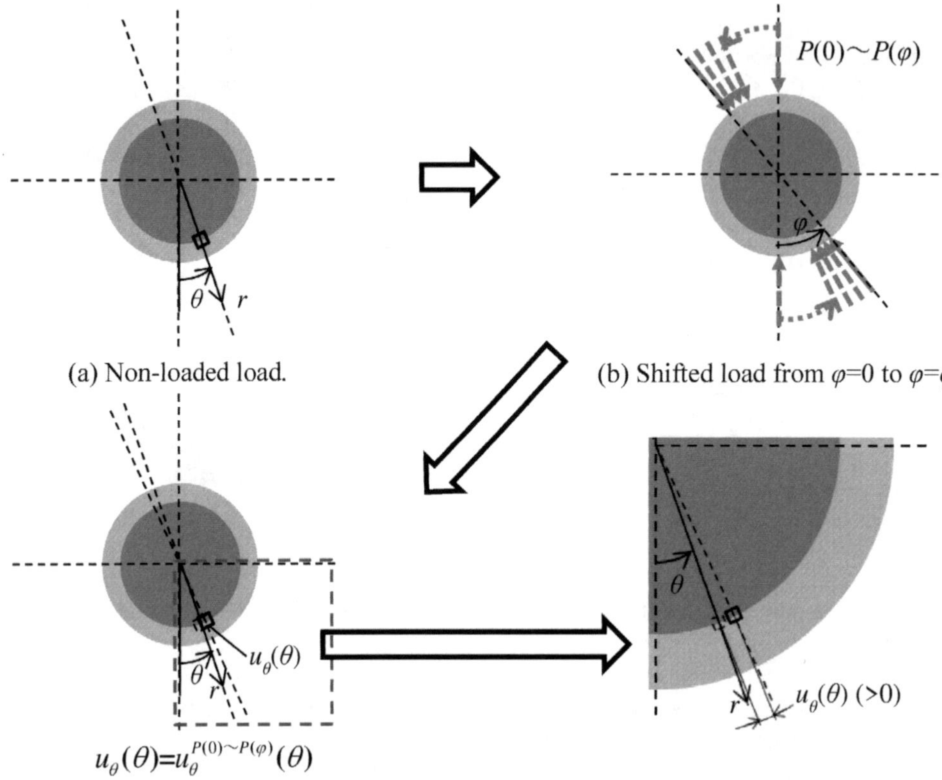

(a) Non-loaded load. (b) Shifted load from φ=0 to φ=φ.

(c) Displacement $u_\theta(\theta)$ at θ=θ due to the load $P(0)\sim P(\varphi)$

(d) Detail of (c).

Figure 3. Interfacial displacement $u_\theta^{P(0)\sim P(\varphi)}(\theta)$ at θ due to the shifted

3.1. Determination of the interfacial creep

In order to determine the interfacial creep, the relative circumferential displacements between the interface of the sleeve and shaft will be focused. In this analysis, since the relative circumferential displacement on the interface of the sleeve is equal to the absolute circumferential displacement on the interface of the sleeve because the rigid shaft is fixed. Therefore the circumferential displacement on the interface of the sleeve will be used to determine the interfacial creep. Assume that $u_\theta^{P(0)\sim P(\varphi)}(\theta)$ is the circumferential displacement on the interface of the sleeve. Figure 3 shows the definition of the interfacial displacement $u_\theta^{P(0)\sim P(\varphi)}(\theta)$ under the shifted load from $\varphi = 0$ to $\varphi = \varphi$.

To simulate the coming out behaviour, the roll rotation is replaced by the shifted load P [16,17] in the circumferential direction on the fixed roll as shown in figure 2. The roller is under a concentrated to load P = 13700 N/mm. As shown in figure 2, the continuous load shifting can be replaced by discrete load shifting with load shift angle φ_0, which is usually used as a standard discretization numerical analysis. Here, a smaller angle φ_0 provides accurate results but large computational time. Therefore, the optimal angle φ_0 should be discussed by investigating u_θ with varying the shift angle, φ_0 = 12°, φ_0 = 8°, φ_0 = 4°, and φ_0 = 0.25°. Figure 4 shows u_θ focusing on the interfacial creep. Although the result for φ_0 = 12° and φ_0 = 8° are different from others, the result for φ_0 = 0.25° and φ_0 = 4° are almost coincide with each other. Since the effect of discrete load shifting is less than a few percent if $\varphi_0 \leq 4°$, it may be concluded that the load shifting angle φ_0 = 4° is the most suitable to reduce large calculation time without losing accuracy. In the following calculation the load shift angle φ_0 = 4° will

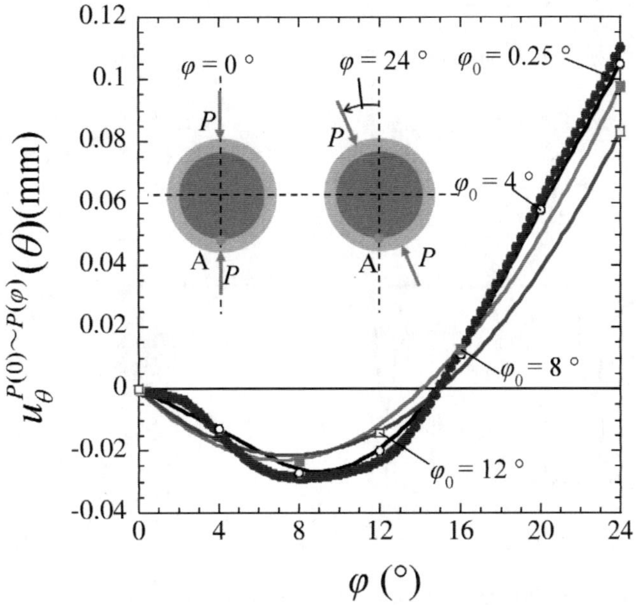

Figure 4. Displacement of the sleeve interface point A when $\theta_0 = 0.25°$, 4°, 8° and 12°.

be used consistently. From figure 4, initially, the displacement of the point A is $u_\theta^{P(0)\sim P(\varphi)} < 0$. This is because the point A is pushed counter-clockwise direction due to the concentrated load on the lower side. Thereafter, the influence of the concentrated load on the upper side is increased, and the point A is pushed clockwise direction. As a result, the displacement of the point A is $u_\theta^{P(0)\sim P(\varphi)} > 0$.

3.2. *Displacement on the interface of the sleeve u_θ*

Figure 5(a) shows the displacement distribution when the loads are applied at $\varphi = 0$ and $\varphi = \pi$ defined as (a) $u_\theta^{P(0)}(\theta)$. and also (b) $u_\theta^{P(0)\sim P(2\pi/15)}(\theta)$. When the concentrated loads $P(\varphi)$ are applied at $\varphi = 0$ and $\varphi = \pi$ as $P(\varphi) = P(0)$, the deformation is symmetric with respect to the loading points $\varphi = 0$ and the displacement is skew-symmetric. In other words, we have

$$-u_\theta^{P(0)}(-\theta) = u_\theta^{P(0)}(\theta) \ (\varphi = 0, 0 \le \theta \le \pi). \tag{1}$$

Figure 5(b) shows the displacement distribution $u_\theta^{P(0)\sim P(2\pi/15)}(\theta)$. In this case, the concentrated loads $P(\varphi)$ are initially applied at $\varphi = 0$ and finally at $\varphi = 2\pi/15$ denoted as $P(\varphi) = P(0) \sim P(2\pi/15)$. In figure 5(b), the deformation is not symmetric anymore with respect to the final loading points $\varphi = 2\pi/15$.

$$-u_\theta^{P(0)\sim P(\varphi)}(-\theta + \varphi) \neq u_\theta^{P(0)\sim P(\varphi)}(\theta + \varphi) \ (\varphi > 0, 0 \le \theta \le \pi) \tag{2}$$

Figure 5 also shows the average value of the displacements $u_{\theta,\text{ave.}}^{P(0)}$ and $u_{\theta,\text{ave.}}^{P(2\pi/15)}$. The average displacement along the interface of the sleeve $u_{\theta,\text{ave.}}^{P(0)\sim P(\varphi)}$ is defined in equation (3).

$$u_{\theta,\text{ave.}}^{P(0)\sim P(\varphi)} = \frac{1}{2\pi}\int_0^{2\pi} u_\theta^{P(0)\sim P(\varphi)}(\theta)d\theta \tag{3}$$

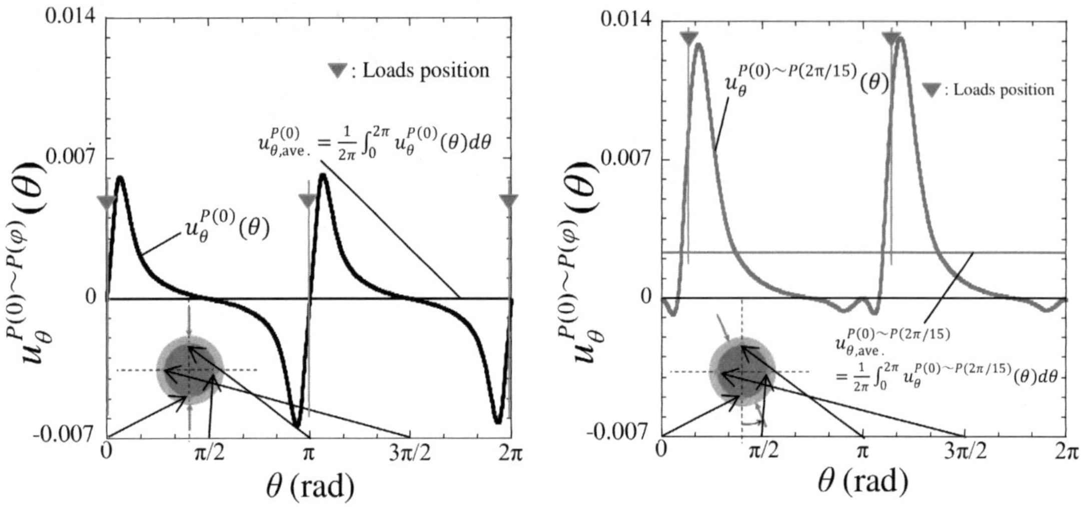

(a) At $P(0)$ (b) At $P(2\pi/15)$

Figure 5. Distribution of displacement of the sleeve interface $u_\theta^{P(0)\sim P(\varphi)}(\theta)$.

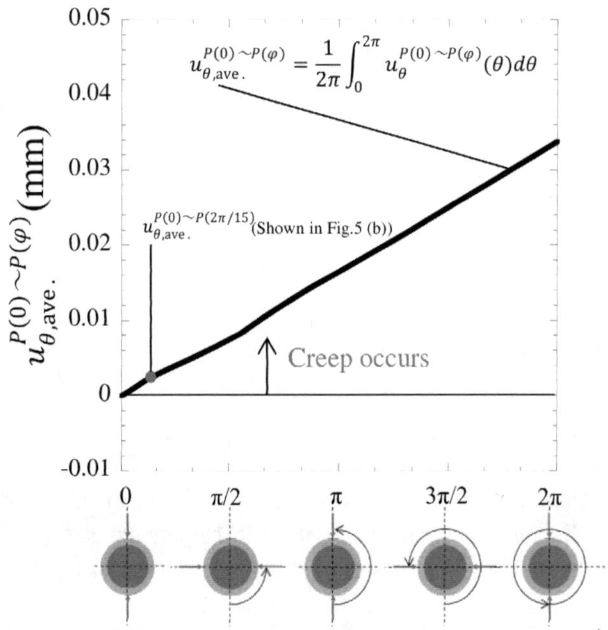

Figure 6. History of average of displacement of the sleeve interface when P is shifted.

Under the initial loads $P(\varphi) = P(0)$, we have $u_{\theta,\text{ave.}}^{P(0)}=0$. This is because the displacement distribution is symmetric. On the other hand, under the shifted loads $P(\varphi) = P(0) \sim P(2\pi/15)$, the average value $u_{\theta,\text{ave.}}^{P(2\pi/15)}$ is not zero anymore because the displacement distribution is not symmetric.

Figure 6 shows the relationship between the average displacement $u_{\theta,\mathrm{ave.}}^{P(0)\sim P(\varphi)}$ and the load shift angle φ. From figure 6, it is found that $u_{\theta,\mathrm{ave.}}^{P(0)\sim P(\varphi)}$ increases with increasing φ. It may be concluded that the interfacial creep occurs as soon as the load shifting starts.

4. Conclusion

- The interface creep phenomenon was realized through quiz-static structural FEM analysis assuming the elastic sleeve and rigid shaft by shifting the load P on the fixed 2D work roll.
- The interface creep phenomenon was discussed by focusing on the relative circumferential displacement on the interface of the sleeve and the shaft. The interface creep occurs due to the collapse of the symmetry of the deformation of the sleeve.
- The interface creep can be regarded as the accumulation of the relative circumferential displacement on the interface of the sleeve and the shaft.
- The average of the relative circumferential displacement on the interface of the sleeve and the shaft. $u_{\theta,\mathrm{ave.}}^{P(0)\sim P(\varphi)}$ increases with increasing the load shift angle φ. That means the interfacial creep occurs as soon as the load shifting starts.

References

[1] Shimoda H, Onodera S., Hori K and Dohi O 1966 *Transactions of the Japan Society of Mechanical Engineering* **32** 689-94
[2] Takigawa H, Hashimoto K, Konno G and Uchida S 2003 Current advances in materials and processes *Report of the ISIJ Meeting* **16** 1150-3
[3] Irie T, Takaki K, Tsutsunaga I and Sano Y 1979 *Testu-to-Hagane* **65** 293
[4] Noda N-A, Sano Y, Takase Y, Shimoda Y and Zhang G 2017 *Sosei-toKako* **58** 66-71
[5] Matsunaga E, Sano Y and Nishida S 1997 Current advances in materials and processes *Report of the ISIJ Meeting* **10** 1078
[6] Matsunaga E, Tsuyuki T and Sano Y 1998 Current advances in materials and processes *Report of the ISIJ Meeting* **11** 362
[7] Soda N 1964 *Bearing, Iwanami Shoten* 196-203
[8] Niwa T 2013 *Ntn Technical Review* **81** 100-3
[9] Murata J and Onizuka T 2004 *Koyo Engineering Journal* **166** 41-7
[10] Miyazaki T, Noda N-A, Ren F, Wang Z, Sano Y and Iida K 2017 *Int J Adhes Adhes.* **77** 118-37
[11] Noda N-A, Miyazaki T, Li R, Uchikoba T and Sano Y 2015b *Int J Adhes Adhes.* **61** 46-64
[12] Noda N-A, Uchikoba T, Ueno M, Sano Y, Iida K, Wang Z and Wang G 2015c *ISIJ International* **55** 2624-30
[13] Wang Z, Noda N-A, Ueno M and Sano Y 2016 *Steel Research International* **88** doi: 10.1002/srin.201600353
[14] Noda N-A, Shen Y, Takaki R, Akagi D, Ikeda T, Sano Y and Takase Y 2017 *Theoretical and Applied Fracture Mechanics* **90** 218–27
[15] Noda N-A, Chen X, Sano Y, Wahab M A, Maruyama H and Fujisawa R 2016 *Materials & Design* **96** 476-89
[16] Noda N-A, Suryadi D, Kumasaki S, Sano Y and Takase Y 2015 *Engineering Failure Analysis* **57** 219-35
[17] Noda N-A, Xu Y, Suryadi D, Sano Y and Takase Y 2016 *Journal of ISIJ International* **56** 303-10

Exact solutions for postbuckling of a graded porous beam

L S Ma[1,2] and Z Y Ou[1]

[1]School of Science, Lanzhou University of Technology, Lanzhou 730050, China

E-mail: lsma@lut.cn

Abstract. An exact, closed-form solution for the postbuckling responses of graded porous beams subjected to axially loading is obtained. It was assumed that the properties of the graded porous materials vary continuously through thickness of the beams, the equations governing the axial and transverse deformations are derived based on the classical beam theory and the physical neutral surface concept. The two equations are reduced to a single nonlinear fourth-order integral-differential equation governing the transverse deformations. The nonlinear equation is directly solved without any use of approximation and a closed-form solution for postbuckled deformation is obtained as a function of the applied load. The exact solutions explicitly describe the nonlinear equilibrium paths of the buckled beam and thus are able to provide insight into deformation problems. Based on the exact solutions obtained herein, the effects of various factors such as porosity distribution pattern, porosity coefficient and boundary conditions on postbuckling behavior of graded porous beams have been investigated.

1. Introduction

The present study is devoted to find the exact solution for postbuckling behavior of graded porous beams under the axial compressive load. Since the overwhelming majority of the nonlinear governing equations for large deflection problems are not soluble analytically, it is not easy to find an exact solution to the nonlinear equations for large deflection beams. Only a few exact solutions have been presented up to the present time. However, in the wake of developments in science and technology, the accurate prediction of the postbuckling behavior of beams is an area of concern.

Graded porous materials are a class of lightweight materials in which a change in their material properties is continuous, that is, these properties change with changes in position along one or more directions of the structure to attain a needed purpose. Due to their unique advantages, the application of structures made of graded porous materials is extensive covers a wide range of aerospace, civil and mechanical fields. Therefore, emerging applications of graded porous materials have led to many conducted researches regarding the mechanical behavior of these materials in the past two decades. Magnucki and Stasiewicz [1] explicitly obtained the critical buckling load of a porous beam subjected to the axial compressive load. Jasion and Magnucka-Blandzi [2] investigated the global and local buckling of sandwich metal foam circular plates and beams and the critical buckling loads for these structures were presented and compared with those obtained by other methods. Magnucka-Blandzi and Magnucki [3] gave some optimal parameters in the effective design for a sandwich metal foam beam and the mass and critical load of the beam were considered at the same time in the analysis. Chen *et al* [4] studied the elastic buckling and bending problems for functionally graded porous beams based on shear deformation theory, two porosity distributions were considered in the analysis. The elastic buckling of the aluminum foam beams was investigated by Grygorowicz *et al* [5] and the formula was proposed to describe the critical buckling load of the beams. Ebrahimi and Zia [6] used the Galerkin's method to investigate large amplitude vibration of a functionally graded porous beam. Recently, Barati and Zenkour [7] examined the postbuckling behavior of grapheme platelet reinforced porous beams and two distributions of grapheme platelets were considered in the analysis. Kitipornchai *et al* [8] investigated the elastic buckling and free vibration of grapheme platelet reinforced porous beam. The nonlinear vibration and postbuckling responses for grapheme platelet reinforced porous beams made of functionally graded materials were investigated by Chen *et al* [9] Shafiei and Kazemi [10] studied the nonlinear buckling behavior of micro- and nano-beams made of functionally graded porous material.

As the aforementioned works show, there are few literatures concerning with an exact solution for postbuckling responses of graded porous beams. Motivated by this consideration, in this study, postbuckling of the classical graded porous beams is investigated. Ma and Lee [11] proposed an exact solution of the postbuckling or bending responses for functionally graded beams and the solution is a function related to the applied load. In the present analysis, their work is extended to the graded porous beams.

In this study, using the physical neutral surface concept, governing equations for the postbuckling response of graded porous beams under axial compressive load are derived based on classical beam theory. The nonlinear equations of equilibrium can be simplified to a single fourth order integral differential equation expressed in term of transverse deformation. The nonlinear equation can be exactly solved, then an exact closed-form solution of the postbuckled configuration is obtained and it is a function related to the axial compressive load. The explicit exact solutions describe the nonlinear buckled paths of the beam and would give ones new insights into these deformation problems. Using the exact solutions, effects of porosity distribution pattern and porosity coefficient on postbuckling responses of graded porous beams have been studied.

2. Basic equations

Consider a graded porous material beam with a rectangular cross-section of area A, height h, and length l as shown in figure 1. The origin of the Cartesian coordinate system, (x, y, z), is at the left end of the beam in present study. The xoy plane is placed in the undeformed midplane of the beam, the x axis is coincident with the centrically axis of the beam, the z axis is perpendicular to the x-y plane and positive direction of the z axis is the one towards the thickness of the cross-section.

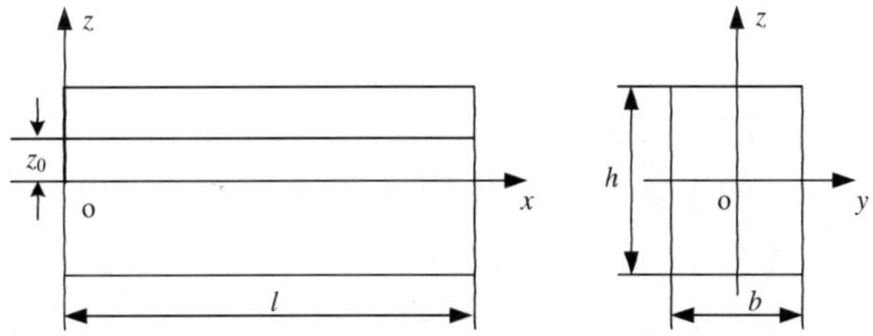

Figure 1. Geometry and coordinates of a beam.

Suppose the mechanical properties of the porous materials vary along the thickness direction of beams. Young's modulus of the porous materials changes according to the two functions, respectively, as follows [12].

$$E(z)=E_1\left[1-e_0\cos(\pi z/h)\right] \tag{1}$$

and

$$E(z)=E_1\left[1-e_0\cos\left(\pi z/2h+\pi/4\right)\right] \tag{2}$$

where, e_0 is the porosity coefficient of the beams, e_0=1-E_0/E_1, E_1 denotes Young's modulus of homogeneous materials beams and is also the maximum value of Young's modulus for the graded porous beams which exist on the upper and lower surfaces corresponding to symmetrical porosity distribution, that is, equation (1), and on the top surface corresponding to asymmetrical porosity distribution, i.e., equation (2), E_0 the minimum value on the midplane for symmetrical porosity distribution and on the bottom plane for asymmetrical porosity distribution. The shear modulus $G(z)=E(z)/[2(1+v)]$, here v is Poisson's ratio and suppose Poisson's ratio is constant along the beam thickness. Figure 2 shows several variation curves of the Young's modulus along thickness of the beam for several values of e_0, (a): symmetrical porosity distribution, (b): asymmetric porosity distribution.

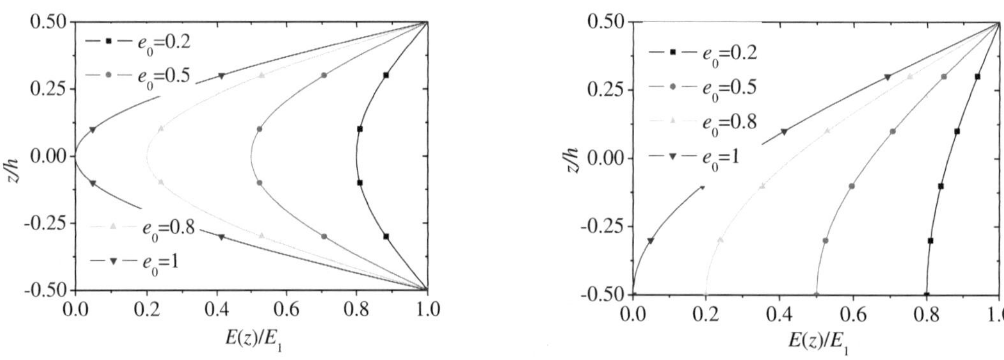

(a) (b)

Figure 2. Variations of the Young's modulus along thickness of the beam for several values of e_0.

When a coordinate system is placed in physical neutral surface of a graded porous beam [13], the term of stretching and bending coupling does not appear in constitutive equations of the beam and this concept can be used to simplify basic equations of the graded porous beam. The physical neutral surface of graded porous beams is represented by $z=z_0$.

$$z_0=\int_{-h/2}^{h/2} zE(z)\mathrm{d}z\left(\int_{-h/2}^{h/2} E(z)\mathrm{d}z\right)^{-1} \tag{3}$$

Therefore, for symmetrical porosity distribution, $z_0=0$, and for asymmetric porosity distribution, $z_0=he_0\left(4/\pi^2-1/\pi\right)/\left(1-2e_0/\pi\right)$. It is obvious that the physical neutral surface is coincident with the geometric midplane for a homogeneous beam or a graded porous beam with symmetrical porosity distribution.

Make use of the physical neutral surface concept and the classical beam theory (CBT), the displacements can be written in the following form

$$U_x(x,z)=u(x)-\left(z-z_0\right)\frac{\mathrm{d}w}{\mathrm{d}x}$$
$$U_z(x,z)=w(x) \tag{4}$$

Here, u and w are the displacements of the physical neutral surface along x and z, respectively. Axial strain is as follows.

$$\varepsilon_x = \varepsilon_x^0 - (z - z_0)\varepsilon_x^1 = \left[\frac{du}{dx} + \frac{1}{2}\left(\frac{dw}{dx}\right)^2\right] - (z - z_0)\frac{d^2 w}{dx^2} \tag{5}$$

In the above, ε_x^0 and ε_x^1 are the strain and curvature in the physical neutral surface, respectively. The constitutive relations are expressed as

$$N_x = \int_A \sigma_x dA = A_x \varepsilon_x^0 + B_x \varepsilon_x^1 = A_x \left[\frac{du}{dx} + \frac{1}{2}\left(\frac{dw}{dx}\right)^2\right]$$

$$M_x = \int_A \sigma_x (z - z_0) dA = B_x \varepsilon_x^0 - D_x \varepsilon_x^1 = -D_x \frac{d^2 w}{dx^2} \tag{6}$$

Here, $A_x = \int_A E(z)\mathrm{d}A$, $B_x = \int_A (z - z_0)E(z)\mathrm{d}A = 0$, $D_x = \int_A (z - z_0)^2 E(z)\mathrm{d}A$.

It is obvious that the term of stretching and bending coupling does not appear in equations (6) based on the physical neutral surface theory.

Through the application of energy principle, the equilibrium equations and boundary conditions can be derived as followed based on CBT.

$$\frac{d}{dx}\left\{A_x\left[\frac{du}{dx} + \frac{1}{2}\left(\frac{dw}{dx}\right)^2\right]\right\} = 0 \tag{7}$$

$$-D_x \frac{d^4 w}{dx^4} + A_x \left[\frac{du}{dx} + \frac{1}{2}\left(\frac{dw}{dx}\right)^2\right]\frac{d^2 w}{dx^2} = 0 \tag{8}$$

$w = 0$ and $\frac{dw}{dx} = 0$ for a clamped end, $w = 0$ and $\frac{d^2 w}{dx^2} = 0$ for a hinged end (9)

The boundary conditions of the axial displacement u can be expressed as

$$u(0)=0 \text{ and } u(l) = -pl/A_x \tag{10}$$

where, p is an axial compressive load.

3. Solution

Integrating equation (7) twice, one obtains:

$$u = -\frac{1}{2}\int_0^x \left(\frac{dw}{d\eta}\right)^2 d\eta + \frac{C_1}{A_x} x + C_2 \tag{11}$$

Using the boundary conditions for the axial displacement, one can obtain:

$$C_1 = \frac{A_x}{2l}\int_0^l \left(\frac{dw}{dx}\right)^2 dx - p, \quad C_2 = 0 \tag{12}$$

The substitution of equations (11) and (12) into equation (8) leads to:

$$-D_x \frac{d^4 w}{dx^4} + \left[\frac{A_x}{2l}\int_0^l \left(\frac{dw}{dx}\right)^2 dx - p\right]\frac{d^2 w}{dx^2} = 0 \tag{13}$$

Equation (13) is the geometrically nonlinear governing equation of a graded porous beam and in the equation, the midplane stretching is taken account of.

The dimensionless variables are defined as follows

$$\xi = x/l,\quad W = w/h,\ F_1 = A_x h^2/D_x\ ,\ P = pl^2/D_x\ , P_0 = pl^2/D_0$$

where $D_0 = \int_A z^2 E_1 \mathrm{d}A$. Hence equation (13) can be transformed into the one in dimensionless form:

$$\frac{\mathrm{d}^4 W}{\mathrm{d}\xi^4} + \lambda^2 \frac{\mathrm{d}^2 W}{\mathrm{d}\xi^2} = 0 \tag{14}$$

where λ^2 is defined by

$$\lambda^2 = P - \frac{F_1}{2}\int_0^1 \left(\frac{\mathrm{d}W}{\mathrm{d}\xi}\right)^2 \mathrm{d}\xi \tag{15}$$

The boundary conditions (9) can be written as

$$W(0) = W(1) = 0 \text{ and } \frac{\mathrm{d}W(0)}{\mathrm{d}\xi} = \frac{\mathrm{d}W(1)}{\mathrm{d}\xi} = 0 \text{ for a beam with clamped ends}$$

$$W(0) = W(1) = 0 \text{ and } \frac{\mathrm{d}^2 W(0)}{\mathrm{d}\xi^2} = \frac{\mathrm{d}^2 W(1)}{\mathrm{d}\xi^2} = 0 \text{ for a simply supported beam} \tag{16}$$

For any given deformed configuration $W(\xi)$, the integral of $W(\xi)$ in equations (14) and (15) is constant, and so a closed form solution for the deformed configuration may be obtained. The closed form solution for the deformed configuration of beams is given by

$$W(\xi) = a\left\{\frac{1-\cos\lambda}{\lambda - \sin\lambda}\left[\sin(\lambda\xi) - \lambda\xi\right] - \cos(\lambda\xi) + 1\right\} \text{ for a beam with clamped ends} \tag{17}$$

$$W(\xi) = a\sin(\lambda\xi) \text{ for a beam with simply supported ends} \tag{18}$$

Here, a is a constant associated with the axial compressive load P and can be expressed as:

$$a = \pm\frac{2}{\sqrt{F_1}}\left(\frac{P}{\lambda^2} - 1\right)^{1/2} \tag{19}$$

The characteristic equation for λ of beams is given by:

$$2 - 2\cos\lambda - \lambda\sin\lambda = 0 \text{ for a beam with clamped ends} \tag{20}$$

$$\sin\lambda = 0 \text{ for a beam with simply supported ends} \tag{21}$$

In the early part of buckling, configuration of a buckled beam is extremely approximate to original straight state of the beam. At the present moment, the axially load P_0 is exactly the critical buckling load P_{0cr} that can be obtained from equation (19) by letting a =0,

$$P_{0cr} = \lambda_1^2 C_{e0} \tag{22}$$

where C_{e0} is a constant related to the porosity coefficient, e_0, and λ_1 denotes the lowest eigenvalue in equation (20) or (21). When e_0 = 0, the dimensionless critical buckling load of a homogeneous beam is obtained, that is, $P_{0cr} = 4\pi^2$ for a clamped beam at two ends, $P_{0cr} = \pi^2$ for a simply supported beam at two ends, which are identical to the solution obtained by Ma and Lee [11].

The variations of the dimensionless critical buckling load P_{0cr} with the porosity coefficient e_0 are given in figure 3. The solid and dashed lines indicate the results of beams with symmetric and asymmetric porosity distribution, respectively. It is seen from figure 3 that as the porosity coefficient e_0 increases, the critical buckling load decreases. Such a trend is observed because an increase in e_0 results in a decrease in the Young's modulus of graded porous beams.

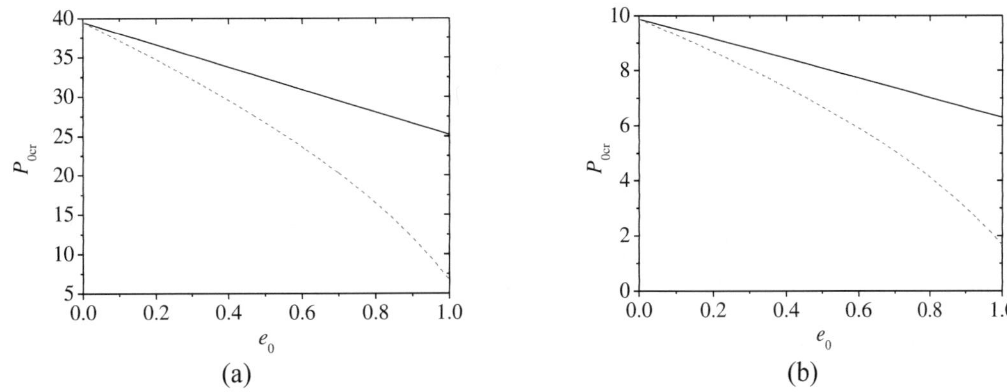

Figure 3. Variation curves of P_{0cr} *vs* e_0. (a): a clamped beam, (b): a simply supported beam.

The typical post-buckling paths of graded porous beams are shown in figure 4. The solid and dashed lines indicate the results of beams with symmetric and asymmetric porosity distribution, respectively. As expected, in figure 4, the variation curves of the post-buckling deflection with axial compressive load for graded porous beams are quite similar to those for pure material beams.

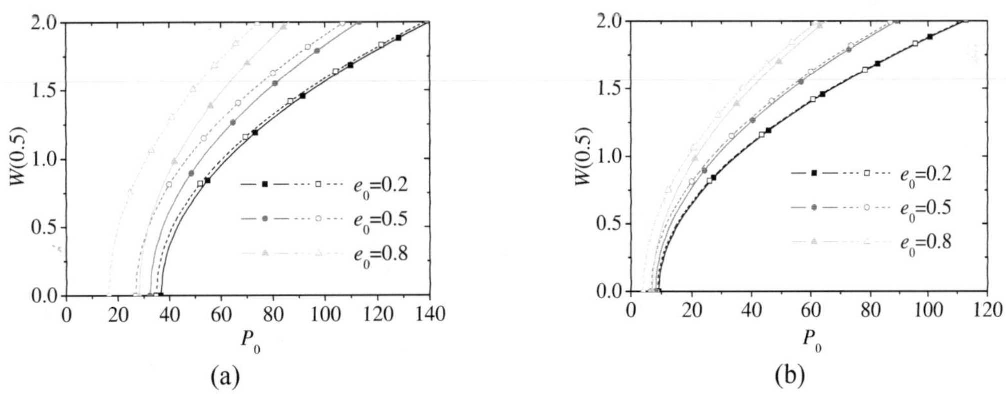

Figure 4. Post-buckling paths of a graded porous beam. (a): a clamped beam, (b): a simply supported beam.

4. Conclusion

An exact solution of the postbuckling responses of graded porous beams under the axial compressive load is obtained. The postbuckled configuration of beams was expressed as a function related to axial compressive load. The nonlinear equilibrium paths of the postbuckled beams can be described explicitly by the exact solutions and thus the exact solutions would give ones new insights into deformation problems and can be used as a standard with which various approximate theories and numerical methods can be verified and improved. One can arrive at some conclusions as followed.

- Under axial compressive load, clamped and simple supported porous beams, in the case of both symmetrical and asymmetric porosity distribution, all exhibit typical bifurcation buckling behavior.
- As the porosity coefficient increases, the critical buckling load decreases, which becomes more pronounced for the case of asymmetric porosity distribution.

Acknowledgments

Many thanks to the National Natural Science Foundation of China (11472123) and the Open Projects of State Key Laboratory for Strength and Vibration of Mechanical Structures at Xi'an Jiaotong University (SV2014-KF-04).

References

[1] Magnucki K and Stasiewicz P 2004 *Theor. Appl. Mech.* **42** 859–68
[2] Jasion P, Magnucka-Blandzi E, Szyc W and Magnucki K 2012 *Thin Walled Struct.* **61** 154–61
[3] Magnucka-Blandzi E and Magnucki K 2007 *Thin-Walled Struct.* **45** 432–8
[4] Chen D, Yang J and Kitipornchai S 2015 *Compos. Struct.* **133** 54–61
[5] Grygorowicz M, Magnucki K and Malinowski M 2015 *Thin-Walled Struct.* **87** 127–32
[6] Ebrahimi F and Zia M 2015 *Acta Astronautica* **116** 117–25
[7] Barati M R and Zenkour A M 2017 *Compos. Struct.* **181** 194–202
[8] Kitipornchai S, Chen D and Yang J 2017 *Mater. Des.* **116** 656–65
[9] Chen D, Yang J and Kitipornchai S 2018 *Compos. Sci. Technol.* **142** 235–45
[10] Shafiei N and Kazemi M 2017 *Compos. Struct.* **178** 483–92
[11] Ma L S and Lee D W 2011 *Eur. J. Mech.* A *Solids* **31** 13–20
[12] Zhang D G and Zhou Y H 2008 *Comput. Mater. Sci.* **44** 716–20
[13] Jabbari M, Joubaneh F, Khorshidvand E and Eslami M R 2013 *Int. J. Mech. Sci.* **70** 50–6

IOP Conf. Series: Materials Science and Engineering **372** (2018) 012028 doi:10.1088/1757-899X/372/1/012028

Experimental study on performance of outdoor ground materials in aspect of surface temperature by constant field experiment in subtropical climate city

T Y Xi[1,3], J H Ding[2,4], X W Lv[1] and Y S Lei[1]

[1]Architecture School of Harbin Institute of Technology, Harbin City, 150006, China
[2]School of Architecture & Urban planning, Shenzhen University, Shenzhen, 518060, China
[3]HeiLongjiang Cold Region Architectural Science Key Laboratory, Harbin, 150000, China

E-mail: a-djh@163.com

Abstract. In order to create a comfortable building thermal environment, it is important to study the outdoor ground materials performance. In this article, we carried out a constant field experiment in Guangzhou, China, studying on the variations of the surface temperature of three common outdoor building materials: concrete, pavement and grass. We put the equipment on six experiment points respectively to measure the ground surface temperature constantly. The result shows that because of the specific heat capacity, both concrete and pavement have an obvious time delay during their temperature decrease when the grass ground has almost no time delay. And when in the same conditions (exposed to sunlight all day), the material with a low specific heat capacity has a more sensitive variation in temperature. The lower the specific capacity is, the steeper the variation trend of the surface temperature will be. So compared with concrete, the pavement brick ground with a low specific heat capacity has a higher surface temperature in daytime and a lower temperature in the late night time. When in different conditions (different time exposed to sunlight), the temperature value is proportional to the time exposed to the sunlight between the same materials. The concrete exposed to sunlight all day has the highest temperature when the shaded one has the lowest. This experiment reveals that both specific heat capacity and the exposed time to sunlight has a strong influence on the surface temperature of outdoor materials. In subtropical region, the materials with a higher specific heat capacity and a less time exposed to sunlight may be more beneficial to the building thermal environment.

1. Introduction

The outdoor thermal environment has become increasingly important on building design. The thermal comfort experience has a directly influence on how much time people spend out the room, especially in hot or cold region, which indirectly effects people's physical health. What's more, according to the Report of the Development of the International Energy in 2017, the proportion of primary energy consumption in China is up to 62%, so if people can spend more time out the house, kinds of energy consumption such as air conditioning, lighting or warm heating can be reduced, which can make contributions to the reduction of world energy. A great attention has been paid on how to create a suitable outdoor thermal environment.

As for how to improve the outdoor thermal environment, various kinds of experiments have been done. Ryozo Ooka *et al* studied the thermal environment of apartment block in Shenzhen trough simulation and field measurement [1]. Uehara Kiyoshi *et al* carried out the wind-tunnel experiments in order to improve the natural ventilation of a street-canyon [2]. Akira Hoyano *et al* clarified characteristics of the summer thermal environment in an actual membrane structure by field measurements and simulations [3]. Xi *et al* had done some researches mainly on the influence of human-built elements and the outdoor thermal environment and thermal environment of piloti [4-7].

Besides the studies above, there are also a lot studies on the specific factors which can influence the building thermal environment. Yue *et al* revealed that the urban water landscape has some effects on the thermal environment [8]. Somsak Chaiyapinunt *et al* had invested different types of glass window to study the influence of it on the thermal comfort and heat transmission [9]. Elisa *et al* studied on the roof and presented an assessment of the effectiveness of cool and green roof on order to improve the thermal comfort [10]. Mohamad *et al* studied the thermal performance of a special wall aiming at examining the energy behavior of the buildings' multi-layer exterior wall structures [11].

As for materials, many studies, related to its effect on the out thermal environment, have been done. But studying on the property of material itself is few in number, let alone the specific data or any constant field experiment. This paper has studied on the performance of ground materials itself via a constant field experiment and present a set of specific data, which offers good references for researchers in further research.

2. Methodology

2.1. Experiment points

The experiment is conducted in the campus of South China University of Technology, Guangzhou. Guangzhou is located in the subtropical region where the winter is short, summer is long. So it is important to study the summer outdoor thermal environment to satisfy the outdoor experience. Because of the typical subtropical climate, the sunlight there is really strong, ground materials receive more solar radiation which can directly influence the ground surface temperature. So it is reasonable to choose Guangzhou city as the experiment site.

In order to compare the different materials performance, six measurement points have been chosen (figure 1). As is shown in the picture, both point 1, 2, 3, 5 are concrete ground. Point 1 is located on an open square where can receive the sunlight all day. Point 2 is located under a piloti shaded from the sun all day. Point 3 is located in building block which is shaded only in the early morning. Point 5 is located under a piloti and only can be exposed to sunlight half the day. And point 4 is pavement brick ground located near a river where can be exposed to sun all day. Point 6 is the grass ground located on an open square, 500 meters far away from other 5 points, exposed to sun all day.

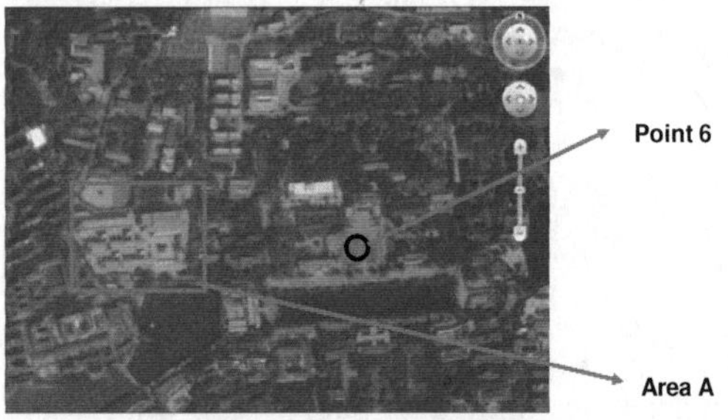

(a) Area A and Point 6

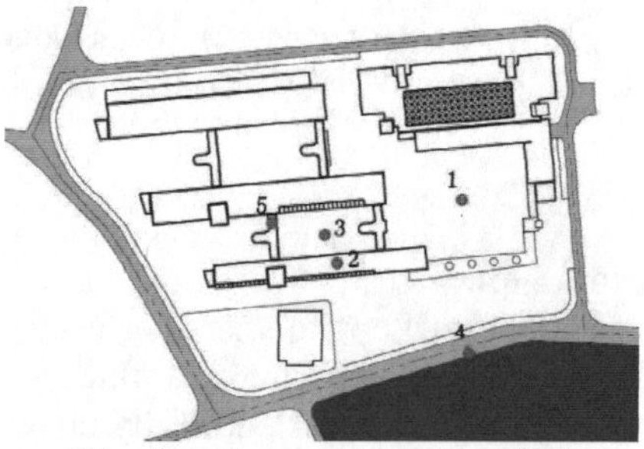

(b) other 5 measurement points

Figure 1. Experiment sites.

2.2. Experiment method

The experiment lasted for 2 whole days, starting on 9:00 am, 4th July, 2009. The equipment is BES-02.

BES-02 is a kind of temperature and humidity collection recorder which can measure the temperature and humidity simultaneously. You can use the computer to preset parameters such as the time to start or to end the measurement. And when the measurement is over, you can down load the measure data through PC. As for the experiment method, we put the equipment on each measurement point respectively to measure the temperature at almost the same time. The surface temperature data of all points is constant during the experiment. In order to exclude the interference of the radiation influence, all sensors were covered by the same materials of the experiment ground. Then a piece of sliver paper was put on it. For example, from point 1 to point 5, the sensors were covered by cement powder and the sensor in point 6 were covered by grass (figure 2). This method to exclude the interference of the radiation influence has been used in many studies such as the experiment held in Guangzhou by Xi [6]. The more specific information about the equipment can be seen in table 1 below.

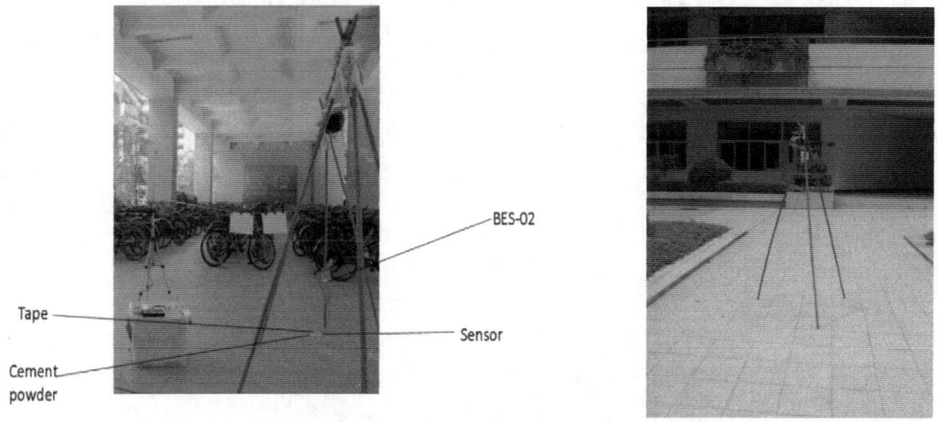

Figure 2. View of experiment.

Table 1. Information of the equipment.

Equipment	Production Place	Operating system	Measurement range	Measurement accuracy
BES-02	China	Windows2000/XP	-30℃-50℃	≤0.5℃

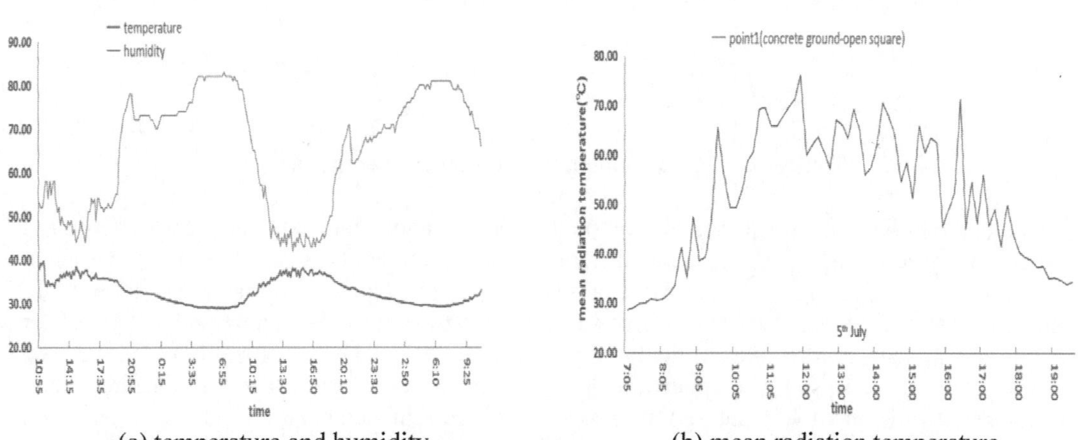

(a) temperature and humidity (b) mean radiation temperature

Figure 3. Environment data.

3. Experiment analysis

What is shown in figure 3 is the simultaneous air temperature and humidity of point 1 in two days and the mean radiation temperature of point 1 in 5th July. The mean radiant temperature (MRT) is defined as the uniform temperature of an imaginary enclosure in which the radiant heat transfer from the human body is equal to the radiant heat transfer in the actual non-uniform enclosure [12]. MRT is an important factor to estimate the building thermal environment. We can see it from the picture that the temperature in Guangzhou can reach almost 40℃ in summer. The humidity of it is also high, the peak value is almost 80%. That is all because Guangzhou is in the typical subtropical climate zone where the temperature and humidity there are usually high. So just as the picture shown below, the value of mean radiation temperature of point 1 is large.

The specific heat capacity values of three materials can be found in table 2 below. We can see that the specific heat capacity of pavement brick is a little bit lower than concrete.

Table 2. Specific heat capacity (J/kgK).

pavement brick	concrete	grass
840	880	calorific potential

All the measured ground temperature was plotted in figure 4. As is shown in figure 4, when compared with the other side of the peak line, all the human-made materials have a time delay when their temperature begins to decrease. Because point 2 is shaded from the sunlight all day, so the temperature fluctuations of it are very small but it still can be found. And compared with the human-made materials, point 6 (grass surface) almost has no time delay.

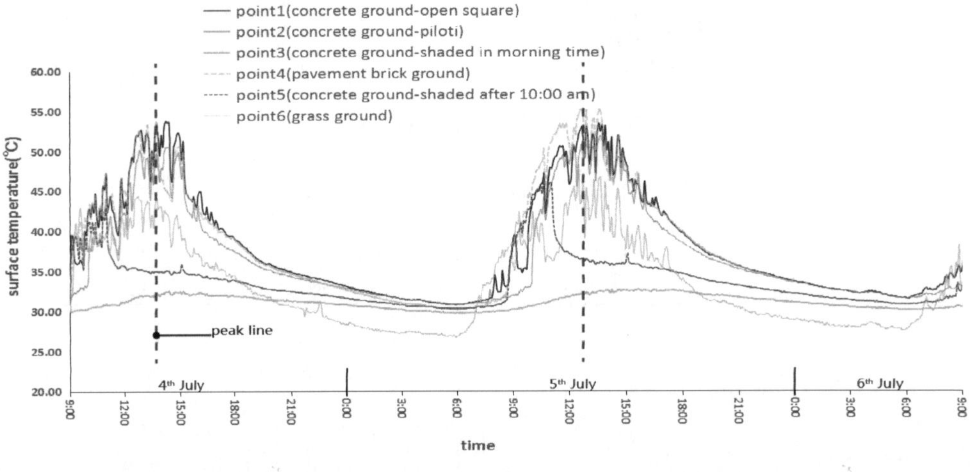

Figure 4. Surface temperature of all measured points.

For a more intuitive comparison, we take out point 1, point 4 and point 6 into one picture (figure 5) to compare the surface temperature of three different kind of materials when they are in the same conditions (exposed to sunlight all day). We can see it clearly from the picture that the material with a low specific heat capacity has a more sensitive variation in temperature. Both the trend of temperature variations and the temperature value of the three points are different. The pavement brick with a lower specific capacity (840 J/kgK) has a higher value and a steeper variation trend than the concrete. What's more, compared with other two points, point 6 has a faster increasing temperature and a relatively stable variation trend.

To compare the temperature variations of the same materials in different conditions (the time exposed to sunlight is different), we make point 2 (under piloti, without any direct sunlight) as the standard. As is shown in figure 6, though the specific heat capacity of the three points is the same, the value of temperature is different which is because the different time exposed to sunlight. Point 2 (shaded all day) has the lowest temperature value when point 1 (exposed all day) has the highest. Point 3 (shaded in the early morning) is a little bit lower than point 1. As for point 3, shaded half the day, has the middle temperature value and an earlier temperature decreasing.

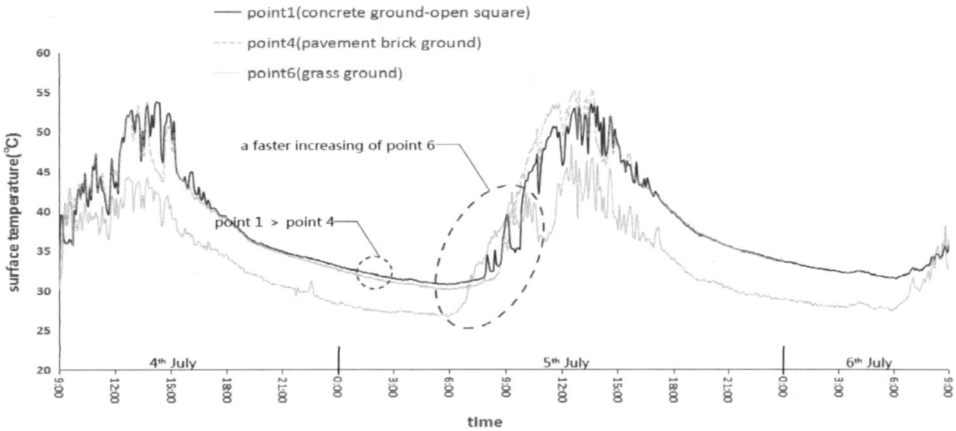

Figure 5. Surface temperature of point 1, 4, 6.

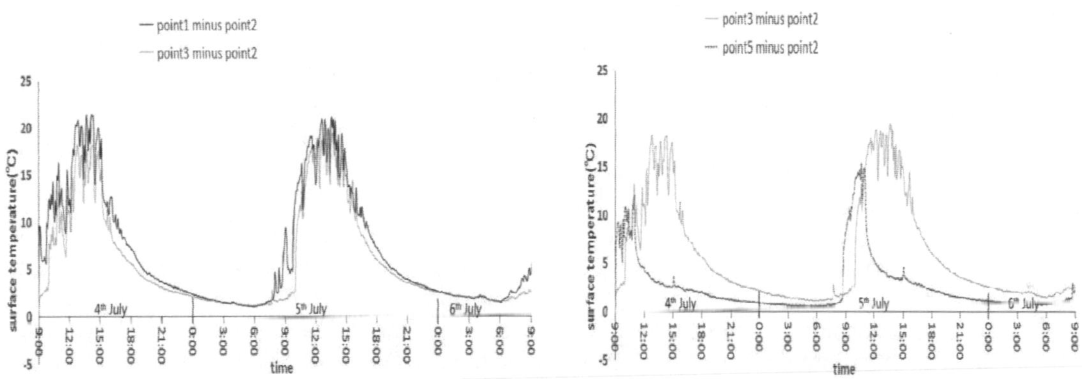

(a) point 1, 3 (b) point 3, 5

Figure 6. Standardized temperature value of point 1, 3, 5.

4. Experiment conclusions

This work studied the surface temperature variations of three common building materials. The result shows that both the specific heat capacity and the exposure to the sunlight have a strong influence on materials' surface temperature. Because of the specific heat capacity, all the human-made materials (concrete, pavement brick) have a time delay.

According to the experiment data, the material with a low specific heat capacity has a more sensitive variation in temperature. Compared with the concrete (exposed all day), the pavement brick (exposed all day) has a higher temperature in daytime and a lower temperature in late night time. And the variation trend of it is much steeper. Besides, the time exposed to the sunlight has a strong influence on the value of surface temperature. For the same materials, the temperature value is proportional to the time exposed to the sunlight which is showed clearly between all the concrete points. What's more, compared with other two materials, the grass ground has almost no time delay and shows a relatively stable variation trend of temperature. That might be an advantage to building a comfortable thermal environment in subtropical region.

The result of this study hopes to have a good influence on the designers on how to use the materials properly in outdoor design. Especially in subtropical region, the materials with a higher specific heat capacity may be more beneficial to the building thermal environment. And in order to get a lower temperature, designers should reduce the time of the materials exposed to the sunlight through designing method.

This experiment was held in Guangzhou, a typical subtropical region. The specific heat capacity and the exposure to sunlight are two major factors that we considered which can influence the surface temperature of the ground. However, besides the two aspects, there are many other factors that also have a strong influence on ground surface temperature such as the convective, conductive and radiation heat transfer. So the present work of this article is a little bit objective to some extent and more studies should be down in the future.

Acknowledgments

This work was financially supported by the National Natural Science Foundation of China (Grant No. 51438005, 51408160) and China Postdoctoral Science Foundation project (2015M571419, 2017T100244). Work partially supported by Harbin Science and Technology Bureau (2015RQQXJ068), Heilongjiang Province Postdoctoral Science Foundation (LBH-Z15084) and HeiLongjiang Province Cold Region Architectural Science Key Laboratory (2016HDJZ-1206).

References

[1] Ooka R, Chen H, Harayama K, Kato S and Li X 2003 Study on outdoor thermal environment of apartment block in Shenzhen, China with coupled simulation of convection, radiation and conduction *Energy Build* **36** 1247-58

[2] Uehara K, Ymamao Y, Oikawa S and Mochida A 2007 Wind-tunnel experiments on improving the natural ventilation of a street-canyon *Journal of Japan Society for Atmospheric Environment* **42** 301-9

[3] Hoyano A and He J 2008 Measurement and simulation of the thermal environment in the built space under a membrane structure *Build. Environ.* **44** 1119-27

[4] Xi T, Li Q, Mochida A and Meng Q 2011 The outdoor thermal environment and thermal comfort of piloti in subtropical climate cities *Journal of Habitat Engineering* **3** 253-60

[5] Xi T, Li Q, Mochida A and Meng Q 2012 Study on the outdoor thermal environment and thermal comfort around campus clusters in subtropical urban areas *Build. Environ.* **52** 162-70

[6] Xi T, Li Q, Mochida A and Meng Q 2011 A study on the influence of various human-built elements on outdoor thermal comfort in subtropical climate cities *4th Int. Conf. on Human-Environment System* 2011 (Sapporo, Japan) pp 85-90

[7] Xi T, Ding J and Jin H 2014 Study on the influence of piloti on mean radiant temperature in residential blocks by 3-D unsteady state heat balance radiation calculation *Journal of Harbin Institute of Technology* **4** 91-5

[8] Yue W Z and Xu L H 2013 Thermal environment effect of urban water landscape *Acta Ecologica Sinica* **33** 1852-9

[9] Somsak C, Bunyarit P, Khemmachart M and Nopparat K 2004 Performance rating of glass windows and glass windows with films in aspect of thermal comfort and heat transmission *Energy Build.* **37** 725-38

[10] Giuseppe E D and D'Orazio M 2015 Assessment of the effectiveness of cool and green roofs for the mitigation of the heat island effect and for the improvement of the thermal comfort in nearly zero energy building *Architecture Science Review* **58** 134-43

[11] Ibrahim M, Biwole P H, Wurtz E and Achard P 2014 A study on the thermal performance of exterior walls covered with a recently patented silica-aerogel-based insulating coating *Build. Environ.* **81** 112-22

[12] International Organization for Standardization 1998 ISO 7726. Ergonomics of the thermal environment - Instrument for measuring physical quantities (Geneva, Switzerland)

A novel finite element analysis of three-dimensional circular crack

X C Ping[1,2,3], C G Wang[1] and L P Cheng[1]

[1]School of Mechanical Engineering, Tianjin University of Science and Technology, China
[2]School of Mechanical Engineering, East China Jiaotong University, China

E-mail: xuechengping@hotmail.com

Abstract. A novel singular element containing a part of the circular crack front is established to solve the singular stress fields of circular cracks by using the numerical series eigen-solutions of singular stress fields. The element is derived from the Hellinger-Reissner variational principle and can be directly incorporated into existing 3D brick elements. The singular stress fields are determined as the system unknowns appearing as displacement nodal values. The numerical studies are conducted to demonstrate the simplicity of the proposed technique in handling fracture problems of circular cracks. The usage of the novel singular element can avoid mesh refinement near the crack front domain without loss of calculation accuracy and velocity of convergence. Compared with the conventional finite element methods and existing analytical methods, the present method is more suitable for dealing with complicated structures with a large number of elements.

1. Introduction

Traditionally, fracture mechanics mainly considers two plane states, i.e., plane stress and plane strain states, these simple two-dimensional (2D) cases make the equations for singular stress fields much easy to obtain by analytical methods [1]. Such essentially 2D theories have been very successful in dealing with fracture and fatigue on the macro-scale where the crack front can often be assumed to be a straight line. In fact, the crack front is usually curved, thereby, the three-dimensional (3D) character of the crack problem must be taken into account. A small number of fracture problems in 3D cracks have analytic solutions in close form due to the difficulty of mathematics [2-5]. For curved cracks in finite specimens, different numerical methods were presented to solve the fracture parameters [6-10]. The finite element method (FEM) is one of the most widely used numerical methods for this problem, in which special elements were developed for stress intensity factor (SIF) evaluation. The shape functions with a $1/r^{1/2}$ singularity at the crack front were used to establish the singular quarter-point element, which make it mainly be used for the crack problems of homogeneous materials [10-12]. Some scholars constructed interpolation functions for special elements by using the leading-order terms of asymptotic solutions [13]. There exist three shortcomings in the special elements with built-in leading-order singularity and conventional elements: the first one is that the numerical results are still dependent on the special element's size; the second one is that these enriched elements usually need transition elements in order to improve the accuracy of the numerical results; thirdly, inter-element compatibility between the special and conventional elements can't be satisfied due to different number of degrees of freedom at nodes. In contrast, special elements with built-in series solutions of asymptotic displacement and stress fields is insensitive to the element sizes, therefore, highly refined meshes are not necessary near the crack tips, and the computational efficiency in the finite element analysis can be improved [14,15]. However, a special element with built-in series solutions for 3D curved crack problems is still needed considering all of the existing special elements are for 2D cracks or 3D straight cracks. For this purpose, a novel singular element containing a part of a circular crack front is developed herein based on the numerical series eigen-solutions as well as Hellinger-Reissner variational principle. Numerical benchmark examples are presented to demonstrate the accuracy and efficiency of the super crack front elements. The first validation problem involves a circular penny-shaped crack in a 3D solid. The second validation problem is a cracked bar under tension and torsion. The last example is an elliptical crack in an infinite solid. The current investigation will pay close attention to the singular stress field along the circular crack front and all possible fracture modes in 3D circular cracks.

2. Establishment of a super corner front element model

Figure 1 shows a part of a circular crack front and the attached Cartesian (x, y, z) and cylindrical polar (r, θ, z) coordinates originating from the center of the circular crack. The key step is the selection an appropriate local (curvilinear) coordinate system (ρ, ϕ, θ), which is convenient to describe the local deformation behavior in the vicinity of the crack front (locus of point o shown in figure 1) . The local coordinate system is, as mentioned above, comprised of ρ, which denotes the radial direction from a point located on the crack front line, ϕ which denotes the angular direction measured counter clockwise from the x-y plane, and θ, which is positive counterclockwise (looking from top) along the circumferential line of the crack tip.

Numerical eigen-solutions of 3D displacement and stress fields solved from a one-dimensional finite element formulation for generalized plane strain problems [16] are resorted to herein. The assumed 3D displacement and stress fields in the crack tip domain have satisfied the governing differential equations and compatibility equations in advance. As a result, the displacement and stress vectors in the ρ-θ coordinate plane of the local coordinate system (ρ, ϕ, θ) (see figure 1) can be written as

$$\boldsymbol{u}(\rho,\theta) = \boldsymbol{U}\boldsymbol{\beta} \tag{1}$$

$$\boldsymbol{\sigma}(\rho,\theta) = \boldsymbol{\Sigma}\boldsymbol{\beta} \tag{2}$$

in which $\boldsymbol{u}(\rho, \theta)$ and $\boldsymbol{\sigma}(\rho, \theta)$ are total asymptotic displacements and stresses near the curved crack front, respectively, $\boldsymbol{U}$ and $\boldsymbol{\Sigma}$ include displacement and stress variations, respectively, due to the existence of n-order stress singularities, and $\boldsymbol{\beta}$ contains unknown multiplicative coefficients to be determined. In the one-dimensional finite element formulation, the angular coordinate θ is related to a natural coordinate ξ.

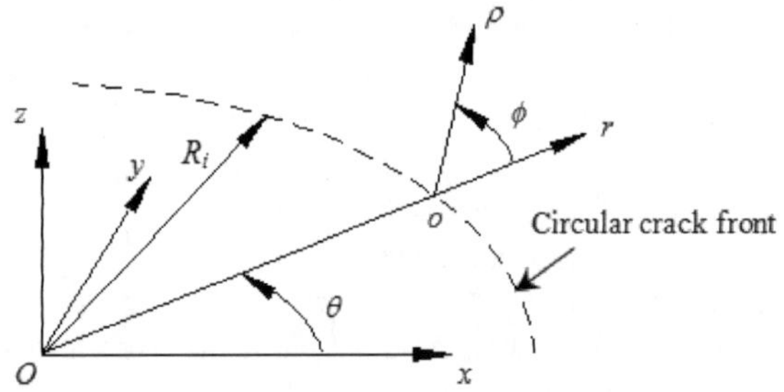

Figure 1. Coordinate system at the crack front.

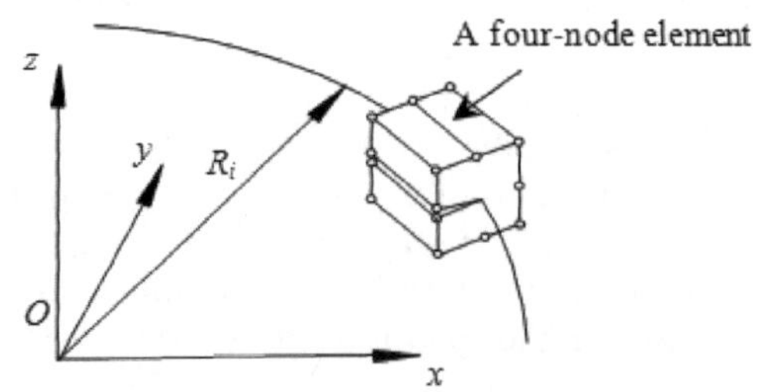

Figure 2. A novel singular element.

In general, the multiplicative coefficients β_i in equations (1) and (2) vary along the crack front for the specific case of a 3D crack, unlike the 2D case where they are assumed to be a constant value for a given geometry and loading. A novel singular element containing a part of curved crack front line is established to address the influence of 3D effects. As shown in figure 2, the novel singular element is composed of some hybrid four-node quadrilateral element. In order to establish the element stiffness matrix of a four-node quadrilateral isoparametric element, a new natural coordinate η is introduced to represent the position of a normal plane along the crack front, and the coefficients $\beta_i(\eta)$ are assumed to be a linear function of η in each element, as a result, the assumed displacement and stress components in local coordinate system (ρ, ϕ, θ) can be written as

$$\boldsymbol{u}(\rho,\phi,\theta)=(1-\eta)/2\cdot\boldsymbol{u}_1(\rho,\theta)+(1+\eta)/2\cdot\boldsymbol{u}_2(\rho,\theta)=(1-\eta)/2\cdot\boldsymbol{U}\boldsymbol{\beta}_1+(1+\eta)/2\cdot\boldsymbol{U}\boldsymbol{\beta}_2=\boldsymbol{Z}_u\boldsymbol{U}_c\boldsymbol{\beta}_c \quad (3)$$

$$\sigma(\rho,\phi,\theta)=(1-\eta)/2\cdot\boldsymbol{\sigma}_1(\rho,\theta)+(1+\eta)/2\cdot\boldsymbol{\sigma}_2(\rho,\theta)=(1-\eta)/2\cdot\boldsymbol{\Sigma}\boldsymbol{\beta}_1+(1+\eta)/2\cdot\boldsymbol{\Sigma}\boldsymbol{\beta}_2=\Sigma_c\boldsymbol{\beta}_c \quad (4)$$

in which $\boldsymbol{\beta}_1$, $\boldsymbol{u}_1(\rho,\theta)$ and $\boldsymbol{\sigma}_1(\rho,\theta)$ are components corresponding to η=-1, and $\boldsymbol{\beta}_2$, $\boldsymbol{u}_2(\rho,\theta)$ and $\boldsymbol{\sigma}_2(\rho,\theta)$ are corresponding to η=1.

To formulate the element stiffness matrix of the novel singular element by the Hellinger-Reissner principle, the hybrid functional for the problem in crack front domain Ω is expressed as the following form [15,17]:

$$\pi=\int_{\Omega}(-\sigma(x,y,z)^{\mathrm{T}}\boldsymbol{S}\,\sigma(x,y,z)/2+\sigma(x,y,z)^{\mathrm{T}}\boldsymbol{\varDelta}\,\boldsymbol{u}(x,y,z))\,\mathrm{d}\,\Omega-\int_{\Gamma}\boldsymbol{t}^{\mathrm{T}}(\boldsymbol{u}(x,y,z)-\tilde{\boldsymbol{u}}(x,y,z))\,\mathrm{d}\,A \quad (5)$$

where Γ denotes the circumferential boundary surfaces of the crack front element, which is composed of outward surfaces of all four-node quadrilateral elements. $\boldsymbol{\varDelta}$ is the matrix differential operator relating strains to displacements. The boundary displacements $\tilde{\boldsymbol{u}}(x,y,z)$ are assumed to be specified. $\boldsymbol{S}$ is the elastic compliance matrix in the constitutive relation between strains and stresses. $\boldsymbol{u}(x,y,z)$, $\sigma(x,y,z)$ and $\boldsymbol{t}(x,y,z)$ are, respectively, the stress, displacement and boundary traction vectors in Cartesian coordinate system (x, y, z).

By using the divergence theorem over Ω, the functional π are simplified as

$$\pi=\frac{1}{2}\int_{\Gamma}\boldsymbol{\sigma}(x,y,z)^{\mathrm{T}}\boldsymbol{n}^{\mathrm{T}}\boldsymbol{u}(x,y,z)\,\mathrm{d}\,A\quad\int_{\Gamma}\boldsymbol{t}(x,y,z)^{\mathrm{T}}\tilde{\boldsymbol{u}}(x,y,z)\,\mathrm{d}\,A \quad (6)$$

In which $\boldsymbol{n}$ is a matrix containing the uniform normal vectors n_x, n_y and n_z. The displacement $\boldsymbol{u}(x, y, z)$, stress $\boldsymbol{\sigma}(x, y, z)$ and traction $\boldsymbol{t}(x, y, z)$ in equation (6) can be transformed from the corresponding components in equations (3) and (4) in the corresponding local coordinate system (ρ, ϕ, θ), i.e.

$$\boldsymbol{u}(x,y,z)=\boldsymbol{Z}_u\boldsymbol{u}(\rho,\phi,\theta)=\boldsymbol{Z}_u\boldsymbol{U}_c\boldsymbol{\beta}_c \quad (7)$$

$$\sigma(x,y,z)=\boldsymbol{Z}_\sigma\boldsymbol{\sigma}(\rho,\phi,\theta)=\boldsymbol{Z}_\sigma\boldsymbol{\Sigma}_c\boldsymbol{\beta}_c \quad (8)$$

$$\boldsymbol{t}(x,y,z)=\boldsymbol{Z}_\sigma\boldsymbol{t}(\rho,\phi,\theta)=\boldsymbol{Z}_\sigma\boldsymbol{n}\boldsymbol{\Sigma}_c\boldsymbol{\beta}_c \quad (9)$$

The coordinate system transformation matrices $\boldsymbol{Z}_u$ and $\boldsymbol{Z}_\sigma$ are from the local coordinate system (ρ, ϕ, θ) to the Cartesian coordinate system (x, y, z).

Based on the hybrid-stress finite element method, the boundary displacements $\tilde{\boldsymbol{u}}(x,y,z)$ are assumed separately from $\boldsymbol{u}(x,y,z)$, and are expressed in terms of the nodal displacements in the displacement-base finite element methods as

$$\tilde{\boldsymbol{u}}(x,y,z)=\boldsymbol{N}\overline{\delta} \quad (10)$$

where the matrix $\boldsymbol{N}$ represents the shape function matrix for a four-node quadrilateral element allowing for 3D displacements at a node, and the vector $\overline{\delta}^{\mathrm{T}}=[q_x\ q_y\ q_z]$ includes the nodal displacements.

Substituting components of equations (7) - (10) into equation (6), leads to

$$\pi = \frac{1}{2}\boldsymbol{\beta}_{\mathrm{c}}^{\mathrm{T}}\boldsymbol{H}\boldsymbol{\beta}_{\mathrm{c}} - \boldsymbol{\beta}_{\mathrm{c}}^{\mathrm{T}}\boldsymbol{G}\bar{\boldsymbol{\delta}} \tag{11}$$

where $\boldsymbol{G}$ and $\boldsymbol{H}$ are two matrix for areal integrals in the outward boundary faces Γ, namely:

$$\boldsymbol{H} = \frac{1}{2}\int_{\Gamma}\left[\left(\boldsymbol{n}\,\boldsymbol{Z}_{\sigma}\boldsymbol{\Sigma}_{\mathrm{c}}\right)^{\mathrm{T}}\boldsymbol{U}_{\mathrm{c}} + \boldsymbol{U}_{\mathrm{c}}^{\mathrm{T}}\left(\boldsymbol{n}\,\boldsymbol{Z}_{\sigma}\boldsymbol{\Sigma}_{\mathrm{c}}\right)\right]\mathrm{d}A \tag{12}$$

$$\boldsymbol{G} = \int_{\Gamma}\left(\boldsymbol{n}\,\boldsymbol{Z}_{\mathrm{c}}\boldsymbol{\Sigma}_{\mathrm{c}}\right)^{\mathrm{T}}\boldsymbol{N}\,\mathrm{d}A \tag{13}$$

Taking the first variation of equation (11) with respect to $\boldsymbol{\beta}_{\mathrm{c}}$ and noting that $\delta\pi = 0$, we determine $\boldsymbol{\beta}_{\mathrm{c}}$ as follows:

$$\boldsymbol{\beta}_{\mathrm{c}} = \boldsymbol{H}^{-1}\boldsymbol{G}\bar{\boldsymbol{\delta}} \tag{14}$$

Substituting $\boldsymbol{\beta}_{\mathrm{c}}$ into equation (11) leads to:

$$\pi = \frac{1}{2}\bar{\boldsymbol{\delta}}^{\mathrm{T}}\boldsymbol{G}^{\mathrm{T}}\boldsymbol{H}^{-1}\boldsymbol{G}\bar{\boldsymbol{\delta}} \tag{15}$$

Invoking the first variation of equation (15) with respect to $\bar{\boldsymbol{\delta}}$, we obtain the following element stiffness matrix for the super crack front element:

$$\boldsymbol{K} = \boldsymbol{G}^{\mathrm{T}}\boldsymbol{H}^{-1}\boldsymbol{G} \tag{16}$$

The implementation of the present finite element model has been carried out in the commercial code Abaqus. Taking advantage of the user element capability of this code, we have written a user-defined element subroutine that includes the proposed element stiffness matrix. In regions where the novel crack front element were not required, the standard 8-node 3D linear brick elements provided in Abaqus were used. In this way, we benefit from the built-in pre- and post-processing features in Abaqus. Consequently, the coefficients $\boldsymbol{\beta}_{\mathrm{c}}$ are retrieved by using equation (14), and the singular displacement and stress fields are finally determined by equations (3) and (4).

3. Numerical examples

Three benchmark problems are studied in this work to verify the proposed method. The first one is a penny-shaped crack in a cylinder, the stress intensity factors corresponding to mode I and III are considered. The second one is a cracked round bar subjected to tensile and torsional loads. The third one considered is an elliptical crack in an infinite cuboid body under tension, the variation of SIFs along the crack front line will be paid close attention to. The convergence of the present method is discussed in the first and second examples to show a dependence of singular stress fields on the number of eigenvalues truncated, and the number of nodes of the super singular hybrid element. In all examples, singular stress fields along the crack front are compared with the existing results.

3.1. A penny-shaped crack in a cylinder

Figure 3 presents a cylinder containing penny-shaped crack. Tensile loading *F* and torsional loading *T* are applied to two end faces of the cylinder. As shown in the figure, the diameter and height of the cylinder are represented by *D* and *H*, respectively. And *a* is the radius of the circular crack. As a consequence of the Saint venant principle, the tensile stress and torsional stress in the horizontal cross-section can be calculated as $\sigma^{\infty}=F/(\pi D^2)$ and $\tau^{\infty}=16T/(\pi D^3)$.

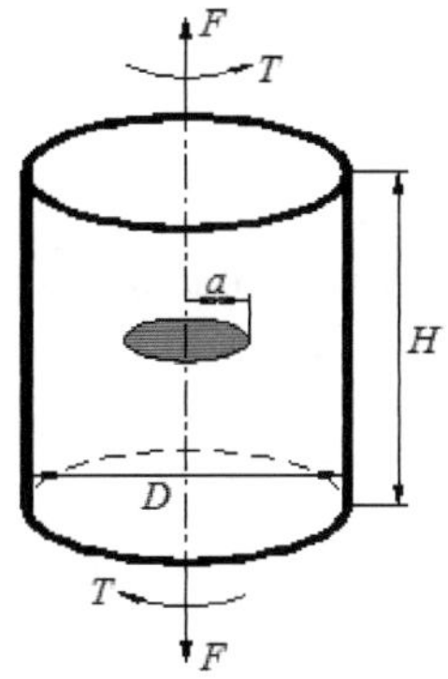

Figure 3. A cylinder containing penny-shaped crack.

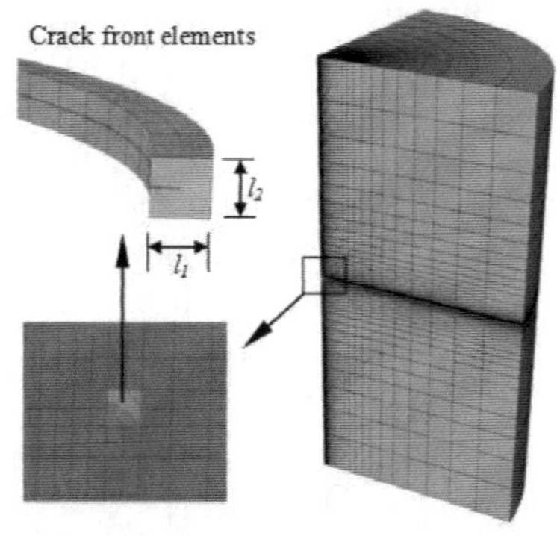

Figure 4. One quarter of the grid model for the penny-shaped crack problem.

Based on the finite element analysis, a finite element model with different mesh densities is established to simulate the crack in an infinite cylinder. D and H is equal to $300a$ in this model. Figure 4 shows one quarter of the cylinder that can clearly illustrate the meshes near the crack fronts. In this case, a certain number of novel 18 node crack front elements are distributed at the crack fronts, around which are traditional 3D eight-node brick elements. The material Poisson's ratio is equal to 0.3.

The von Mises stress contours of the cylinder with penny-shaped crack is shown figure 5. Figures 5(a) and 5(b) are the results of the tensile loading and the torsional loading, respectively. No function for post-processing of contours is an existing limitation of user-defined elements in Abaqus. As shown in figure 5, contours do not represent the stresses in the super crack front element. As a result, the program of the user-defined element adds a callable Fortran code in order that the results of singular stresses can be exported to a log file generated by Abaqus. According to the definition of modes I-III fracture modes, the SIFs can be numerically estimated by equations (17) - (19):

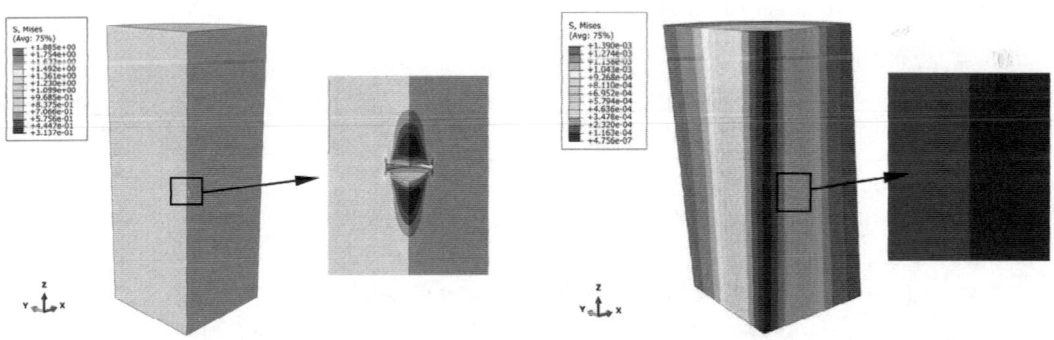

Figure 5. The von Mises stress contours of the cylinder with penny-shaped crack.

$$K_{\mathrm{I}} = \lim_{\rho \to 0} \sqrt{2\rho} \cdot \sigma_{\phi}(\rho, 0, \theta) \tag{17}$$

$$K_{\mathrm{II}} = \lim_{\rho \to 0} \sqrt{2\rho} \cdot \sigma_{\rho\phi}(\rho, 0, \theta) \tag{18}$$

$$K_{\mathrm{III}} = \lim_{\rho \to 0} \sqrt{2\rho} \cdot \sigma_{\phi\theta}(\rho, 0, \theta) \tag{19}$$

Convergence property is researched to validate the effectiveness of the present finite element model. As shown in figure 6, The SIF values calculated from the super crack front element model are

IOP Conf. Series: Materials Science and Engineering **372** (2018) 012029 doi:10.1088/1757-899X/372/1/012029

exhibited. For the penny-shaped crack in an infinite cylinder, the normalized SIFs are $K_I/(\sigma^\infty a^{0.5})=0.6366$ for tension and $K_{III}/(\sigma^\infty a^{0.5})=0$ for torsion. This figure demonstrates that when the element size of the novel crack front elements in a plane perpendicular to the crack front is less than $0.05a\times0.05a$ and the number of elements along the crack front is larger than 30, the present solutions converge to the reference solutions. On the other hand, the influence of the number of terms in series solutions ($2N+M$) on SIFs are listed in table 1. It shows that when the dimension of vector $\boldsymbol{\beta}$ increases to 26, the present solutions converge to reference values. In terms of an 18-node super crack front element, at least 24 terms from the series representation are needed [15], and the present model still obeys this convergence condition.

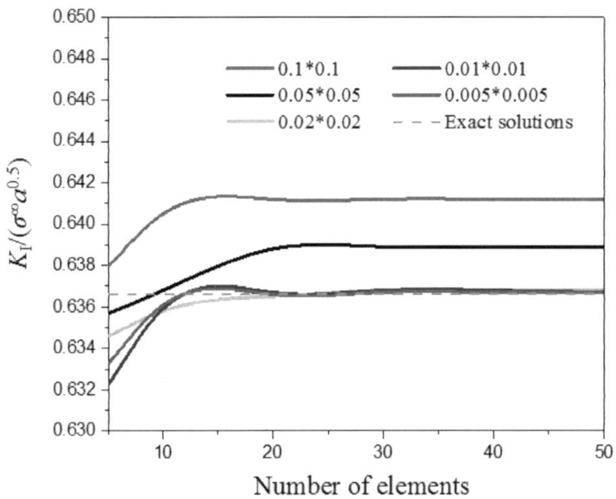

Figure 6. Convergence of the SIF values calculated from the super crack front element model for the penny-shaped crack.

Table 1. The effects of the number of terms in series solutions on SIFs for cracks.

Dim(β)	8	14	20	26	32	38	44
Results for the penny-shaped crack problems							
$K_I/(\sigma\ a^{0.5})$	0.6634	0.6566	0.6410	0.6372	0.6373	0.6371	0.6378
Results for the circumferential crack problems							
$K_I/(\sigma\ a^{0.5})$	0.9517	1.0004	1.0987	1.1227	1.1231	1.1225	1.1226
$K_{III}/(\sigma\ a^{0.5})$	-1.0048	-1.0045	-1.0045	-1.0057	-1.006	-1.0008	-1.0009

3.2. A circumferential crack in a cylinder

Figure 7 presents a cylinder containing circumferential crack. The cylinder is subjected to tensile loading F and torsional loading T. D and H denote the diameter and height of the cylinder. a represents the depth of the crack for the circumferential crack. Just like the penny-shaped crack problem, the tensile stress and torsional stress in a horizontal cross-section can be calculated as $\sigma^\infty=F/(\pi D^2)$ and $\tau^\infty=16T/(\pi D^3)$.

Both D and H are equal to $300a$ in the finite element model. Figure 8 shows one quarter of the finite element model that can clearly illustrate the meshes near the crack fronts. A certain number of super 18 node crack front elements are distributed at the crack fronts as before. The material Poisson's ratio is equal to 0.3.

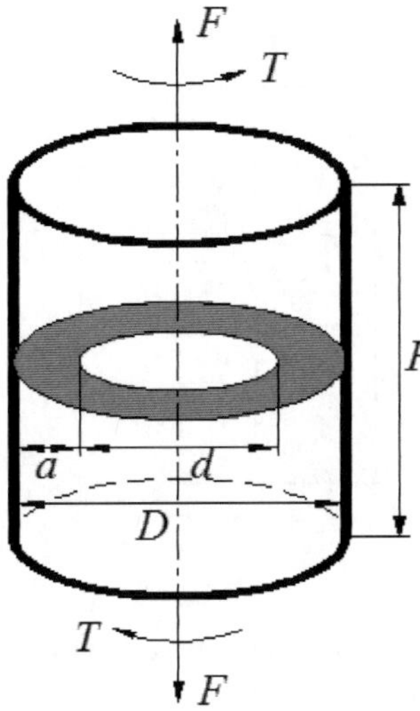

Figure 7. A cylinder containing circumferential crack.

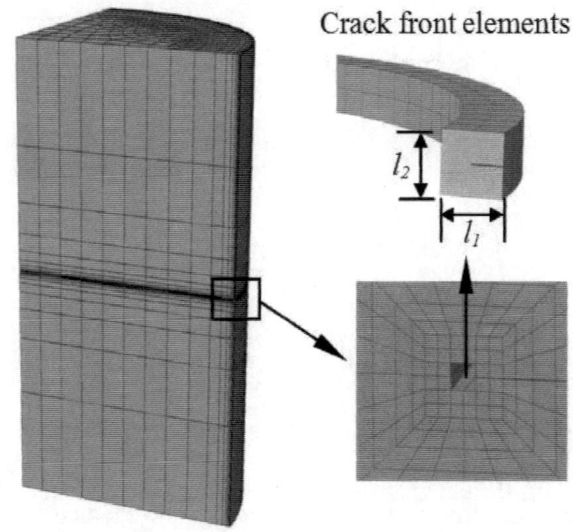

Figure 8. One quarter of the grid model for the circumferential crack problem.

Figure 9 shows the von Mises stress contours of the cylinder with circumferential cracks, and figures 9(a) and 9(b) are results corresponding to the tensile loading and the torsional loading, respectively. According to equations (17)-(19), the SIFs can be numerically estimated. Figure 10 presents the convergence of the present solutions for the circumferential crack problem. the normalized SIFs are $K_I/(\sigma^{\infty}a^{0.5})$=1.1225 for tension and $K_{III}/(\sigma^{\infty}a^{0.5})$=-1.007 for torsion. This figure indicates that an element size of 0.1 $a \times$ 0.1 a in the face perpendicular to the crack front are enough to acquire converged solutions provided that 30 or more elements are used. Compared with the penny-shaped crack problem in section 3.1, the present problem is easier to get a convergent solution. This suggests that larger the curvature radius is, the less the numerical solutions are affected by the element size. As shown in table 1, the present solutions also converge to reference results when the dimension of vector $\boldsymbol{\beta}$ increases to 26.

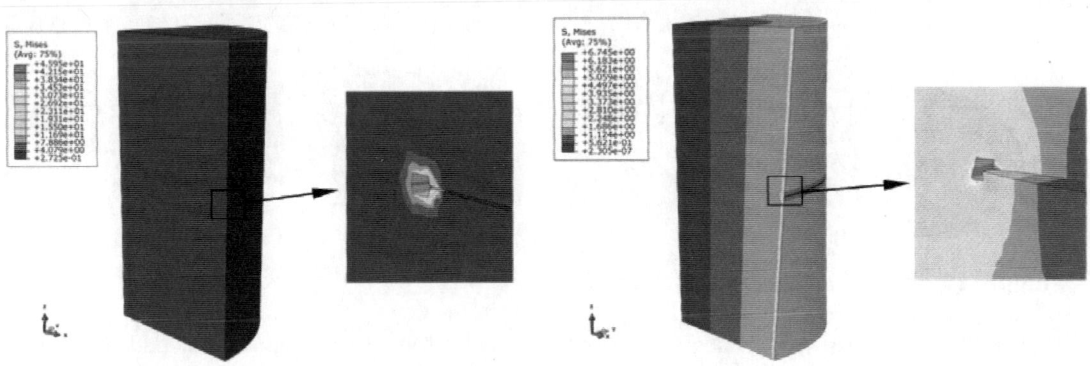

(a) (b)

Figure 9. The von Mises stress contours of the cylinder with a circumferential crack.

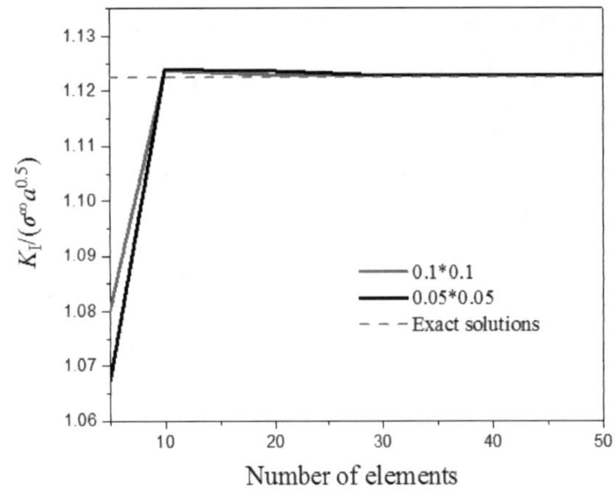

Figure 10. Convergence of the SIF values calculated from the super crack front element model for the circumferential crack problem.

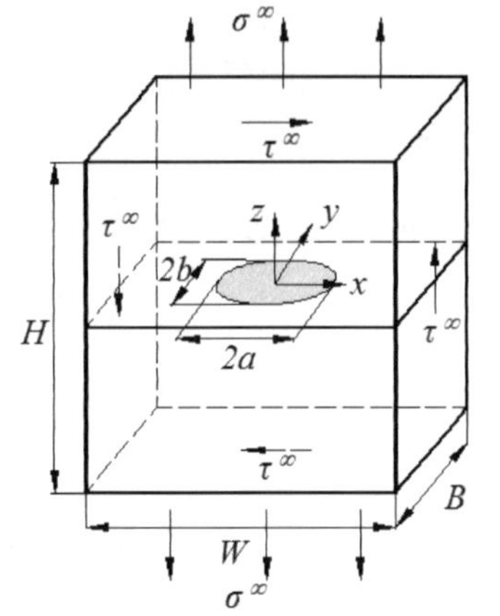

Figure 11. An four prism containing an embedded elliptical crack.

3.3. An elliptical crack in an infinite four prism

Figure 11 shows an four prism containing an embedded elliptical crack, and far-field tension and shear loading are applied to the prism. In the finite element analysis, $2H$, $2W$ and $2B$ are equal to 300 a, and the ratio of the elliptical crack's minor and major axes, b/a, is a variable. In the case of uniform tension stress σ^∞ and a shear loading τ^∞, Kassir and Sih [18] gave the accurate analytical expressions of SIFs for an elliptical crack with arbitrary ratio of a/b. In the same manner, the novel crack front element is used to solve the singular stress fields. The element size of the super crack front element is set to $0.02a\times0.02a$, and the number of elements along the crack front is 160. The dimension of $\boldsymbol{\beta}$ is 26.

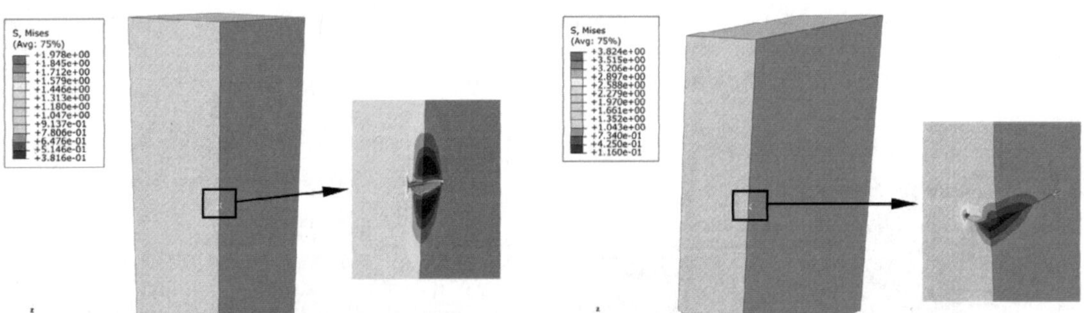

Figure 12. The von Mises stress contours of the four prism with an embedded elliptical crack.

Figure 12 presents the von Mises stress contours of the four prism with an embedded elliptical crack, in which figures 12(a) and 12(b) are the results of the tensile loading and the shear loading, respectively. When the prism is in a far-field tension loading condition, pure mode I fracture exists ahead of the crack front. As shown in figure 13(a), the normalized SIFs, $K_I/(\sigma^\infty b^{0.5})$, and the angle θ for different ratios of a/b is plotted. The present results, the exact solutions and the conventional FEM solutions have good consistency. In the case of conventional elements, very refined elements are required to calculate the numerical asymptotic solutions for singular stress fields. Therefore, the present super crack front element can increase the computational efficiency. If the prism is subjected to pure shear loadings τ, the stress states ahead of the crack front is under modes I and III fracture.

According to equations (18) and (19), the normalized SIFs, $K_{II}/(\tau^{\infty}b^{0.5})$ and $K_{III}/(\tau^{\infty}b^{0.5})$, and the angle ϕ for different ratios of *a/b* are plotted in figures 13(b) and 13(c). The present solutions are also in very good agreement with the exact solutions [18]. We can't obtain converged solutions for mixed mode fracture with conventional elements although we have tried many ways to solve this problem.

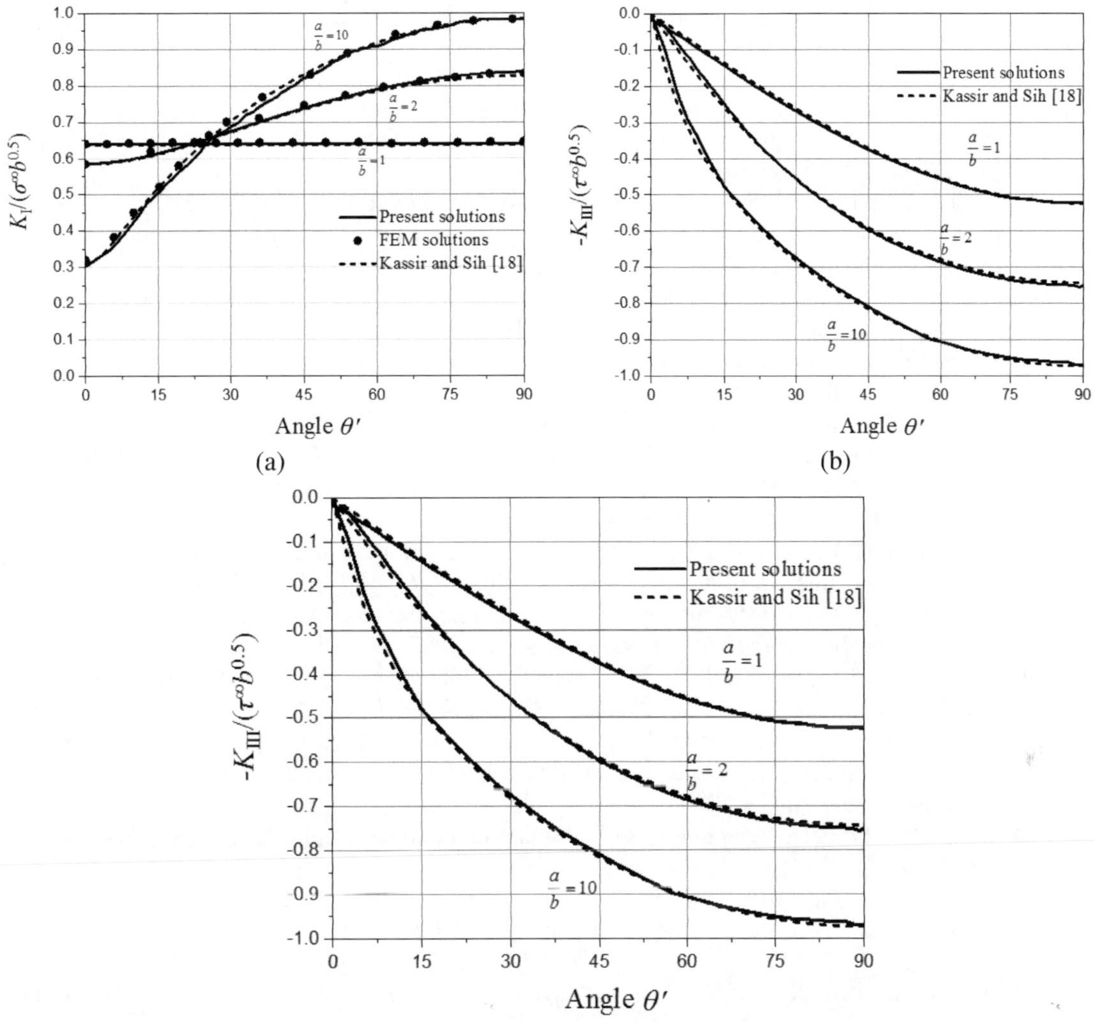

Figure 13. Normalized SIFs for different ratios of *a/b* of elliptical.

4. Conclusion

For the circular crack problems, numerical series solutions of singular displacement and stress fields from an *ad hoc* finite element eigen-analysis method are used to generate a novel 3D crack front element. The built-in series solutions include not only the leading-order terms but also a certain number of high-order terms.

When the curvature radius of a crack front is sufficiently large compared with the element size, the present method is insensitive to the sizes of novel crack front elements in a normal plane of the crack front, so coarse meshes near the crack fronts can be used to acquire converged results. For the engineering analysis, the present novel crack front element can improve the computational efficiency by decreasing the number of elements. Therefore, the novel crack front element is much easier to use compared with conventional FEMs.

Acknowledgments
The authors acknowledge the support of National Natural Science Foundation of China under grant No 51365013. The Personnel training plan of Tianjin city in China for the key young and middle-aged innovation talents is also acknowledged.

References

[1] Broberg K B 1999 *Cracks and Fracture* (Burlington: Academic Press)
[2] Sneddon I N 1946 The distribution of stress in the neighbourhood of a crack in an elastic solid *Proc. R. Soc. A.* **187** 229–60
[3] Gao H and Rice J R 1987 Somewhat circular tensile cracks *Int. J. Fract.* **33** 155–274
[4] Chen M C, Noda N A and Tang R J 1999 Application of finite-part integral to planar interfacial fracture problems in 3D bimaterials *J. Appl. Mech* **66** 885-90
[5] Nik Long N M A, Koo L F and Eshkuvatov Z K 2011 Computation of energy release rates for a nearly circular crack *Math. Prob. Engin.* **2011**(3) 1-17
[6] Ayhan A O 2011 Simulation of three-dimensional fatigue crack propagation using enriched finite elements *Comput. Struct.***89**(9) 801-12
[7] Sukumar N, Chopp D and Moran B 2003 Extended finite element method and fast marching method for three-dimensional fatigue crack propagation *Engin. Fract. Mech.* **70** 29–48
[8] Yu P and Guo W 2013 An equivalent thickness conception for evaluation of corner and surface fatigue crack closure *Engin. Fract. Mech.* **99**(1) 202–13
[9] Koguchi H, Yokoyama K and Luangarpa C 2015 Variation of stress intensity factor along a small interface crack front in singular stress fields *Int. J. Solids. Struct.* **71** 156–68
[10] Ariza M P, Sáez A and Domínguez J 1997 A singular element for three-dimensional fracture mechanics analysis *Engin. Analys. Boundary. Ele.* **20**(4) 275–85
[11] Pageau S S and Biggers S B 1997 Enrichment of finite elements with numerical solutions for singular stress fields *Int. J. Num. Methods. Engin.* **40**(14) 2693–713
[12] Liu G R, Nourbakhshia N and Zhang Y W 2011 A novel singular ES-FEM method for simulating singular stress fields near the crack tips for linear fracture problems *Engin. Fract. Mech.* **78**(6) 863–76
[13] Chen E P 1985 Finite element analysis of a bimaterial interface crack *Theore. Appl. Fract. Mech.* **3**(3) 257–62
[14] Barut A, Guven I and Madenci E 2001 Analysis of singular stress fields at junctions of multiple dissimilar materials under mechanical and thermal loading *Int. J. Solids. Struct.* **38**(50-51) 9077–109
[15] Chen M C and Sze K Y 2001 A novel hybrid finite element analysis of bimaterial wedge problems *Engin. Fract. Mech.* **68**(13) 1463–76
[16] Sze K Y and Wang H T 2000 A simple finite element formulation for computing stress singularities at bimaterial interfaces *Finite. Ele. Analys. Design.* **35**(2) 97–118
[17] Zhang J and Katsube N 1995 A hybrid finite element method for heterogeneous materials with randomly dispersed elastic inclusions *Finite. Ele. Analys. Design.***19**(94) 45-55
[18] Kassir M K and Sih G C 1966 Three-dimensional stress distribution around an elliptical crack under arbitrary loadings *J. Appl. Mech.* **33**(3) 601-11

Application of improved wavelet total variation denoising for rolling bearing incipient fault diagnosis

W Zhang[1] and M P Jia[1,2]

[1]School of Mechanical Engineering, Southeast University, Nanjing 211189, China

E-mail: mpjia@seu.edu.cn

Abstract. When incipient fault appear in the rolling bearing, the fault feature is too small and easily submerged in the strong background noise. In this paper, wavelet total variation denoising based on kurtosis (Kurt-WATV) is studied, which can extract the incipient fault feature of the rolling bearing more effectively. The proposed algorithm contains main steps: a) establish a sparse diagnosis model, b) represent periodic impulses based on the redundant wavelet dictionary, c) solve the joint optimization problem by alternating direction method of multipliers (ADMM), d) obtain the reconstructed signal using kurtosis value as criterion and then select optimal wavelet subbands. This paper uses overcomplete rational-dilation wavelet transform (ORDWT) as a dictionary, and adjusts the control parameters to achieve the concentration in the time-frequency plane. Incipient fault of rolling bearing is used as an example, and the result shows that the effectiveness and superiority of the proposed Kurt-WATV bearing fault diagnosis algorithm.

1. Introduction

Rolling bearing is one of the most significant part of the mechanical transmission system. Incipient fault diagnosis is conducive to the formulation of reasonable maintenance plan, which can realize the maximization of production efficiency and change the disadvantages of the traditional Run-to-Breakdown maintenance way. When rolling bearing is in the incipient stage of failure, the fault feature is relatively weak. Hence, it is difficult to extract fault information due to the influence of vibration and noise of signal transmission path and other components.

In recent years, domestic and foreign scholars have done a great deal of researches on the incipient fault extraction in rolling bearings. Although the signal processing methods [1-5] show outstanding performance in extracting incipient fault feature, the methods themselves have some shortcomings. Wavelet transform [1], empirical mode decomposition [2,3] and some other methods [4,5] are all based on the construction of the best band-pass filter to obtain fault information, but in-band noise can not be removed, which affects the extraction of faults related information from the strong background noise.

Sparse representation has become a hot topic in the field of signal processing in recent years. The purpose of signal sparse representation is to express a signal with as few atoms as possible in a given overcomplete dictionary, to obtain a more concise representation way [6]. Liu *et al* [7] proposed a time-frequency atomic matching pursuit algorithm to analyze vibration signals and extract rolling bearing fault characteristics. Shi *et al* [8] applied an adaptive time-frequency decomposition method based on matching pursuit and Gaussian line-frequency dictionaries to analyze transient vibration signals. Yang *et al* [9] combined the basic pursuit and wavelet packet dictionary to extract bearing fault features. They point out that basic pursuit can sparsely represent the rolling bearing fault features at fine resolution and enhance signal-to-noise ratio. Feng and Chu [10] applied several typical methods of sparse signal decomposition, including frame methods, best orthogonal basis, matching pursuit and basis pursuit, to extract the embedded meaningful information. Qin *et al* [11] combined basis pursuit with various transformation bases to decompose different components of mixture signals, such as impacts, harmonics and modulation components, which can effectively diagnose rotor and bearing faults. Feng and Liang [12] used iterative threshold method to suppress background noise and extracted the fault feature of planetary gearbox.

Wavelet thresholding often introduces artifacts such as spurious noise spikes and pseudo-Gibbs oscillations [13]. Total variation (TV) term is added into the optimization objective function to alleviate the spurious noise spikes and pseudo-Gibbs artifacts [14]. The existing sparse representation method based on wavelet dictionary can not adjust the time-frequency resolution. Therefore, the frequency resolution will decrease as the frequency increases, and the appropriate time-frequency division can not be selected according to the signal characteristics. The proposed algorithm utilizes ORDWT dictionary to achieve a better time-frequency distributions. In addition, the kurtosis is introduced to determine the optimal wavelet subbands, and the denoised results show that the impact features are more obvious. Lastly, experimental results indicate that the proposed algorithm can be effectively used to incipient rolling bearings fault diagnosis.

2. Kurtosis-base sparse model and algorithmic framework

2.1. Sparse fault diagnosis model

When the rolling bearings appear local damage, the relative motion between components produce periodic impact, so the fault signal model of rolling bearing contains periodic impact feature. In the incipient stage of the rolling bearings failure, the amplitude of impact component is not only small, but also the impact features are easily drowned in the strong noise environment. Usually collected vibration signal contains periodic impact components, mechanical frequency components and noise components. Rolling bearing fault signal model can be written as follows:

$$y = x + f + v \tag{1}$$

where x is periodic impacts, f is mechanical frequency, v is noise. The total variation of signal x is defined as $TV(x) = \|Dx\|_1$, where D is first-order difference matrix,

$$D = \begin{bmatrix} -1 & 1 & & & \\ & -1 & 1 & & \\ & & \cdots & \cdots & \\ & & & -1 & 1 \end{bmatrix} \tag{2}$$

In order to extract x from the observed signal y, we construct the following optimization problem by using the sparse priori knowledge of the periodic impact component under the wavelet dictionary representation:

$$\hat{w} = \arg\min_{w} \left\{ F(w) = \frac{1}{2}\|Wy - w\|_2^2 + \sum_j \lambda_j \varphi(w_j) + \beta \|DW^T w\|_1 \right\} \tag{3}$$

The first term in equation (3) is the fidelity term, which means signal reconstruction error. The last two terms are penalty function. The function $\varphi(w_j)$ can be expressed in the form of l_1 norm. The estimate of the signal x is then given by the inverse wavelet transform of $\hat{w}$: $\hat{x} = W^T \hat{w}$. The regularization parameters are $\lambda_j > 0$ and $\beta > 0$.

2.2. Overcomplete rational dilation wavelet dictionary

The selection and construction of the basis functions in redundant dictionaries is one of the important factors that determine the effectiveness of signal sparse representation. When the basis function is as close as possible to the internal structure of the signal, it is possible to obtain the sparse representation of the signal through a sparse solution algorithm, so as to reveal the effective components in the signal. For a simple signal, pre-constructed dictionaries can describe their sparsity more effectively. Continuous wavelet, discrete wavelet, complex wavelet, lifting wavelet, wavelet packet and multiwavelet transform are widely used in fault diagnosis [15]. As a new branch of wavelet series, ORDWT have attracted more attention recently in the field of fault diagnosis [16]. Compared with the traditional binary wavelet, ORDWT is a new type of wavelet tight frame, which has better translational invariance, adjustable time-frequency distribution and flexible oscillations of flexible wavelet atoms. ORDWT is composed of an iterative two-channel filter, as shown in figure 1. It can be seen that the upsampling is done only in the low-pass path. The dilation factor of ORDWT is q/p, where p and q are satisfied $1 < q/p < 2$, and p and q are coprime.

The frequency response of filters $H(w)$ and $G(w)$ is:

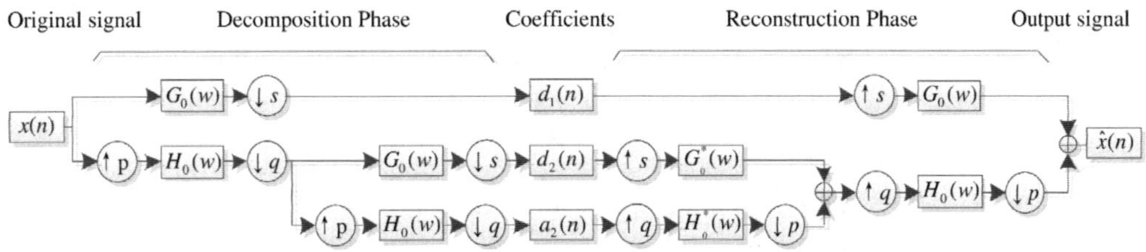

Figure 1. Filter-banks for ORDWT.

$$H(w)=\begin{cases}\sqrt{pq}, & w\in\left[0,\left(1-\frac{1}{s}\right)\frac{\pi}{p}\right)\\ \sqrt{pq}\theta_H\left(\frac{w-a}{b}\right), & w\in\left[\left(1-\frac{1}{s}\right)\frac{\pi}{p},\frac{\pi}{q}\right)\\ 0, & w\in\left[\frac{\pi}{q},\pi\right]\end{cases} \quad (4)$$

$$G(w)=\begin{cases}0, & w\in\left[0,\frac{s-1}{s}\pi\right)\\ \sqrt{s}\theta_G\left(\frac{w-pa}{pb}\right) & w\in\left[\frac{s-1}{s}\pi,\frac{p}{q}\pi\right)\\ \sqrt{s} & w\in\left[\frac{p}{q}\pi,\pi\right]\end{cases} \quad (5)$$

where $a=\left(1-\frac{1}{s}\right)\frac{\pi}{p}$, $b=\frac{1}{q}-\left(1-\frac{1}{s}\right)\frac{1}{p}$, $\theta_H=1/2(1+\cos w)\sqrt{2-\cos w}$, $w\in[0,\pi]$; $\theta_G=1/2(1-\cos w)\sqrt{2+\cos w}$, $w\in[0,\pi]$.

The variables p, q, s are the controlling parameters of ORDWT bases. The function $\theta_H(w)$ and $\theta_G(w)$, originating from Daubechies' orthonormal wavelet with 2 vanishing moments, are used to construct the transition bands of $H(w)$ and $G(w)$. According to equations(4) and (5), the frequency response curve of the filter can be obtained, as shown in figure 2. A complete description of ORDWT can be found in [17].

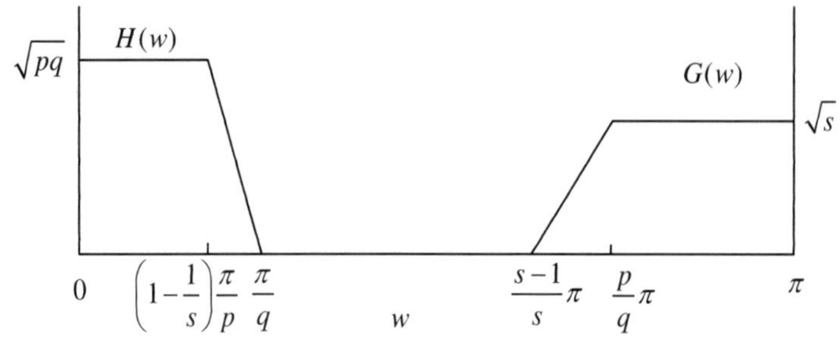

Figure 2. Frequency responses of the filter.

2.3. Optimization algorithm

ADMM is suitable for solving distributed convex optimization problems [18]. In order to solve the WATV denoising problem in equation (3), ADMM is used to solve equation (3):

$$\arg\min_{u,w} g_1(w) + g_2(u) \text{ subject to } u = w \tag{6}$$

where $g_1(w) = \frac{1}{2}\|Wy - w\|_2^2 + \sum_j \lambda\varphi(w_j)$ and $g_2(u) = \beta\left\|DW^T w\right\|_1$ are strictly convex. The augmented Lagrangian is given by

$$L(w,u,\mu) = g_1(w) + g_2(u) + \frac{\mu}{2}\|u - w - d\|_2^2 \tag{7}$$

ADMM transforms the joint optimization problem (7) of w and u into two simple sub-problems to solve. First, w and u are alternately minimized, and thus have the following iterative scheme:

$$w = \arg\min_{w}\{g_1(w) + \frac{\mu}{2}\|u - d - w\|_2^2\} \tag{8a}$$

$$u = \arg\min_{u}\{g_2(u) + \frac{\mu}{2}\|u - d - w\|_2^2\} \tag{8b}$$

$$d = d - (u - w) \tag{8c}$$

Combining the quadratic terms, equation (8a) may be written as follows:

$$w = \arg\min_{w} \sum_j \left\{\frac{1}{2}(p_j - w_j)^2 + \lambda_j\varphi(w_j)\right\} \tag{9}$$

where $p = (Wy + \mu(u-d))/(\mu+1)$. It can be further expressed as:

$$w_j = \theta\left(p_j; \lambda_j\right) \tag{10}$$

where θ is the threshold function, which can be defined as $\theta(y;\lambda) = \arg\min_{x\in R}\left\{\frac{1}{2}(y-x)^2 + \lambda\varphi(x)\right\}$. That is, $\theta(y;\lambda,a) = 0$ for all $|y| < \lambda$.

According to the proximity operators, equation (8b) can be transformed into the problem of TV denoising [19]. Hence, (8b) can be implemented as

$$u = v + W(tvd(W^T v, \beta/\mu) - W^T v) \tag{11}$$

where $tvd(W^T v, \beta/\mu) = \arg\min_x \left\{ \frac{\beta}{\mu} \|Dx\|_1 + \frac{1}{2} \|x - W^T v\|_2^2 \right\}$.

Therefore, solving problem equation (3) by equations (8a)-(8c), consists of wavelet threshold denoising equation (9) and the TV denoising equation (11), and see [20] for detail.

2.4. Kurt-WATV for incipient fault feature extraction

Kurtosis is a time statistics that reflects the distribution characteristics of vibration signals. It is very sensitive to impact feature, especially for incipient fault diagnosis. Kurtosis is defined as:

$$Kurt = \frac{E\left\{(x - \bar{x})^4\right\}}{\sigma^4} \tag{12}$$

where $\bar{x}$ and σ are the mean and standard variance of signal x, respectively. $E\{\cdot\}$ is mathematical expection. Therefore, the kurtosis is used to select the optimal wavelet subbands. Table 1 shows the proposed Kurt-WATV algorithm framework.

Table 1. Kurt-WATV algorithm framework.

1. Input:	$y, \lambda_j, \beta, \mu, p, q, s$
2. Construct a dictionary	Based on given p, q, s constructed wavelet dictionary W
3. Initialization	$u = Wy, d = 0$
4. Repeat	$p = (Wy + \mu(u - d))/(\mu + 1)$
	$w_j = \theta(p_j; \lambda_j)$
	$v = d + w$
	$u = v + W(\text{tvd}(W^T v, \beta/\mu) - W^T v)$
	$d = d - (u - w)$
Until convergence	
5. Select the optimal wavelet subbands	$w_{opt} = \max\left\{kurt[w_1(n)], kurt[w_2(n)], \ldots, kurt[w_{J+1}(n)]\right\}$
6. Reconstruct signal	$x = W^T w_{opt}$

3. Experiment

In order to further demonstrate the performance of proposed Kurt-WATV algorithm, rolling bearing with incipient fault are used to diagnosis. The collected signal is provided by the Center of Intelligent Maintenance Systems. Figure 3 shows an experimental rig that comprises a motor and a shaft with four Rexnord ZA-2115 double-row bearings. The bearing parameters are shown in table 2. The motor speed is 2,000 r/min, and the sampling rate is 20 kHz. More information can be seen in [21] According to the structural parameters of rolling bearing (table 2), the fault feature frequency of outer race is equal to 236.4 Hz. The collected vibration signal from bearing 1 with outer fault is used to analyze.

Table 2. Parameters of the rolling bearing.

Pitch diameter d/mm	Ball diameter e/mm	Ball number Z	Contact angle α /°
71.5	8.4	16	15.17

IOP Conf. Series: Materials Science and Engineering **372** (2018) 012030 doi:10.1088/1757-899X/372/1/012030

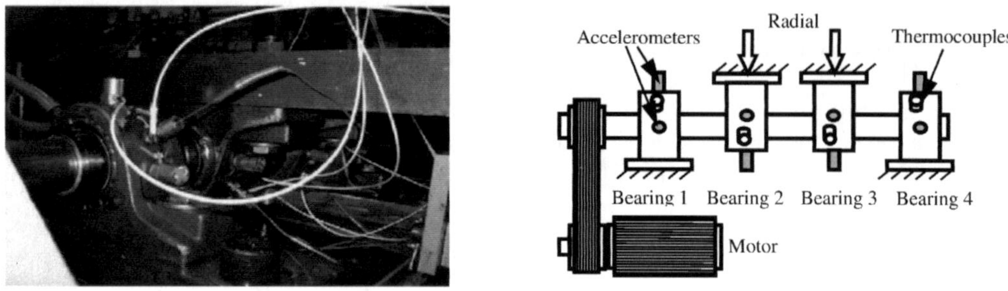

(a) (b)

Figure 3. Experimental rig: (a) Bearing test rig; (b) Schematic presentation of sensor placement.

Figure 4 shows the raw vibration signal at the incipient stage of failure, figure 4(a) shows the time-domain waveform, in which the impulses are submerged in strong background noise. Figure 4(b) shows the envelop spectrum, in which the fault feature frequency is not clear. So we need to further processing of the raw vibration signal.

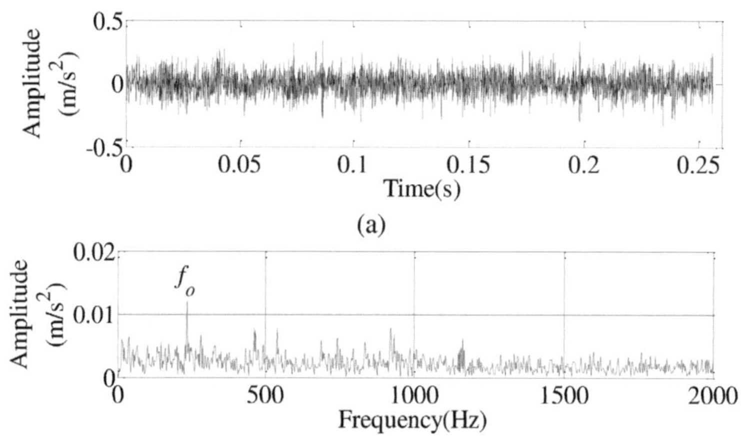

Figure 4. Experimental signal: (a) time-domain waveform; (b) envelop spectrum.

The proposed Kurt-WATV algorithm is employed enhance fault feature and weaken heavy noise. The range and constraint condition of the parameter space are set as follows:

$$T_1 = \{(p,q,s,J) \mid p,q,s,J \in Z^+, 2 \le p \le 9, q = p+1, 1 \le s \le 3, J = 20\} \tag{13}$$

The kurtosis of each wavelet subband is calculated and the result is shown in figure 5. The controlling parameters and wavelet subbands are chosen according to the kurtosis maximization principle. The optimal parameters are p=6, q=7, s=2, and the w_{16} wavelet subband is chosen to reconstructed signal, which reveal the hidden transient fault impulses. Figures 6(a) and 6(b) display the time-domain waveform and envelop spectrum using the proposed algorithm respectively. From figure 6(a), we can see the obvious periodic impacts. In the envelope spectrum, it can be seen that the fault feature frequency is attenuated while the other non-fault feature frequencies are weakened. Therefore, the proposed algorithm can successfully extract weak periodic impulses from strong background noise.

In order to further verify the superiority of the proposed algorithm, wavelet thresholding methods are used to compare with the proposed algorithm in experimental case. Figure 7(a) displays the time-domain waveform of hard-thresholding denoising, which indicates that the periodic impacts are not obvious. Figure 7(b) shows the envelop spectrum of denoised signal. From the results, we can see the amplitude of the fault feature frequency is not clear. There are some noise spikes in hard-thresholding denoising because the wavelet coefficients exceed the threshold. Figure 8 shows the time-domain waveform and envelop spectrum obtained after soft-thresholding denoising. Although the prominent peaks in figure 8(b) correspond to the feature frequency, the spectrum is again clouded with many unwanted small peaks due to background noise. The proposed algorithm in this paper is improved base on the method mentioned in [14], the denoised result obtained via the method in [14] is presented in figure 9. In the envelope spectrum, figure 9(b), the fault feature frequency of the periodic impact can not be obtained. We compare the result of the proposed algorithm with the result obtained using the method in [14]. The results demonstrated that Kurt-WATV is superior to the method proposed in [14]. It is mainly because the wavelet dictionary in [14] is a undecimated wavelet, and the time-frequency resolution is not adjustable. In this paper, time-frequency distribution characteristics can be varied by adjusting a set of controlling parameters. In addition, the optimal wavelet subband is selected using kurtosis index, which is equivalent to further performing band-pass filtering.

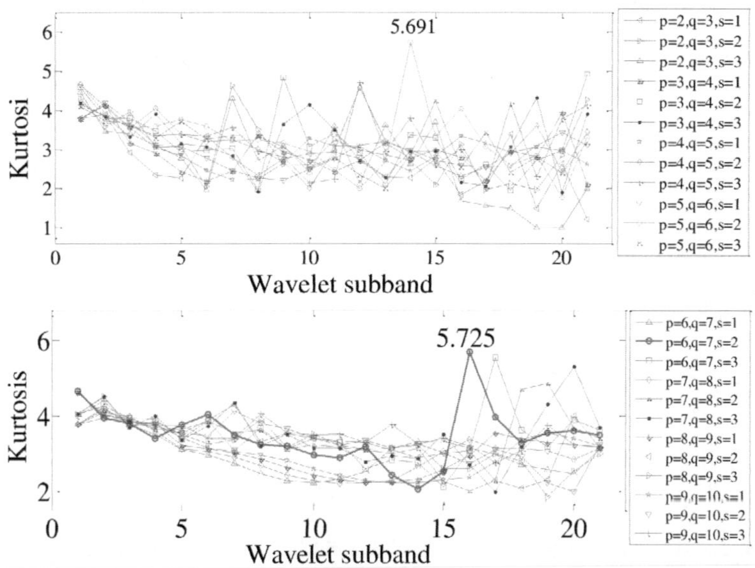

Figure 5. Kurtosis of the wavelet subbands by using parameter space T_1.

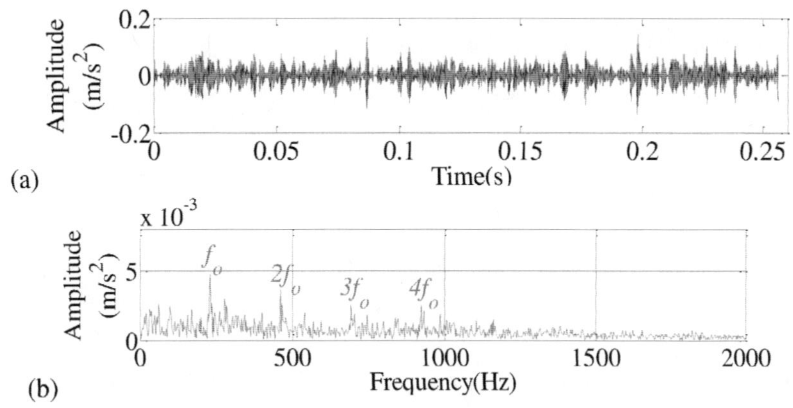

Figure 6. The denoised results using the proposed algorithm: (a) time-domain waveform and (b) envelope spectrum.

IOP Conf. Series: Materials Science and Engineering **372** (2018) 012030 doi:10.1088/1757-899X/372/1/012030

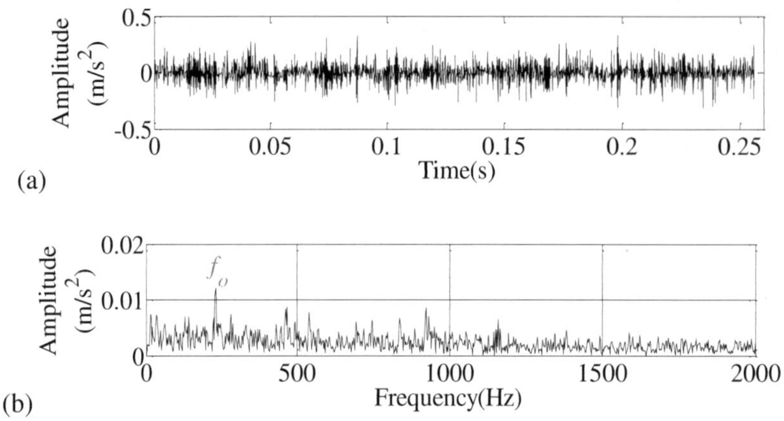

Figure 7. The denoised results obtained by hard-thresholding denoising: (a) time-domain waveform; (b) envelope spectrum

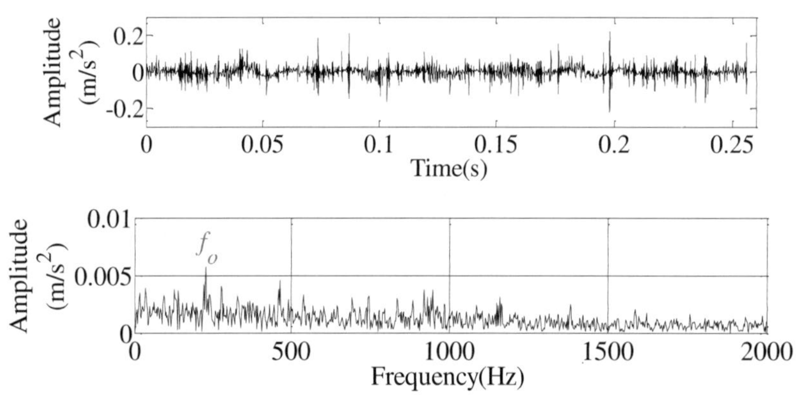

Figure 8. The denoised results obtained by soft-thresholding denoising: (a) time-domain waveform; (b) envelope spectrum

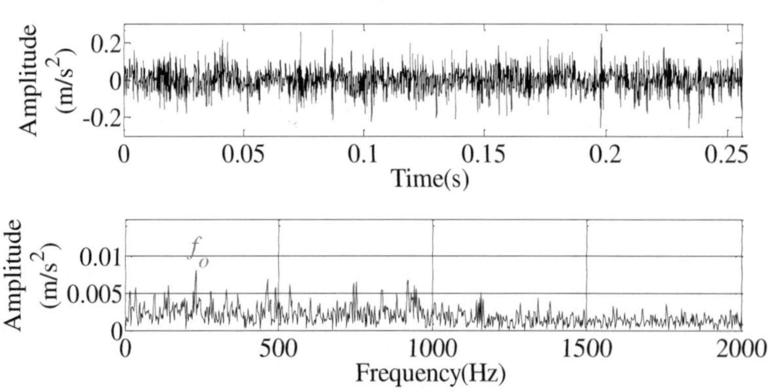

Figure 9. The denoised result obtained via the method in [14]: (a) time-domain waveform; (b) envelope spectrum

4. Conclusion

Aiming at the difficulty of extracting the incipient fault features of rolling bearing, a sparse optimization objective function is constructed. The regularization term of the objective function contains a TV term, which can effectively avoid noise spikes and pseudo-Gibbs phenomenon. ADMM is used to solve the constructed optimization objective function, and kurtosis is used as the choice of the optimal wavelet subbands, which is equivalent to a band-pass filter to extract the resonance frequency band of related fault features. The experimental signal is used to verify the the proposed algorithm. The denoised results indicate that the proposed Kurt-WATV can effectively detect fault feature frequency of rolling bearing at the incipient fault stage. The proposed Kurt-WATV, however, is limited by the lack of theoretical support for determining the parameters λ and β. Hence, future study in this area is parameter optimization.

Acknowledgments
The authors very appreciate the support from the National Natural Science Foundation of China (Nos. 51675098).

References

[1] Wang D, Guo W and Wang X J 2013 *Appl Soft Comput* **13** 4097-104
[2] Wang H C, Chen J and Dong G M 2014 *Mech Syst Signal PR* **48** 103-19
[3] Lv Y, Yuan R and Song G B 2016 *Mech Syst Signal PR* **81** 219-34
[4] Park C S, Choi Y C and Kim Y H 2013 *Mech Syst Signal PR* **38** 534-48
[5] Antoni J 2007 *Mech Syst Signal PR* **21** 108-24
[6] Stéphane M 2009 *A Wavelet Tour of Signal Processing: The Sparse Way* 3rd Ed (Burlington, MA: Academic Press)
[7] Liu B, Ling S F and Gribonval R 2002 *NDT&E INT* **35** 255-62
[8] Shi D F, Tsung F and Unsworth P J 2004 *Mech Syst Signal PR* **18** 127-41
[9] Yang H Y, Mathew J and Ma L 2005 *Mech Syst Signal PR* **19** 341-56
[10] Feng Z P and Chu F L 2007 *J Sound Vib* **302** 138-51
[11] Qin Y, Mao Y F and Tang B P 2013 *J Sound Vib* **332** 5217-35
[12] Feng Z P and Liang M 2014 *J Sound Vib* **333** 5196-211
[13] Coifman R R and Donoho D L 1995 *Wavelets & Statistics* **2** 125-50
[14] Ding Y and Selesnick I W 2015 *IEEE Signal Proc Lett* **22** 1364-8
[15] Yan R Q, Gao R X and Chen X F 2014 *Signal Process* **5** 1-15
[16] Chen B Q, Zhang Z S, Sun C, Li B, Zi Y Y and He Z J 2012 *Mech Syst Signal PR* **33** 275-98
[17] Ding Y and Selesnick I W 2015 *IEEE Signal Proc Let* **9** 1364-8
[18] Boyd S, Parikh N, Chu E, Peleato B and Eckstein 2011 *J Found Trends in Mach Learn* **3** 1-122
[19] Condat L 2013 *IEEE Signal Proc Lett* **11** 1054-7
[20] Pustelnik N, Pesquet J C and Chaux C 2012 *IEEE Trans Signal Process* **2** 968-73
[21] Qiu H, Lee J, Lin J and Yu G 2006 J Sound Vib **289** 1066-90

IOP Conf. Series: Materials Science and Engineering **372** (2018) 012031 doi:10.1088/1757-899X/372/1/012031

The study of micro-inextensible piezoelectric cantilever plate

L H Chen[1,2], J W Xu[1] and W Zhang[1]

[1]Department of Mechanics, Beijing University of Technology, Beijing, 100124, China

E-mail: chenlihua@bjut.edu.cn

Abstract. In this paper, a micro-inextensible piezoelectric cantilever plate is analyzed and its nonlinear dynamic behaviour is studied. The nonlinear oscillation differential equation is established by using Hamilton's principle with the application of strain gradient theory to consider the size effect, and inextensible theory to consider the large deformation and rotation effect of cantilever plate. Based on MATLAB software, using the Runge-Kuta method, we can obtain the response of the nonlinear oscillation differential equation. The influences of the strain gradient length scale parameter and voltage on the dynamic response of micro piezoelectric cantilever plate are investigated separately. The results confirmed an increase of the stiffness of the system by using the strain gradient theory and the amplitude of the vibration is reduced. The vibration of the system can be controlled by applying an active voltage. The effect of external excitation frequency on nonlinear dynamic behaviour is considered by using Poincare surface of section and diagrams of waveforms, phase and bifurcation.

1. Introduction

Piezoelectric materials have both positive and negative piezoelectric effects, thus it can be applied to vibration control, piezoelectric harvester and other engineering fields. With the development of micro-electromechanical technology (MEMS), the piezoelectric element itself is bound to be miniaturized. The deformation of materials in the micron dimension has been shown to be size dependent [1-4]. Therefore, the strain gradient theory is introduced to explain the size effect. Cantilever beams and plates can be assumed to be inextensible under lateral loads. Cantilever plates are important structures and extensive studied in structural engineering, such as bridges, wings and so on. Therefore, the analysis of mechanical properties of cantilever plates can explain many phenomena in practical engineering. There are plenty of studies on inextensible cantilever beams, but few on the plate.

The Kirchhoff hypothesis assumes that the deflection of the midplane is smaller than the plate thickness and the midplane remains unstrained during bending. When the transverse deflection of a plate is large, the midplane is strained and assumptions of Kirchhoff are no longer applicable. Von Kármán strains were derived to systematically account for moderate nonlinearities of thin plates, by assuming large deflection and small rotation. However, even the cantilever plates have a larger displacement when they are under lateral load, the stresses and strains of the midplane are still very small. The nonlinearities of the inextensible plate are mainly derived from the large rotation angles. Consequently, it is inappropriate to follow Von Kármán's approach where the nonlinearities of rotations are neglected to describe the bending of a cantilever plate. Dowell [5-7] derived the nonlinear modal equations of motion using Hamilton's principal for inextensible plates based on the full formulation of Novozhilov [8]. It is suitable to describe the deformation of cantilever plates with large deflection and rotations.

According to the experimental results, it is necessary to account for size effects to study microstructures. The strain gradient theory introduces the characteristic length dimension of the material in the constitutive relation, and considers the influence of strain gradient higher order tensor on the variable energy density function. It can describe and explain the size effect phenomenon of the mechanical properties of micro components. Mindlin [9] developed a general higher-order stress theory, which includes higher-order strain gradients. In 1992, Aifantis and Altan [10] introduced the second order strain tensor in the constitutive equation, and propose a theory of gradient elasticity. In 2015, Lei J [11] presented a size-dependent functionally graded (FG) micro-plate model based on a simplified strain gradient theory. The results prove that the inclusion of the size effects results in an increase in the micro-plate's stiffness. In 2016, Ansari and Gholami [12] considered the influences of the magneto electro thermos mechanical coupling to investigate the free vibration with various edge conditions using the Mindlin plate theory and von Karman hypothesis.

Inextensible plates have been studied in some literatures; no research on micro inextensible plates with strain gradient has been reported yet. The piezoelectric effect is also considered on this basis. In this paper, the inextensible theory and strain gradient theory are utilized to establish a size-dependent nonlinear cantilever piezoelectric plate model. The effects of the strain gradient length scale parameter and voltage on the dynamic response of micro piezoelectric cantilever plate are analyzed. The complex nonlinear behavior of the structure is revealed under different external excitation frequencies.

2. Potential energy of micro-inextensible piezoelectric cantilever plate

Figure 1 illustrates the cantilever plate which is studied herein:

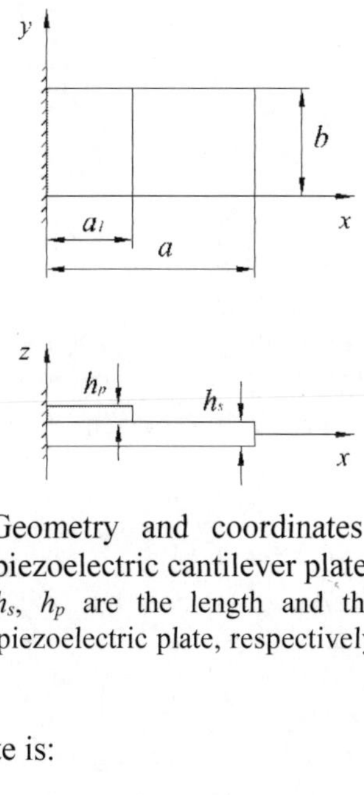

Figure 1. Geometry and coordinates of a micro inextensible piezoelectric cantilever plate.
note: a, a_l, h_s, h_p are the length and thickness of the baseplate and piezoelectric plate, respectively b is the width of the plate.

The strain at each point in the plate is:

$$\varepsilon_{xx} = \hat{\varepsilon}_{xx} + z\kappa_{xx}, \quad \varepsilon_{yy} = \hat{\varepsilon}_{yy} + z\kappa_{yy}, \quad \varepsilon_{xy} = \hat{\varepsilon}_{xy} + z\kappa_{xy} \tag{1}$$

$\hat{\varepsilon}_{xx}, \hat{\varepsilon}_{yy}, \hat{\varepsilon}_{xy}$ are the strain of midplane, $\kappa_{xx}, \kappa_{yy}, \kappa_{xy}$ are the curvature of midplane.

Midplane strains can be considered to:

$$\hat{\varepsilon}_{xx} = \frac{\partial u}{\partial x} + \frac{1}{2}(\frac{\partial w}{\partial x})^2, \quad \hat{\varepsilon}_{yy} = \frac{\partial v}{\partial y} + \frac{1}{2}(\frac{\partial w}{\partial y})^2, \quad \hat{\varepsilon}_{xy} = \frac{\partial u}{\partial y} + \frac{\partial v}{\partial x} + \frac{\partial w}{\partial x}\frac{\partial w}{\partial y} \tag{2}$$

where w is the plate transverse deflection in the z-direction and u, v are the in-plane deformations in x

and y-directions.

For the inextensible plate, the following hypotheses are made:

$$\hat{\varepsilon}_{xx} = \hat{\varepsilon}_{yy} = \hat{\varepsilon}_{xy} = 0 \tag{3}$$

According to (2.2-2.3), we can obtain expression:

$$\frac{\partial u}{\partial x} = -\frac{1}{2}(\frac{\partial w}{\partial x})^2,\quad \frac{\partial v}{\partial y} = -\frac{1}{2}(\frac{\partial w}{\partial y})^2,\quad \frac{\partial u}{\partial y} + \frac{\partial v}{\partial x} = -(\frac{\partial w}{\partial x})(\frac{\partial w}{\partial y}) \tag{4}$$

Following Dowell [6], the curvature is considered as:

$$\kappa_{xx} = -\frac{\partial^2 w}{\partial x^2}[1 + \frac{1}{2}(\frac{\partial w}{\partial x})^2 + \frac{1}{2}(\frac{\partial w}{\partial y})^2]$$

$$\kappa_{yy} = -\frac{\partial^2 w}{\partial y^2}[1 + \frac{1}{2}(\frac{\partial w}{\partial x})^2 + \frac{1}{2}(\frac{\partial w}{\partial y})^2] \tag{5}$$

$$\kappa_{xy} = -2\frac{\partial^2 w}{\partial x \partial y}[1 + \frac{1}{2}(\frac{\partial w}{\partial x})^2 + \frac{1}{2}(\frac{\partial w}{\partial y})^2]$$

The micro piezoelectric plate is a two layer structure, the stress of the base and the piezoelectric layer should be calculated separately, and the results can be expressed as

$$\sigma_{xx}^{S} = \frac{E_S}{1-\upsilon_S^2}(\varepsilon_{xx} + \upsilon_S \varepsilon_{yy}),\quad \sigma_{yy}^{S} = \frac{E_S}{1-\upsilon_S^2}(\varepsilon_{yy} + \upsilon_S \varepsilon_{xx}),\quad \tau_{xy}^{S} = \frac{E_s}{2(1+\upsilon_S)}\gamma_{xy}$$

$$\sigma_{xx}^{P} = \frac{E_P}{1-\upsilon_P^2}(\varepsilon_{xx} + \upsilon_P \varepsilon_{yy}) - e_{31}E_3,\quad \sigma_{yy}^{P} = \frac{E_P}{1-\upsilon_P^2}(\varepsilon_{yy} + \upsilon_P \varepsilon_{xx}) - e_{32}E_3,\quad \tau_{xy}^{P} = \frac{E_P}{2(1+\upsilon_P)}\gamma_{xy} \tag{6}$$

where $E_3 = \frac{V_0}{h_p}$, E and υ are young's modulus and Poisson's ratio for each layer of material, respectively. The subscripts S and P respectively represent the base and the piezoelectric layer. The total potential energy of the system is:

$$U = \bar{U} + \hat{U} \tag{7}$$

in which $\bar{U}$ is the potential energy dependent on strain and $\hat{U}$ is the potential energy dependent on strain gradient. $\bar{U}$ can be considered as:

$$\bar{U} = \iiint_{V_S} \frac{1}{2}(\sigma_{xx}^{S}\varepsilon_{xx} + \sigma_{yy}^{S}\varepsilon_{yy} + \tau_{xy}^{S}\gamma_{xy})dxdydz + \iiint_{V_P} \frac{1}{2}(\sigma_{xx}^{P}\varepsilon_{xx} + \sigma_{yy}^{P}\varepsilon_{yy} + \tau_{xy}^{P}\gamma_{xy})dxdydz \tag{8}$$

The expression of strain gradient is proposed by Mindlin [9]:

$$\eta_{ijk} = \varepsilon_{jk,i} \qquad (i,j,k = x,y,z) \tag{9}$$

For the inextensible plate, from (1), (3) and (5), the expression η_{ijk} are:

$$\eta_{xxx} = -z\frac{\partial \kappa_{xx}}{\partial x},\quad \eta_{yxx} = -z\frac{\partial \kappa_{xx}}{\partial y},\quad \eta_{xxy} = \eta_{xyx} = -z\frac{\partial \kappa_{xy}}{\partial x},\quad \eta_{yxy} = \eta_{yyx} = -z\frac{\partial \kappa_{xy}}{\partial y},\quad \eta_{xyy} = -z\frac{\partial \kappa_{yy}}{\partial x}, \tag{10}$$

$$\eta_{yyy} = -z\frac{\partial \kappa_{yy}}{\partial y}, \quad \eta_{zxx} = \kappa_{xx}, \quad \eta_{zxy} = \kappa_{xy}, \quad \eta_{zyy} = \kappa_{yy}$$

others $\eta_{ijk} = 0, (j\,\mathrm{I}\; k = z)$.

For microstructures, $\hat{U}$ is expressed by the strain gradient obtained above:

$$\hat{W}_{\xi} = L_1^{\xi}\eta_{iik}\eta_{kjj} + L_2^{\xi}\eta_{ijj}\eta_{ikk} + L_3^{\xi}\eta_{iik}\eta_{jjk} + L_4^{\xi}\eta_{ijk}\eta_{ijk} + L_5^{\xi}\eta_{ijk}\eta_{kji}, \quad (\xi = S, P) \tag{11}$$

from L_l^{ξ} to L_5^{ξ} are the material parameters, the reference values for each material parameter is obtained by Ramezani [13]:

$$L_2^{\xi} = \frac{1}{2}l_{\xi}^2\lambda_{\xi}, \quad L_4^{\xi} = l_{\xi}^2\mu_{\xi}, \quad L_1^{\xi} = L_3^{\xi} = L_5^{\xi} = 0, \quad (\xi = S, P) \tag{12}$$

l_{ξ} is the strain gradient length scale parameter, λ_{ξ} and μ_{ξ} are lame constants:

$$\lambda_{\xi} = E_{\xi}\upsilon_{\xi} / (1-2\upsilon_{\xi})(1+\upsilon_{\xi}), \quad \mu_{\xi} = E_{\xi} / 2(1+\upsilon_{\xi}), \quad (\xi = S, P) \tag{13}$$

Then we obtain the high order stress tensor χ:

$$\chi_{ijk}^{\xi} = \chi_{ijk}^{\xi} = \frac{\partial \hat{W}_{\xi}}{\partial \eta_{ijk}} \quad (\xi = S, P) \tag{14}$$

where:

$$\begin{aligned}
&\chi_{xxx}^{\xi} = (L_4^{\xi} + L_5^{\xi})(-2z\frac{\partial \kappa_{xx}}{\partial x}), \quad \chi_{xxy}^{\xi} = \chi_{xyx}^{\xi} = (L_4^{\xi} + L_5^{\xi})(-2z\frac{\partial \kappa_{xy}}{\partial x}), \quad \chi_{xyy}^{\xi} = (L_4^{\xi} + L_5^{\xi})(-2z\frac{\partial \kappa_{yy}}{\partial x}),\\
&\chi_{yxx}^{\xi} = (L_4^{\xi} + L_{55}^{\xi})(-2z\frac{\partial \kappa_{xx}}{\partial y}), \quad \chi_{yxy}^{\xi} = \chi_{yyx}^{\xi} = (L_4^{\xi} + L_5^{\xi})(-2z\frac{\partial \kappa_{xy}}{\partial y}), \quad \chi_{yyy}^{\xi} = (L_4^{\xi} + L_5^{\xi})(-2z\frac{\partial \kappa_{yy}}{\partial y}),\\
&\chi_{zxx}^{\xi} = -2L_4^{\xi}\kappa_{xx}, \quad \chi_{zxy}^{\xi} = \chi_{zyx}^{\xi} = -2L_4^{\xi}\kappa_{xy}, \quad \chi_{zyy}^{\xi} = -2L_4^{\xi}\kappa_{yy},
\end{aligned} \tag{15}$$

Following the equation above, the energy based on the strain gradient can be obtained:

$$\hat{U} = \iiint_{V_S}\frac{1}{2}(\chi_{ixx}^{S}\cdot\eta_{ixx} + \chi_{iyy}^{S}\cdot\eta_{iyy} + \chi_{ixy}^{S}\cdot\eta_{ixy})dxdydz + \iiint_{V_P}\frac{1}{2}(\chi_{ixx}^{P}\cdot\eta_{ixx} + \chi_{iyy}^{P}\cdot\eta_{iyy} + \chi_{ixy}^{P}\cdot\eta_{ixy})dxdydz \tag{16}$$

$(i = x, y, z)$

3. Nonlinear dynamic model of micro piezoelectric plate

The kinetic energy of the system can be expressed as the sum of the kinetic energy of the base plate and the piezoelectric layer:

$$T_S = \iiint_{V_S}\frac{\rho_S}{2}(\dot{w}^2)dV_S, \quad T_P = \iiint_{V_P}\frac{\rho_P}{2}(\dot{w}^2)dV_P, \quad T = T_S + T_P \tag{17}$$

ρ_S and ρ_P are the density of the material of the base plate and the piezoelectric layer, respectively. $\dot{w}$ is the derivative of w to time.

The external work of the system divided into three parts:

$$W_q = \int q \cdot w dxdy, \quad W_S = \int r_S \cdot \dot{w} \cdot w dxdydz, \quad W_P = \int r_P \cdot \dot{w} \cdot w dxdydz \tag{18}$$

$q=q_0 \sin\omega t$ is the distributed load, r_s and r_p are the damping coefficient of the material of the base plate and the piezoelectric layer, respectively.

The total external power:

$$W = W_S + W_P + W_q \tag{19}$$

Based on the boundary conditions of a clamped boundary (x=0) and three free boundaries (other sides) , using the combination of beam functions, the form of the solution is assumed to be:

$$w(x,y,t) = w_1(t) \cdot W_1(x,y) + w_2(t) \cdot W_2(x,y) + w_3(t) \cdot W_3(x,y) \tag{20}$$

where:

$$W_1(x,y) = X_1 \cdot Y_1, \quad W_2(x,y) = X_1 \cdot Y_2, \quad W_3(x,y) = X_2 \cdot Y_1$$

$$X_1 = \cosh\frac{k_1 x}{a} - \cos\frac{k_1 x}{a} - \beta_1(\sinh\frac{k_1 x}{a} - \sin\frac{k_1 x}{a}), \quad X_2 = \cosh\frac{k_2 x}{a} - \cos\frac{k_2 x}{a} - \beta_2(\sinh\frac{k_2 x}{a} - \sin\frac{k_2 x}{a}),$$

$$Y_1 = 1, \quad Y_2 = \sqrt{3}(1-2y/b), \qquad k_i^4 = \omega^2\frac{\rho A}{EJ}(i=1,2), \quad \beta_i = \frac{\cosh k_i b - \cos k_i b}{\sinh k_i b - \sin k_i b}(i=1,2)$$

EJ is the bending stiffness of section, A is the cross-sectional area. X_i are the shape functions of cantilevered beam and Y_i are the shape functions of free-free beam. Bring the expression of the assumed solution (20) into (7), (17) and (19), according to Hamilton principle: $0 = \delta\int_{t_1}^{t_2}(U-T)dt + \int_{t_1}^{t_2}\delta W dt$, t_1 and t_2 represent the initial and final moments of the system movement, respectively. we obtain the ordinary differential motion equation:

$$\begin{aligned}
\ddot{w}_1 &= a_1 w_1 + a_2 w_2 + a_3 w_3 + a_4 w_1^2 + a_5 w_2^2 + a_6 w_3^2 + a_7 w_1^3 + a_8 w_2^3 + a_9 w_3^3 + a_{10} w_1 w_2 + a_{11} w_1 w_3 \\
&+ a_{12} w_2 w_3 + a_{13} w_1 w_2^2 + a_{14} w_1 w_3^2 + a_{15} w_2 w_1^2 + a_{16} w_2 w_3^2 + a_{17} w_3 w_1^2 + a_{18} w_3 w_2^2 + a_{19} w_1 w_2 w_3 \\
&+ a_{20}\dot{w}_1 + a_{21}\dot{w}_2 + a_{22}\dot{w}_3 + a_{23}V_0 + a_{24}w_1^2 V_0 + a_{25}w_2^2 V_0 + a_{26}w_3^2 V_0 + a_{27}w_1 w_2 V_0 + a_{28}w_1 w_3 V_0 \\
&+ a_{29} w_2 w_3 V_0 + a_{30} q_0 \sin\omega t
\end{aligned}$$

$$\begin{aligned}
\ddot{w}_2 &= b_1 w_1 + b_2 w_2 + b_3 w_3 + b_4 w_1^2 + b_5 w_2^2 + b_6 w_3^2 + b_7 w_1^3 + b_8 w_2^3 + b_9 w_3^3 + b_{10} w_1 w_2 + b_{11} w_1 w_3 \\
&+ b_{12} w_2 w_3 + b_{13} w_1 w_2^2 + b_{14} w_1 w_3^2 + b_{15} w_2 w_1^2 + b_{16} w_2 w_3^2 + b_{17} w_3 w_1^2 + b_{18} w_3 w_2^2 + b_{19} w_1 w_2 w_3 \\
&+ b_{20}\dot{w}_1 + b_{21}\dot{w}_2 + b_{22}\dot{w}_3 + b_{23}V_0 + b_{24}w_1^2 V_0 + b_{25}w_2^2 V_0 + b_{26}w_3^2 V_0 + b_{27}w_1 w_2 V_0 + b_{28}w_1 w_3 V_0 \\
&+ b_{29} w_2 w_3 V_0 + b_{30} q_0 \sin\omega t
\end{aligned} \tag{21}$$

$$\begin{aligned}
\ddot{w}_3 &= c_1 w_1 + c_2 w_2 + c_3 w_3 + c_4 w_1^2 + c_5 w_2^2 + c_6 w_3^2 + c_7 w_1^3 + c_8 w_2^3 + c_9 w_3^3 + c_{10} w_1 w_2 + c_{11} w_1 w_3 \\
&+ c_{12} w_2 w_3 + c_{13} w_1 w_2^2 + c_{14} w_1 w_3^2 + c_{15} w_2 w_1^2 + c_{16} w_2 w_3^2 + c_{17} w_3 w_1^2 + c_{18} w_3 w_2^2 + c_{19} w_1 w_2 w_3 \\
&+ c_{20}\dot{w}_1 + c_{21}\dot{w}_2 + c_{22}\dot{w}_3 + c_{23}V_0 + c_{24}w_1^2 V_0 + c_{25}w_2^2 V_0 + c_{26}w_3^2 V_0 + c_{27}w_1 w_2 V_0 + c_{28}w_1 w_3 V_0 \\
&+ c_{29} w_2 w_3 V_0 + c_{30} q_0 \sin\omega t
\end{aligned}$$

The parameter a_i, b_i, c_i (i=0,1,...30), are related to the geometric size and material parameter, as space is limited, the detailed derivation won't be described here.

4. The effect of voltage on the amplitude

By using Runge–Kutta numerical integration method and MATLAB the influences of the scale effect and voltage on the amplitude of piezoelectric plate are studied. The result shows the necessity of introducing the scale effect and the control effect of the voltage. First, the length scale parameter l is introduced to reflect the effect of the strain gradient. The response of the plate is calculated at l=0 and l=0.5h_S, respectively, under the same external excitation. It can be seen from figure 2(a) that considering the length parameter, the stiffness of the plate will be obviously increased, then the deflection- thickness ratio is reduced. This is because the thickness of the plate is close to the length scale parameter, and the effect of the scale is obvious, so the strain gradient needs to be considered.

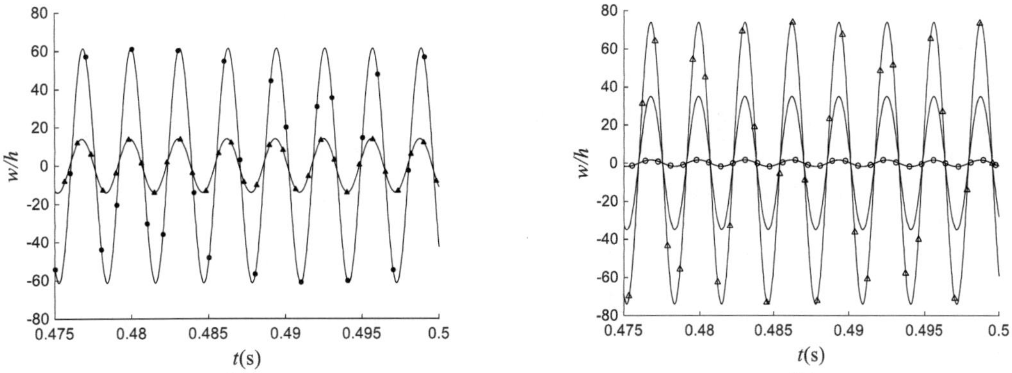

(a) (b)

Figure 2. The influence of l and V_0, l=0(-•-) and l=0.5 h_S (-▲-) in (a), V_0=-2V(-o-) ,V_0=0V (—) and V_0=2V(-△-) in (b).

The geometric parameters and material parameters of the micro piezoelectric cantilever laminates are shown in table 1 below.

Table 1. Parameters.

physical quantity	value	physical quantity	value
a	2 mm	υ_S	0.278
a_1	1 mm	υ_P	0.35
b	1 mm	e_{31}	-20C · m^2
h_S	4 μm	e_{32}	-20C · m^2
h_P	2 μm	ε_{33}	6500 ε_0
E_S	210 Gpa	ρ_S	2330 kg/m^3
E_P	1.4 Gpa	ρ_P	1800 kg/m^3

Next, the impact of the voltage is studied, where $V=V_0\sin\omega t$. Both positive and negative voltages are applied to the piezoelectric plate respectively.

From figure 2(b), it can be seen that the application of both positive and negative voltages has a diametrically opposite effect on the amplitude of the plate compared to the plate without loaded voltage. Therefore, the application of voltage can play a controlling effect.

5. Complex nonlinear behavior

This section mainly studies the nonlinear dynamic behavior of micro piezoelectric plate. The nonlinear dynamic equation (21) mentioned above is solved by Runge-Kutta method. With the help of MATLAB software, the nonlinear vibration waveform diagram (ratio of deflection to thickness and time), phase diagram (velocity and deflection), Poincare cross section (velocity and deflection) and bifurcation diagram (deflection and excitation force frequence) of the micro piezoelectric inextensible cantilever under the influence of external vibration are obtained using numerical simulation. The results of these numerical simulations are observed and analyzed. the length scale parameter l=0.5h_S.

The frequency of the applied voltage is adjusted to $\omega_V = 2\omega_F$, where ω_V and ω_F are the frequency of the external voltage and the external excitation force, respectively. Changing the frequency of external excitation, we get a series of waveforms of deflection- thickness ratio and phase diagrams of displacement and velocity, and bifurcation diagram with frequency variation of ω_F. From figure 3(a), we can find the micro piezoelectric plate has been maintaining a single periodic harmonic vibration before ω_F=2200 rad/s. When ω_F=2200 rad/s, the system becomes bifurcate, the single period vibration becomes two times period, then changes into three and four times period. After ω_F=2250 rad/s. The system goes into a stable time period. When the external excitation frequency increases to 2550 rad/s, the micro piezoelectric plate gradually appeared 2T—3T—4T—3T—2T of periodic motion, shown in the figure 3(b). The vibration returns to the single periodic vibration after 2900 rad/s. With the increase of frequency, the micro piezoelectric cantilever plate returns to single periodic harmonic vibration.2T—3T—4T of period occurred again between ω_F=3600 rad/s and ω_F=3700 rad/s. The amplitude of vibration increases obviously, when the system appears bifurcation. The waveform, phase diagram and Poincare cross section of period—1 to period—4 are shown in figure 4.

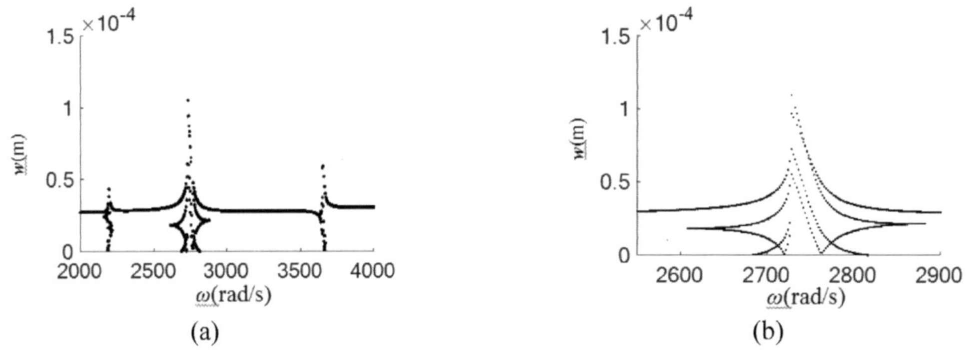

Figure 3. Bifurcation diagram of ω (rad/s) and w (m), (b) is the partial enlarged drawing of (a).

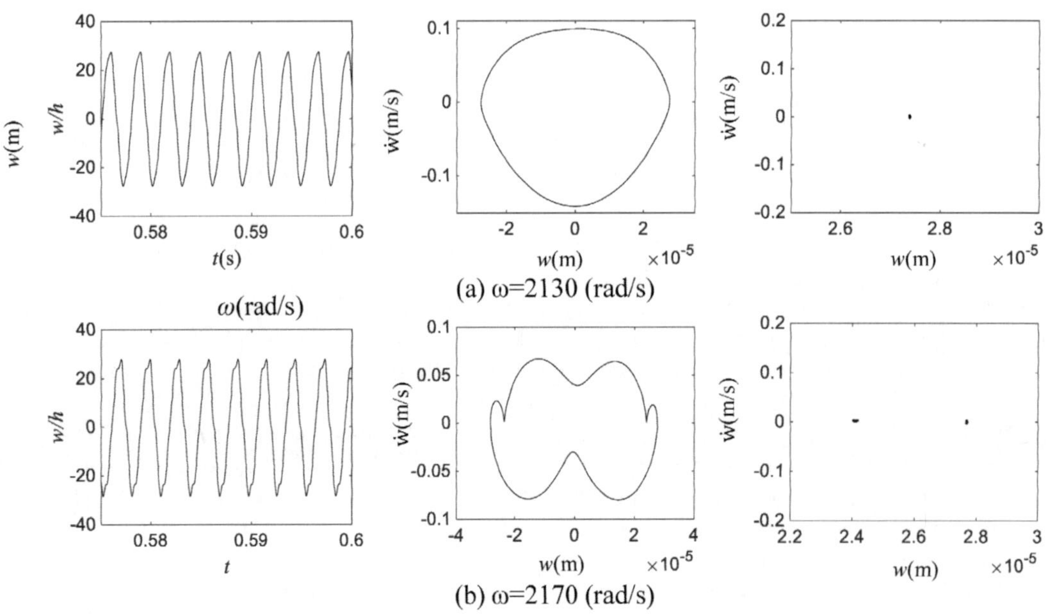

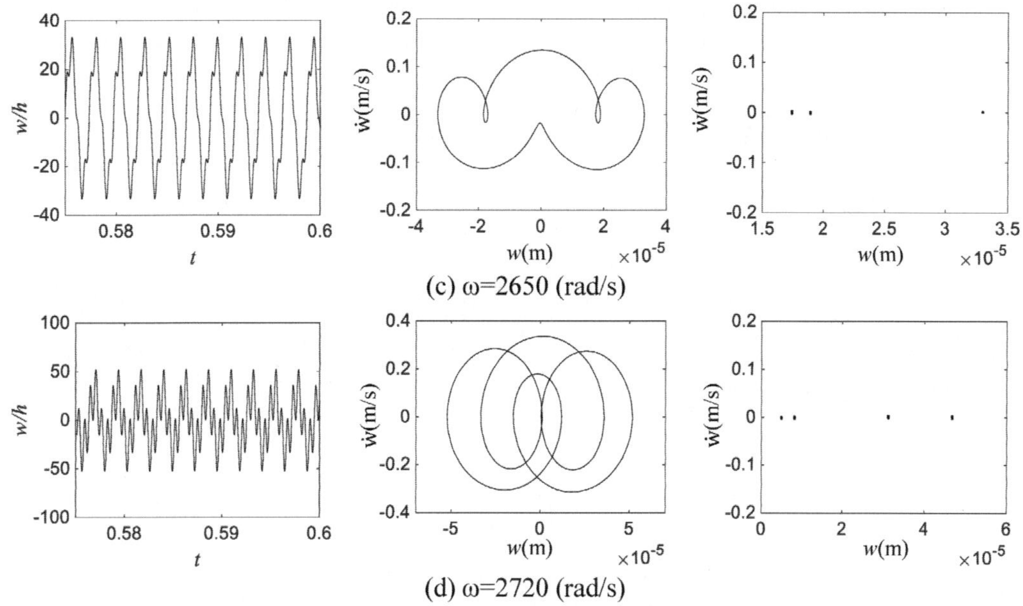

Figure 4. The waveform, phase diagram and poincare cross section.

6. Conclusion

In this paper, a nonlinear dynamic model of a micro cantilever piezoelectric plate is established, considering the inextensible midplane, the scale effect and the mechatronic coupling effect. When the geometric size of the microstructure approaches the material scale parameters, the strain gradient theory can fully reflect the influence of the scale effect. Therefore, it is necessary to consider the size effect of the micro structure. The positive and negative voltages are applied to enlarge and restrain the response so that the vibration response is controlled. The complex nonlinear dynamic behavior of the plate is generated by changing the external excitation frequency. The amplitude increases sharply, when the system appears bifurcation. This phenomenon can be used as a theoretical basis for the application of the piezoelectric plate in the field of energy harvester.

Acknowledgments

This paper is supported by the national natural science foundation of China, project No. 11472019. We really appreciate the assistance of the committee of the national natural science foundation of China.

References

[1] Lam D C C, Yang F, Chong A C M, Wang J and Tong P 2003 Experiments and theory in strain gradient elasticity *J Mech Phys Solids* **51** 1477-508

[2] Mcfarland A W and Colton J S 2005 Role of material microstructure in plate stiffness with relevance to microcantilever sensors *J Micromech Microeng* **15** 1060-7

[3] Poole W J, Ashby M F and Fleck N A 1996 Micro-hardness of annealed and work-hardened copper polycrystals *Scr Mater* **34** 559-64

[4] Stölken J S and Evans A G 1998 A microbend test method for measuring the plasticity length scale *Acta Mater* **46** 5109-15

[5] Tang D, Gibbs S C and Dowell E H 2015 Nonlinear aeroelastic analysis with inextensible plate theory including correlation with experiment *AIAA J* **53** 1299-308

[6] Tang D, Zhao M and Dowell E H 2014 Inextensible beam and plate theory: computational analysis and comparison with experiment *J Appl Mech* **81** 061009

[7] Dowell E and Mchugh K 2016 Equations of motion for an inextensible beam undergoing large

deflections *J Appl Mech* **83** 051007
[8] Novozhilov V V 1953 *Foundations of the Nonlinear Theory of Elasticity* (Graylock Press)
[9] Mindlin R D 1964 Micro-structure in linear elasticity *Arch Ration Mech Anal* **16** 51-78
[10] Altan S B and Aifantis E C 1992 On the structure of the mode iii crack-tip in gradient elasticity *Scr Metall Mater* **26** 319-24
[11] Lei J, He Y, Zhang B, Liu D, Shen L and Guo S 2015 A size-dependent fg micro-plate model incorporating higher-order shear and normal deformation effects based on a modified couple stress theory *Int J Mech Sci* **104** 8-23
[12] Ansari R and Gholami R 2016 Nonlocal free vibration in the pre- and post-buckled states of magneto-electro-thermo elastic rectangular nanoplates with various edge conditions *Smart Mater Struct* **25** 095033
[13] Ramezani S 2013 Nonlinear vibration analysis of micro-plates based on strain gradient elasticity theory *Nonlinear Dyn* **73** 1399-421

IOP Conf. Series: Materials Science and Engineering **372** (2018) 012032 doi:10.1088/1757-899X/372/1/012032

Elasto-limited plastic analysis of structures for probabilistic conditions

M Movahedi Rad[1,2]

[1]Department of Structural and Geotechnical Engineering, Széchenyi István University, Egyetem Tér1, H-9026 Győr, Hungary

E-mail: majidmr@sze.hu

Abstract. With applying plastic analysis and design methods, significant saving in material can be obtained. However, as a result of this benefit excessive plastic deformations and large residual displacements might develop, which in turn might lead to unserviceability and collapse of the structure. In this study, for deterministic problem the residual deformation of structures is limited by considering a constraint on the complementary strain energy of the residual forces. For probabilistic problem the constraint for the complementary strain energy of the residual forces is given randomly and critical stresses updated during the iteration. Limit curves are presented for the plastic limit load factors. The results show that these constraints have significant effects on the load factors. The formulations of the deterministic and probabilistic problems lead to mathematical programming which are solved by the use of nonlinear algorithm.

1. Introduction

The basic aim of plastic limit analysis is to calculate the plastic limit load factors and the stresses, strain rates at the plastic limit condition of the structures. The strong preference of this principle is due to the fact that, without a study of the entire loading history, the principal directly determine the value of the upper and lower boundaries of the plastic limit load factors. At the plastic limit analysis stresses can be solved using a static equilibrium equation with the limited plastic load and, meanwhile, satisfy the yield condition at any part in the structure.

2. Loading and limit state of elasto-plastic structures

Let us consider a body what is subjected to multiparameter quasi-static load on its surfaces. This loading can be given in separable form:

$$Q_i(x_i,t) = \sum_j m_i(t)\, Q_{i0}^j(x_i). \quad (j = 1,2,\dots,n) \tag{1}$$

Here the indepemdent group of forces $Q_{i0}^j(x_i)$; $(j = 1,2,\dots,n)$ do not depend on time. The time dependency is given by the associated load multipliers $m_j(t)$. The loading process is represented by a curve in loading space $m_1, m_2, \dots, m_j$. These multipliers can vary independently, but they are limited by the following equation:

$$L(m_j) = 0, \tag{2}$$

Equation (2) defines all the possible combinations of the external loads which can occur during the loading process. In case of two-parameter loading the loading process and the loading domain are presented in figure 1. Here the formulation $L(m_1, m_2) = 0$ means all the combined values of the load intensities $m_1$and $m_2$where the structure is in plastic limit state. This equation is represented by a closed curve in coordinate system m_1, m_2 and it can be proved that it is covex Kaliszky [1].

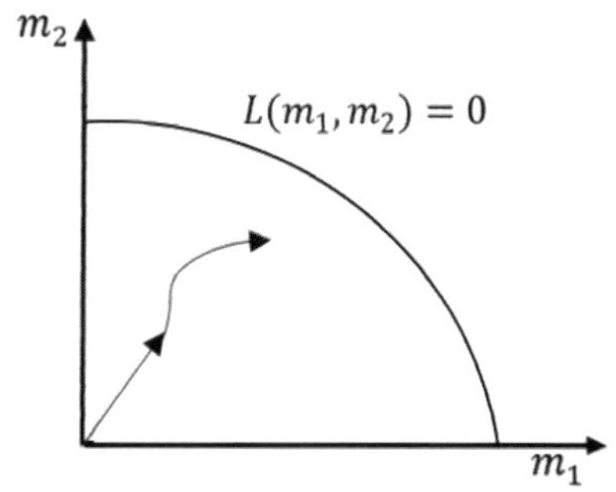

Figure 1. loading process and loading domain.

In the literature there are several computational methods which provide appropriate tools to determine the limit of the accumulated permanent deformations (see e.g. Maier [2]; Polizzotto [3]; König [4]; Weichert and Maier [5]; Levy *et al* [6]; Simon and Weichert [7]; Ponter [8]; Corradi [9]; Capurso *et al* [10]; Kaneko and Maier [11]; Tin-Loi [12] and Liepa *et al* [13]). Also an appropriate computational procedure is presented which based on the fact that the permanent deformations and residual stresses are not independent, the complementary strain energy of the residual stresses can approximately describe the plastic behaviour of structure and the amount of the plastic deformations, respectively (Kaliszky and Lógó [14]; Movahedi and Lógó [15]; Lógó *et al* [16] and Movahedi [17]). Therefore if the complementary strain energy of the residual stresses is bounded, in the same time it can limit the amount of the permanent deformations of the structure.

In the case of truss element the complementary strain energy of the residual stresses can be expressed as follow:

$$C_p = \frac{1}{2E}\sum_{k=1}^{n}\frac{l_k}{A_k}(S_k^r)^2; \tag{3}$$

in the case of residual moments:

$$C_p = \frac{1}{2E}\sum_{k=1}^{n}\frac{1}{I_k}\int_0^{l_k}[M_k^r(s)]^2 ds. \tag{4}$$

Here $l_k, (k = 1, 2, \dots, n)$ denotes the length of the bars or beam members, A_k and I_k are the area and moment of inertia of the bar and beam elements, respectively, S_k^r: the residual normal force of the bar members, $M_k^r(s)$: the residual moment of the beam members, E:the Young's modulus. The integral expression in equation (4) can be defined in case of the residual moments M_{k1}^r and M_{k2}^r acting at the ends of the beam members as:

$$R_k^r = \frac{1}{3}[(M_{k1}^r)^2 + (M_{k1}^r)(M_{k2}^r) + (M_{k2}^r)^2], \tag{5}$$

for sake of simplicity, R_k^r denotes the residual force of the element, D_k *is* the area or the moment of inertia for each member. The general form of equation(3) and equation (4) can be written as follows:

$$C_p = \frac{1}{2E}\sum_{k=1}^{n}\frac{l_k}{D_k}(R_k^r)^2. \tag{6}$$

Applying the expressions above the amount of the plastic deformations can be limited if a permissible energy value C_{p0} is introduced. Permissible energy value can not exceed by the complementary strain energy of the residual stresses, the inequality can be written as follows:

$$\frac{1}{2E}\sum_{k=1}^{n}\frac{l_k}{D_k}(R_k^r)^2 \leq C_{p0}. \tag{7}$$

2.1. Probabilistic problem

Reliability methods aim at evaluating the probability of failure of a system whose modelling takes into account randomness. The probability function for a random variable Xcan be defined by the following integral:

$$P_f = \int f(X)dx. \tag{8}$$

Since uncertanties have significant role in engineering, the constraints on the complementary strain energy of the residual forces are considered randomly and for the reason of simplicity normal distribution function applied. Mean value: C_{p0} and standard deviation: S. The probability of the failure function can be constructed as follow:

$$P_{f,target} = \int f(C_{p0}, S)dx. \tag{9}$$

Probabilistic bound can be defined with introducing relability index β:

$$\beta_{target} - \beta_{calc} \leq 0; \tag{10}$$

where $\beta_{target}$is working as a constraint for β_{calc}. Reliability indexes can be defined as follows:

$$\beta_{target} = -\Phi^{-1}(P_{f,target}); \tag{11}$$

$$\beta_{calc} = -\Phi^{-1}(P_{f,calc}); \tag{12}$$

in equations (11) and (12) Φ is cumulative distribution function of the normal distribution.

3. Elasto-limited plastic analysis

Consider any statically admissible internal force $\boldsymbol{Q}^s$fields which satisfy equilibrium equations and define to them a statically admissible load multiplier $\boldsymbol{m}_s$, the following equilibrium equation can be constructed:

$$\boldsymbol{G}\boldsymbol{Q}^s + \boldsymbol{m}_s\boldsymbol{P}_0 = \boldsymbol{0}; \tag{13}$$

$\boldsymbol{G}$: equilibrium matrix and $\boldsymbol{P}_0$: base load vector.

Equation (14) defines the yields condition:

$$-\boldsymbol{Q}^p \leq \boldsymbol{Q}^s \leq \boldsymbol{Q}^p; \tag{14}$$

$\boldsymbol{Q}^p$: limit plastic force. The following equation defines elastic fictitious internal force:

$$\boldsymbol{Q}^e = \boldsymbol{T}^{-1}\boldsymbol{G}\boldsymbol{K}^{-1}\boldsymbol{m}_s\boldsymbol{P}_0; \tag{15}$$

Here $\boldsymbol{T}$: flexibility matrix and $\boldsymbol{K}$: stiffness matrix. The static principle of limit analysis states that: any statically admissible stable load multiplier $\boldsymbol{m}_s$ of the elasto-plastic structures is less than or equal to the collapse load multiplier $\boldsymbol{m}_p$, i.e.:

$$\boldsymbol{m}_s \leq \boldsymbol{m}_p. \tag{16}$$

Since the exact value of $\boldsymbol{m}_p$ is equal to the maximum value of $\boldsymbol{m}_s$, the problem based on the static principle can be define as follows:

For *deterministic problems* residual plastic behavior of structures are limited with applying permissible energy value $\boldsymbol{C}_{p0}$:

$$\left.\begin{array}{c} \boldsymbol{m_s = max} \\ \boldsymbol{GQ^s + m_sP_0 = 0} \\ \boldsymbol{-Q^p \le Q^s \le Q^p} \\ \boldsymbol{Q^e = T^{-1}GK^{-1}m_sP_0} \\ \boldsymbol{\frac{1}{2E}\sum_{k=1}^{n}\frac{l_k}{D_k}(R_k^r)^2 \le C_{p0}} \end{array}\right\} \tag{17}$$

In the case of *probabilistic problems* bound can be defined with introducing relability index $\boldsymbol{\beta}$:

$$\left.\begin{array}{c} \boldsymbol{m_s = max} \\ \boldsymbol{GQ^s + m_sP_0 = 0} \\ \boldsymbol{-Q^p \le Q^s \le Q^p} \\ \boldsymbol{Q^e = T^{-1}GK^{-1}m_sP_0} \\ \boldsymbol{\beta_{target} - \beta_{calc} \le 0} \end{array}\right\} \tag{18}$$

4. Numerical examples and results

4.1. Example 1

Consider the skeletal frame illustrated in figure 2. The structure is subjected to horizontal loads $P_1 \geq 0$ and vertical loads $P_2 \geq 0$. The elastic modulus $E = 21 \cdot 10^4\ N/mm^2$ and allowable stress $f_y = 210N/mm^2$. The cross-section area of the beams and columns $A = 2.85 \cdot 10^3 mm^2$ and $3.9 \cdot 10^3\ mm^2$, respectively. The moment of inertia of the beams and columns $I = 19.43 \cdot 10^6 mm^4$ and $38.91 \cdot 10^6\ mm^4$, respectively.

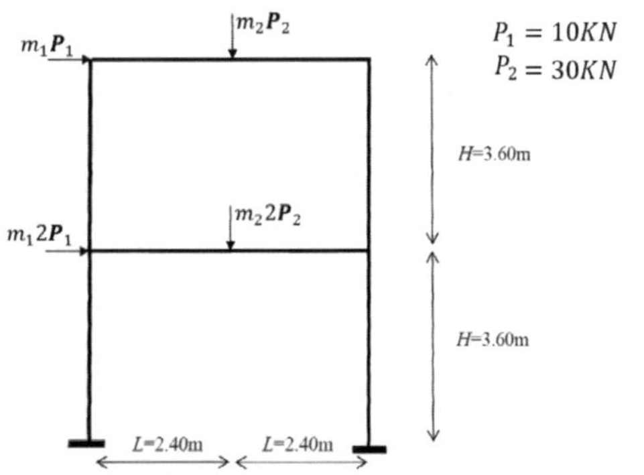

Figure 2. Skeletal frame structure as numerical example.

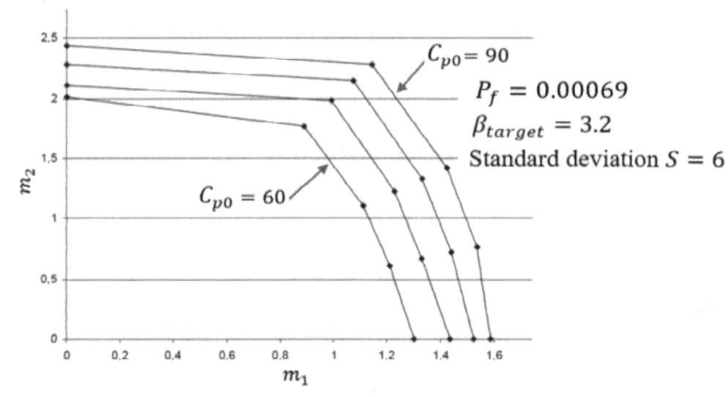

Figure 3. Loading domains of the structure.

The results of the numerical example in case of two-parameter (m_1, m_2) loading domains for different permissible energy value C_{p0} are presented in figure 3 for probabilistic conditions.

4.2. Example 2

The second numerical example shows a long steel pile structure in cohesion less soil. In this example two loading parameter (m_1, m_2) are taken into consideration for different cross-sections (figure 4). The elastic modulus $E = 21 \cdot 10^4\ N/mm^2$and allowable stress $f_y = 210\ N/mm^2$.

Figure 5 shows loading domains in case of two-parameter (m_1, m_2) for different cross-sections. Permissible energy value $C_{p0} = 50$. In this example elasto-limited plastic analysis are considered for deterministic and probabilistic problems. Standard deviation $S = 3$ for probabilistic problems.

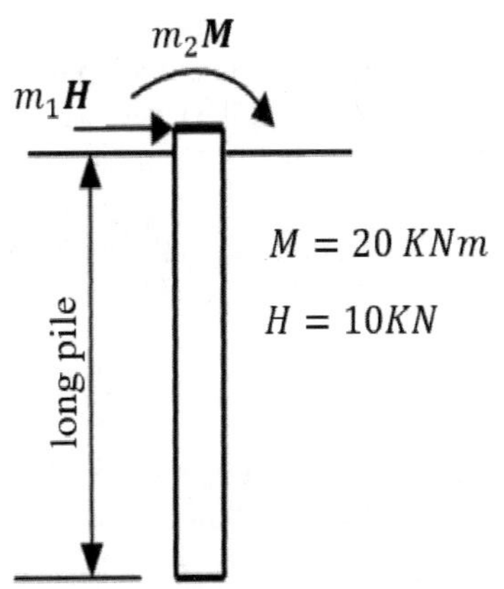

Figure 4. Long steel pile as numerical example.

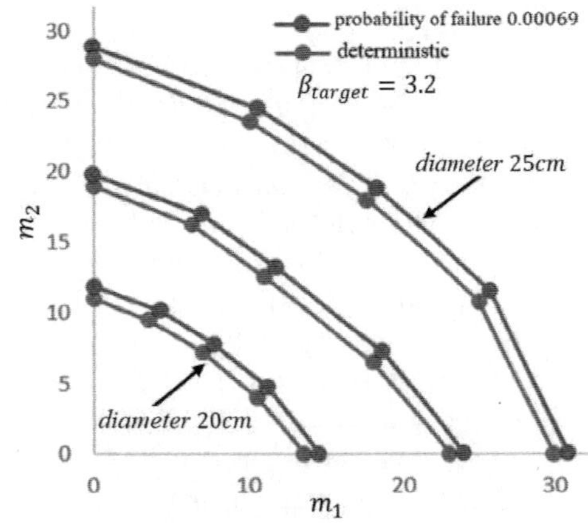

Figure 5. Loading domains.

5. Conclusions

In this research elasto-limited plastic analysis of two different typical type of steel structures are taken into consideration for deterministic and probabilistic solutions. In both examples, residual plastic behavior of structures are restricted with applying permissible energy value. Loading domains are presented for different given permissible energy values. The numerical examples shows that the bounds on the residual plastic deformation of structures and expected probability of failure have significant effects on the load parameters and loading domains.

Acknowledgment

The research described in this paper was financially supported by the Hungarian Human Resources Development Operational Programme (EFOP-3.6.1-16-2016-00017).

References

[1] Kaliszky S 1989 Plasticity theory and engineering applications *Akadémia Kiadó, Budapest* (Amsterdam: Elsevier Science Publisher)

[2] Maier G 1969 Complementary plastic work theorems in piece wise-linear elastoplasticity, *International Journal of Solids and Structures* **5** 261–70

[3] Polizzotto C 1982 A unified treatment of shakedown theory and related bounding techniques *Solid Mechanics Archives* **7** 19-75

[4] König J A 1987 Shakedown of elastic-plastic structures *Fundamental Studies in Engineering* (Warsaw, Poland: Elsevier Publishing Company)

[5] Weichert D and Maier G 2002 *Inelastic Behaviour of Structures under Variable Repeated Loads* (Vienna: Springer)

[6] Levy N H, Einav I and Hull T 2009 Cyclic shakedown of piles subjected to two-dimensional lateral loading *International Journal for Numerical and Analytical Methods in Geomechanic* **33**(10) 1339-61

[7] Simon J W and Weichert D 2012 Shakedown analysis of engineering structures with limited kinematical hardening *International Journal of Solids and Structures* **49** 2177–86

[8] Ponter A R S 1972 An upper bound on the small displacements of elastic, perfectly plastic structures *Journal of Applied Mechanics* **39**(4) 959–63

[9] Corradi L 1977 Deformation bounding techniques in elastoplastic analysis in: Cohn M Z and Maier G, eds. *Engineering Plasticity by Mathematical Programming* (New York: Pergamon Press) pp 495-516

[10] Capurso M, Corradi L and Maier G 1987 Bounds on deformations and displacements in shakedown theory *Materiaux et Structures sous Chargement Cyclique* (Palaiseau, France)

IOP Conf. Series: Materials Science and Engineering **372** (2018) 012032 doi:10.1088/1757-899X/372/1/012032

213–44

[11] Kaneko I and Maier G 1981 Optimum design of plastic structures under displacements constraints *Computer Methods in Applied Mechanics and Engineering* **27**(3) 369–92

[12] Tin-Loi F 2000 Optimum shakedown design under residual displacement constraints *Structural and Multidisciplinary Optimization* **19**(2) 130–9

[13] Liepa L, Blaževičius G, Merkevičiūtė D and Atkočiūnas J 2016 Structural shakedown: a new methodology for estimating the residual displacements *Journal of Civil Engineering and* Management **22**(8) 1055-65

[14] Kaliszky S and Lógó J 1997 Optimal plastic limit and shakedown design of bar structures with constraints on plastic deformation *Engineering Structures* **19**(1) 19-27

[15] Movahedi Rad M and Lógó J 2011 Plastic behavior and stability constraints in the reliability based shakedown analysis and optimal design of skeletal structures *Asian Journal of Civil Engineering (Building and Housing)* **12**(4) 395-413

[16] Lógó J, Movahedi Rad M, Knabel J and Tauzowski P 2011 Reliability based design of frames with limited residual strain energy capacity *Periodica Polytechnica Civil Engineering* **55**(1) 13-20

[17] Movahedi Rad M 2017 Reliability based analysis and optimum design of laterally loaded piles *Periodica Polytechnica Civil Engineering* **61**(3) 491-7

IOP Conf. Series: Materials Science and Engineering **372** (2018) 012033 doi:10.1088/1757-899X/372/1/012033

The application of decommissioned GEO satellites to CAPS

S Y Fu[1,2], Z R Wang[1], H L Shi[1] and L H Ma[1]

[1]National Astronomical Observatories, Chinese Academy of Sciences, 20A Datun Road, Chaoyang District, Beijing, 100012, P R China
[2]University of Chinese Academy of Sciences, 19A Yuquan Road, Shijingshan District, Beijing, 100049, P R China

E-mail: zhaorui_w@nao.cas.cn

Abstract. To ensure the reliable service of geostationary earth orbiting (GEO) communication satellites during the period of in-orbit, the hardware design life of each system usually has some redundancies in contrast to the limited fuel used to keep the satellite position and attitude. After the brief analysis of the life of the satellite subsystems, the feasibility of turning the decommissioned GEO communication satellites into slightly inclined geosynchronous orbiting (SIGSO) satellites is proved. In addition, the role and the actual usage of SIGSO satellites in Chinese Area Positioning System (CAPS) are analysed and discussed, including the effect on the improvement of Position Dilution of Precision (PDOP) of the navigation constellation and the application to satellite communication system, thus the potential value of satellite material and devices is exploited.

1. Introduction

In recent years, with the development of space technology, more and more satellites have been launched into space and are widely used in communication, navigation, remote sensing, meteorology and space exploration, etc. Among them, GEO have become the preferred orbit of most satellites due to its wide coverage, continuous visibility and it's geosynchronous. Limited orbital space has become a scarce resource. Currently, the life of GEO satellites in orbit is generally 8 to 15 years, and the GEO satellites from some countries even can work in orbit for more than 15 years [1]. When a GEO satellite ends service or enters the end of life, in order to guarantee continual customer service, new satellite will be launched before the satellite decommissions. According to international convention, the original decommissioned satellites must shut down all the power and be pushed to the grave orbit about 300 km above geostationary orbit, become the space debris.

Decommissioning of satellite usually because of the lacking of carried fuel, it's difficult to continue maintain the position and attitude of satellite for a long time, when the structure and electronic equipment such as solar cells, signal transponders of the satellite remained intact, they can still work for a long time. In references [2], a scheme of maneuvering the end-of-life GEO communication satellites to SIGSO (slightly inclined geosynchronous orbiting) satellites was introduced. It has realized the full and reasonable utilization of the residual fuel and the electronic equipment such as transponders on the satellites, and has created the second life cycle of the end-of-life GEO satellites. Meanwhile, we use these SIGSO satellites and GEO communication satellites to constitute the space satellite constellation of CAPS (Chinese Area Positioning System), to transpond the carrier signal modulated ranging code and navigation message launched from the ground station, and to realize satellite positioning [3].

This work briefly introduces the influence factors of satellites life, analyzes and proves the feasibility of using decommissioned satellites as SIGSO satellites, and introduces the technology and application of SIGSO satellites in CAPS.

2. Satellite life and main influencing factors

The working life is one of the most important performance indexes of the satellites. Users tend to prefer the longer working life of the satellites, but the working life is restricted by various factors. The extension of satellite life will generally require higher or additional requirements for the satellite subsystems, which will increase the complexity and mass of the subsystem, resulting in the total mass and total cost of the satellite rising. Therefore, the extension of design life is at the cost of increasing satellite mass and development cost [5].

For the components of the satellites, the probability of failure is a function of time, and the typical failure rate curve is the Bathtub curve, as shown in Figure 1. The shape of the curve is high at both sides and low in the middle, with obvious stages, which can be divided into three phases: early life failures, random failures and wear out failures. The early life failures often occurs in the development stage before the satellites are launched, and the failure rate is high and decrease rapidly with the increase of working time. It is possible to eliminate faults by modifying design, improving production, screening of components and so on, so that the early life failure isn't as part of system failure in normal circumstances. The wear out failures generally occurs at the end of the equipment life and increases with the extension of time. When the device or equipment is developed, the life time can be extended by means of design method and production process, so that the device will not work to the wear out failures period before the satellites decommission. The failure rate in the random failures period is low and stable, which is generally regarded as the effective life of the equipment. The failures of this period is mainly due to working conditions, external environment conditions, error operation and other factors causing the equipment stress to exceed a certain intensity [6].

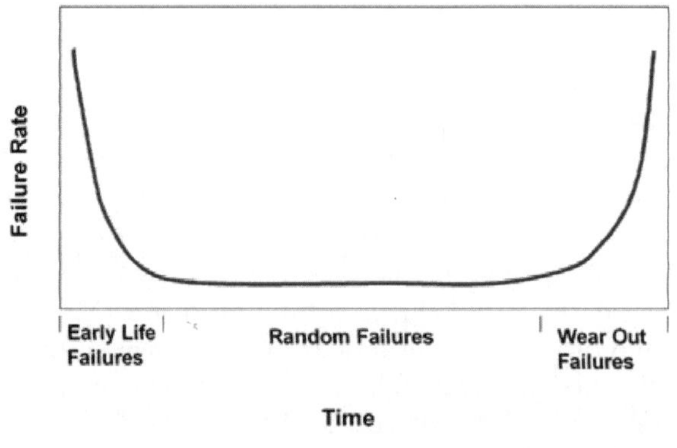

Figure 1. Bathtub curve of satellite equipment.

According to the operating situation of GEO satellites, the factors that affect the life of on-orbit satellites are mainly the space environment, equipment performance, the consumption on satellites and the ground misoperation.

The various perturbing forces of GEO satellites will cause the orbit and attitude disturbance, so it is necessary to constantly consume propellant fuel for attitude control and position maintenance. If satellite with large communication antennas and solar panels is affected by solar radiation pressure and earth's non-spherical gravitation, it will drift on east-west direction. The effect of lunisolar gravitation makes satellites drift on south-north direction, which is the most important factor that causes the change of the orbital inclination angle of the satellites [7]. On the other hand, space radiation will cause the problems such as flexible parts damage, reduced energy supplies, the device performance degradations, and star misoperation and so on. The space geomagnetic storm will affect the performance of the electromagnetic device and the electromagnetic environment of the satellites. In addition, the performance of solar cells and other devices will be reduced by temperature changes, solar radiation, and space electronics, etc. [6].

Due to the components on satellites work in random failure period, the failure rate is low and constant, and in the process of design and development, means of improving the satellite reliability such as the environment protection design, redundancy key components design and derating design are used, so the satellite life will not end except the occasional single point failures. The in-orbit situation of GEO satellite also shows that propellant used for attitude and orbit controlling is often carried according to design life due to restriction of launching mass. It will gradually reduce with use, eventually runs out and the satellites are out of control, so the propellant fuel becomes the main factors restricting the service life of the satellites.

3. Reuse of decommissioned GEO communication satellites

As mentioned above, when GEO communication satellites invalid at the end of life, the remaining fuel on satellites is not enough to maintain orbit position and satellite attitude for a long time. But other components, such as solar power and signal transponders, are still in good condition, even some spare parts are not used which can continue to work for many years. If GEO satellites are decommissioned at this time and pushed into grave orbit, it does waste the equipment and residual fuel on satellites, furthermore, more and more space debris also causes destruction to the valuable space environment around the GEO orbit. Therefore, the reuse of decommissioned GEO satellites has significant social and economic benefits.

The drift of GEO satellite orbit position is divided into two directions: north-south and east-west, usually the fuel consumption of north-south station-keeping maneuvers accounts for about 90%, which is about 10 times of fuel consumption of east-west station-keeping maneuvers. So if only maintaining position of satellites in east-west direction, and leaving uncorrected on north-south direction under the influence of lunisolar gravitation, they will gradually become SIGSO (slightly inclined geosynchronous orbiting) satellites. In this way the use time of the residual fuel of satellites will be extended by 10 times. After stopping control in north-south direction, the north-south drift of satellites leads to orbital inclination angle increasing gradually. The inclination angle increases about 0.75°~0.96° every year [9].

On the other hand, the performance of solar cells will decline gradually about 2% ~ 3% per year due to long-time use and the complex space environment, and the battery will charge and discharge repeatedly because of the satellite eclipse about 90 times every year. These can lead to a decline in the capacity of electric power supply of GEO satellites at the end of life. The normal work of most of the equipment can be guaranteed by turning some transponders off.

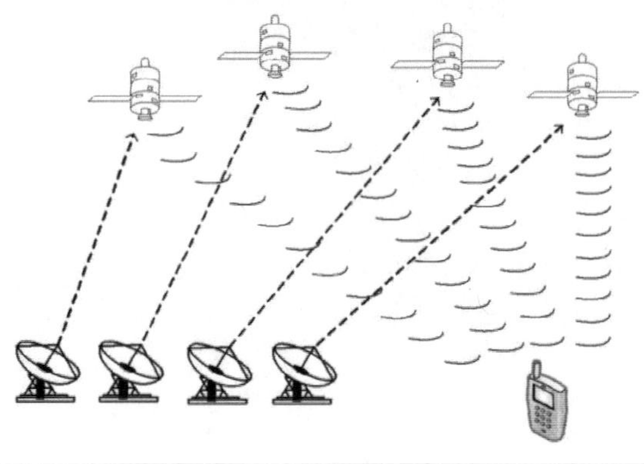

Figure 2. CAPS schematic.

4. Application of SIGSO satellites to CAPS

The CAPS (Chinese Area Positioning System) system is a transponding satellite navigation system based on communication satellites, which uses communication satellites to transpond the navigation signals generated by the ground station for the first time. The system can provide users with position, velocity and time service, and the system configuration is shown in figure 2.

4.1. Improvement of the CAPS navigation constellation by SIGSO satellites

For satellite positioning systems, satellite space geometry has important influence on the positioning accuracy, system availability and continuity, in general, position dilution of precision (PDOP) is used to describe the quality of satellites' spatial layout [9]. The positioning RMS error can be expressed as:

$$\sigma_{(x,y,z)} = \mathrm{PDOP} \cdot \sigma_{\rho}$$

Where, $\sigma_{(x,y,z)}$ is the positioning error of the user receiver and σ_{ρ} is the user equivalent range error. It can be seen that PDOP values are amplification of user range error. Therefore, it is necessary to minimize the PDOP value in the navigation system constellation design to improve the positioning precision.

Since 2003, CAPS has leased commercial communication satellite transponders for experiments. But most communication satellites are GEO satellites, the leased satellites are all in the equatorial plane, so it's unable to achieve three-dimensional positioning. However the SIGSO (slightly inclined geosynchronous orbiting) satellites drifted from decommissioned GEO satellites fundamentally solved this problem. The drift quantity in north-south direction of SIGSO satellites greatly improve the constellation space layout, GEO satellites and SIGSO satellites constellation has completely realize three-dimensional positioning. To verify the contribution of the amount and orbit inclination angle of SIGSO satellites to PDOP value, the PDOP values every 10 minutes are calculated by using digital simulation, the tested fields include China and surrounding areas with 1° latitude × 1° longitude grid, the results are as follows [11,12].

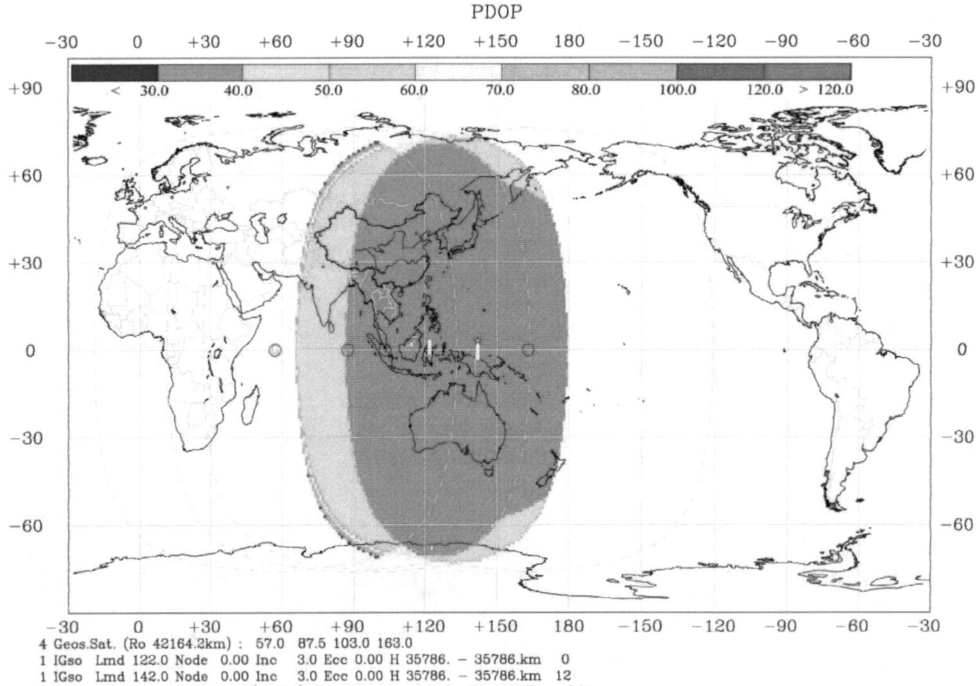

Figure 3. PDOP value of the constellation with 4 GEO and 2 SIGSO (3° inclination angle) satellites.

6 GEO satellites located at 57°E, 87.5°E, 103°E, 122°E, 142°E and 163°E orbits are used. First, we drift 2 of them which located at 122°E and 142°E into SIGSO satellites with 3° inclination angle, the daily mean PDOP values of China and surrounding areas are shown in Figure 3. It can be seen from the figure that the average daily PDOP values in China and surrounding areas are all less than 50. Except for parts of Xinjiang and Tibet, PDOP values of most areas are under 40. Although it is still large, it's possible to realize three-dimensional positioning. Next, we set the GEO satellite at 103°E as SIGSO satellite, constitute 3 GEO + 3 SIGSO constellation, average daily PDOP values are shown in Figure 4, PDOP values of Chinese territory are less than 20, it is obvious that the addition of SIGSO satellites can observably optimize the geometry of constellation. Finally, the orbital angle of 3 SIGSO satellites are increased to 5°, as shown in Figure 5, it can be seen that with the increasing of inclination angle of SIGSO satellites gradually, the geometry of the constellation will be continually optimized, PDOP value reduced to about 10.

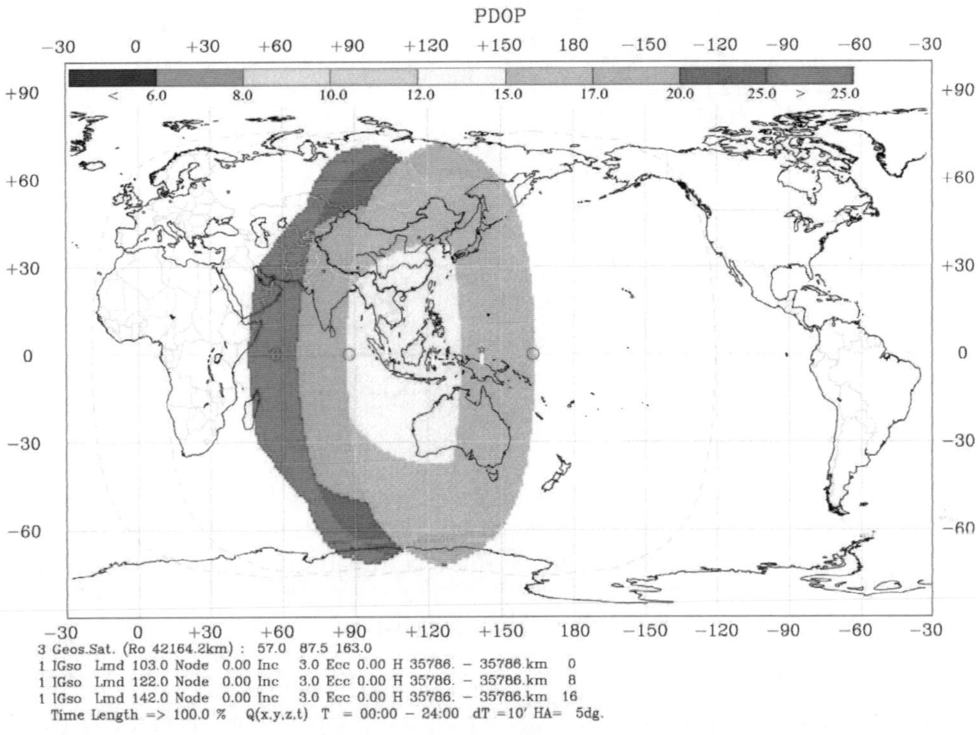

Figure 4. PDOP value of the constellation with 3 GEO and 3 SIGSO (3° inclination angle) satellites.

In CAPS constellation, two decommissioned satellites are controlled to be SIGSO satellites. They are the APSATR-1 satellite located at 138°E and the APSATR-1A satellite located at 134°E, respectively. When these two satellites are decommissioned, they have no any malfunctions, the 24 C-band transponders can still work, and even the six backup transponders are never used. Moreover, the remaining fuel can supply for 1 year, if we adopt the work mode of SIGSO satellites described previously, i.e. only maintain the east-west station-keeping maneuvers and allow to drift freely in north-south direction, then the fuel consumption of two satellites will be reduced 90%, and satellites can continue to be used about 10 years. It can greatly improve the service life and utilization of satellites and the equipment.

In 2004, the APSATR-1 satellite located at 138°E was drifted into the orbit at 142°E, meanwhile, stopped north-south station-keeping maneuvers, and drifted freely to be SIGSO satellite. Until 2014, the orbital inclination angle of APSATR-1 satellite has been drifted to 9°, but the fuel exhausted, the satellite was officially invalid and pushed into the grave orbit, while the equipment such as transponders can still work normally. During the 10 years of reutilization of APSATR-1 satellite, it constituted the space constellation of the CAPS system together with the GEO satellites, and assisted by the pressure measurement technology, the CAPS system successfully realized the three-dimensional positioning. At the same time, a number of experiments and research projects have been completed.

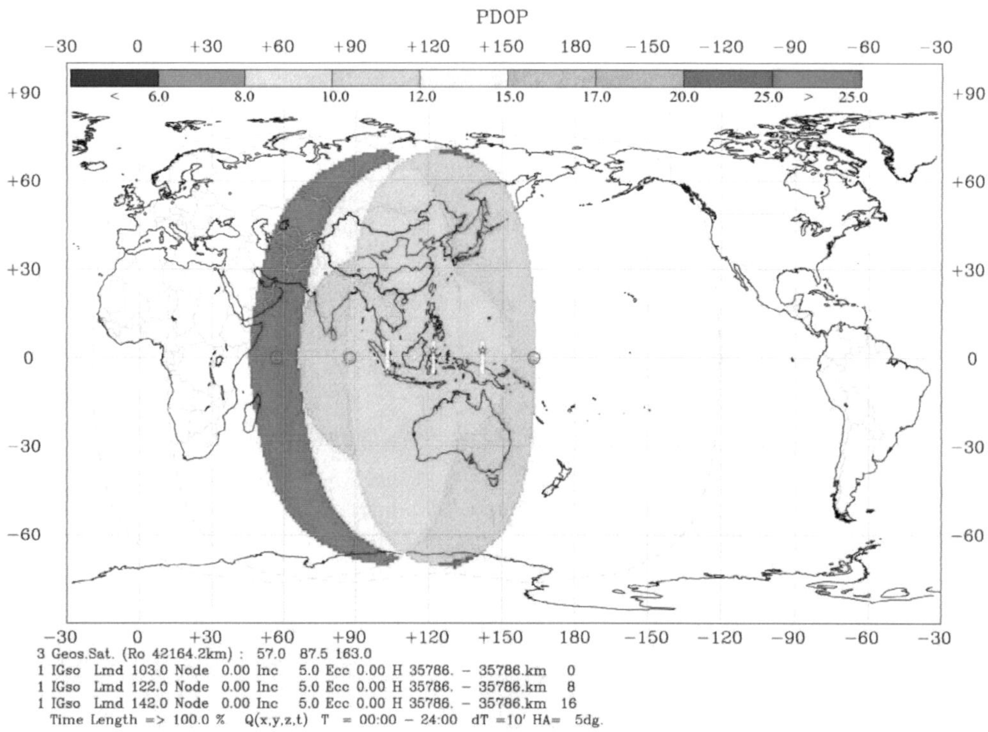

Figure 5. PDOP value of the constellation with 3 GEO and 3 SIGSO (5° inclination angle) satellites.

4.2. Application of SIGSO satellites to CAPS satellite communication

After the decommissioned GEO satellites are re-used as the SIGSO satellites, due to the change of orbit, the original business ceased or is undertaken by other GEO satellites. All 24 signal transponders on the SIGSO satellite are idle. Except two transponders are used for navigation and positioning applications, and some transponders are closed to reduce the power supply load of solar cells and batteries, there are still many transponders can be used continually for satellite communications business. CAPS system is an integration of navigation and communication system based on communication satellites, therefore, SIGSO satellites with abundant frequency bandwidth and transponders can provide a large number of precious resources for communication, and make the CAPS system combine the advantages of navigation system and communication system, so it can provide the location and time information communication when position, velocity and time are determined. The low rate (50 bps~1 kbps) information transmission based on SIGSO satellites can contain a large number of terminals to realize communication, and can make up the covers insufficient defects of the current communication methods in remote areas, mountains and sea areas. Combined with satellite positioning, it is widely used in vehicle management, disaster relief, satellite search and rescue, combat command, etc., with the advantages of fast, cheap and reliable.

The communication with SIGSO satellite also faces some new problems. For example, the information transmission rate of CAPS communication system is less than 1kbps, forming a new low rate satellite communication system. At the same time, the SIGSO satellite is different from the GEO satellite, the earth trace of satellite has "8" shape. In order to track the SIGSO satellites, it needs to increase the antenna beam width. Considering the factors as antenna gain and cost performance, the antenna diameter of communication center station is about 10~15 meters. In addition, to prevent the interference of the wide beam uplink signal to the adjacent satellites, the power limited communication signal needs to adopt the spread spectrum technology, and the amplification gain is about 40 dB [12].

Based on the technology mentioned above, the CAPS satellite communication system has successfully taken several experiments and obtained good results. The testing field of CAPS communication experiment in Tianjin is shown in figure 6. This indicates the feasibility of applying SIGSO satellites to satellite communication. The long-term stable operation of the CAPS system has also verified the reliability of the decommissioned GEO satellites, and the research and experiment of the integrated navigation and communication system have application value and social value.

Figure 6. Testing field of CAPS communication experiment.

4.3. Application of SIGSO satellites to the maintenance of frequency and orbital resources

With the development of space technology around the world, more satellites are competing for limited GEO orbital position, which is a very important space resource. Orbital positions need to be occupied by satellites in order to retain the right to use orbit. As a result, the decommissioned GEO satellite can be used as an effective satellite in orbit with its fewer services and longer running time. These satellites can maintain national satellite frequency and orbital resources while maximizing the use of the equipment. Since 2004, the decommissioned APSATR-1 satellite stayed at 142°E orbit for a long time, serviced for the CAPS navigation and communication experiments. In 2012, when the fuel was running out, APSATR-1 satellite was moved into 163°E orbit, until completely ran out of fuel in 2014. It retained orbit position for the new GEO satellite. Also, the APSATR-1A satellite was moved into 115.5°E, 125°E, 130°E and 51.5°E orbit position for a short time, to protect the orbital and frequency resources. In 2015, CAPS bought decommissioned satellite MEASAT-2 at 148°E position now. It will be able to use until about 2025, servicing for CAPS navigation and communication project.

5. Conclusions

CAPS (Chinese Area Positioning System) is a new navigation and communication system based on communication satellites. The experiment results and project acceptances have proved the feasibility of CAPS navigation and communication technology after years of effort and overcoming difficulties, and have had solid foundation. In CAPS, SIGSO (slightly inclined geosynchronous orbiting) satellites transformed from decommissioned GEO communication satellites are important component. CAPS use SIGSO satellites to achieve positioning and communication functions innovatively, so that greatly improve the service efficiency and life of decommissioned satellites. During the process of using, CAPS develops the unique advantages of SIGSO satellites.

In order to extend the service life of decommissioned GEO satellites, we only maintain the position in east-west direction and make them drift freely in north-south direction at an annual rate of about 0.85° orbital inclination into SIGSO satellites. The SIGSO satellites can provide the navigation constellation of CAPS with the north-south drift distances which can improve the spatial constellation geometry, reduce the PDOP value, and improve the positioning accuracy. At the same time, decommissioned satellites have abundant C-band full frequency transponders. On the one hand, they can provide more options for navigation, for example, three frequency or multiple frequency positioning, changing frequency and hopping frequency navigation, etc., so that improve the anti-interference ability of the satellite navigation system significantly. On the other hand, a large number of redundant transponders can be used in satellite communications, and they can also be used for short message transmission, voice communication and so on at the time of navigation and positioning services. In addition, extending the life of GEO communications satellites can help preserve the valuable GEO orbital position and frequency resources, so as not to lose the use priority because of lacking a satellite in orbit for a long time.

The reuse of decommissioned GEO communication satellites not only reduces the cost of the CAPS system construction, but also maximizes the use of the residual fuel and transponders aboard the satellites. Years of practical use have demonstrated that the SIGSO satellites can do play a unique role in CAPS system.

The research and application of decommissioned GEO satellites proved that the whole service life of the important equipment and devices often is not limited by material strength and service life of most devices, but is only restricted by a few conditions of use. So we can change the restraint condition, it is possible to make the material and most devices to extend the service life and make the material and devices play more role.

Acknowledgments
This work was supported by a grant from National Natural Science Foundation of China (11573041).

References

[1] Pang H W and Feng W Q 2006 Space environmental factors influencing on GEO satellite's life and reliability and their evaluation, verification and guarantee technologies *Spacecraft Environment Engineering* **23(2)** 63-6
[2] Shi H L, Ai G X, Han Y B, Ma L H, Chen J B and Geng J P 2009 Multi-life cycles utilization of retired satellites *Sci. China* G **52(3)** 323-7
[3] Shi H L, Han Y B, Ma L H, Pei J, Yin Z Q and Ji H F 2010 *Satellite Communications* ed Diodato N (Rijeka:Sciyo) chapter 16 pp 323-65
[4] Ai G X, Shi H L, Wu H T, Li Z G and Guo J 2009 The principle of the positioning system based on communication satellites *Sci. China* G **52(3)** 472-88
[5] Zhu Y L 2007 An analysis of influence of satellite design lifetime *Spacecraft Engineering* **16(4)** 9-18
[6] Li J C, Yang Y A and An J W 2006 Geostationary satellite's end-of-life predication based on propellant-remaining estimation *Chin. J. Space Sci.* **26(3)** 193-6
[7] Ji H F 2013 *On Principle and Algorithm for a Navigation Constellation Composed of SIGSO Communication Satellites* (Beijing: University of Chinese Academy of Sciences)
[8] Ma L H, Jing Y X, Ji H F and Zhang L R 2011 Analysis of evolution of orbits of GEO satellites near end of life based on the STK software *Astronomical Research & Technology* **8(4)** 47-353
[9] Liu Y X, Shang H B and Wang S 2015 Research on GEO satellite station keeping method using electric propulsion *Journal of Deep Space Exploration* **2(1)** 80-7
[10] Han Y B, Ma L H, Qiao Q Y, Yin Z Q, Shi H L and Ai G X 2009 Functions of retired GEO

communication satellites in improving the PDOP value of CAPS *Sci. China* G **52(3)** 423-33

[11] Ma L H, Zhang L R, Han Y B and Qiao Q Y 2011 Analysis of optimizing CAPS navigation performance with two retired GEO satellites *GNSS World of China* **36(1)** 13-7

[12] Ji H F, Ma L H, Ai G X and Shi H L 2013 The principle of a navigation constellation composed of SIGSO communication satellites *Research in Astron. Astrophys.* **13(4)** 479-89

[13] Cui J X, Shi H L, Chen J B and Pei J 2009 The transmission link of CAPS navigation and communication system *Sci. China* G **52(3)** 402-11

IOP Conf. Series: Materials Science and Engineering **372** (2018) 012034 doi:10.1088/1757-899X/372/1/012034

Anti-buckling design of variable stiffness composite cylinder under combined loading based on the multi-objective optimization method

Y Zheng[1,2,3] and J Chen[1,2,3]

[1]Department of Mechanics, Huazhong University of Science and Technology, Wuhan 430074, China
[2]Hubei Key Laboratory for Engineering Structural Analysis and Safety Assessment, Wuhan 430074, China

E-mail: zyc276565153@126.com/ jpchen@hust.edu.cn.

Abstract. Variable stiffness composite structures take full advantages of composite's design ability. An enlarged design space will make the structure's performance more excellent. Through an optimal design of a variable stiffness cylinder, the buckling capacity of the cylinder will be increased as compared with its constant stiffness counterpart. In this paper, variable stiffness composite cylinders sustaining combined loadings are considered, and the optimization is conducted based on the multi-objective optimization method. The results indicate that variable stiffness cylinder's loading capacity is increased significantly as compared with the constant stiffness, especially when an inhomogeneous loading is considered.

1. Introduction

Due to the excellent mechanical property and lightweight feature, composites are widely used in different engineering fields, such as the offshore platform in the ocean, the pipeline in deep sea, the pressure vessels in industry, the rocket launchers and space vehicles, and the aircraft cabin and wings, etc. Their large design freedom provides large design potential. In order to take full advantages of composite structures, composite structures can be continuous tailoring and with variable thickness or fiber angles, different stacking sequences, resulting in the variable stiffness (VS) composite structures. The design of VS composite structure has attracted many attentions in recent years [1-4]. Having been reported since the late 1980s, it is shown that automated fiber placement (AFP) technology can not only reduce labor costs and scrap materials, but also improve its quality and performance [5-10]. Continuously variable fiber orientations are steered by AFP machines, resulting in VS composite structures and giving the designers large designability to extend the design space and fully exploit the directional properties of composite materials [11]. As a result, VS design can offer a significantly improved performance compared with constant stiffness (CS) counterparts at the same weight.

In 1994, Hyer *et al* [12] optimized the fiber path of a flat plate by reducing the variation of the angle between adjacent elements. The laminated plate was fabricated and the buckling loading was verified by experiments. Tatting [13] is the first one who studied design optimization of VS composite cylinders for buckling. Rouhi [14-16] *et al* studied the optimization of maximum buckling load of composite cylinders subject to bending moment, the buckling behavior of a VS composite cylinder

IOP Conf. Series: Materials Science and Engineering **372** (2018) 012034 doi:10.1088/1757-899X/372/1/012034

under axial compression and a test about VS composite has been done. These studies only considered buckling of structures under single loading.

There have been some studies conducted for multi-objective optimization of laminated composite structures. For example, Lee *et al* [17] presented a work which aimed to minimize the weight of multilayered composite plates and minimize their maximum displacement. The design variables include the type of fiber, thickness and the fiber orientations of each layer. Vosoughi and Nikoo [18] developed a hybrid method for maximizing the fundamental natural frequency and thermal buckling temperature of laminated composite plates.

The kind of combined loadings is the common situation in engineering, which should be involved in the optimization of a composite cylinder. In general, the VS composite shell structures are prone to buckling and their anti-buckling ability is crucial. In this paper, two kinds of combined loading, i.e., the bending and external pressure, and the bending and torque are considered respectively for the optimal design of VS composite cylinders. Multi-objective particle swarm optimization is utilized for obtaining the final Pareto-optimal solutions, which is the optimal compromise between the two loadings.

2. Design of a VS composite cylinder

The common composite material cylinder is the filament winding composite cylinder in which fiber is wrapped in a certain winding angle to make the stiffness constant along the cylindrical surface. The AFP technique can be used for manufacturing VS composite structures, leading to the increase of interest on the optimal design of composite structures with VS. VS composite structures can be formed in three ways: the curved fiber ply, the non-even thickness ply and the ply with different fiber volume fractions. The AFP technique changes the fiber-laying of composite structures, through the curve lay, thus greatly influencing the material's properties. The following figure shows a VS composite cylinder.

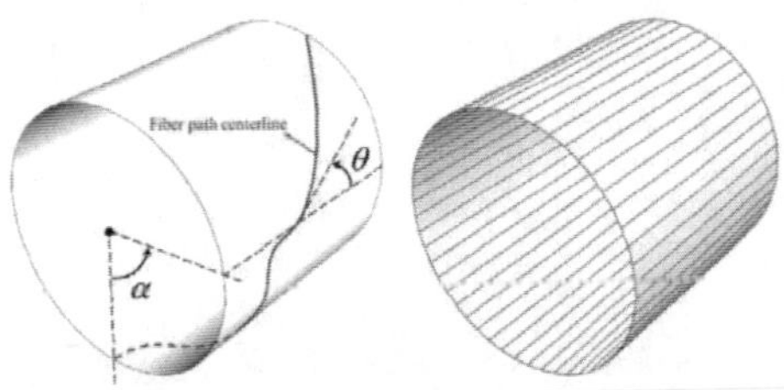

Figure 1. A VS composite cylinder, (a) Curve fiber path and (b) The cylinder is divided into narrow strips.

Figure 1(a) schematically shows the curve fiber path. Fiber angle is changing along the circumferential direction while maintaining the same along the axial direction. The cylinder structure is formed by finding the fiber angle in each narrow strip as shown in figure 1(b). In order to describe the variation of the orientation angle, 4 steps from the keel to the crown of the cylinder is defined. Each segment is divided into M narrow bands in the finite element model. All elements in one narrow band possess the same orientation angle in a ply. The symmetry about the vertical axis is assumed so that five design variables per VS ply: φ_1, φ_2, φ_3, φ_4 and φ_5, should be considered for 4 segments, as shown in figure 2. Interpolation between five design variables approaches with satisfaction. More steps will take more computational cost and the results will remain the same. In each segment, the orientation angle of the fiber paths are linear variation along the circumferential direction:

$$\theta_{i,k} = \varphi_i + \frac{\varphi_{i+1} - \varphi_i}{\alpha_{i+1} - \alpha_i}(\alpha_k - \alpha_i) \quad i = 1,2,3,4 \quad k = 1,2,\cdots,M . \tag{1}$$

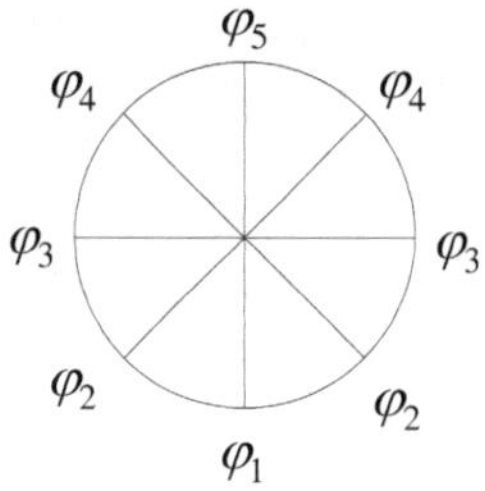

Figure 2. Circular cylindrical design nodes.

Where $[\alpha_i, \alpha_{i+1}]$ denotes the lower boundary and the upper boundary of the i-th segment, $\theta_{i,k}$ is the fiber orientation angle in the k-th narrow band, α_k is the circumferential angle of the k-th narrow band, φ_i is the fiber orientation angle of the i-th segment lower boundary, and φ_{i+1} is the upper boundary.

A composite cylinder is studied in this section whose diameter is 0.457 m and length is 0.457 m. Sixteen layers are arranged in the form of $[0°/+\theta/90°/-\theta/-\theta/90°/+\theta/0°]_s$, where $\theta \in [0°, 90°]$ is the ply angle that varies in the circumferential direction. The thickness of each layer is 0.127 mm. The plies are made of AS4D/9310 carbon/epoxy materials for which the material properties are given in table 1.

Table 1. AS4D/9310 Material properties of carbon fiber-epoxy material.

Property	AS4D/9310
E_1 (GPa)	134
E_2 (GPa)	7.71
$G_{12} = G_{13}$ (GPa)	4.31
G_{23} (GPa)	2.76
$\upsilon_{12} = \upsilon_{13}$	0.301
υ_{23}	0.396
V_f	0.55
Thickness (mm)	0.127

When two sides of the composite cylinder are simply supported under bending loading, buckling will take place and the mode. Two cylinders' buckling modes which are quasi-isotropic (QI) composite and VS composite are shown in the figure 3.

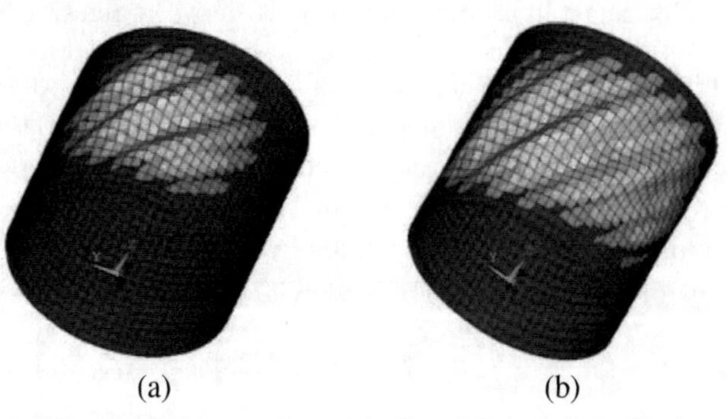

(a) (b)

Figure 3. The buckling modes, (a) Quasi-isotropic cylinder and (b) Variable stiffness cylinder.

It indicates that buckling failure occurs in the compression area. The difference between two cylinders is the area of compression. Compression area in VS cylinder's mode is larger than QI composite because multiple design orientation angles make the cylinder sustaining loading uniformly after optimized. Its critical bending moment promotes from 1.058e5 N*m to 1.330e5 N*m. More elements sustain compression resulting in critical bending moment's promoting.

3. Design optimization of VS composite cylinders under combined loadings

In engineering, a common situation is that the cylinder is under a combined loading condition. Considering the buckling of cylinders, most emphasis is placed on axial compressive force and bending. How to design a cylinder under a variety of load conditions to ensure the performance such as buckling and the critical strength is of important. In order to enhance the different kinds of cylinders' loading capacities simultaneously, designs of VS cylinders under two kinds of combined loadings are considered. One is bending and external pressure combination and the other one is bending and torsion. The cylinder may be failure when sustaining different combined loadings. The critical loadings are chosen as the objectives. The five design variables are those mentioned in section 2. To solve such a multi-objective optimization problem, the multi-objective particle swarm optimization (MOPSO) is applied to obtain the final Pareto-optimal solutions.

3.1. MOPSO

In traditional particle swarm optimization (PSO) method, candidate solutions are represented by a number of particles. These particles are attracted by the global optimum position of the crowd $\boldsymbol{gbest}$ and the personal ones $\boldsymbol{pbest}_i$. In each step, the velocity and position of a particle will be updated by equation (2):

$$\begin{aligned} \boldsymbol{v}_i(t+1) &= \omega \boldsymbol{v}_i(t) + c_1 rand_1(\ )(\boldsymbol{pbest}_i - \boldsymbol{X}_i(t)) + c_2 rand_2(\ )(\boldsymbol{gbest} - \boldsymbol{X}_i(t)) \\ \boldsymbol{X}_i(t+1) &= \boldsymbol{X}_i(t) + \boldsymbol{v}_i(t+1) \end{aligned} \quad (2)$$

where $\boldsymbol{v}_i(t)$ and $\boldsymbol{X}_i(t)$ are the velocity and position, respectively, of particle i at step t; ω is the inertia factor which is 0.9 at the beginning of the iteration and decreases to 0.4 along the iteration; c_1 and c_2 are acceleration coefficients which are set to 2.05; $rand_1(\ )$ and $rand_2(\ )$ are two independent random numbers following a uniform distribution $U(0,1)$. A maximum velocity value v_{max} is assigned for $\boldsymbol{v}_i(t)$. As the iteration proceeds, all particles tend to move towards $\boldsymbol{gbest}$ until the optimal is obtained.

In MOPSO, different updating rules are applied as compared to PSO, i.e., $\boldsymbol{pbest}_i$ and $\boldsymbol{gbest}$ are updated through a dominating principle. A solution $\boldsymbol{X}_{(1)}$ dominating another solution $\boldsymbol{X}_{(2)}$ means that it is not worse than $\boldsymbol{X}_{(2)}$ in all objectives, and strictly better than $\boldsymbol{X}_{(2)}$ for at least one of the objectives. So, three kinds of dominating cases exist: $\boldsymbol{X}_{(1)}$ dominates $\boldsymbol{X}_{(2)}$, $\boldsymbol{X}_{(2)}$ dominates $\boldsymbol{X}_{(1)}$, and $\boldsymbol{X}_{(1)}$ and $\boldsymbol{X}_{(2)}$ are non-dominated to each other. When $\boldsymbol{X}_{(1)}$ and $\boldsymbol{X}_{(2)}$ are non-dominated in a step of MOSPO, they should be saved and gathered in a set called an external archive. In the next step, $\boldsymbol{gbest}_i$ for particle i is selected randomly from the external archive. When updating $\boldsymbol{pbest}_i$, if $\boldsymbol{pbest}_i$ and $\boldsymbol{X}_i(t+1)$ are non-dominated to each other or $\boldsymbol{pbest}_i$ dominates $\boldsymbol{X}_i(t+1)$, $\boldsymbol{pbest}_i$ will be saved, otherwise $\boldsymbol{pbest}_i$ is replaced by $\boldsymbol{X}_i(t+1)$.

3.2. Formulation of VS composite cylinders' design under combined loading

In this paper, two kinds of combined loadings are considered. The first case is the combination of bending M and external pressure P. And the second case is bending M and torsion T. The five ply angles are the design variables. The multi-objective optimization problem can be described as:

$$\begin{aligned} &\max_{\boldsymbol{\varphi}} \quad M_{cr}, F_{cr} \\ &s.t. \quad F.I. \leq 1 \\ &\boldsymbol{\varphi} = \{\varphi_1, \varphi_2, \varphi_3, \varphi_4, \varphi_5\} \\ &F_{cr} = P_{cr}\, or\, T_{cr} \end{aligned}, \tag{3}$$

in which, M_{cr}, P_{cr} and T_{cr} represent the loading capacity for bending, external pressure and torsion respectively. $F.I.$ is the Tsai-Wu criterion coefficient and it is used to avoid strength failure.

In problem (3), for a candidate solution $\boldsymbol{\varphi}$, there is a negative correlation between M_{cr} and F_{cr}. To determine the correlation between them, a number of M_{cr} or F_{cr} need to be considered, and one of them is regarded as the independent variable and the other one is the dependent variable. The problem (3) is transformed as:

$$\begin{aligned} &\max_{\boldsymbol{\varphi}, F_{cr}} \quad M_{cr}, F_{cr} \\ &s.t. \quad F.I. \leq 1 \\ &\boldsymbol{\varphi} = \{\varphi_1, \varphi_2, \varphi_3, \varphi_4, \varphi_5\} \\ &F_{cr} = P_{cr}\, or\, T_{cr} \end{aligned} \quad \text{or} \quad \begin{aligned} &\max_{\boldsymbol{\varphi}, M_{cr}} \quad M_{cr}, F_{cr} \\ &s.t. \quad F.I. \leq 1 \\ &\boldsymbol{\varphi} = \{\varphi_1, \varphi_2, \varphi_3, \varphi_4, \varphi_5\} \\ &F_{cr} = P_{cr}\, or\, T_{cr} \end{aligned}. \tag{4}$$

Problem (4) is solved by MOPSO mentioned in section 3.1.

4. Examples

In this section, the multi-objective optimization of VS composite cylinders under two kinds of combining loadings is calculated. For each condition, three kinds of fiber paths will be considered: QI composite, CS composite and VS composite. The differences among them are that $\varphi_1 = \varphi_2 = \varphi_3 = \varphi_4 = \varphi_5 = 45°$ for QI, $\varphi_1 = \varphi_2 = \varphi_3 = \varphi_4 = \varphi_5$ for CS and φ_i is independent to the others for VS.

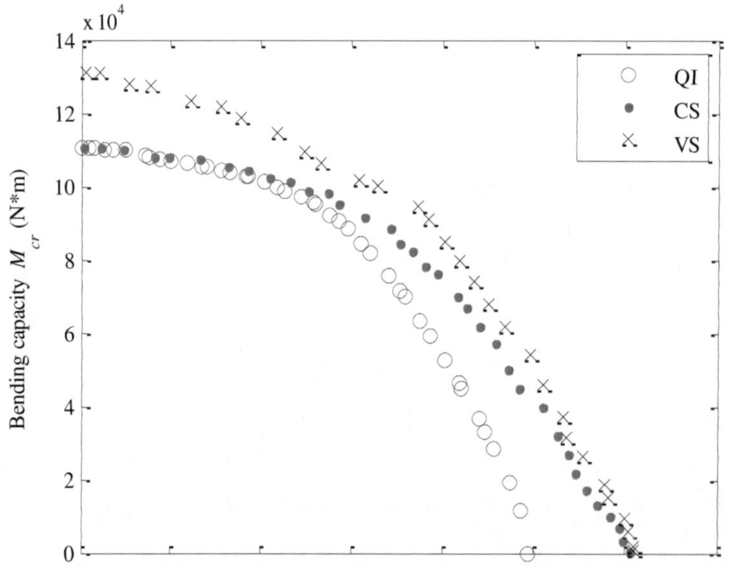

Figure 4. The optimal compromise between bending capacity and external pressure capacity.

4.1. Bending and external pressure case

The Pareto-optimal solutions for the combination of bending and external pressure are shown in figure 4.

It indicates that the Pareto front of VS is the best among QI, CS and VS, i.e., VS's loading capacity is superior to CS and QI when sustaining bending and external pressure simultaneously. Performance of CS is the middle, and that of QI is the worst. The enlarged design space enhances composite cylinder's combining loading capacity. In the single bending case, CS's bending capacity is almost the same as QI's. This is because that a QI composite cylinder possesses almost the best anti-bending performance under the linear fiber path. VS's bending capacity is superior to CS's and QI's because curve fiber path makes composite inhomogeneous and optimization of curve fiber path can increase VS composite cylinder's loading capacity. In the single external pressure case, the optimal solution of VS composite cylinder is close to CS, owing to the fact that external pressure is a kind of homogeneous loading and the curve fiber path is degenerated to the linear path.

4.2. Bending and torsion

The Pareto-optimal solutions are shown in figure 5 for the bending and torsion combined loading.

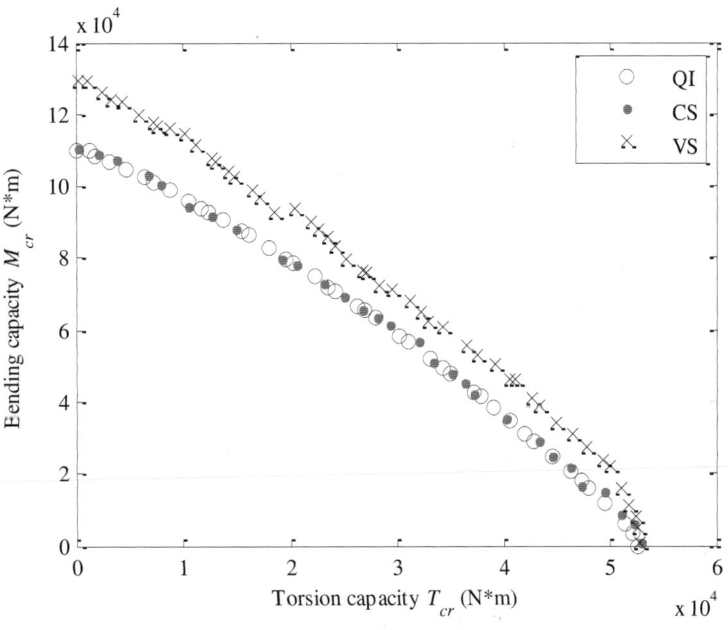

Figure 5. The optimal compromise between bending capacity and torsion capacity.

It is found that VS's combining loading capacity is superior to CS and QI. For the single torsion, solutions of VS, CS and QI are almost the same, implying that the three type structures provide the same anti-torsion performance. For the single bending or bending-torsion, VS structures' advantage is obvious.

5. Conclusions

Loading capacity of a VS composite cylinder under combined loading is studied by conducting multi-objective design optimization. It is shown that the VS composite cylinder whose fiber path is curve is superior to CS and QI ones. Due to the curve fiber path, a VS composite cylinder possesses inhomogeneous feature. When sustaining inhomogeneous loadings, its loading capacity will be enhanced greatly as compared with its CS or QI counterpart. For homogeneous loading condition, CS composite cylinder provides almost the best solution.

IOP Conf. Series: Materials Science and Engineering **372** (2018) 012034 doi:10.1088/1757-899X/372/1/012034

References

[1] Jing Z, Fan X L and Sun Q 2015 *Compos Struct* **121** 225-36
[2] Jing Z, Fan X L and Sun Q 2015 *Compos B: Eng* **69** 181-90
[3] Jing Z, Sun Q and Silberschmidt V V 2016 *Compos Struct* **141** 240-52
[4] Jing Z, Sun Q and Silberschmidt V V 2016 *Compos Struct* **154** 106-28
[5] Pinckney R L 1991 *Proceedings of First NASA Advanced Composites Technology Conference* (NASA Langley Research Center) p 385-97
[6] Anderson R L and Gran C 1991 *Proceeding of First NASA Advanced Composites Technology Conference* (NASA Langley Research Center) p 817-30
[7] Kisch R A 2006 *Proceedings of International SAMPE Symposium and Exhibition* (Long Beach CA (USA)) p 1537-47
[8] Measom R and Sewell K 1996 *Proceedings of First NASA Advanced Helicopter Society* (American Helicopter Society 52nd Annual Forum) p 611-22
[9] Grant C and Benson V 1993 *Proceedings of Third NASA Advance Composites Technology Conference* (NASA Langley Research Center) p 648-52
[10] Buchanan D, Alexander R and Thomas J 1999 *Proceedings of the 40th AIAA/ASCE/AHS/ASC SDM Conference* (St. Louis, MO (USA)) p1999
[11] Sabido A, Bahamonde L, Harik R and Tooren M J L V 2017 *Compos Struct* **160** 804-12
[12] Hyer M W, Rust R J and Waters W A 1994 *Technical Report NASA-CR-197045*, NASA
[13] Tatting B F 1998 Analysis and design of variable stiffness composite cylinders (Virg. Tech, Blacksburg, Virginia)
[14] Rouhi M 2015 *Composite: Part B* **69** 249-55
[15] Rouhi M and Ghayoor H 2016 *Composite Structures* **150** 115-23
[16] Rouhi M, Ghayoor H, Fortin-Simpson J, Zacchia T T, Hoa S V and Hojjati M 2017 *Composite Structures* **184** 146-52
[17] Lee D S, Morillo C, Bugeda G, Oller S and Onate E 2012 *Compos Struct* **94** 1087-96
[18] Vosoughi A R and Nikoo M R 2015 *Thin-Walled Struct* **95** 408-15

IOP Conf. Series: Materials Science and Engineering **372** (2018) 012035 doi:10.1088/1757-899X/372/1/012035

A three degree of freedom manipulator used for store separation wind tunnel test

R Wei[1,2], B-H Che[1], C-B Sun[1], J Zhang[1] and Y-Q Lu[1]

[1]China Aerodynamics Research & Development Center, Mianyang, China

E-mail: wrywrr@mail.ustc.edu.cn

Abstract. A three degree of freedom manipulator is presented, which is used for store separation wind tunnel test. It is a kind of mechatronics product, have small volume and large moment of torque. The paper researched the design principle of wind tunnel test equipment, also introduced the transmission principle design, physical design, control system design, drive element selection calculation and verification, dynamics computation and static structural computation of the manipulator. To satisfy the design principle of wind tunnel test equipment, some optimization design are made include optimizes the structure of drive element and cable, fairing configuration, overall dimension so that to make the device more suitable for the wind tunnel test. Some tests are made to verify the parameters of the manipulator. The results show that the device improves the load from 100 Nm to 250 Nm, control accuracy from 0.1°to 0.05°in pitch and yaw, also improves load from 10 Nm to 20 Nm, control accuracy from 0.1°to 0.05°in roll .

1. Introduction

The Store Separation Wind Tunnel Test is a wind tunnel test technology that studies the separation characteristics of aircraft external store. The main test device is the multi degree of freedom traverse system [1-3].

Many wind tunnels in the world are equipped with a multi degree of freedom traverse system [4]. The Royal Aeronautical Research Center has a five degree of freedom traverse system [5], with three linear displacement degrees of freedom of X, Y and Z and two Angular degrees of freedom of roll and yaw. The limitation is that there are only two Angular displacement degree of freedom, can not be completed at the same pitch and yaw angle position simulation. France ONERA Modane Wind Tunnel has a four degrees of freedom [6] traverse system, with X-directional linear displacement freedom and pitch, yaw, roll three angular displacement of freedom, when used with the tail boom strut installed Together, the limitation is that the position simulation can only be done within a certain range on one side of the master model.

Many wind tunnels in China are equipped with multi-degree-of-freedom traverse system, including the China Aerodynamics Research and Development Center's FL-12, FL-14, FL-24 and FL-26 wind tunnels and the AVIC Aerodynamics Research Institute's FL-2 Wind Tunnel, etc [7-9]. This paper investigates the six degrees of freedom traverse system (figure 1) equipped by the FL-12 wind tunnel in 1987. The advantage of this system is that it has a wide range of motion, many positions can be simulated, and the disassembly and assembly are convenient during the test. However, the device was affected by the current mission requirements and mechanical and electrical technology constraints; the design load is small, the design carrying weight of 2 kg, pitch moment of 100 Nm. At this stage, the weight of the aircraft external store wind tunnel model has reached the level of 5 kg, and the maximum pitch moment reaches the magnitude of 200 Nm. Due to the lack of design load, the range of locations that the existing devices can simulate in the experiment is limited, and part of the experiment contents can not be completed. Therefore, it is necessary to redesign the six-degree-of-freedom traverse system of the FL-12 wind tunnel to meet the requirements of the Store Separation Wind Tunnel Test at the present stage and the next stage.

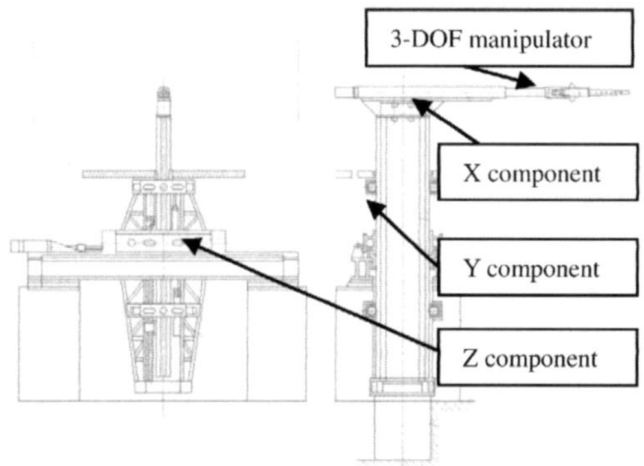

Figure 1. Existing traverse system diagram.

After checking the parts loss and checking calculation, the conclusion is: the installation foundation of the existing traverse system and the X, Y, Z linear displacement components meet the conditions of continued use, which need to re-design is the angular displacement components of the existing traverse system. The function of this part is to support the aircraft external store model to complete the rotation in the angular degree of freedom of pitch, yaw and roll. It is called as three degree of freedom manipulator. [12]

2. Research content

Based on the experience and user feedback of the existing three-degree-of-freedom manipulator (figure 2), it was found that the problems that can be improved are not only the load capacity of the device, but also the wind tunnel test device, which is also unsatisfactory in the applicability of the wind tunnel. To this end, this paper devoted to the design principles of wind tunnel test device, according to the conditions of use of wind tunnel test equipment, the following design principles are proposed:

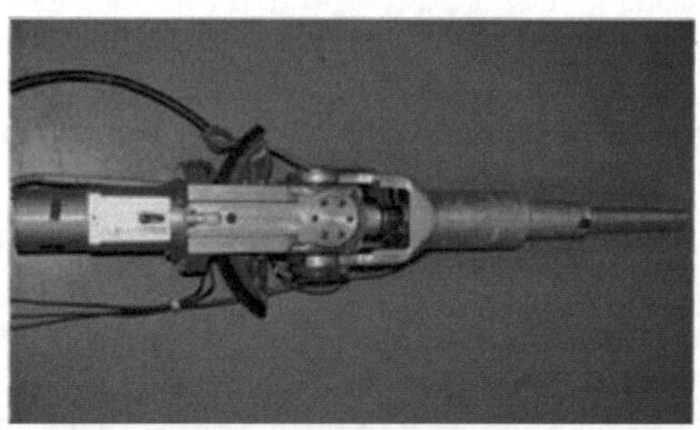

Figure 2. Existing three degrees of freedom manipulator data photos.

- The device must be easy to install and disassemble, the parts can be disassembled and saved as a whole component;
- The Device must be easy to control and calibration, supporting a dedicated control software

and calibration tooling;

- The aerodynamic shape of the device should be smooth to reduce the impact on the flow field quality of the wind tunnel;
- Wind tunnel blockage ratio of the device can not be too large, for reducing the impact of wind tunnel test section of the pressure;
- Shielding the power cable to reduce the signal interference of the balance measuring element;
- Try not to expose the cables in the wind tunnel to avoid the swaying lead to the connectors loose.

After the investigation of the model of the aircraft external store at the present stage, the aerodynamic load of a new type missile is selected as the estimation case, and the load capacity requirement of the three-degree-of-freedom robot is proposed. Device overall technical indicators are listed in table 1.

Table 1. 3-DOF manipulator technical indicators.

Project		Existing Ability	Design Requirements
Pitch	Load	100nm	250nm
	Accuracy	0.1°	0.05°
	Speed	10°/S	4°/S
	Angle Range	±45°	±45°
Yaw	Load	100nm	250nm
	Accuracy	0.1°	0.05°
	Speed	10°/S	4°/S
	Angle Range	±45°	±45°
Roll	Load	10Nm	20Nm
	Accuracy	0.1°	0.05°
	Speed	10°/S	10°/S
	Angle Range	±180°	±180°
Weight	\		60kg
Blockage Ratio	\		1.2%

According to the above technical requirements, the research and development of the device has the following difficulties:

- At the same time reduce the blockage ratio of the device and enhance the load capacity will be mutually constrained. In order to reduce the blockage ratio of the device, the overall size of the device will be limited, and the components of the motor, the sensor and the driving arm may not be oversized. At the same time, the thrust provided by the small-size component may not meet the device load requirements.
- In order to reduce the impact of the device on the quality of the flow field, its shape must be smooth and rectified, and the cable buried. The requirements for the layout of the driving elements, the planning of the cable and the mode of mechanical transmission are higher.
- Control accuracy requirements also have brought difficulties to the design. In addition to the selection of control components higher requirements, the deformation of various parts, tolerances, processing methods and weight control is more stringent.

3. Structural design and optimization

In the transmission principle design, three sets of programs are designed: the worm gear in program one can provide a larger load, shown in figure 3, the advantage is large torque, self-locking function, the disadvantage is the size is too large, the blockage ratio does not meet the Requirements, solid worm impede the built-in cable; program two is to parallel arrangement the electric cylinder ball hinge form, shown in figure 4, the advantage is small axial size, simple structure, the disadvantage is the radial size is too large, the blockage ratio Does not meet the requirements, and parallel arrangement of the electric cylinder in the pitch and yaw direction of the two angles caused by the movement of the coupling lead difficulty to the control of decoupling; program three is arranged in axis series connection of the electric cylinder, shown in figure 5, the advantage is diameter size is small, in the pitch and yaw angle of the two directions without the coupling, the disadvantage is that the series connection of the electric cylinder will cause the device axial size larger than the other program.

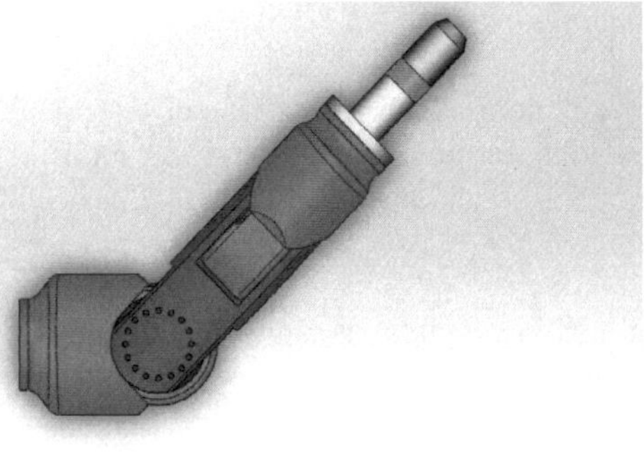

Figure 3. Sketch map of program 1.

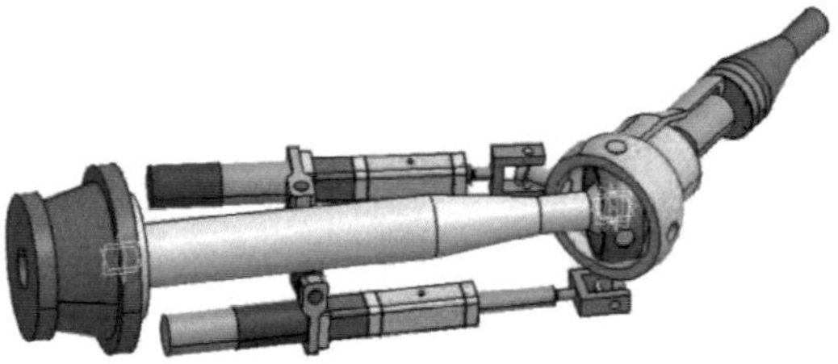

Figure 4. Sketch map of program 2.

Figure 5. Sketch map of program 3.

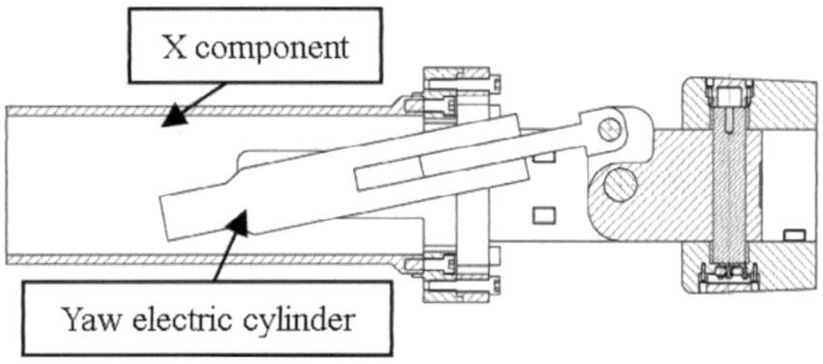

Figure 6. Sketch map of improved program 3.

According to the design requirements of the three-degree-of-freedom manipulator, program three is finally selected as the transmission scheme of the manipulator, and then, in order to reduce the axial size, an improved program three is proposed as shown in figure 6. After the research, the installation base of the X component is a hollow structure. After structural optimization, the yaw electric cylinder and the supporting components are arranged in the X component. The total length of the device with the improved program three is 1140 mm, optimized structure reduces the 240 mm in length, and the radial size can be controlled at about 210 mm. According to the numerical model, the maximum blockage area of the device is about 0.12 m^2 and the maximum blockage ratio is 0.12 m^2 /10.72 m^2 (10.72 m^2 is the cross-sectional area of wind tunnel test section) = 1.12%, which meets the blockage ratio requirements.

The detailed design of the three degree-of-freedom manipulator is shown in figure 7, from the rear end to the front followed by a yaw component, pitch component and roll component, of which yaw angle axis, pitch angle axis and roll angle axis are continuously orthogonal. In order to meet the requirements of the smooth shape of the device, a conical fairing is designed. Through the internal structure optimization, the pitch and roll components are arranged in the conical fairing. The material of the fairing is made of high-strength aluminum alloy 7075-T6, which has the advantages of light weight and high strength, which not only reduces the weight of the pitch component but also effectively controls the deformation of the device under load.

Transmission structure of the device is shown in figure 8, the pitch component work by the electric cylinder to push the device along the main axis of rotation, in order to achieve the angle deflection, the angle sensor is arranged on the pitch axis, the acquisition angle is the pitch angle of the device. The

IOP Conf. Series: Materials Science and Engineering **372** (2018) 012035 doi:10.1088/1757-899X/372/1/012035

yaw component and drives the same way with the pitch component and are arranged in series connection. The advantage of this kind of structure is that the device has the advantages of small radial dimension, small backlash and no long-distance transmission, and the device has no motion coupling in the two directions of pitch and yaw. Rolling component work by the DC servo motor with harmonic gear reducer drive model probe rod to achieve rotation, the ring angle sensor is arranged in the probe rod, the acquisition angle is the body rolling angle, which is characterized by direct and reliable drive mode the angle of the encoder feedback signal is also the device roll angle, without the need for angle conversion.

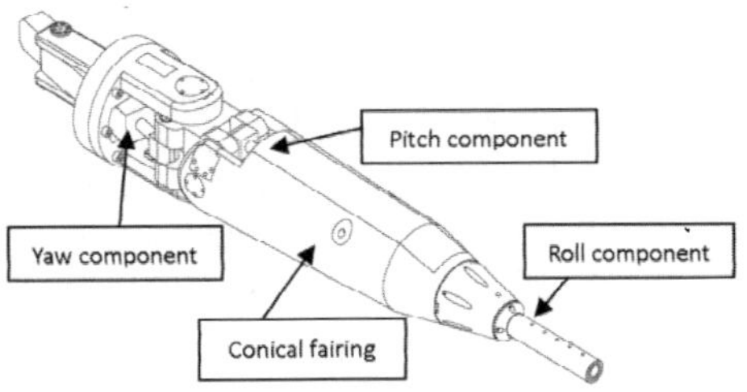

Figure 7. Outline of the 3-DOF manipulator.

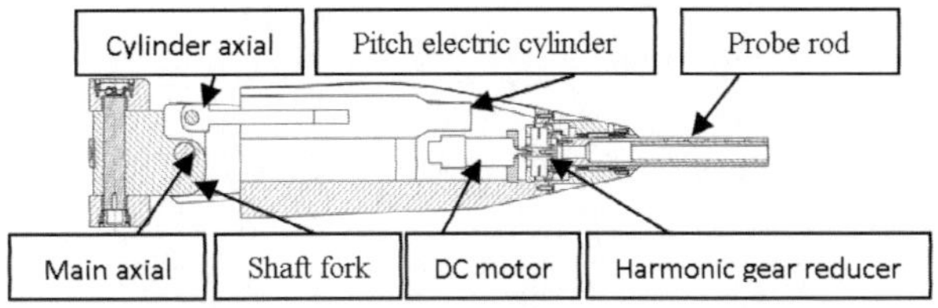

Figure 8. Transmission structure diagram.

Inside the conical fairing, the roller motor and pitch electric cylinder are dislocation to optimize the axial dimension of the device, as shown in figure 9. In order to meet the requirement of cable embedded, cable troughs are reserved in the interior of the conical fairing and the probe rod, which not only ensures the stable signal transmission but also ensures the smooth aerodynamic shape of the conical fairing and reduces the adverse effects on the flow field quality.

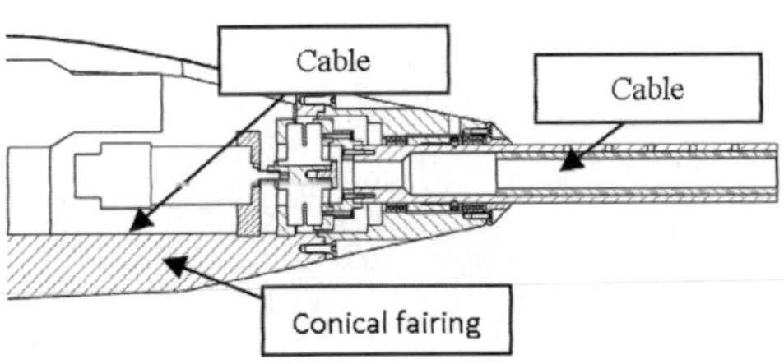

Figure 9. Structure inside the conical fairing.

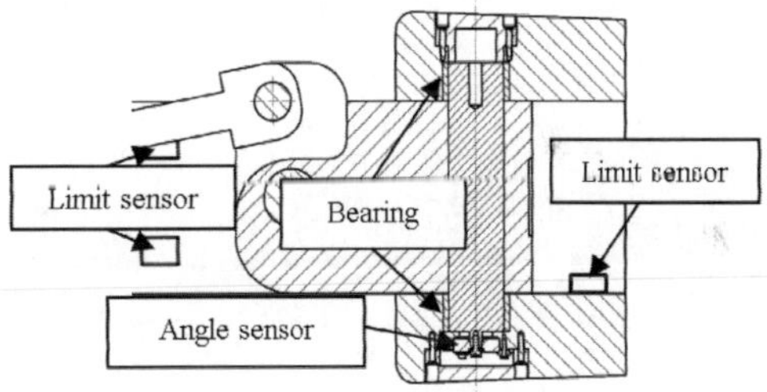

Figure 10. Shaft fork sketch map.

Schematic diagram of the shaft fork structure is shown in figure 10, the angle sensor is arranged on the corresponding axis, the acquisition accuracy of the sensor is 0.0005°, the main axis of rotation angle signal feedback in the closed-loop control. Metal-based self-lubricating bearings (JDB) was selected, which can achieve oil-free lubrication, the friction coefficient is about 0.05, and can protect the device for a long time smooth operation. Photoelectric limit sensor will ensure the safety of the device, when the shaft fork beyond the angle range can trigger the limit sensor, through the control of computer software to stop running to prevent parts collide with each other.

4. Driver selection and check

4.1. Pitch, yaw direction of the components selected calculation

Yawing electric cylinder and pitch electric cylinder drive in the same form, so choose the larger load pitch electric cylinder to do the calculation. The device in the pitch direction of the technical indicators: the load is 250 Nm, the speed is 4 °/s, positioning accuracy is 0.05°. Because of the complexity of the transmission of the electric cylinder, the simple statics calculation can not completely simulate the whole load and the movement state of the electric cylinder within the specified movement range of the device. Therefore, the ADAMS software is used to calculate the dynamics of the device, and the sensor is set at the position of the pitch electric cylinder, by analysing the thrust, displacement and speed curve of the electric cylinder, the parameters of the cylinder can be determined.

According to the three-dimensional model of the device, the yaw joint of the three-degree-of-freedom manipulator is set as the base point O, the pitch joint is the relay point A, the front end of the probe rod is the reference point B, then the transmission principle diagram is drawn as figure 11.

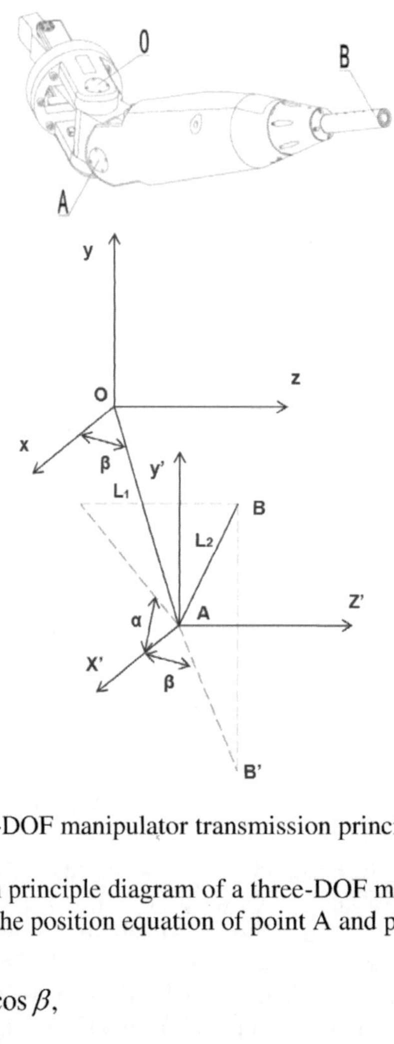

Figure 11. 3-DOF manipulator transmission principle diagram.

According to the transmission principle diagram of a three-DOF manipulator, the dynamic equation of the device is established, and the position equation of point A and point B is obtained:

- $A:(l_1 \cdot \cos\beta, 0, l_1 \cdot \sin\beta)$
  $B:(l_1 \cdot \cos\beta + l_2 \cdot \cos\alpha \cdot \cos\beta,$
- $l_2 \cdot \cos\beta \cdot \sin\alpha,$
  $l_1 \cdot \sin\beta + l_2 \cdot \cos\alpha \cdot \sin\beta)$

Enter the dynamic equation into ADAMS, select -45° to +45° as the calculation range, the calculation of the settings taking the gravity of the parts, device load, friction and start characteristics into account, the pitch mechanism operating speed is set at 4°/s, the load is set at 250 Nm, and the displacement curve, thrust curve and speed curve of the electric cylinder of the device are obtained by calculation.

It can be seen from the displacement curve in figure 12 that the displacement of the electric cylinder has a good linear relationship with the pitch angle. When the pitch angle of the device runs from -45° to 0° and then to + 45°, the total displacement of the electric cylinder is 38 mm + 34 mm = 72 mm, Therefore, the travel requirements of the electric cylinder should be more than 72 mm.

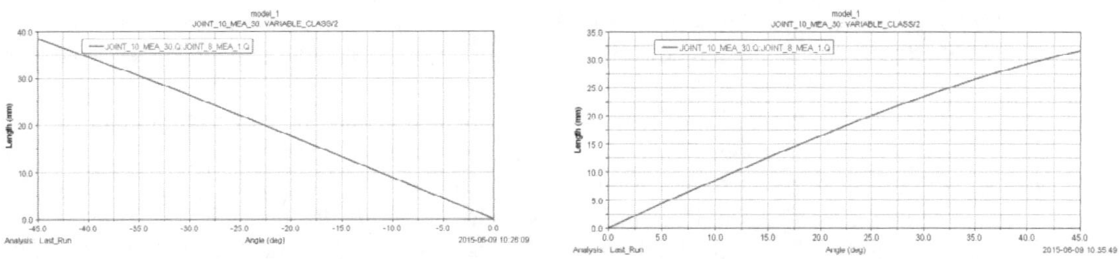

Figure 12. Electric cylinder displacement calculation result.

From the thrust diagram of figure 13 can be seen that although the electric cylinder affected by the start-up characteristics, friction torque, compression torque and arm changes and other factors, but its thrust in the device operating range is still gentle changes, no step-growth, The calculation results show that the maximum thrust required for the cylinder is 9800 N.

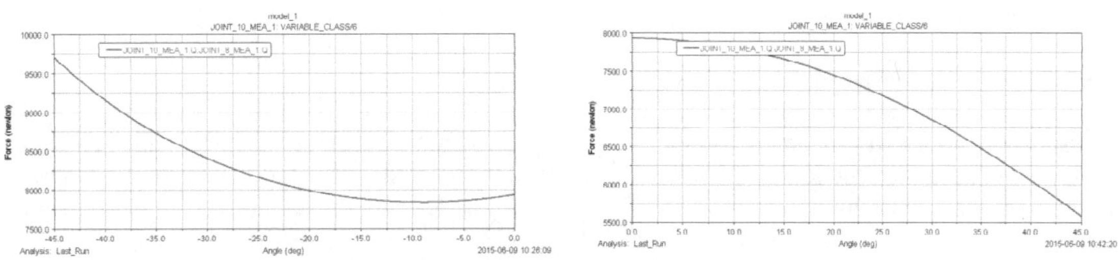

Figure 13. Electric cylinder thrust calculation result.

The speed of the device at different positions of the cylinder can be obtained from the speed diagram figure 14, and it can be seen that the speed requirement of the electric cylinder should be greater than 3.9 mm/s.

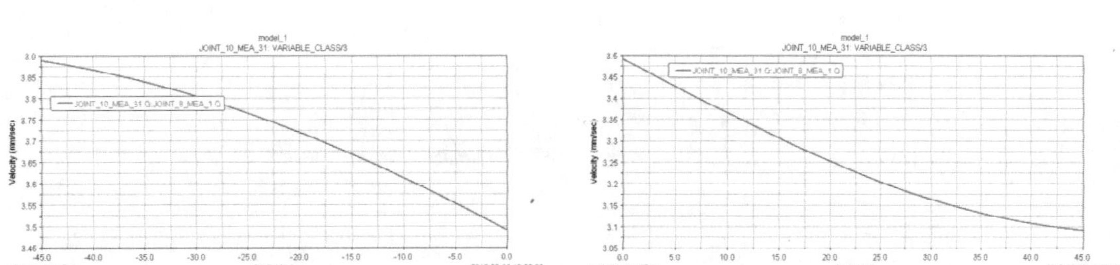

Figure 14. Electric cylinder velocity calculation result.

According to the dynamic calculation results, the selected electric cylinder has a continuous thrust of 12000 N, a maximum displacement of 100 mm, a maximum speed of 4 mm/s and a control precision of 0.01 mm. The accuracy of the mechanism in pitch direction is 0.016° <0.05°, which meets the design requirements.

4.2. Roll direction of the components selected calculation

The technical indicators of the rolling components are: the load is 20 Nm, the speed is than 10 °/s, the accuracy is 0.1°. The drive components are selected by the above technical indicators.

Considering the internal structure of the device size, a 40 mm diameter motor is selected, the power of which is 150 W, the speed is 7000 rpm, the rated load is 0.184 Nm, the maximum load is 2.28 Nm.

When selecting the reducer, firstly considers the reduction ratio and secondly considers the reducer size, and thirdly considers the rotation speed. According to the selected motor reduction ratio calculation: 20 Nm/0.184 Nm = 109.29, considering the efficiency 80% and the overload allowance of 20%, the reduction ratio is suitable at about 160-170.

A harmonic reducer is selected, the deceleration ratio of which is 160, the outer diameter is 70 mm, grease lubrication efficiency is about 80%, and the reducer maximum input speed is 750 rpm, the maximum output torque is 49 Nm, which meet the load requirements.

Consider the control accuracy requirement, the resolution of the motor encoder should be greater than: 360 °/(0.1° × 160) = 22.5. The final selected encoder has 500 frames per turn to meet the requirements.

Calculate obtained the required starting torque is 60 Nm at maximum angular acceleration and the maximum overload torque is 2.28 Nm × 160 = 364 Nm, start-up characteristics to meet the requirements.

According to the selected technical indicators of the motor and reducer accounting device, the speed of the rolling mechanism is obtained as 262 °/s> 10°/s to meet the design requirements; the controllable precision of the rolling mechanism is 0.0045° <0.1°, satisfying Design requirements; Rolling body load capacity is 23.5 Nm> 20 Nm, which meets the design requirements.

5. Control system design

Device control structure is shown in figure 15. When the device is running, a command is given to the corresponding motor at an angle through the computer-controller-driver, and the angle signal is feedback by the angle encoder located on the main axial to make the closed-loop control of the angle. At the same time, the limit sensor at the front of the model and the photoelectric sensor signal mentioned above are also sent back to the controller. If the device exceeds the preset angle range or the risk of collision between the model and the aircraft, the software will automatically determine and send the command to stop the device.

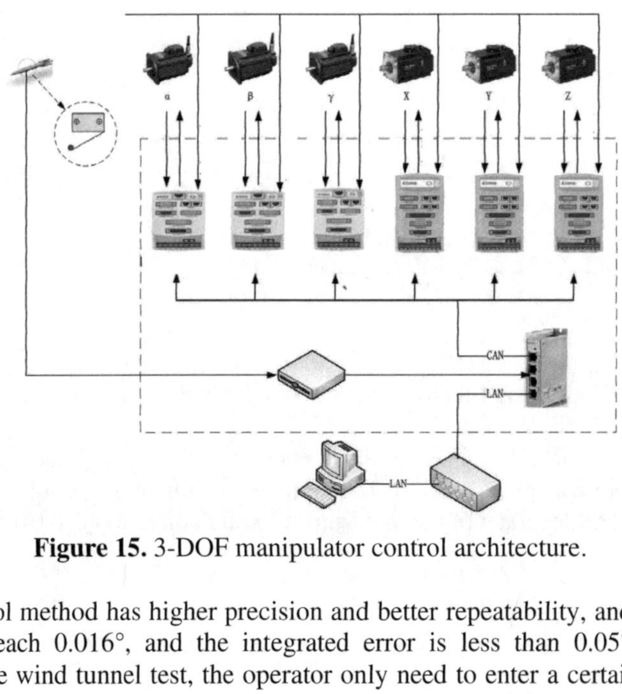

Figure 15. 3-DOF manipulator control architecture.

This kind of control method has higher precision and better repeatability, and the control accuracy in pitch angle can reach 0.016°, and the integrated error is less than 0.05°. Its high degree of automation, during the wind tunnel test, the operator only need to enter a certain point of the control software values, the software will automatically adjust the angle of the feedback signal to the device angle control values, during which no need to intervention.

6. Design verification

6.1. Stress calculation

According to the device bearing capacity, give an example of two key parts of the strength calculation. The shaft fork during work is under repeated thrust, choose aluminum alloy 7075-T6 as the material of the shaft fork, the mechanical properties of the parameters are listed in table 2. In order to ensure there is no risk of fracture during device work, the safety factor require is 5 [10-11].

Table 2. Mechanical properties of used material.

Material	Yield Strength / MPa	Poisson's Ratio	Density / (kg · m $^{-3}$)
7075-T6	5.05×10^2	0.33	2.81×10^3
30CrMnSiA	8.35×10^2	0.28	7.25×10^3

The result of stress calculation for the shaft fork is shown in figure 16. It can be seen from the calculation results that the maximum stress of the part after loading is 23.2 MPa, which is lower than the yield strength of material properties 505 MPa, and the safety factor of 21.7.which meets the safety factor requirements.

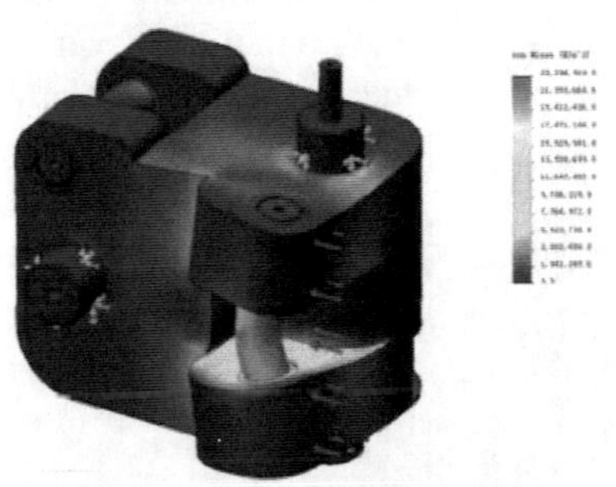

Figure 16. Bushing stress calculation result.

Figure 17. Electric cylinder stress calculation result.

Yaw electric cylinder and its supporting parts, under working conditions need repeatedly withstand thrust, choose alloy steel 30CrMnSiA as its material, its mechanical properties are listed in table 2, the safety factor require is 5.

The stress calculation results of the yaw electric cylinder and the supporting parts are shown in figure 17. It can be seen that the maximum stress of the part is 42 MPa, lower than the yield strength of the material of 835 MPa and the safety factor is 19.8, which meets the safety factor requirements.

6.2. Collision calculation

In order to control the overall size, the gap between the parts of the device is small, and with the device movement, there are some risk of collision between the electric cylinder and the other parts, so the three-dimensional software is used for check the minimum gap between the parts. Figure 18 illustrates several typical locations where the part clearance is less than 10 mm. The clearances of the parts in the first three pictures are more than 3 mm. There is no special requirement on the machining tolerances of the parts. The clearance of the parts in the latter picture is less than 3 mm, which indicates that there is a collision risk between the electric motor and the cover of the conical fairing. Therefore, the shape requirements of the corresponding parts and the position tolerance requirements are improved, to ensure that they will not collide with the electric cylinder.

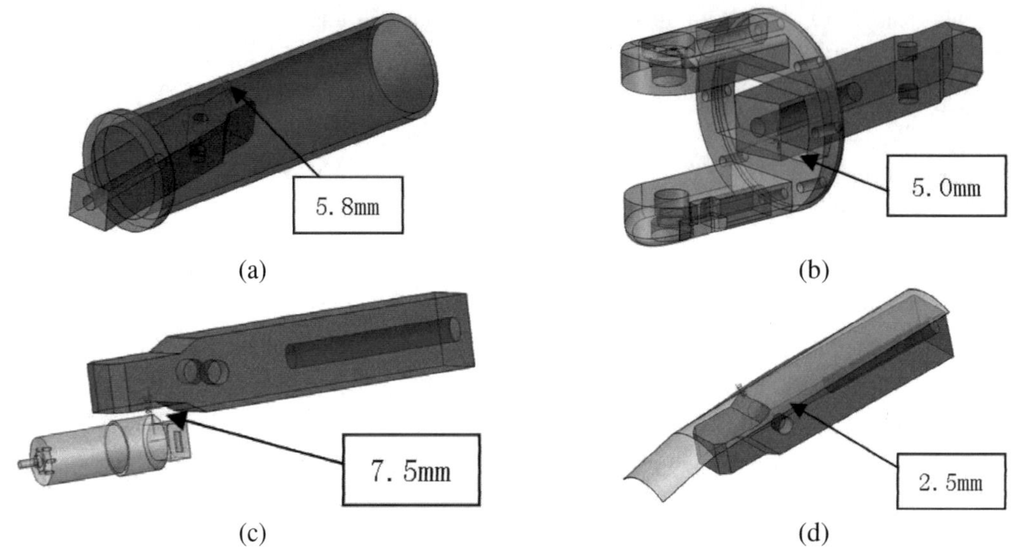

Figure 18. Clearance calculation results.

7. Test verification

When the three degrees of freedom manipulator is assembled, the ground commissioning and debugging within the wind tunnel are made, shown in figure 19. Then the main performance indicators of the three-degree-of-freedom manipulator were tested check in the wind tunnel. The verification results show that the load capacity and angle control accuracy of the three-degree-of-freedom manipulator meet the technical indicators.

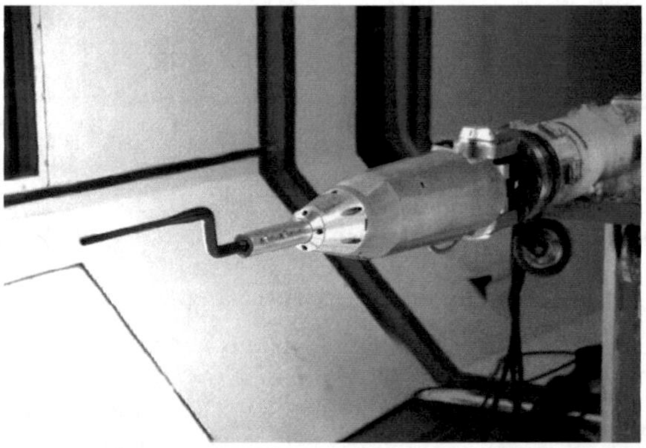

Figure 19. Three degrees of freedom manipulator wind tunnel debugging.

7.1. Load verification

Exert moment load Respectively on the three degrees of freedom of the manipulator to observe the operation of the device, if there are no jitter, or can not maintain a given position, then the test can pass. A certain position is selected on the probe rod at the tip of the manipulator and a moment load is applied by adding weights. Calculation of torque is according to the formula:

- $M=F\times L$

M is the moment, in unit of Nm;

F is the gravity of the weight, in units of N, wherein the weight standard is according to GB / T 4167-2011 M2 level;

L is the distance from the axis of load application point, the measurement error does not exceed

0.2%;

Within the design load range, the manipulator is operating normally in the three degrees of freedom directions. The load is checked according to the evaluation criteria mentioned above. The test data are shown in table 3.

Table 3. Load capacity test results.

Project	Load (Nm)	Test Result
Pitch Direction	100	Pass
	150	Pass
	200	Pass
	220	Pass
	250	Pass
Yaw Direction	100	Pass
	150	Pass
	200	Pass
	220	Pass
	250	Pass
Roll Direction	5	Pass
	10	Pass
	15	Pass
	18	Pass
	20	Pass

7.2. Accuracy verification

The angle of the three degrees of freedom of the manipulator is measured by the electronic inclinometer (the accuracy is 1") and the angle checking tool in figure 20, and compared with the angles given by the software to evaluate the angle positioning accuracy. Tolerances on shape and position of the angle checking tool shall be Level 6 in GB / T 1182-2008.

Figure 20. Pitch direction accuracy check.

After testing, the maximum error of the three directions are no more than 3', that is less than 0.05°, the accuracy verification is passed. Tables 4-6 show the test data.

Table 4. Angle test results of pitch direction.

Nominal Angle (°)	Measured Angle	Software Angle	Error

0	0°1′22″	—	—
5	5°2′15″	5°0′7″	7″
10	10°0′10″	9°58′48″	1′12″
15	15°1′25″	15°0′3″	3″
20	20°3′15″	20°1′53″	1′53″
25	25°2′15″	25°0′53″	53″
30	30°2′30″	30°1′8″	1′8″
35	35°3′45″	35°2′23″	2′23″
40	40°3′15″	40°1′53″	1′53″

Table 5. Angle test results of yaw direction.

Nominal Angle (°)	Measured Angle	Software Angle	Error
0	0°0′ 30″	—	—
5	5°1′ 0″	5°0′ 30″	30″
10	10°0′ 55″	10°0′ 25″	25″
15	14°59′ 55″	14°59′ 25″	35″
20	19°59′ 15″	19°58′ 45″	1′ 15″
25	24°59′ 45″	24°59′ 15″	45″
30	29°59′ 10″	29°58′ 40″	1′ 20″
35	34°59′ 20″	34°58′ 50″	1′ 10″
40	39°59′ 35″	39°59′ 5″	45″

Table 6. Angle test results of roll direction.

Nominal Angle (°)	Measured Angle	Software Angle	Error
0°	-0°0′45″	—	—
-5°	-5°1′20″	-5°0′35″	35″
-10°	-10°1′55″	-10°1′10″	1′10″
-15°	-15°0′45″	-15°0′0″	0″
-20°	-20°1′25″	-20°0′40″	40″
-25°	-24°59′15″	-24°58′30″	1′30″
-30°	-30°2′40″	-30°1′55″	1′55″
-35°	-35°2′45″	-35°2′0″	2′0″
-40°	-40°1′35″	-40°0′50″	50″

8. Conclusion

- In this paper, the design of small-size and large-load three-degree-of-freedom manipulator is the research content. Through the investigation of the requirements of the aircraft external store model, the technical indicators of the device are proposed. At the same time, design principle of wind tunnel test device is researched, covering the device blockage ratio, aerodynamic shape and cable layout.
- In the design of transmission principle, an improved program three is selected with orthogonal three degrees of freedom. The advantage is that the angle changes will not be coupled to each other and the control process does not need to be decoupled. The angle sensors are all located on the rotating axial. The collected angle signals No need to convert and can directly feedback to the closed-loop control, with less conversion process, higher reliability.
- According to the research contents, the structural design and optimization of the three-DOF manipulator, control system design, design verification and test verification are carried out. The results are shown in table 7. The results show that all the technical indicators of the device

meet the design requirements.

- The three-DOF manipulator has the ability to complete the store separation wind tunnel test at present and in the future. However, the accuracy of the model positioning accuracy caused by factors such as the shape and position tolerance of the part and the deformation by loading of the device also needs further research and exploration, to improve the data quality of store separation test.

Table 7. Three degrees of freedom manipulator developed results.

Project		Existing ability	Design requirements	verification results
Pitch	Load	100 Nm	250 Nm	pass
	Accuracy	0.1°	0.05°	pass
	Speed	10°/s	4°/s	pass
Yaw	Load	100 Nm	250 Nm	pass
	Accuracy	0.1°	0.05°	pass
	Speed	10°/s	4°/s	pass
Roll	Load	10Nm	20 Nm	pass
	Accuracy	0.1°	0.05°	pass
	Speed	10°/s	10°/s	pass
Weight	\		60 kg	56 kg
Size (mm)	1070×295×295		\	1140×Φ210
Blockage Ratio	\		1.2%	1.12%

References

[1] Andrew G 1999 *Design & Development of A New Captive Trajectory Simulation Model Support System* American Institute of Aeronautics and Astronautics AIAA-99-2165

[2] Navair A C and Md P R 2009 *Lessons Learned in 30 Years of Store Separation Testing* American Institute of Aeronautics and Astronautics AIAA-2009-0098

[3] Carter R and Lind R 2012 *Trajectory Optimization for Guided Store Separation* American Institute of Aeronautics and Astronautics AIAA-2012-4686

[4] Veazey D T and Hopf J C 1998 *Comparison of Aero Dynamic Data Obtained in the Arnold Engineering Development Center Wind Tunnels 4T and 16T* American Institute of Aeronautics and Astronautics AIAA-1998-2874

[5] Coulton D G 1991 Recent developments in data acquisition and control systems at the aircraft research association limited *IEEE Instrumentation in Aerospace Simulation Facilities* **1991** 196-205

[6] Garcon F 2001 *Recent Developments in Captive Trajectory Systems of the ONERA Modane Wind Tunnels* American Institute of Aeronautics and Astronautics AIAA-2001-0579

[7] Huang X H and Pang X D 2008 Development of a new captive trajectory simulation system in the 1.2m transonic wind tunnel *Journal of Experiments in Fluid Mechanics* **2008** 95-8

[8] Huang X H and Luo X F 2008 Design & development of a new captive trajectory simulation system in FL-24 wind tunnel *Acta Aerodynamica Sinica* **2008** 145-9

[9] Cui C X, Xing H Q, Zhang R, *et al* 2007 FL-2 wind tunnel CTS test technology *1st Experimental Aerodynamics Conference* **2007** 172-6

[10] Chen W H and Wang C Q 2005 Finite element analysis on main body of a wind tunnel *Journal of Experiments in Fluid Mechanics* **2005** 90-3

[11] China PLA General Armament Department Military Training Materials Editorial Board 2003 *Aerodynamic and Structural Design of High and Low Speed Wind Tunnel* (Beijing: National Defense Industry Press) pp 10-2

[12] Wei R, Che B H, Zhang J, *et al* 2016 The design of a 3-DOF robot arm used for captive trajectory simulation in wind tunnel test *Journal of Experiments in Fluid Mechanics* **30** 91-7

IOP Conf. Series: Materials Science and Engineering **372** (2018) 012036 doi:10.1088/1757-899X/372/1/012036

Study on magnetic force of electromagnetic levitation circular knitting machine

X G Wu[1,2], C Zhang[1], X S Xu[1], J G Zhang[1], N Yan[1] and G Z Zhang[1]

[1]School of mechanical engineering and automation, Wuhan Textile University, Wuhan, Hubei 430073, China

E-mail: 2006wist@163.com

Abstract. The structure of the driving coil and the electromagnetic force of the test prototype of electromagnetic-levitation (EL) circular knitting machine are studied. In this paper, the driving coil's structure and working principle of the EL circular knitting machine are firstly introduced, then the mathematical modelling analysis of the driving electromagnetic force is carried out, and through the *Ansoft Maxwell* finite element simulation software the coil's magnetic induction intensity and the needle's electromagnetic force is simulated, finally an experimental platform is built to measure the coil's magnetic induction intensity and the needle's electromagnetic force. The results show that the theoretical analysis, the simulation analysis and the results of the test are very close, which proves the correctness of the proposed model.

1. Introduction

With the successful application of EL technology in train, more and more areas are exploring how to introduce the high-tech, maglev knitting technology is one of them [1], the so-called maglev knitting technology using electromagnetic force to drive the needle up and down movement to replace the traditional mechanical connecting rod drive. The driving mechanism of the EL circular knitting machine adopts hybrid drive (EPMH) mode of electromagnetic and permanent-magnet (PM), the electromagnetic field excited by the driving coil, PM and needles are fixedly connected together. In the working process, the control card gives the driving coil electricity in a certain phase sequence, thus the drive coil produces a certain sequence of magnetic field, so the PM and the needle can do regular movement to knit in the magnetic field. By using the electromagnetic drive technology, the EL circular knitting machine can eliminate the middle part of the complex mechanical transmission and realize zero transmission of the needle selecting process compared with the traditional circular knitting machine's drive mechanism of the mechanical triangle cam (A triangular shaped cam that drives the needle motion in a certain phase sequence), the direct drive mode will not produce rigid impact and shortcomings the friction and wear, energy loss and noise, which greatly improves the knitting efficiency of circular machine.

2. Knitting part and working principle

As shown in figure 1, the knitting part of the EL circular knitting machine includes needle, drive coil, sinker, cam, gear, a base and a support member, which are arranged on the corresponding position of the base. In the process of work, the coils drive the needle up and down when the coils are energized, at the same time, the gear drives cam rotation, the rotation of the cam push sinker to do reciprocating motion of front and back (the sinker is installed in the groove of the fixed disc and forms a cam with the cam, so when the cam rotates, the sinker is driven to reciprocate) , the combination of the two movements completes the knitting process.

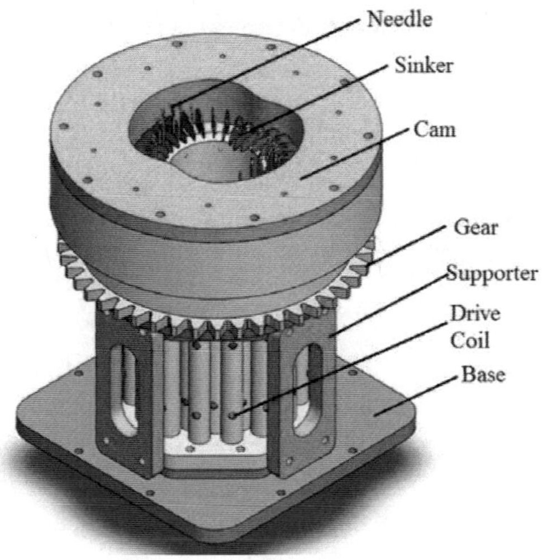

Figure 1. Knitting part of Magnetic levitation knitting circular

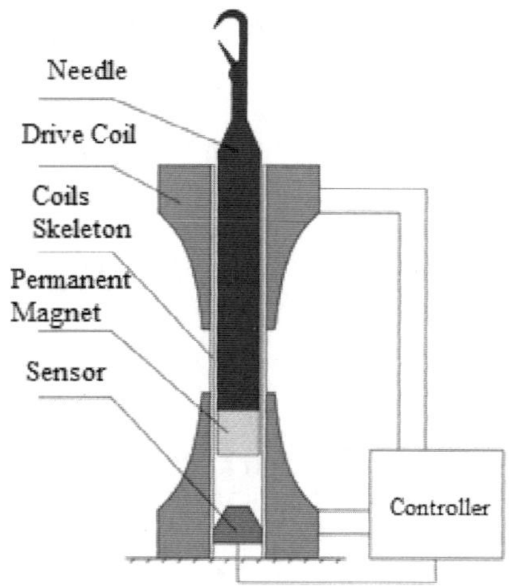

Figure 2. Working principle of Magnetic levitation knitting

Magnetic drive knitting device core component comprises drive coil, PM, coil frame, needle, sensor and controller, the principle of its specific structure is shown in figure 2. From figure 2 we know that as long as the fixed drive electromagnetic coil is electrified, a certain direction of magnetic field will be produced in the vertical direction of the centre, the magnetic field will produce an interaction with the PM magnetic field, thus PM and the needle will be driven up and down. This is an open loop system, which can't control the time and size of the electricity, so adding a sensor and controller to form a closed loop system, the displacement sensor to detect changes in needle, needle position feedback to the controller, the controller will change the size and direction of the coil current, the size and direction of the coil current will directly reflect the change of the magnetic force and the direction of the magnetic force, the change of the magnetic force of the PM will change the direction of up and down movement of the needle and speed, therefore, precise weaving action will be completed through. the needle reciprocating motion.

3. Theoretical model of electromagnetic force

3.1. PM magnetic field

As shown in figure 2, the force of the conducting wire is equal to the one of the PM in the opposite direction when the PM moves during the knitting process in EPMH mode (Newton's 3rd law) and the force will be equal to the force of the PM in the opposite direction, therefore, in the process of calculating the magnetic force, the force of the conducting wire in the magnetic field can be obtained by the theory of ampere force, so that the form and size of the force during the movement of the PM can be obtained. According to the *Ampere* molecular circulation hypothesis: the smallest unit that makes up a magnet is a ring current, and if these molecules are oriented in a circular arrangement, they will show N and S poles macroscopically [2].

Expanding a single energized ring into space can create a PM field mathematical model. Let a cylindrical PM radius a, high h, magnet bottom surface centre of the circle as the origin of coordinates, the bottom surface of the *xoy* plane, z-axis positive direction for the magnetization, the establishment of geometric model shown in figure 3.

If the magnet material and magnetization direction unchanged, and saturation magnetization, the magnetization density *Js* can be regarded as a constant. Calculate the magnetic field strength of any point $P(r\,cos(\theta), r\,sin(\theta), z - z_0)$ in the space, where *r* and θ are the polar coordinates of the point *Q* of *P* on the *xoy* plane. Select the arbitrary thickness of the PM dz_0 thin layer circulation, the *z-axis* axial height of z_0. While *a* and α are the polar coordinates of the current element ***dl*** in the plane of dz_0, and ***R*** is the radius vector of the current element ***dl*** to the point *P*.

According to the superposition characteristics of the magnetic field, the total magnetic field strength at any point *P* in the space of the PM can be given by the following equation (1).

$$\boldsymbol{B} = \int_0^h d\boldsymbol{B} = \frac{\mu_0 J_s a}{4\pi}\int_0^h \oint \frac{dl \times \boldsymbol{R}}{R^3}$$
$$= \frac{\mu_0 J_s a}{4\pi}\int_0^h \int_0^{2\pi} \frac{\begin{vmatrix} i & j & k \\ -\text{asin}(\alpha) & \text{acos}(\alpha) & 0 \\ rcos(\theta) - \text{acos}(\alpha) & rsin(\theta) - \text{asin}(\alpha) & z - z_0 \end{vmatrix}}{R^3} d dz_0 \tag{1}$$

$R = AP - a = (rcos(\theta) - acos(\alpha), rsin(\theta) - asin(\alpha), z - z_0)$ in the equation (1), μ_0 the vacuum permeability.

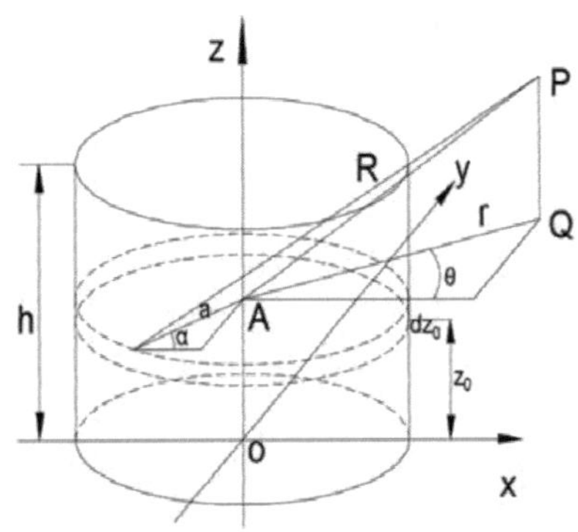

Figure 3. Magnetic field model of cylindrical PM

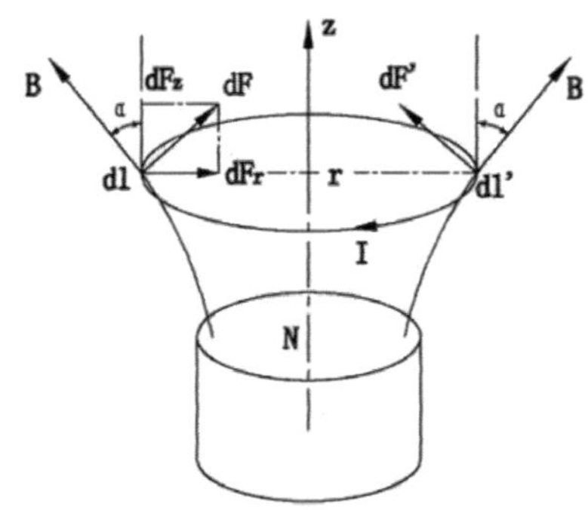

Figure 4. Force model of single electric coil

3.2. Electromagnetic force model

A single coil energized wire in the magnetic field force model as shown in figure 4 above. From the *Ampere* formula, the electric current ***Idl*** receives the electromagnetic force is the following equation (2).

$$d\boldsymbol{F} = Id\boldsymbol{l} \times \boldsymbol{B} \tag{2}$$

Assuming the current direction of the coil is as shown in the figure, the direction of any point in the coil can be judged by the left-hand rule. The decomposition force F is along the z-axis direction F_z and along the coil radial direction F_r. As the strength of the magnetic field at the concentric ring is equal in magnitude and different in direction, so the F_z is equal in size and direction, the F_r points the same point to the centre of the circle and its size is equal. Further analysis shows that F_z is provided by B_x and B_y, and F_r is provided by B_z. Axial force only needs to be known during needle movement. therefore, it is only necessary to study the effect of B_x and B_y on the conducting wire.

The electromagnetic force of the single coil current element obtained by substituting the equation (1) into the equation (2), which is represented by the following equation (3).

$$dF = \frac{\mu_0 J_s a}{4\pi} Idl \int_0^h \int_0^{2\pi} \frac{\begin{vmatrix} i & j & k \\ -\operatorname{asin}(\alpha) & \operatorname{acos}(\alpha) & 0 \\ rcos(\theta)-\operatorname{acos}(\alpha) & rsin(\theta)-\operatorname{asin}(\alpha) & z-z_0 \end{vmatrix}}{R^3} d\alpha dz_0 \quad (3)$$

$dl = (-rsin(\theta), rcos(\theta), 0)d\theta$ in the equation (3), substitute it into equation (3) and integrate $d\theta$ to obtain the electromagnetic force on the whole single coil, which is represented by the following equation (4).

$$F = \frac{\mu_0 J_s a I}{4\pi} \int_0^{2\pi} \int_0^h \int_0^{2\pi} \frac{D}{R^3} d\alpha dz_0 d\theta \quad (4)$$

Which D is as follows equation (5).

$$D = \begin{vmatrix} i & j & k \\ -rsin(\theta) & rcos(\theta) & 0 \\ \operatorname{acos}(\alpha)(z-z_0) & \operatorname{asin}(\alpha)(z-z_0) & a^2 - ar(\sin(\theta)\sin(\alpha) + \cos(\theta)\cos(\alpha)) \end{vmatrix} \quad (5)$$

As mentioned earlier, there is only need to calculate the axial force for the PM, so the z axial force F_{sz} is shown in the following equation (6).

$$F_{sz} = -\frac{\mu_0 J_s a I}{4\pi} \int_0^{2\pi} \int_0^h \int_0^{2\pi} \frac{ar(\cos(\theta)\cos(\alpha)+\sin(\theta)\sin(\alpha))}{R^3} d\alpha dz_0 d\theta \quad (6)$$

This is a single-coil force model in the magnetic field, for any multi-turn coil, the force can be calculated by the force can be superimposed features. Assume that the inner diameter of the coil is r, the outer diameter is $f(z)$ and $f(z)$ is the function of the outer diameter with respect to the z axis, the coordinates of the coil in the axial direction are z_1 and z_2. Also consider the existence of gaps between the coils, the axial density ρ_z and radial density ρ_r is Introduced. In turn, the axial force and the radial force of the single coil are integrated. Finally, the total force of the coil in the z axis of any shape can be expressed as the following equation (7).

$$F_z = \rho_z \rho_r \int_r^{f(z)} \int_{z_1}^{z_2} F_{sz}\, dzdr \quad (7)$$

If any certain force F_z is given, the shape of the corresponding coil can be calculated.

4. Simulation and measurement of magnetic field and electromagnetic force

4.1. Simulation of magnetic field and electromagnetic force

According to the previous experimental study [3], knitting, the needle can smoothly hook yarn to provide conditions for the electromagnetic force is about 85mN in the process of weaving needle, the presupposition condition into equation (7) obtained a curved coil as shown in figure 5. In order to increase the needle stroke, the coil bipolar symmetrical layout, the opposite direction, through 12 V DC. Through the *Ansoft Maxwell* magnetic field finite element simulation software, the magnetic induction intensity simulation cloud chart under the actual working condition is shown in figure 6.

At the same time, the electromagnetic force of the PM and the needle during the movement is simulated, the simulation model is shown in figure 7, where *Region* is the boundary solution area and *Band* is the movement area. The electromagnetic force simulation curve of the PM during the movement is as shown in figure 8.

4.2. Measurement of magnetic field and electromagnetic force

Under the same power-on condition, the magnetic field induction strength of the drive coil and the magnetic force of the needle in the working stroke were measured with a Gauss meter and a dynamometer, respectively. The measurement interval is 1 mm, and the magnetic field strength curve and the magnetic force curve in the working stroke are shown in figures 9 and 10 respectively.

IOP Conf. Series: Materials Science and Engineering **372** (2018) 012036 doi:10.1088/1757-899X/372/1/012036

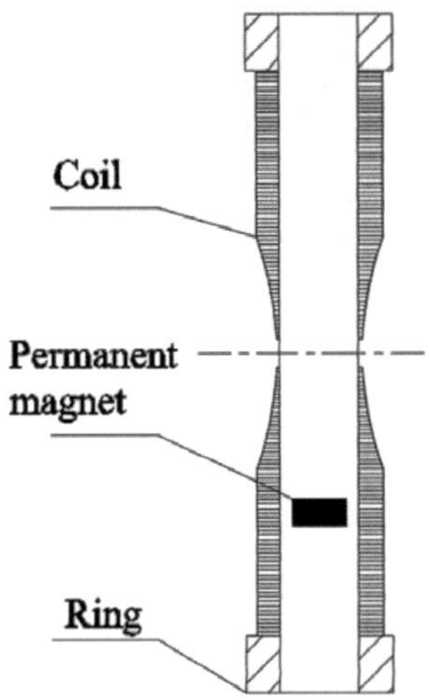

Figure 5. Coil structure model

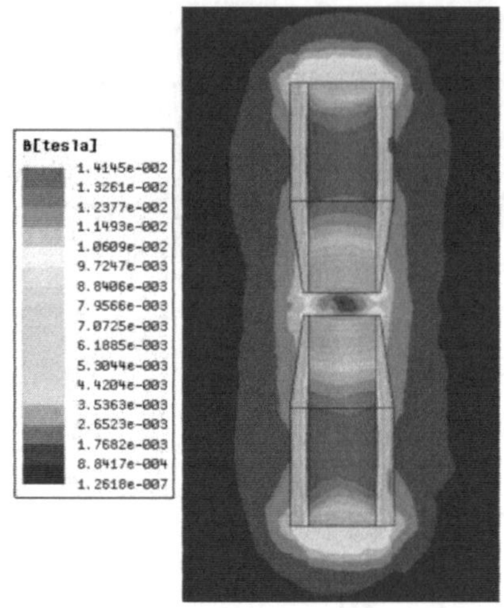

Figure 6. Simulation of magnetic field strength of coil

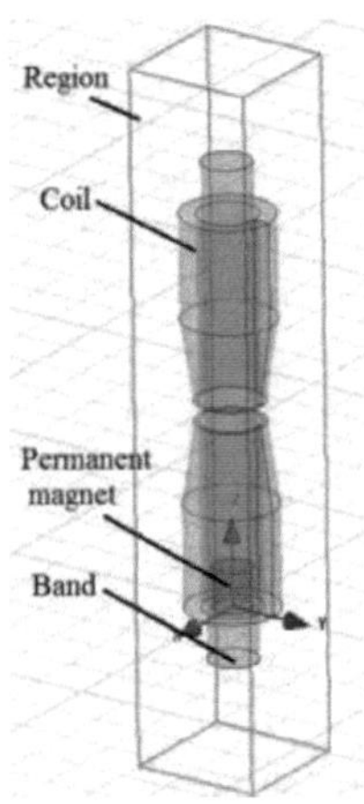

Figure 7. Magnetic force simulation model

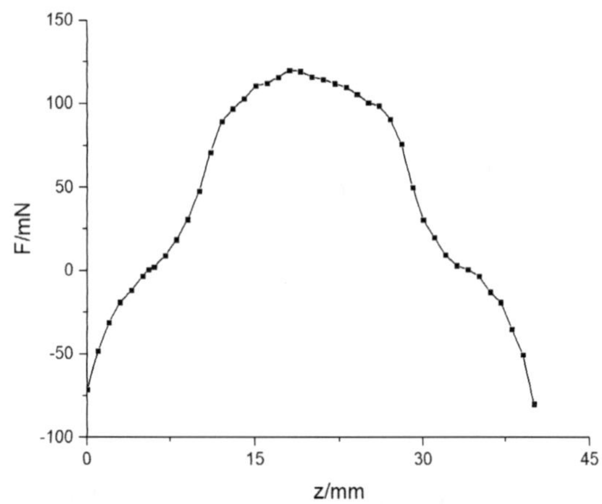

Figure 8. Simulation magnetic force diagram

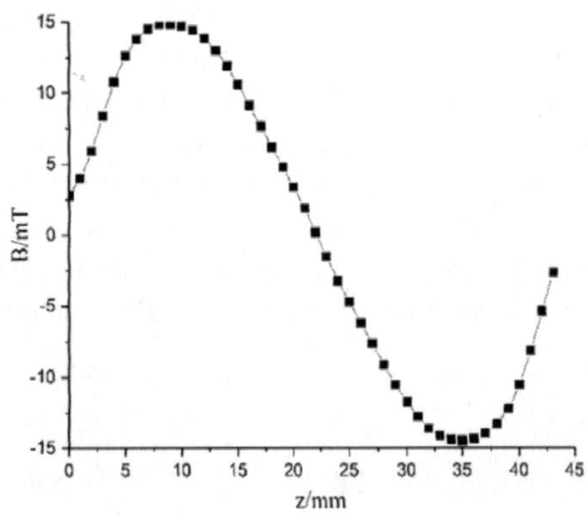

Figure 9. Magnetic induction intensity of measurement

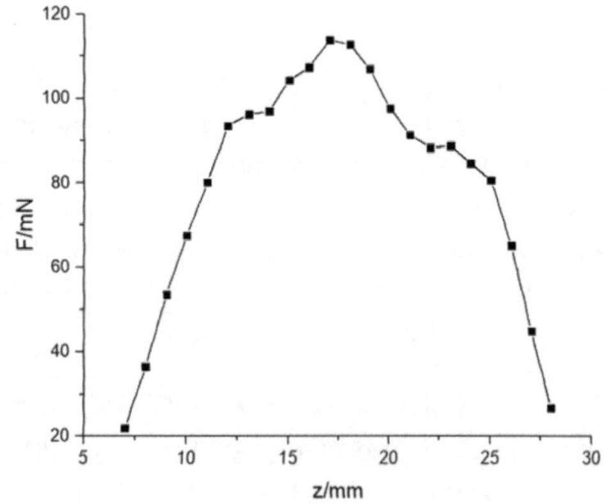

Figure 10. Magnetic force of measurement

Comparing figures 9 and 6, the simulation results of the magnetic induction intensity of the bipolar symmetric driving coil are in good agreement with the measured results, the magnetic fields of the two coils arranged symmetrically in the up and down direction are opposite and mainly concentrate in the middle of the coil with an average of 14 mT. Comparing figure 10 with figure 8, we can see that the electromagnetic force simulation result and the measurement result are similar, and the electromagnetic force is stable within the working stroke (position from 10 mm to 25 mm), about 95 mN, which meets the driving requirement.

5. Summary

This paper introduces a direct-drive circular knitting machine based on electromagnetic drive. Based on the basic law of electromagnetic field, a theoretical model of electromagnetic force is established and a kind of curved coil is determined. *Ansoft Maxwell* software was used to simulate the magnetic field and magnetic force of the established coil, the magnetic field and magnetic force were also measured. The results show that the simulation results of the magnetic field and magnetic force of the coil are in good agreement with the measured results, which can meet the demand of stable and accurate drive motion for knitting.

Acknowledgments

This research is a Textile Industry Federation guidance project and was financially supported by the National Science Foundation of China (51175384) and the Key Laboratory of Digital Textile Equipment Foundation of Hubei Province (DTL2017003 and DTL2017002).

References

[1] Wan D Y, Wu X G and Zhang C 2017 Study on needle electromagnetic force and coil profile optimization of magnetic suspension drive *J. Knitting industry* **8** 9-12
[2] Zhao K H and Chen X M 2003 *Electromagnetics* (Beijing: Higher Education Press) pp 85-137
[3] Wu X G, Zhang C and Yuan B 2017 Research on magnetic force coupling and magnetic force coupling of magnetic suspension driven needles *J. Knitting industry* **1** 13-7

Eddy current evaluation of metallic coating thicknesses over nonmagnetic metals

B A Abu-Nabah[1,2]

[1]Department of Mechanical Engineering, American University of Sharjah, Sharjah, UAE

E-mail: babunabah@aus.edu

Abstract. The use of titanium and nickel alloys in metal coating technologies have been gaining rapid popularity due to their high strength characteristics and excellent corrosion resistance behaviour operating at different temperature ranges. Phase-sensitive eddy current technique set by the International Organization for Standardization to assess metal coating thicknesses over nonmagnetic metals delivers ±30% uncertainty in ±25.4 μm lift-off variation from that used over metal-coated calibration blocks. This study takes the first step towards adopting AECC spectroscopy as an alternative forward measurement technique, which offers a reduced sensitivity to lift-off and sample conductivity deviations from those used over the calibration blocks in a broad range of inspection frequencies. To map the depth-dependent conductivity profile from the frequency-dependent AECC measurement, a new AECC-based inversion algorithm is developed to estimate metallic coating thicknesses over nonmagnetic metals. This work is validated following the plane-wave approximation and COMSOL simulations over different coating thicknesses relevant to the industry. Results indicate that AECC spectroscopy can potentially deliver ±3% uncertainty in metallic coating estimation in ±25.4 μm lift-off range. That is a significant improvement over existing eddy current measurement methods.

1. Introduction

Joining processes of dissimilar metals using metal spraying technologies [1,2] and weld overlay [3] are considered practical solutions in the design and manufacturing of critical components in Aviation and Oil and Gas industries. Their cost-effective solution to protect the functionality and life expectancy of their products make them even more attractive in the service sector of these industries. The growing demand for applying these technologies requires the use of a practical nondestructive evaluation approach to evaluate metallic coating thicknesses over nonmagnetic metals.

Phase-sensitive eddy current technology can be used to assess such nonmagnetic layered metal structures following ISO 21968 [4]. Impedance-based eddy current technology [5] following Dodd and Deeds exact solution [6,7] can also be used as alternative solution. The use of time-based pulsed eddy current technology has also demonstrated its potential capabilities to assess such structures [8]. Regardless of the eddy current technique used to assess these coating thicknesses, the calibration mandates building a coil-specific characteristic curve for a coating/substrate conductivity combination measured at a specific lift-off value, i.e., a known separation distance between the probe and the sample. Even with spring loading the probe against the sample to assure lift-off consistency between measurement and calibration, the measurement delivers an uncertainty in metal coating thickness estimation of approximately 13% [5,8].

Recent improvements in high-frequency eddy current conductivity spectroscopy delivers ± 0.1% uncertainty in apparent eddy current conductivity (AECC) up to 80-100 MHz in a ± 5% conductivity and ± 25 μm lift-off ranges [9-15]. Not only it allows the sample conductivity to deviate from those used over the calibration blocks, but it also allows enough lift-off variation consistent with the lift-off variations from spring loading the probe against the sample. The capabilities of AECC-based technology was initially built to assess 1-3% near-surface conductivity variations due to surface enhancement methodologies such as shot peening and low plasticity burnishing [16-18]. This study takes the first step towards building an AECC-based capability to assess metallic coating thicknesses over nonmagnetic metals.

2. AECC-based model

AECC over metallic-layered structure employs the analysis of an electromagnetic wave over depth-dependent inhomogeneous conductivity profiles. The previously mentioned AECC deals with the electric conductivity of a homogeneous medium that would produce the same complex coil impedance over an inhomogeneous medium at a given frequency [19]. The electromagnetic plane-wave approximation [20] has demonstrated its capabilities in capturing the AECC spectrum over small conductivity variations that are smooth and continuous (≤ 1% variation) as shown in figure 1. In other words, the plane-wave approximation assumes an infinitely large coil.

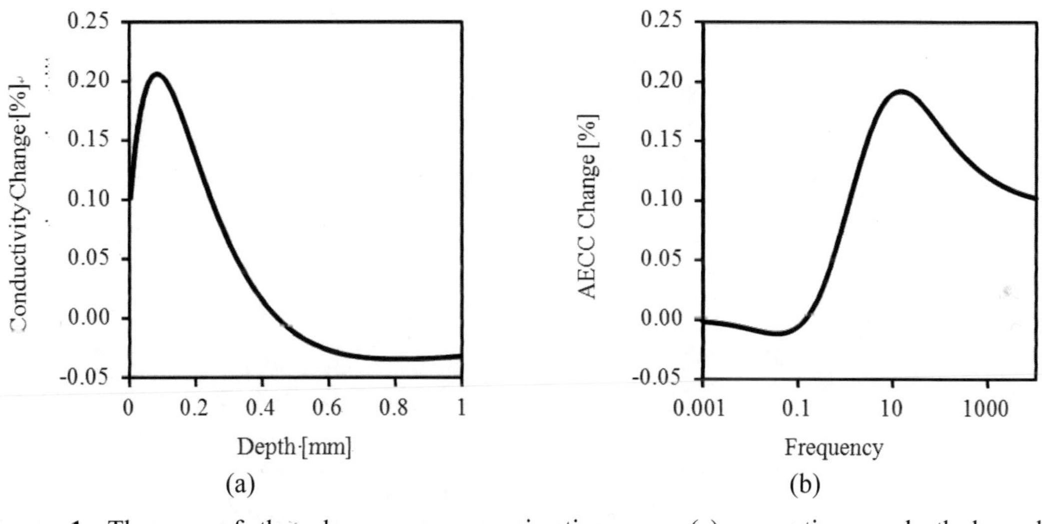

Figure 1. The use of the plane-wave approximation over (a) a continuous depth-dependent conductivity profile and (b) its corresponding AECC spectrum.

2.1. Forward model

The plane-wave approximation is considered as a direct approach to assess AECC spectrums. This study uses the same approach over step conductivity profiles, which represent conductivity profiles in coating/substrate metal structures. Figure 2 demonstrates the conductivity profile of a coating/substrate metal structure with relatively small conductivity variation. However, the challenge here is to validate the plane-wave approximation over relatively large conductivity variations.

For the case of titanium alloy over stainless steel, the coating will have a conductivity that is approximately 50% lower than the substrate. To validate the plane-wave approximation over this conductivity range with coating thicknesses (t_c) relevant to the industry, the indirect approach of AECC spectroscopy is followed. Linear system calibration is implemented using two calibration blocks to be measured with (l = 50.8 μm) and without (l = 0.0 μm) lift-off bracketing the coil impedance measured or simulated over the coated sample [15,20]. To cover the conductivity range of

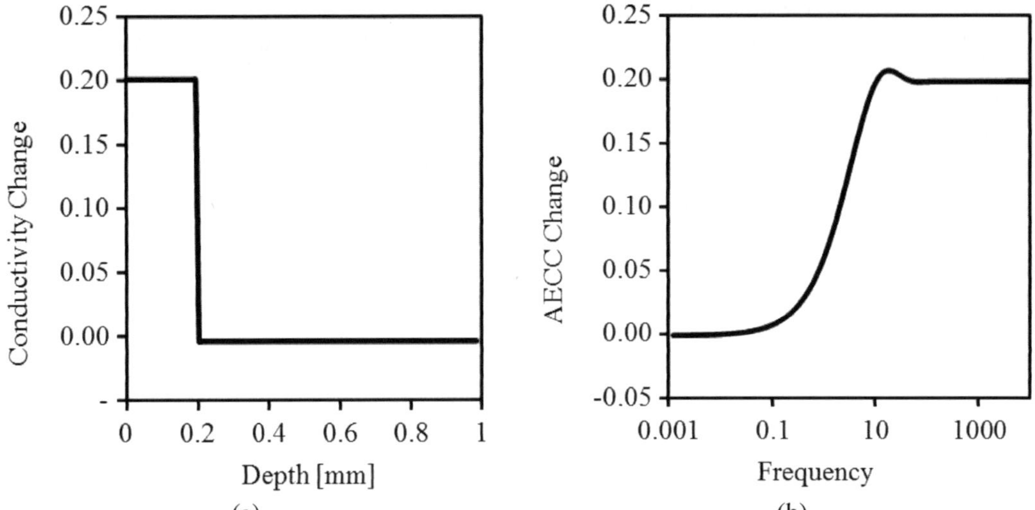

Figure 2. The use of the plane-wave approximation over (a) a step depth-dependent conductivity profile and (b) its corresponding AECC spectrum.

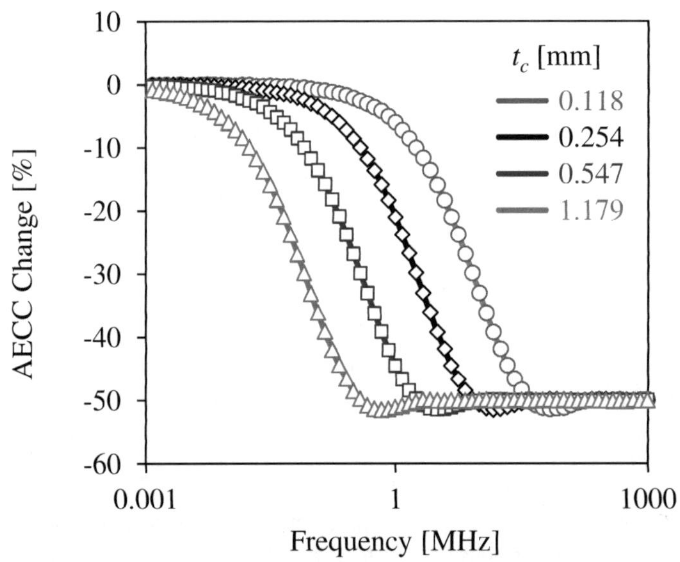

Figure 3. AECC spectrums using the plane-wave approximation (solid lines) and COMSOL simulations (open symbols) over different titanium coating thicknesses on stainless steel substrate.

interest with accuracy, 10 homogenous calibration blocks were simulated with and without lift-off. In this section, the coil impedance was simulated over the coated sample at 25.4 μm lift-off. A relatively large coil design was selected with 50-mm outer diameter to resemble the plane-wave approximation in an axisymmetric COMSOL model. The substrate was selected to be 10 times thicker than the eddy current standard penetration depth for the lowest frequency used in this study. This is considered to resemble the semi-infinite substrate used in the plane-wave approximation. A follow up study will address the challenges encountered upon reducing the coil outer diameter. System calibration and measurement simulations were implemented using the commercially available software COMSOL. Figure 3 shows the AECC change using the plane-wave approximation (solid lines) for different Ti-6Al-4V (σ_c= 1.125% IACS) coating thicknesses over semi-infinite SS304 (σ_s= 2.250% IACS) substrate. Using the indirect approach of AECC assessment, COMSOL simulations (open symbols) show an excellent agreement with the plane-wave approximation. This indicates that AECC spectroscopy lends itself for coating thickness estimation over such metallic layered structures. In figure 3, the conductivity of SS304 substrate was used as a reference for AECC change simulations. The simulations at low frequencies will only see the conductivity of SS304. As the frequency increases, the effect of Ti-6Al-4V will reduce the AECC change. When the frequency is high enough, the AECC change will only see the conductivity of Ti-6Al-4V delivering -50% AECC change relative to the conductivity of the substrate.

2.2. Inverse model

Existing simple [20] and iterative [9] AECC-based inversion models only work for smooth and continuous depth-dependent conductivity profiles. In the case of step conductivity profiles, increasing the number of iterations increases the numerical instability after the 0th iteration, i.e., after applying the simplistic approach, as shown in figure 4. This indicates that existing inversion models are not suitable for step conductivity profiling applications. In the proposed model, it is assumed that the coating and substrate conductivities are known or measured separately. Accordingly, a best-fit analysis is implemented between a measured or COMSOL simulated AECC spectrum over an unknown coating thickness with the results of from the plane-wave approximation using the secant method. In this case, the whole AECC spectrum and two initial coating thickness points i.e., initial thickness guesses, bracketing the thickness of interest are needed to use the plane-wave approximation until its AECC spectrum converges to the measured or simulated AECC spectrum using the secant method.

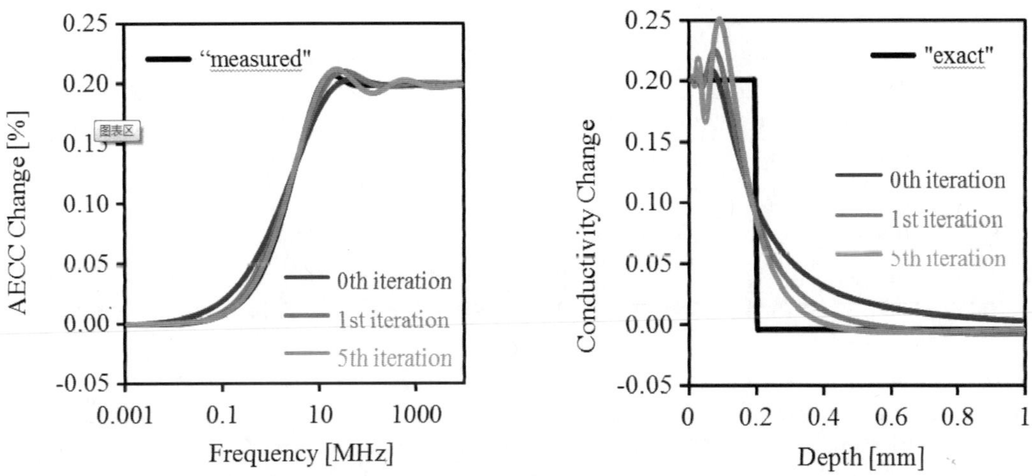

(a) (b)

Figure 4. A demonstration of existing AECC-based inversion models over (a) continuous and (b) step depth-dependent conductivity profiles.

3. Results and discussion

Figure 5 demonstrates the convergence of coating thickness estimation using the proposed AECC-based inversion model on a 0.2-mm titanium alloy coating thickness over a stainless steel substrate. It is worth mentioning here that it only takes few iterations to reach the accurate coating thickness. The estimated coating thickness was taken from a point measured or simulated at 25.4 μm lift-off, and delivered the accuracy shown in figure 5. To illustrate the robustness of the proposed method, figure 6 shows the average coating thickness estimates in a ±25.4 μm lift-off range. This is a significant improvement over other eddy current methods which only delivers a 13% uncertainty at fixed lift-off.

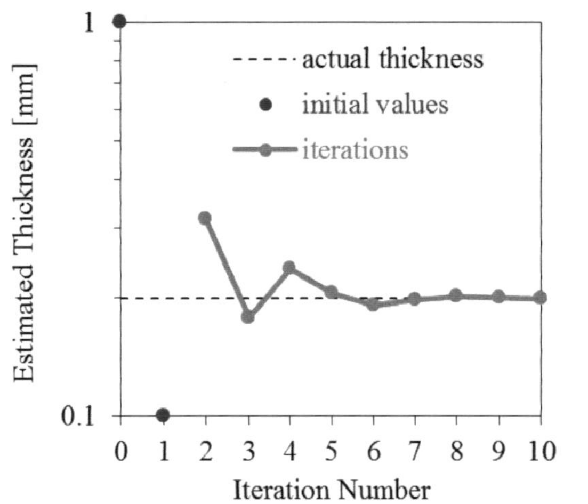

Figure 5. The convergence of coating thickness estimation using the secant method.

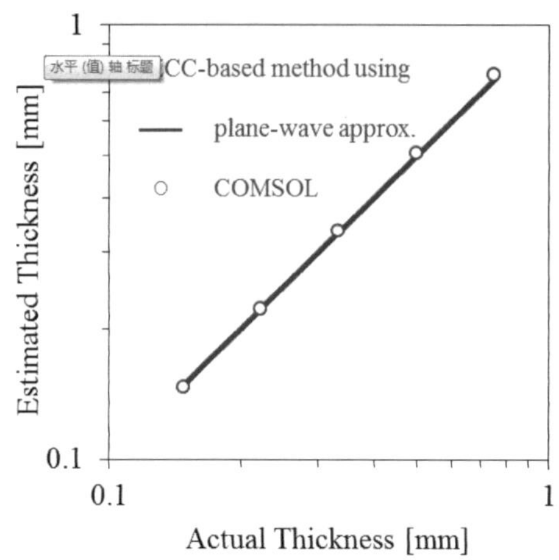

Figure 6. The lift-off effect on AECC-based estimates of coating thicknesses.

Table 1. Analysis of accuracy, precision and total uncertainty of the proposed method.

t_c	Lift-off (l) [μm] 0.0	12.7	25.4	38.1	50.8	Analysis			
	t_m					$(t_m)_{avg}$	B_x	P_x	w_x
[mm]	[mm]	[mm]	[mm]	[mm]	[mm]	[mm]	[mm]	[mm]	[%]
0.148	0.1460	0.1453	0.1486	0.1473	0.1467	0.1468	-0.0012	0.0036	2.56
0.222	0.2212	0.2202	0.2205	0.2216	0.2226	0.2212	-0.0008	0.0027	1.25
0.333	0.3330	0.3330	0.3328	0.3392	0.3328	0.3342	0.0012	0.0078	2.37
0.500	0.5054	0.5031	0.5031	0.5079	0.5124	0.5064	0.0064	0.0108	2.51
0.750	0.7674	0.7531	0.7672	0.7712	0.7682	0.7654	0.0154	0.0197	3.33

Table 1 illustrates a complete analysis of measured or estimated coating thicknesses (t_m) due to a ±25.4 μm lift-off range. Bias (B_x), precision (P_x) and total uncertainty (w_x) are included in this analysis. Since only 5 lift-off values are considered, Student's t distribution at 95% confidence level is used in assessing the measurement precision. The average overall uncertainty equals to approximately 2.40% relative to the actual thickness (t_c).

4. Conclusions

AECC spectrums over conductive layered structures were simulated using two different approaches, namely the direct and indirect approaches. The simple direct approach uses the plane-wave approximation, which offers high accuracy in estimating AECC spectrums over coating/substrate metals. The indirect approach using COMSOL simulations with a relatively large coil design was used to validate the plane-wave approximation over 50% coating/substrate conductivity variations with coating thicknesses of interest to the industry. It was demonstrated that existing simplistic and iterative AECC-based inversion models only work for smooth and continuous conductivity profiles and are unfit to capture rectangular conductivity profiles in metallic coatings over nonmagnetic metals. The proposed AECC-based inversion model leverages the accuracy offered using the plane-wave approximation by applying best-fit analysis using the secant method to deliver the best agreement with the indirectly measured or simulated AECC spectrums. The proposed method delivers a ±3%

uncertainty in estimating metal coating thickness over nonmagnetic metals measured in a ±25.4 μm lift-off range. This is a one order of magnitude improvement over the applicable ISO standard for the same lift-off range.

References

[1] Kawakita J, Katanoda H, Watanabe M, Yokoyama K and Kuroda S 2008 Warm spraying: An improved spray process to deposit novel coatings *Surf. Coatings Technol.* **202** 4369–73
[2] Molak R M, Araki H, Watanabe M, Katanoda H, Ohno N and Kuroda S 2014 Warm spray forming of Ti-6Al-4V *J. Therm. Spray Technol.* **23** 197–212
[3] Sudha C, Shankar P, Subba Rao R V, Thirumurugesan R, Vijayalakshmi M and Raj B 2008 Microchemical and microstructural studies in a PTA weld overlay of Ni–Cr–Si–B alloy on AISI 304L stainless steel *Surf. Coatings Technol.* **202** 2103–12
[4] ISO 21968 2005 Non-magnetic metallic coatings on metallic and non-metallic basis materials -- Measurement of coating thickness -- Phase-sensitive eddy-current method
[5] Moulder J C, Uzal E and Rose J H 1992 Thickness and conductivity of metallic layer from eddy current measurements *Rev. Sci. Instrum.* **63** 3455–65
[6] Dodd C V and Deeds W E 1968 Analytical solutions to eddy-current probe-coil problems *J. Appl. Phys.* **39** 2829–38
[7] Luquire J W, Deeds W E and Dodd C V 1970 Alternating current distribution between planar conductors *J. Appl. Phys.* **41** 3983–91
[8] Tai C, Rose J H and Moulder J C 1996 Thickness and conductivity of metallic layers from pulsed eddy-current measurements *Rev. Sci. Instrum.* **67** 3965–72
[9] Abu-Nabah B A and Nagy P B 2006 Iterative inversion method for eddy current profiling of near-surface residual stress in surface-treated metals *NDT E Int.* **39** 641–51
[10] Abu-Nabah B A and Nagy P B 2007 Iterative inversion method for eddy current evaluation of near-surface residual stress profile in surface-treated metals *Review of Progress in Quantitative Nondestructive Evaluation* **894** 310–8
[11] Abu-Nabah B A and Nagy P B 2007 High-frequency eddy current conductivity spectroscopy for residual stress profiling in surface-treated nickel-base superalloys *NDT E Int.* **40** 405–18
[12] Abu-Nabah B A and Nagy P B 2007 High-frequency eddy current conductivity spectroscopy for near-surface residual stress profiling in surface-treated nickel-base superalloys *Review of Progress in Quantitative Nondestructive Evaluation* **894** 302–9
[13] Abu-Nabah B A and Nagy P B 2007 Lift-off effect in high-frequency eddy current conductivity spectroscopy *NDT E Int.* **40** 555–65
[14] Abu-Nabah B A and Nagy P B 2008 Recent improvements in high-frequency eddy current conductivity spectroscopy *Review of Progress in Quantitative Nondestructive Evaluation* **975** 392–9
[15] Abu-Nabah B A 2017 Reduction of lift-off effect in high-frequency apparent eddy current conductivity spectroscopy *Meas. Sci. Technol.* **28** 55107
[16] Abu-Nabah B A and Nagy P B 2008 The feasibility of eddy current conductivity spectroscopy for near-surface cold work profiling in titanium alloys *Review of Progress in Quantitative Nondestructive Evaluation* **975** 1228–35
[17] Abu-Nabah B A, Yu F, Hassan W T, Blodgett M P and Nagy P B 2009 Eddy current residual stress profiling in surface-treated engine alloys *Nondestruct. Test. Eval.* **24** 209–32
[18] Abu-Nabah B A, Hassan W T, Ryan D, Blodgett M P and Nagy P B 2010 The effect of hardness on eddy current residual stress profiling in shot-peened nickel alloys *J. Nondestruct. Eval.* **29** 143–53
[19] Blodgett M P and Nagy P B 2004 Eddy current assessment of near-surface residual stress in shot-peened nickel-base superalloys *J. Nondestruct. Eval.* **23** 107–23
[20] Yu F and Nagy P B 2004 Simple analytical approximations for eddy current profiling of the near-surface residual stress in shot-peened metals *J. Appl. Phys.* **96** 1257–66

IOP Conf. Series: Materials Science and Engineering **372** (2018) 012038 doi:10.1088/1757-899X/372/1/012038

The investigation of strength and plasticity mechanism of low-temperature annealed ultrafine grained stainless steel

Q H Pang[1,2], W J Li[1,2,3], M Y Cai[1,2], H Qi[1,2], C C Zhang[1,2] and J N Wu[1,2]

[1]School of Materials and Metallurgy, University of Science and Technology Liaoning, Anshan Liaoning, 114051, China
[2]State Key Laboratory of Metal Material for Marine Equipment and Application, Anshan Liaoning, 114009, China

E-mail: liweijuan826@163.com

Abstract. The properties of steel material can be greatly improved when its grain size becomes finer, so grain refinement is the most commonly used to improve the strength, ductility and impact on the toughness of steel at the same time. Compared to ordinary fine-grained steel, the average grain size of ultra-fine grained steel is less than 2 μm and it has different strength and plasticity mechanism. In this paper, effect of low-temperature annealing on the microstructure and properties of ultra-fine grained stainless steel in internal friction was studied by utilizing transmission electron microscope (TEM) and universal tensile testing machine. At the same time, the strength and plasticity mechanism of ultra-fine grained stainless steel was revealed by internal friction experiments. The results show that with the increase of annealing time and annealing temperature, both the tensile strength and yield strength decreases, while the elongation increases continuously. The mechanical properties of tested steel are more sensitive to the temperature. In the internal friction curve of tested steel, the peak of P_1 which result from the interaction between solid solution atoms and dislocations increases first and then decreases, while the peak of P_2 which result from the interaction between dislocation and carbide precipitates decreases first and then increases. The peak of P_3 which result from the structure, area and movement of grain boundary decreases continuously. The strength and plasticity mechanism of tested steel at room temperature is the result of interaction between the grain boundary and dislocation.

1. Introduction

When the average grain size of material is between 100 nm and 2 μm, we ordinarily call it ultrafine-grained material. Due to its high fracture strength, high specific strength, high specific stiffness and excellent wear resistance, it has been more and more widely applied in transportation, aerospace, electrical and electronic fields [1-3]. The methods of ECAP (equal channel angular pressing) [4], HPT (high pressure torsion) [5], electrodeposition [6], cyclic extrusion-upsetting [7] and powder metallurgy [8] etc can prepare ultrafine grained material, but these methods are difficult to produce ultrafine-grained materials as components over 30 cm diameter or length. At present, the forging (or rolling) with large deformation and then recrystallization annealing is used to achieve large size ultrafine grained material [1,9,10], so the regulation of its thermal stability is the key factor to the success preparation of ultrafine-grained material. Numerous studies [11-14] have shown that the key influence factors of thermal stability in ultrafine grained material are the heating temperature and heating time.

The heating temperature directly affects the activation energy of grain boundary diffusion. When the ambient temperature exceeds a certain temperature (critical temperature), the grain size will increase exponentially with the increase of temperature. On the other hand, the heating time affects the degree of grain growth. Fereshteh E [15] prepared nanoscale nickel base alloy with grain sizes of 11 nm, 59 nm and 278 nm at 250°C, 400°C and 500°C annealing 90 min, respectively.

Compared with ordinary steel, the strength and hardness of ultrafine-grained steel increased significantly, but its elongation decreased significantly. Even for steel with good plasticity under the conventional grain size, its elongation at break was less than 5% [6,16]. For instance, the grain size of ferrite in carbon steel with 0.55% carbon content was 1.7 μm~30 nm. Its strength increased gradually with the grain size refined, but the plasticity decreases obviously [17]. For IF steel whose the average grain size is in 12 μm~0.45 μm, the stress-strain curve changed from continuous yielding to discontinuous yielding with the grain refined. When the grain size was less than 1 μm, the elongation after fracture decreased suddenly [18]. When the average grain size of ferrite in medium carbon steel was less than 150 nm, there was no necking phenomenon before fracture, and the fracture would suddenly occurr [19]. Therefore, the research of strength and plasticity mechanism in ultrafine-grained material is a very important research content, and further investigation is required.

The change of point, line and surface defect in the material inside will directly determine the plastic deformation behavior during uniaxial tensile test. In this paper, the effect of different annealing processes on microstructure, mechanical properties and internal friction of ultrafine-grained stainless steel was studied. According to the spectrum of internal friction, the relationship between microcosmic state (such as solid solution atoms, dislocation and grain boundary) and mechanical properties was analyzed. Meanwhile, the strength and plasticity mechanism of ultrafine-grained stainless steel was revealed.

2. Experimental work

Experimental materials are ultrafine-grained cold rolled stainless steel provided by a factory, with its chemical composition as shown table 1. Samples were heated to 300°C, 350°C, 400°C and 500°C, and held 20 min, 30 min, 60 min and 120 min respectively, to anneal. Finally, the samples were cooled down slowly to room temperature in an insulated blanket.

Table 1. Chemical composition of tested steels (wt%).

No.	C	Si	Mn	P	S	Ni	Cr	Mo	V	Nb
Tested steel	0.055	0.4	1.63	≤0.035	≤0.03	8.45	17.3	0.12	0.08	0.04
316L	≤0.03	≤1.00	≤2.00	≤0.035	≤0.03	10.0-14.0	16.0-18.0	2.0-3.0	—	—

The samples for transmission electron microscope were cut from the annealed sample and grinded to 50 μm thick foils, and then twin-jet electropolished with 5% perchloric acid alcohol solution (volume fraction) at 35~45 V. The microstructures were analyzed using JEM-2100 transmission electron microscope (TEM). The tensile load was applied with a constant strain rate of 5×10^{-3} m/sec. From the engineering stress strain curve, the yield strength was calculated by 0.02% offset method.

Internal friction experiments were performed in MFP-1000 multifunctional internal friction instrument. The internal friction-temperature spectrum was measured by the free attenuation method, and free attenuation frequency was 2.78 Hz. The actual friction curve can be obtained by the deduction of background internal friction [20], and the relaxation activation energy of actual internal friction peak are measured according to equation (1).

$$H = RT_m \ln(K_B T_m / hf_m) + T_m \Delta S \tag{1}$$

Where T_m means peak temperature, f_m means peak frequency, R means the gas constant, K_B means the boltzmann constant, h means the Panck constant, ΔS means entropy change whose value is 1.1×10^{-4} Ev/k.

3. Experimental results

3.1. Mechanical properties

Figure 1 shows the stress-strain curve of ultrafine-grained stainless steel annealed at 400°C and soaked different time. Initial tested steel has the maximum tensile strength and yield strength, and the minimum elongation after fracture as illustrated in figure 1(a). Their values are respectively 1403 MPa, 893 MPa and 11%. With the extension of tempering time, the strength and yield strength of tested steels both decrease, and the decrease degree of tensile strength is greater than that of yield strength. While the elongation after fracture steeply increases at the beginning, and then the raise of them becomes progressively slower as illustrated in figure 1(b).

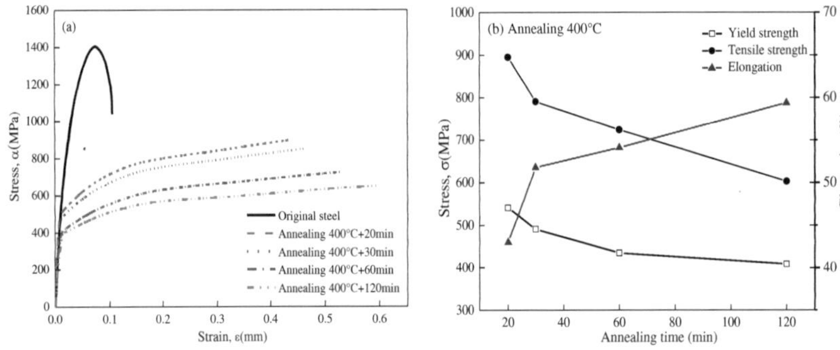

Figure 1. Tensile stress-strain curves of tested steels with different annealing time.

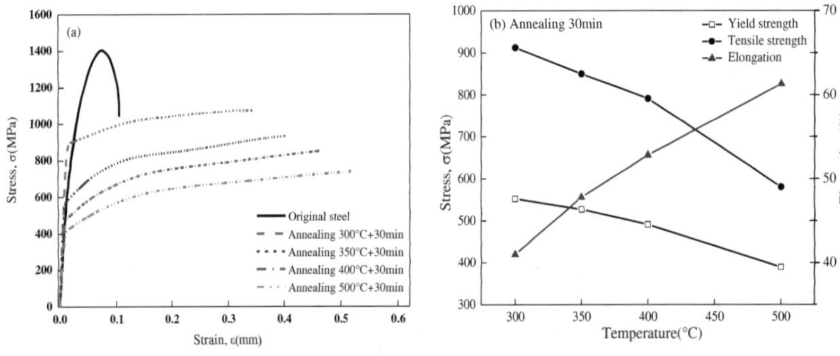

Figure 2. Tensile stress-strain curves of tested steels with different annealing temperatures.

Figure 2 shows the stress-strain curve of ultrafine-grained stainless steel annealed at different temperature and soaked 30 min. As shown in figure 2, the stress-strain curves of tested steel are continuous yield under different annealing temperatures, and there is no obvious yield platform. The change regularity of their strength and elongation is that the tensile strength and yield strength decrease gradually with the increase of annealing temperature, while the elongation increases gradually. Compared with the stress-strain curves of tested steel at different annealing time, the curve of tested steel is more dispersed at different annealing temperatures. This indicates that the mechanical property of tested steel is more sensitive to the temperature change than soaked time in the low temperature annealing process.

3.2. Microstructure characterization

Figure 3 shows the microstructure of tested steels with different annealing processes obtained from the TEM images. It can be seen that the microstructure of initial stainless steel was mainly made of fine polygon austenite structure, and the boundary of that was clear as illustrated in figure 3(a). There are low dislocation density and some high angle boundaries in the austenite, due to the contrast image of adjacent grains in TEM image. The average grain size of initial stainless steel microstructure is (100±20) nm measured by Image Pro Plus. At this size level of the grain, the tested steel exhibited "the brittle fracture behavior" during the uniaxial tension test. It is because when the grain size is very small, the number of grain boundaries increases significantly. Moreover, the simultaneous slip probability of grain boundaries in the same direction may decrease, so the mobility of grain boundaries will be weaken. Due to the decreasing of grain diameter, the removable distance of dislocation is decreasing, which may limit the mobility of dislocation. Therefore, the pinning effect of grain boundary on dislocation is enhanced [14,19]. Figures 3(b)-3(e) show the microstructure of tested steels annealed at 300°C, 350°C, 400°C and 500°C respectively, and soaked 30 min. There was no obvious change in the microstructure and phase constituents of tested steels, but the grain size of them obviously increased. The effective average grain size is respectively (0.4±0.03) μm, (0.6±0.03) μm, (0.9±0.03) μm and (1.5±0.03) μm measured by Image Pro Plus. A few dislocation lines and dislocation tangle can be observed near the grain boundary. Simultaneously, the grain boundary in local area became blurred and a small amount of annealing twins appeared. This illustrates that the grain boundary migration and grain growth occured in low temperature annealing process.

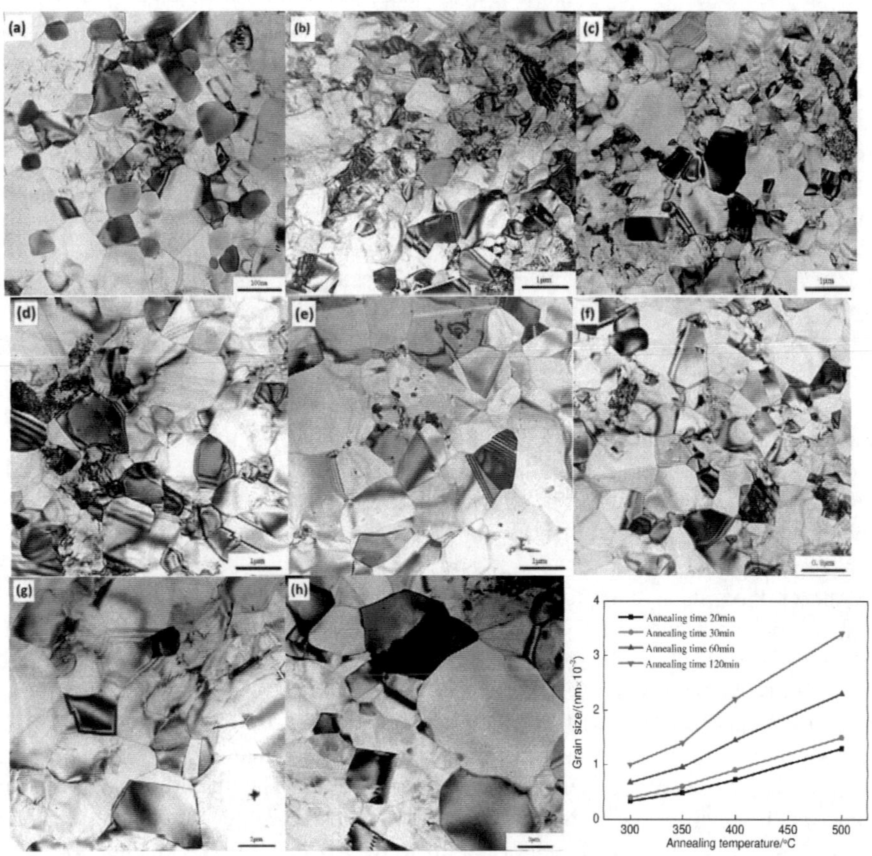

Figure 3 TEM image and grain size of the tested steel with different annealing process: (a) Original steel; (b) 300°C +30 min; (c) 350°C +30 min; (d) 400°C +30 min; (e)500°C +30 min; (f)400°C +20 min; (g) 400°C +60 min; (h) 400°C +120 min.

Figures 3(f)-3(h) show the microstructure of tested steels annealed at 400°C and soaked 20 min, 30 min, 60 min and 120 min. Their effective average grain size are respectively (0.7±0.03) μm, (0.9±0.03) μm, (1.3±0.03) μm and (2.2±0.03) μm. It can be seen from the statistical results of effective average grain size, the change of grain size is more sensitive to the change of annealing temperature. After annealing treatment, the strength of tested steel decreased with the increase of grain size, and the yield strength meeted the Hall-Petch relationship. Moreover, the plasticity of tested steel significantly increased, and this is in accordance with the results of the literature [14,16-19]. It indicates that the strengthening and toughening mechanism of classic fine grain strengthening will no longer be applied to ultrafine grain materials, so it is necessary to reveal the strength and plastic mechanism of ultrafine-grained materials in the micro level of grain boundary and dislocation.

3.3. Internal friction-temperature spectrum

Figure 4(a) shows the internal friction spectrum of tested steel annealed at 400°C with different annealing time. The background internal friction was deducted from internal friction-temperature spectrum as shown in figure 4(b), so it can be seen that the actual P_1 peak increased first and then decreased with the extension of tempering time. When the annealing times were 60 min and 120 min, the values of P_1 peak were maximum and minimum, respectively. According to the calculated internal friction peak activation energy and characteristic parameters [20-22], P_1 peak is the Snoek peak which is caused by the interaction between solid solution atoms and dislocation. In the initial stage of annealing, solid solution atoms migrated to the dislocations, so the concentration of it gradually increased near dislocations. The continuous diffusion of solid solution atoms in this process would lead to the enhancement of relaxation phenomenon of P_1 peak. When the segregation of solid solution was saturated, the value of P_1 peak was maximum. With the extension of tempering time, the pinning effect of a large number of solid solution atoms on the dislocation began to appear, so the interaction between the solid solution atoms and dislocations was enhanced to result in the decrease of P_1 peak value. Figure 4(c) is the actual curve of P_2 peak, the peak height change of it is different from the P_1 peak. It was shown that the P_2 peak height decreased first and then increased with the extension of tempering time. It is because that P_2 peak is the SKK peak which is caused by the common interaction between dislocations, or dislocation and precipitated carbide [23,24]. When the solid solution atoms began to saturate at the position of dislocations, the second phase would precipitate to pin the dislocations. It resulted in the decrease of the P_2 peak height. With the extension of annealing time, the pinning effect weakened due to the second phase precipitation coarsening, and the interaction between dislocations would play a leading role at this time. It led to the rise of P_2 peak height. The P_3 peak is Kê peak which is caused by the interaction of grain boundary structure, grain boundary area and grain boundary movement [25]. With prolonging annealing time, the grain size of tested steel constantly increased, and the number of grain boundaries decreased. So the P_3 peak height had reduced at all times as illustrated in figure 4(d). It can be shown from figure 4 that the peak temperature of P_1, P_2 and P_3 peak moved right with the prolonging annealing time, except for the peak temperature of P_2 peak left shift with annealing 120 min.

Figure 5 shows the free attenuation internal friction curve of tested steel annealed 60 min at different temperature. Each internal friction curve has three internal friction peaks (the P_1, P_2 and P_3 peak), and the curves of P_1, P_2 and P_3 peak at different temperatures were obtained after deducting the background as shown in figures 5(b)-5(d). As shown in figure 5(b), at first the higher the annealing temperature is, the higher the value of P_1 peak height is. It is because that the annealing temperature is higher, solid solution atoms may get more thermal excitation energy. The time used for the aggregation of solid solution atoms near dislocations is shorter. Thus it will cause the effect of solid solution atoms on the dislocation mobility to take place earlier. When the annealing temperature is more than 400°C, the value of P_1 peak height begins to decrease. The reason is due to the reduction of dislocation density or the excessive concentration of solid solution atoms near the dislocations, the pinning effect on the dislocation may increase, which reduces the mobility of dislocations. The curve of P_2 peak height first decreases and later increases with the increase of annealing temperature as

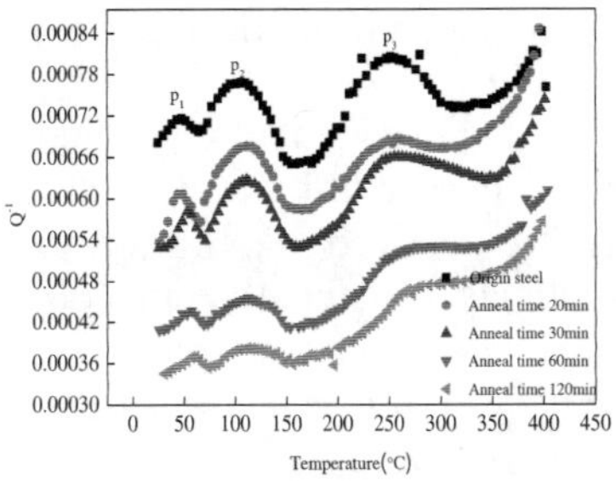

(a) different annealing time

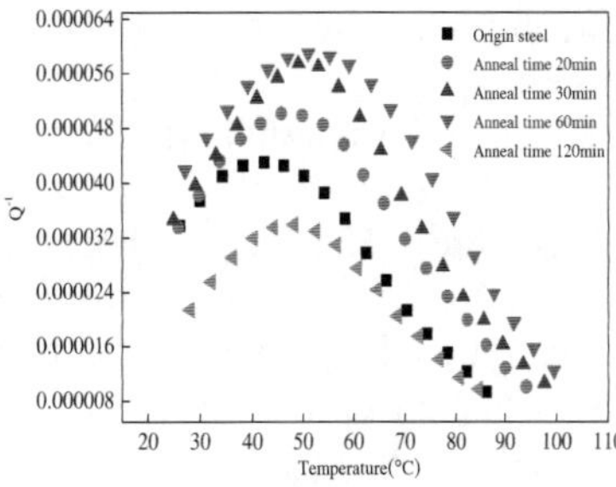

(b) the P_1 peak

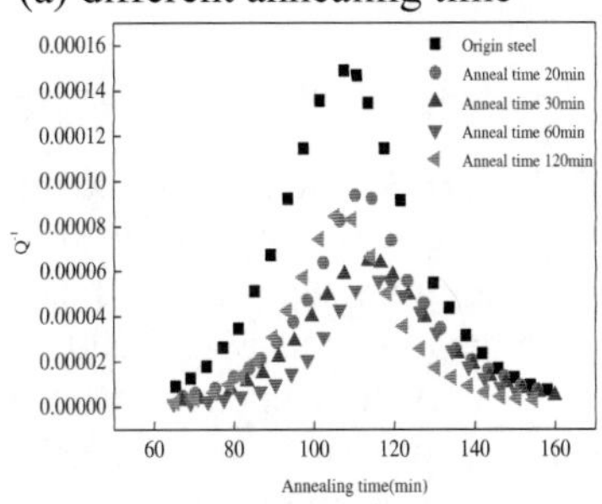

(c) the P_2 peak

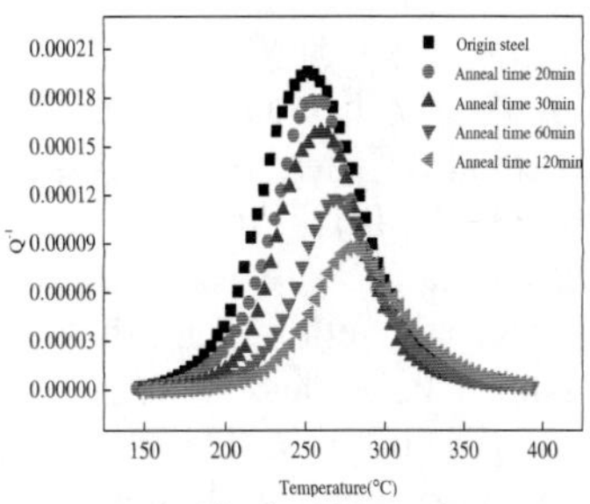

(d) the P_3 peak

Figure 4. The curves of free damping internal friction with different annealing time.

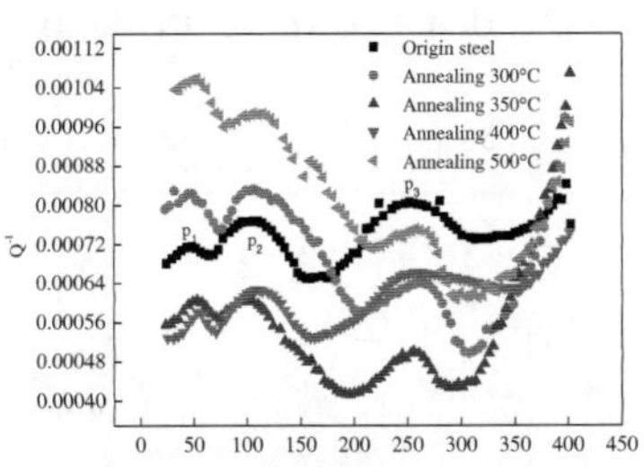

(a) different annealing temperature

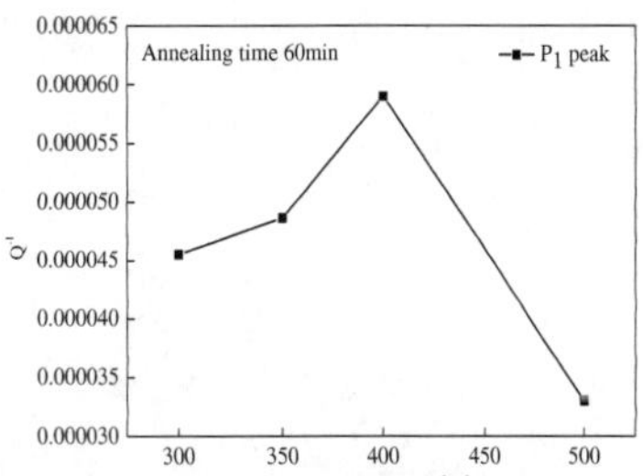

(b) the P_1 peak

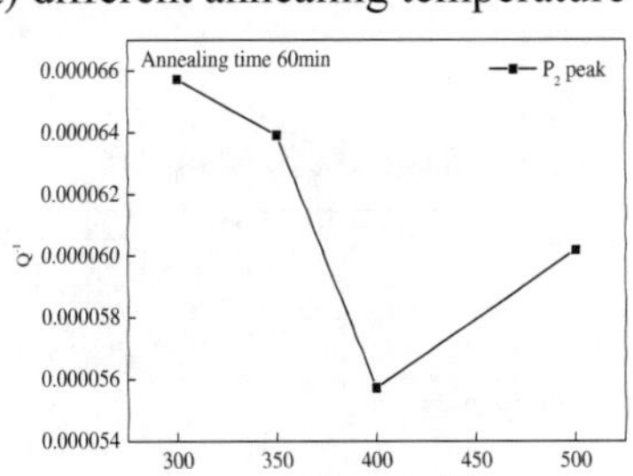

(c) the P_2 peak

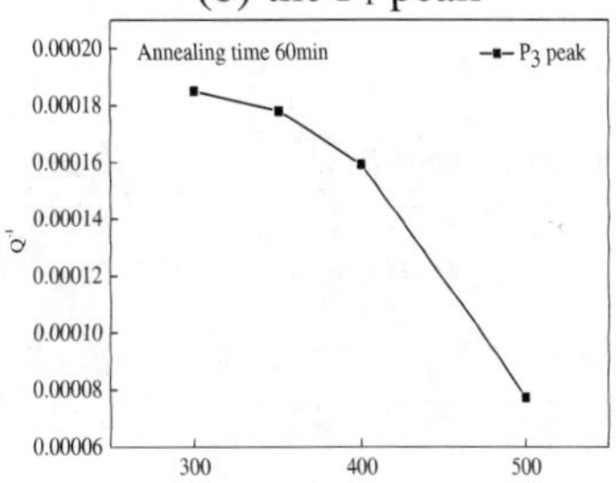

(d) the P_3 peak

Figure 5. The curves of free damping internal friction at different annealing temperature.

shown in figure 5(c), when the peak value reached the minimum at 400°C. That is because that the carbide which pinned dislocation movement precipitated faster with the higher annealing temperature. As annealing the temperature continues to rise, it may miss the optimum temperature of carbide precipitation "tip temperature". So the carbide precipitation slows down, the carbides begin coarsening, which results in the pinning force on dislocations decrease. It can be seen from figure 5(d) that the curve of P_3 peak height assumes first slow decreasing and then drastic declining with the increase of annealing temperature.

In order to investigate the internal friction caused by grain boundary relaxation more accurately, the P_3 peak of tested steel under different grain sizes was measured by forced vibration method. The vibration frequencies were 1 Hz, 2 Hz, 4 Hz and 6 Hz, respectively. Figure 6 is the internal friction curve of tested steels whose average grain size is respectively 100 nm, 400 nm, 600 nm, 1300 nm and 2800 nm at 2 Hz vibration frequency. Peak temperature of the internal friction curve is 604.96°C, 612.96°C, 616.06°C, 626.58°C and 641.26°C, respectively. Using transform frequency method and Arrhenius formula [26] to calculate the activation energy of internal friction peak:

$$\tau = \tau_0 e^{H/(k_B T)} \tag{2}$$

where, τ is relaxation time; τ_0 is pre exponential factor.

The calculated activation energies are respectively 197.18 kJ/mol, 201.03 kJ/mol, 207.76 kJ/mol, 219.30 kJ/mol and 231.81 kJ/mol. It can be seen from figure 6(a) that the starting temperature which internal friction peak began to rise was determined by the grain size of tested steel. For tested steel with 100 nm grade internal friction peak, the starting temperature was nearly 100°C lower than that of 2800 nm grade tested steel. This is because that the activation energy of internal friction peak increases with the increase of grain size. When the grain size of tested steel is 2800 nm, it needs higher temperature to meet the demand of larger activation energy. Moreover, the peak height of internal friction peak decreases with the increase of grain size, and that of 2800 nm grade tested steel decreases sharply. It is because that the grain size level has exceeded the scope of ultrafine grain steel. Overall, about the internal friction caused by grain boundary sliding, the grain size is smaller, the activation energy at the beginning of relaxation is smaller, and the distance of grain boundary movement is shorter.

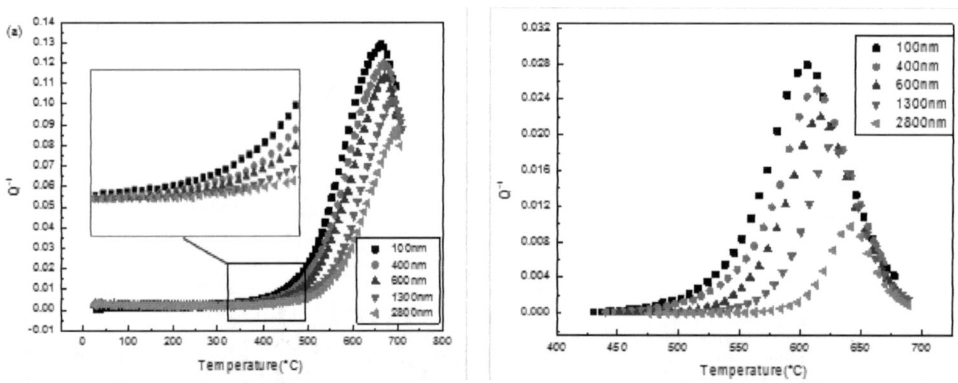

Figure 6. The curves of free damping internal friction with different grain size: (a) the curves of forced vibration internal friction; (b) the P_3 peak.

4. Analysis and discussion

The strength and plasticity of metal materials is mainly affected by dislocation movement. When the solid solution atom, the second phase or the grain boundary pins the dislocation to make it impossible to move freely, the macroscopic mechanical properties reflected by this micro-mechanism of interaction is the strength increase of material. Conversely, the dislocation is easier to move, and the plasticity of material is better [26-28]. The strength and plasticity mechanism of ultrafine-grained stainless steel in this paper are as follows: with the increase of grain size, the tensile and yield strength decreases, and the elongation increases gradually as shown in figures 1 and 2. It seems to be contrary to the classic "fine-grained strengthening mechanism" for ultrafine grained metals. Due to a substantial increase in the number of grain boundary, the interaction between dislocations and dislocation, or grain boundary and dislocation, or the movement of dislocation and grain boundary will produce different synergy mechanism.

Figures 4(b) and 5(b) shows the peak height of P_1 peak first increases and then decreases with the increase of annealing time and annealing temperature. In essence, different annealing processes change the grain size of tested steel, when the grain size is less than 900 nm, with the increase of grain size, the migration distance of solid solution atom to dislocation may increase. It leads to an increase in the peak height of P_1 peak. When the grain size reaches a critical value between 900~1300 nm, the migration of solid solution atom to dislocation will reach critical value, and the peak height no longer increases with the increase of the grain size. At this time, the segregation of solid solution atom plays a leading role in the dislocation pinning, and the mobility of dislocation became weakened. This leads to the decrease of P_1 peak height. The peak height of P_2 peak first decreases and then increases with the increase of annealing time and annealing temperature as shown in figures 4(c) and 5(c). It indicates that when the grain size is less than 900 nm, the number of dislocation and carbides in the unit grain may increase with the increase of grain size. The interaction between dislocation and dislocation may increase, which leads to the decrease of P_2 peak height. However, when the grain size is larger than 900 nm, the movement distance of dislocation increases with the increase of grain size. Due to the coarsening of the second phase carbide, the effect of pinning dislocation is weakened, which led to the P_2 peak height rising. For the internal friction (P_3 peak) caused by the grain boundary relaxation, the peak height decreases with the increase of grain size, and the peak temperature shifts to the right as shown in figures 4(d) and 5(d). There are two main factors affecting the grain boundary relaxation [25]: One factor is the boundary itself, including the geometry configuration of grain boundary, such as the Orientation and mismatch relation on both sides of grain boundary. The other factor is external factors, such as the adsorption of impurities, etc. All of these may affect the mobility of grain boundary. Mori T's research [29] shows the relaxation time of grain boundary increases with the increase of grain size, that means, the distance of grain boundary sliding in the relaxation time increases with the increase of grain diameter. This is in accordance with the regularity shown in figure 6, the critical temperature when the internal friction of grain boundary begins to increase rapidly increases with the increase of grain size. Therefore the larger the grain size, the greater the activation energy required for grain boundary sliding, namely, the greater the sliding resistance.

According to present study the strength and plasticity mechanism of ultrafine-grained stainless steel does not only depend on the interaction between dislocations, but also the interaction between grain boundary and dislocation might effect as well. When the grain size is below 900 nm, the dislocation and grain boundary are easy to be pinned and hard to move. The strength of material is high and the plasticity is poor. As the grain is coarsening (more than a critical size within 900~1300 nm), although the resistance of grain boundary sliding increases, the increase of the grain boundary number and grain diameter may lead to the weakening of pinning effect on dislocation. In addition, the distance of grain boundary sliding increases with the increase of grain diameter, these make the movement of grain boundary become enhanced. So the strength of material decreases and the plasticity increases.

5. Conclusions

With the increase of annealing temperature and annealing time, the tensile strength and yield strength of tested steel decrease, and the elongation of tested steel increases gradually. Moreover, the mechanical property of tested steel is more sensitive to the temperature change than soaked time in the low temperature annealing process. In order to obtain better plasticity in stainless steels 400°C and 60 min annealing is recommended. The P_1 peak of tested steel first rises and then decreases, the P_2 peak first decreases and then increases, and the P_3 peak decreases constantly. In essence, the strength and plasticity mechanism of ultrafine-grained stainless steel is not only the interaction between dislocations, but also the interaction between grain boundary and dislocation should be considered.

Acknowledgments

Supported by National Nature Science Foundation of China (No. 51274121); Supported by State Key

Laboratory of Marine Equipment made of Metal Material and Application (No. SKLMEA-USTL-201708); National Nature Science Foundation of USTL (No. 2017QN11).

References

[1] Yin Y Q, Wu C L, Xie P, Zhu K, Tian S L, Han M and Chen J H 2016 *J. Acta Metall. Sin.* **12** 1527-35
[2] Yang G, Liu Z D, Gong Z H, Wang L M, Lin Z J and Cheng S C 2006 *J. Iron steel Res.* **18** 32-6
[3] Muley S V, Vidvans A N, Chaudhari G P and Udainiya S 2016 *Acta Biomater* **30** 408-19
[4] Tang L L, Zhao Y H, Islamgaliev R K, Chi Y A, Valiev R Z, Lavernia E J and Zhu Y T 2016 *KnE. Mater. Sci.* **1** 168-71
[5] Song Y P, Chen M M, Gao D S and Wang W 2017 *J Hot Working Technol.* **3** 6-9
[6] Zhang Q, Liu Y, Liu Y S, Ren Y H, Wu Y X, Gao Z P, Wu X L and Han P D 2017 *J. Mater Sci Eng. A* **701** 196-202
[7] Wang Q, Mu Y, Lin J, Zhang L and Roven H J 2017 *J. Mater. Sci. Eng.* **699** 26-30
[8] Wang X 2013 Research on mechanical properties of ultra-fine grain AZ31 magnesium alloy prepared by powder metallurgy (Harbin: Harbin Institute of Technology)
[9] Qian T, Karaman I and Marx M 2014 *Adv. Eng. Mater.* **16** 1323-39
[10] Morris D G, Gutierrez-Urrutia I and Muñoz-Morris M A 2008 *J Mater. Sci.* **43** 7438-44
[11] Lan H, Liu W J and Liu X 2008 *Chin. J. Mater. Res.* **22** 279-86
[12] Etienne A, Radiguet B, Genevois C, Breton J M, Valiev R and Pareige P 2010 *Mater. Sci. Eng.* **527** 5805-10
[13] Park K T, Kim Y S, Lee J G and Dong H S 2000 *Mater. Sci. Eng.* **293** 165-72
[14] Xu D M, Li G Q, Wan X L, Xiong R L, Xu G, Wu K M, Somani M C and Misra R D K 2017 *Mater. Sci. Eng. A* **688** 407-15
[15] Ebrahimi F and Li H 2007 *J Mater Sci.* **42** 1444-54
[16] Zheng Z G, Xie C J, Sun T and Yuan S 2014 *Mater. Mech. Eng.* **50** 77-83
[17] Rodríguez-Baracaldo R, Benito J A and Cabrera J M 2010 *J. Mater. Sci.* **4517** 4796-804
[18] Gao S, Chen M, Joshi M, Shibata A and Tsuji N 2014 *J. Mater. Sci.* **49** 6536-42
[19] Haddad M, Ivanisenko Y, Courtois-Manara E and Fecht H J 2015 *Mater. Sci. Eng.* **620** 30-5
[20] Li W J, Zhang H Y, Fu H, Zhang J and Qi X Y 2015 *Acta Metall Sin.* **51** 385-92
[21] Wang H, Li W, Hao T, Jiang W B, Fang Q F, Wang X P, Zhang T, Zhang J, Wang K and Wang L 2017 *Mater. Sci. Eng.* **695** 193-8
[22] Zeng B, Wu J, Shen B and Zhang H 2011 *Shanghai Met.* **33** 18-21
[23] Ge T S and Wang Q M 1955 *Acta Phys. Sin-Ch Ed.* **11** 387-402
[24] Xu L Y, Ji J, Huang J, Zhu Z H, Wu Y W and Wang H B 2016 *Phys Testing Chem. Anal. Part A* **52** 156-8
[25] Ge T S 2014 *Solid Internal Friction Theoretical Basis* (Beijing: Science Press) pp 375-425
[26] Shi D K 1988 *Dislocation and Material Strength* (Xi'an: Xi 'an jiaotong univ Press) pp 253-70
[27] Yu M H 2011 *New System of Strength Theory* (Xi'an: Xi 'an jiaotong univ Press) pp 62-83
[28] Hussein A M, Rao S I, Uchic M D, Parthasarathay T A and Ei-awady J A 2017 *J. Mech. Phys. Solids* **99** 146-62
[29] Mori T, Koda M, Monzen R and Mura T 1983 *Acta Metall.* **31** 275-83

IOP Conf. Series: Materials Science and Engineering **372** (2018) 012039 doi:10.1088/1757-899X/372/1/012039

Shear properties of pultruded fiber reinforced polymer composite materials

J H Seo[1], S H Kim[1,2], D M Ok[1], D J An[3] and S J Yoon[1,4]

[1]Department of Civil Engineering, Hongik University, 94, Wausan-ro, Mapo-gu, Seoul, 04066, Korea
[2]Department of Architectural Engineering, Gachon University, 1342 Seongnam-daero, Sujeong-gu, Seongnam-si, Gyeonggi-do, 13120, Korea
[3]Woojoo Enginerring, 38, Heungan-daero, Dongan-gu, Anyang-si, Gyeonggi-do, 14059, Korea

E-mail: sjyoon@hongik.ac.kr

Abstract. This paper focuses on the mechanical properties of PFRP composite materials. Especially, relationship between shear property and the other mechanical properties of PFRP composite materials is investigated through comparison between experimental and theoretical results. The shear property of PFRP composite specimen is calculated from the theoretical equations which were suggested in previous studies. In addition, comparison between the shear property determined by the tensile test and the shear property calculated from theoretical equations is conducted and discussed. It was found that the theoretically predicted shear modulus of elasticity considering contiguity is close to the shear modulus of elasticity obtained by the 45° off-axis tensile test.

1. Introduction

The pultruded fiber reinforced polymer plastic (PFRP) is one of the most actively studied composite materials for the structural member in construction industries [1]. Pultrusion is a continuous manufacturing process for mass production with good mechanical properties. The PFRP is considered as an orthotropic (transversely isotropic) material because fibers are placed to the longitudinal direction of the member [2].

The performance of an engineering material is often measured by the mechanical properties and static behavior such as tensile, compressive, shear or dynamic behavior in both normal and adverse test environments. This information of performance is important to take the proper material in a given application and designing a structure with the material [3].

The stiffness properties can be determined by experimental measurements, but an experimental set of measurements determines the properties of a fiber-matrix system produced by a single manufacturing process. When any change in the system variables occurs, additional measurements are required. These experiments may become time consuming and cost prohibitive [4]. Therefore, it is necessary to study the prediction of material properties of fiber reinforced composites.

This study focuses on the mechanical properties of PFRP composite materials. Especially, relationship between shear property and the other mechanical properties of PFRP composite materials is investigated through comparison between experimental and theoretically predicted results.

IOP Conf. Series: Materials Science and Engineering **372** (2018) 012039 doi:10.1088/1757-899X/372/1/012039

2. Theoretical calculation method tensile test of PFRP composite

To determine the mechanical properties of PFRP structural members, PFRP composite coupon specimens are used in the test. Width of PFRP composite coupon specimens is 600 mm and thickness of PFRP composite coupon specimens is 5 mm. Weight ratio of fiber is 60% and weight ratio of matrix is 40%. Typical Poisson's ratio of fiber and matrix is adopted from the reference [5]. Properties of PFRP composite coupon specimens are given in table 1.

Table 1. Properties of PFRP composite.

Properties of PFRP composite coupon specimen			
Width (mm)	600	Modulus of elasticity of fiber (E_f, GPa)	72.50
Thickness (mm)	5	Modulus of elasticity of matrix (E_m, GPa)	4.00
Weight ratio of fiber (%)	60.00	Shear modulus of elasticity of fiber (G_f, GPa)	30.21
Weight ratio of matrix (%)	40.00	Shear modulus of elasticity of matrix (G_m, GPa)	1.50
Volume ratio of fiber (V_f, %)	41.19	Poisson's ratio of fiber (ν_f)	0.22
Volume ratio of matrix (V_m, %)	58.81	Poisson's ratio of matrix (ν_m)	0.33

To determine the tensile strength and modulus of elasticity in longitudinal direction and transverse direction, tensile tests in longitudinal direction and transverse direction are conducted. There are several test methods of shear test of composite materials such as ±45° off-axis tensile shear test, Iosipescu shear test, V-notched rail shear test, etc. The ±45° off-axis tensile shear test is one of the most popular testing methods for the pultruded composite structural profile. To determine the shear strength and shear modulus of elasticity, 45° off-axis tensile tests are conducted. The test information and stress-strain curve are detailed in table 2 and summary of test results is given in table 3.

Table 2. Test specimen dimension and stress-strain curve.

Test	Specimen dimension (Length*width*thickness)	Stress-strain curve
Tensile test in longitudinal direction (ASTM D 3039/D 3039 [6])	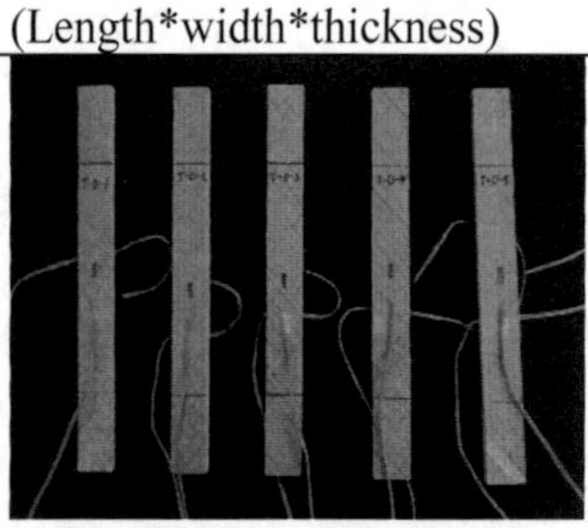 250mm*25mm*5mm	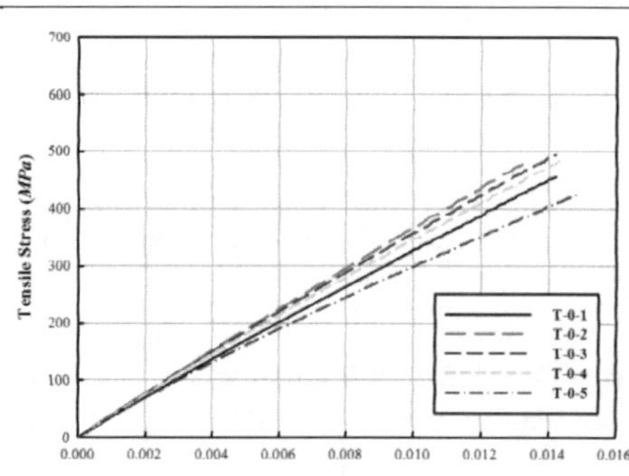

Tensile test in transverse direction (ASTM D 3039/D 3039 [6])

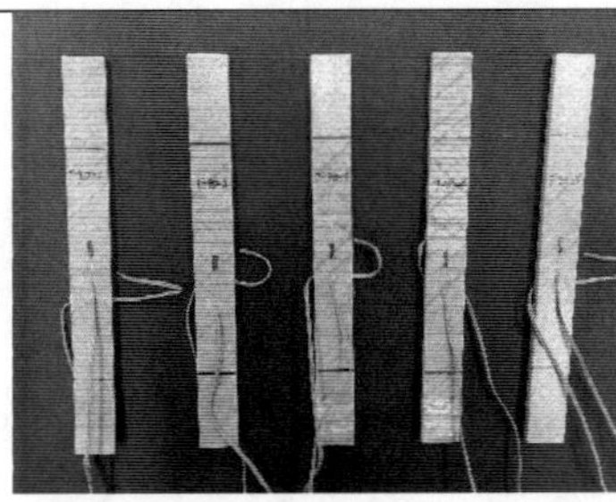

250mm*25mm*5mm

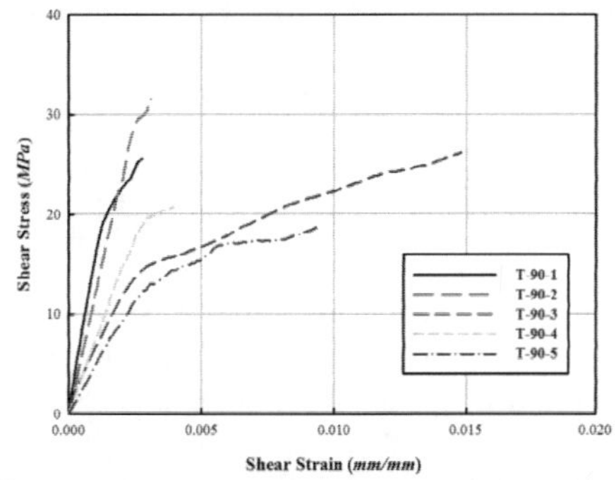

Tensile test in 45° off-axis direction (ASTM D 3518/D 3518 [7])

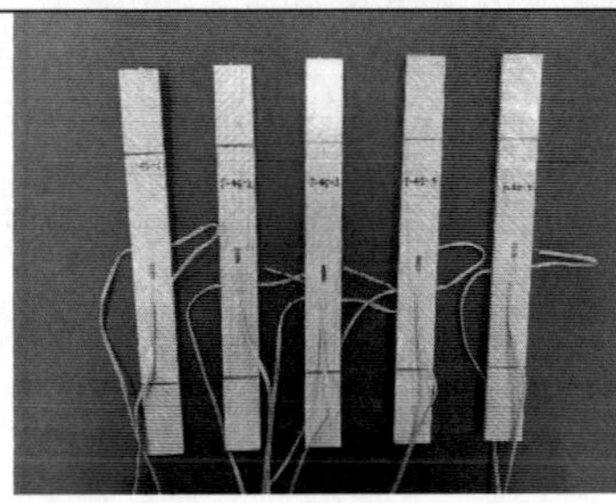

250mm*25mm*5mm

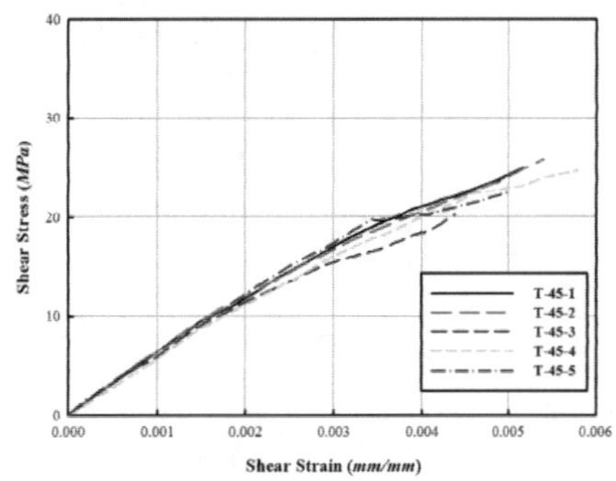

Table 3. Summary of average test results.

Test	Maximum stress (MPa)	Modulus of elasticity (GPa)	Poisson's ratio (ν_{12})
Tensile test in longitudinal direction	496.30	36.39	0.33
Tensile test in transverse direction	23.86	9.89	-
Tensile test in 45° off-axis direction	24.50	5.13	-

3. Theoretical prediction of the shear modulus of elasticity

In theoretical analysis, the shear property of PFRP composite specimen is calculated from the theoretical equation which was suggested in previous studies.

According to the rule of mixtures [8], the shear property of PFRP composite materials is determined in the mechanics of materials approach by presuming that the shearing stresses on the fiber and on the matrix are the same. The shear modulus of elasticity is determined by equation (1).

$$G_{12} = \frac{G_f G_m}{(V_m G_f + V_f G_m)} \quad (1)$$

where, G_{12} is the shear modulus of elasticity, G_f is the shear modulus of elasticity of fiber, G_m is the shear modulus of elasticity of matrix, V_f is the volume ratio of fiber, and V_m is the volume ratio of matrix.

According to Jones's study [8], the shear modulus of orthotropic composite materials was suggested as equation (1). The shear modulus of the PFRP composite used in tensile test is calculated from equation (1) and calculated result is given in table 4.

Table 4. Theoretically predicted result by equation (1).

G_f (GPa)	G_m (GPa)	V_f (%)	V_m (%)	G_{12} (GPa)
30.21	1.50	41.19	58.81	4.42

In addition, Jones's study suggested as equations which considered the effect of actual fiber arrangement. In the fabrication of fiberous composite materials, the fibers are often somewhat randomly placed rather than being packed in a regular array. Therefore, the theoretical analysis for the modulus of composite materials with regular arrays must be modified to account for the fact that fibers are contiguous, i.e., that fibers touch each other rather than being entirely surrounded by matrix material. If *C* denotes degree of contiguity, then *C*=0 corresponds to no contiguity (isolated fibers) and *C*=1 corresponds to perfect contiguity (all fibers are in contact) as shown in figure 1 [8].

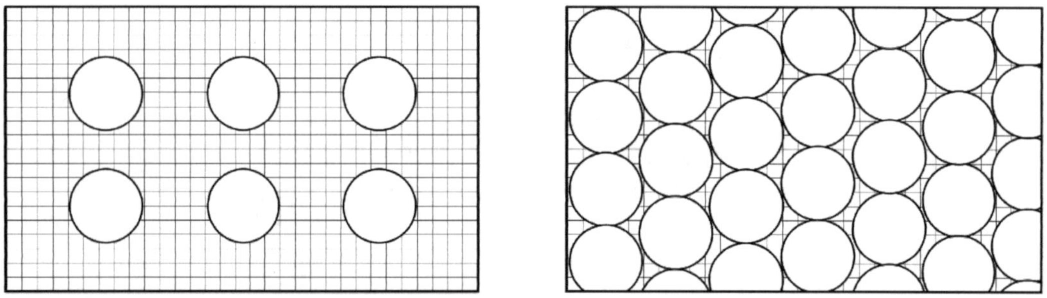

(a) Isolated fibers matrix contiguous (*C*=0) (b) Isolated matrix fibers contiguous (*C*=1)

Figure 1. Extremes of fiber contiguity [8].

For the elasticity approach in which the contiguity is considered, the modulus of elasticity in the transverse direction was suggested as equation (2).

$$E_{22} = 2\left[1 - \nu_f + (\upsilon_f - \upsilon_m)V_m\right] \times \left[(1-C)\frac{K_f(2K_m + G_m) - G_m(K_f - K_m)V_m}{(2K_m + G_m) + 2(K_f - K_m)V_m} + C\frac{K_f(2K_m + G_f) + G_f(K_m - K_f)V_m}{(2K_m + G_f) - 2(K_m - K_f)V_m}\right] \quad (2)$$

where, E_{22} is the modulus of elasticity in the transverse direction, G_f is the shear modulus of elasticity of fiber, G_m is the shear modulus of elasticity of matrix, V_f is the volume ratio of fiber, V_m is the volume ratio of matrix, ν_f is the Poisson's ratio of fiber, ν_m is the Poisson's ratio of matrix, K_f is the transverse plane strain bulk modulus of fiber, K_m is the transverse plane strain bulk modulus of matrix, and *C* is the degree of contiguity.

In equation (2), the parameters K_f and K_m are calculated by equation (3) and equation (4).

$$K_f = \frac{E_f}{2(1-\upsilon_f)} \quad (3)$$

$$K_m = \frac{E_m}{2(1-\upsilon_m)} \quad (4)$$

From a practical point of view, *C* would be determined by comparison of theoretical curves of E_{22} versus (V_f or V_m) for various values of *C* with experimental results. Therefore, *C* of the test specimen was determined by comparison with the tensile test result in transverse direction and equation (2). The result of comparing with the tensile test in transverse direction result and equation (2) is given in table 5. Mechanical properties of PFRP composite specimen obtained by the tensile tests are applied to the parameters in equation (2).

Table 5. Contiguity from equation (2).

E_{22} (GPa)	G_f (GPa)	G_m (GPa)	V_f (%)	V_m (%)	ν_f	ν_m	C
9.89	30.21	1.50	41.19	58.81	0.22	0.33	0.32

In addition, equation (5) was suggested in Jones's study for calculating the shear modulus of elasticity.

$$G_{12} = (1 - C)G_m \frac{2G_f - (G_f - G_m)V_m}{2G_m + (G_f - G_m)V_m} + CG_f \frac{(G_f + G_m) - (G_f - G_m)V_m}{(G_f + G_m) + (G_f - G_m)V_m} \tag{5}$$

where, G_{12} is the predicted shear modulus of elasticity, G_f is the shear modulus of elasticity of fiber, G_m is the shear modulus of elasticity of matrix, V_f is the volume ratio of fiber, V_m is the volume ratio of matrix, ν_f is the Poisson's ratio of fiber, ν_m is the Poisson's ratio of matrix, and C is the degree of contiguity.

The shear modulus of the PFRP composite used in tensile test is calculated from equation (5). The result of calculation is given in table 6. Properties of PFRP composite used in tensile tests are applied to the parameters in equation (5).

Table 6. Predicted shear modulus of elasticity by equation (5).

G_f (GPa)	G_m (GPa)	V_f (%)	V_m (%)	ν_f	ν_m	C	G_{12} (GPa)
30.21	1.50	41.19	58.81	0.22	0.33	0.32	5.15

4. Comparison

Experimental results and theoretical are compared to determine the reliability of the theoretical analysis. Comparison of experimental and theoretical results is given in table 7.

Table 7. Comparison of the shear modulus of elasticity.

Description	G_{12} (GPa)	Difference (%)	Remark
Tensile test in 45° off-axis direction	5.13	-	Tested
Calculated result by equation (1)	4.42	-13.84	Predicted
Calculated result by equation (5)	5.15	0.39	Predicted

As a result of comparison with the theoretical and experimental results of the shear modulus of elasticity, the difference obtained by equation (1) is –13.84%, the difference obtained by equation (5) is 0.39%. In the case of equation (5), the predicted shear modulus of elasticity showed a reliable difference compared with experimental results.

In the results of comparison between calculation result of equation (1) and equation (5), the result of equation (5) showed 13.45% less difference than the result of equation (1). Therefore, in the theoretical analysis for the shear modulus of PFRP composite materials, if the arrangement of fibers is considered it may reduce the difference between experimental and theoretical results. Also, it is thought that the prediction of shear modulus of PFRP composite materials is possible using equation (5).

5. Conclusion

In this study, the mechanical properties of PFRP composite materials are investigated. Especially, relationship between the shear property and the other mechanical properties of PFRP composite materials is investigated through comparison between experimental and theoretical results. Results obtained in this study is as follows:

To determine the tensile strength and the modulus of elasticity in longitudinal direction and transverse direction, the tensile test in longitudinal direction and transverse direction is conducted. To determine the shear strength and the shear modulus of elasticity, the tensile test in 45° off-axis direction is conducted. As a result of the tensile tests, the modulus of elasticity in longitudinal direction is 36.39 GPa, the Poisson's ratio is 0.33, the modulus of elasticity in transverse direction is 9.89 GPa, and the shear modulus of elasticity is 5.13 GPa. Standard deviation and coefficient of variation on the test results are evaluated and it was found that they are reliable.

In the theoretical analysis, the shear property of PFRP composite specimen is calculated from the theoretical equations which were suggested in previous studies. As a result of comparison between the theoretical and experimental results of the shear modulus of elasticity, the difference obtained by equation (1) is –13.84% and the difference obtained by equation (5) is 0.39%. In the case of equation (5), the theoretical shear modulus of elasticity showed a reliable difference when compared with experimental results. However, from the result by equation (1), the theoretical shear modulus of elasticity showed significant difference compared with experimental results.

In the results of comparison between predicted result by equation (1) and equation (5), the result of equation (5) showed 13.45% less difference than the result of equation (1). Therefore, in the theoretical analysis for the shear modulus of PFRP composite materials, considering the arrangement of fibers is considered to reduce the difference between the experimental and the theoretical results. In addition, the shear modulus of PFRP composite materials can be predicted from material properties of fiber and matrix of PFRP composites.

Acknowledgment

This work was supported by the National Research Foundation of Korea Grant funded by the Korean Government (NRF-2017R1A6A3A01076282)

References

[1] Yoo J H and Yoon S J 2017 Evaluation of safety factor on the buckling of pultruded glass fiber reinforced polyester composite plate *J of Korean Soc Adv Comp Struc* **8** 30-6
[2] Lee Y G, Park S Y, Park J S, Nam J H, An D J and Yoon S J 2011 Structural behavior of PFRP connection with single bolt *Proc 18th Int Conf Comp Material (ICCM18)*
[3] Mallick P K 2008 *Fiber-reinforced Composites: Materials, Manufacturing, and Design* 3rd Edition (Boca Raton: CRC Press)
[4] Hohan R, Bejan L and Taranu N 2012 Effect of contiguity on shear elastic modulus of fiber reinforced polymeric composites *Construction Architecture Section* **58** 9-22
[5] Schwartz M M 2009 *Smart Materials* (Boca Raton: CRC Press)
[6] ASTM D 3039/D 3039 M-00 2000 *Standard Test Method for Tensile Properties of Polymer Matrix Composite Materials* (West Conshohocken: ASTM International)
[7] ASTM D 3518/D 3518 M-94 2001 *Standard Test Method for In-Plane Shear Response of Polymer Matrix Composite Materials by Tensile Test of a ±45° Laminate* (West Conshohocken: ASTM International)
[8] Jones R M 1999 *Mechanics of Composite Materials* (Philadelphia: Taylor and Francis)

IOP Conf. Series: Materials Science and Engineering **372** (2018) 012040 doi:10.1088/1757-899X/372/1/012040

Failure modes of single and multi-bolted joint in the pultruded fiber reinforced polymer composite members

S Y Kim[1], J H Yoo[1], H K Kim[1], K Y Shin[1] and S J Yoon[1,2]

[1]Department of Civil Engineering, Hongik University, 94 Wausan-ro, Mapo-gu, Seoul 04066, Korea

E-mail: sjyoon@hongik.ac.kr

Abstract. In this paper, we discussed the structural behavior of bolted lap-joint connections in pultruded FRP structural members. Especially, bolted connections in pultruded FRP members are investigated for their failure modes and strength. Specimens with single and multiple bolt-holes are tested in tension under bolt-loading conditions. All of the specimens are instrumented with strain gages and the load-strain responses are monitored. The failed specimens are examined for the cracks and failure patterns. The purpose of this paper is to predict the failure strength by using the ratio of the results obtained by the experiment and the finite element analysis. In the study, several tests are conducted to determine the mechanical properties of pultruded FRP materials before the main experiment. The results are used in the finite element analysis for single and multiple bolted lap-joint specimens. The results obtained by the experiment are compared with the results obtained by the finite element analysis.

1. Introduction

Fiber reinforced polymer (FRP) composites have high specific strength and stiffness compared with those of conventional structural materials such as concrete and steel. FRP composites have several advantages such as non-magnetic and non-conductive properties, high resistance to corrosion and chemical attack. Due to the excellent physical, chemical, and mechanical properties of FRP composites, it is beginning to be widely used in the aerospace, shipbuilding, automotive, and leisure industries, as well as civil engineering and construction. However, although FRP is recognized as an alternative of steel structural member, it is very difficult to predict the structural behavior due to complex failure mode depending on the kind of composite material and connection method. In addition, most of the composite material suitable for use as construction materials have a disadvantage in that the reinforcing fibers are arranged in the axial direction of the members, hence the strength of the bolted connections is considerably low. The pultrusion process shown in figure 1 is a cost effective composite manufacturing technique in which continuous reinforcing fibers with other additional fabric layers are pulled from creels and are passed through a resin tank where fibers are impregnated with polymer resin. The saturated fibers are drawn through a preforming and heating die in which polymerization into a hardened form takes place. The pultruded FRP (PFRP) is considered as an orthotropic (transversely isotropic) material because fibers are placed to the longitudinal direction of the member [1]. In order to use the PFRP as the structural material, basic research on the structural behavior of the bolted joint should be preceded.

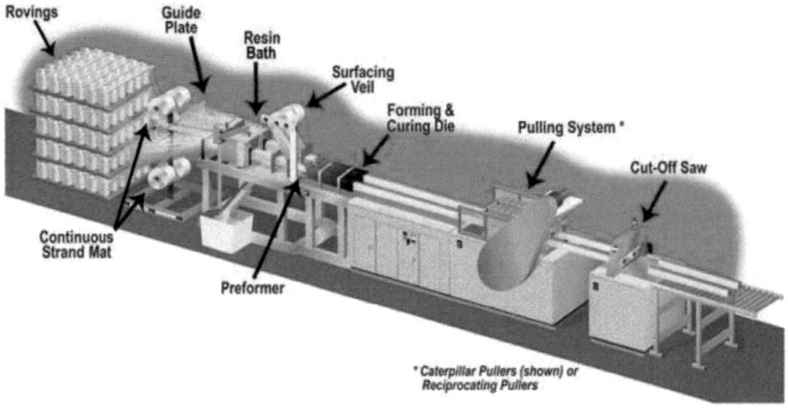

Figure 1. Pultrusion process [2].

A study on the bolted connection of PFRP is conducted by Abd-El-Naby on the effect of FRP shear strength and bolt-hole shape, which varies with the fiber volume ratio in the loading direction [3]. In addition, Prabhakaran studied using the design equation of the load and resistance factor design method (AISC-LRFD) of the American Institute of Steel Construction for FRP connections connected by bolts [4]. However, since failure modes are limited to net-tension failure and block shear failure, there are limitations in using the failure modes as basic data for various failure modes of composites. In addition, Choi considers various effects of stress distribution, surface friction coefficient, and torque on the circular hole, through the finite element analysis [5]. However, additional experiment is necessary. In Lee's study, it is confirmed through the lap-joint test of the PFRP that a number of bolts and the arrangement of bolts cause the change of strength [6]. In the experimental program E-glass fiber and polyester resin composite with 55% fiber volume ratio produced by the pultrusion process with structural profiled cross-section are used.

In this study, it is attempted to find the design method and guideline for connection of the most effective FRP connection method. The FRP plate width, the number of bolts, the spacing between the bolts, and the size of the bolt holes are arbitrarily determined, and the finite element analysis for the joints is conducted. The load - displacement behavior is investigated by performing the tensile test. The fracture mode of the member is confirmed, and the stress distribution is investigated through the strain gages attached to the surface of test specimen.

2. Failure mode of bolted connection

As discussed by Bank, the pultruded parts used in the connection are made of orthotropic materials, and therefore the orientation of the individual parts in the connection is critical, unlike in steel-bolted connections, where the base material is isotropic and not orientation dependent. Failure modes of bolted lap joints are mainly net-tension failure, shear-out failure, cleavage failure, block-shear failure, and bearing failure. Each failure mode is shown in figure 2. First, net-tension failure is caused by the cross-sectional area reduced by the bolt hole. Therefore, the tensile strength is reduced due to the reduction, so the plate is broken along the bolt line. Second, shear-out failure is a phenomenon that the fracture of the member touching the neck part of the bolt progresses further and then it becomes weak to the shear resistance and is broken along the shear section. Third, the cleavage failure is a phenomenon that the bolt-hole section of the member that receives the load from the neck part of the bolt does not resist the load and is broken as if it is cracked. Fourth, block shear failure is a phenomenon in which shear fracture and net section tensile fracture occur at the same time. Finally, bearing failure is that the cross-section of the bolt hole of the member that comes into contact with the neck part of the bolt is gradually broken when the member is subjected to a load and a loss occurs in the cross-section. According to Bank, bearing failure may be need to induce for the safety of the structure [7].

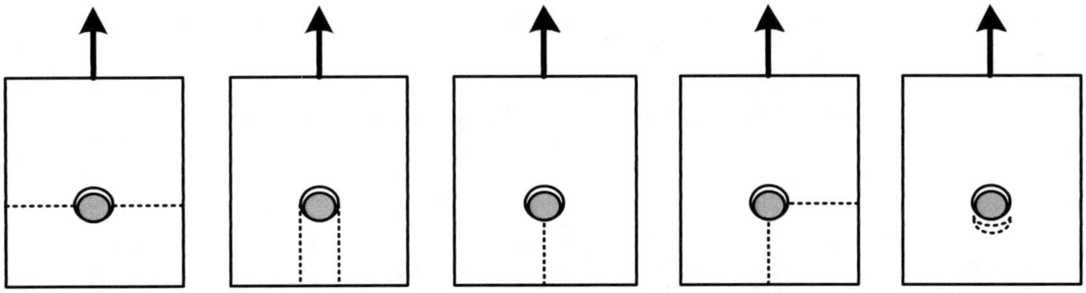

Figure 2. Failure modes of single-bolt lap-joint in-plane connection [7]..

3. Mechanical properties of PFRP materials

The purpose of this study is to provide basic data for the design of bolted joints of PFRP. Before performing the tensile test of the PFRP specimen with bolted connection, it is necessary to investigate the material properties of the PFRP. Therefore, the axial tensile test, the compression test in the direction perpendicular to the member axis, and the shear strength test were performed.

Because of the specimen length limitation in the structural shape, it is difficult to make the test specimen suggested in ASTM D 3916. Therefore, prismatic bar shape specimen with rectangular cross-section is made with reference to existing research results [8]. The test results are given in table 1.

Table 1. The axial tensile test result.

Specimen number	Thickness (mm)	Width (mm)	Length (mm)	Tensile strength (MPa)	Tensile modulus of elasticity (GPa)	Poisson`s ratio
No. 1	6	25.4	254	584.3	38.6	0.24
No. 2	6	25.4	254	553.8	37.1	0.16
No. 3	6	25.4	254	664.5	34.5	0.09
No. 4	6	25.4	254	592.9	35.4	0.24
No. 5	6	25.4	254	571.9	32.3	0.25
Average	6	25.4	254	583.0	35.7	0.24

Table 2. The compression test result in the direction perpendicular to the member axis.

Specimen	Thickness (mm)	Width (mm)	Length (mm)	Compressive strength (MPa)	Compressive modulus of elasticity (GPa)	Poisson`s ratio
No. 1	6	20	80	109.7	12.3	0.07
No. 2	6	20	80	123.6	11.0	0.08
No. 3	6	20	80	137.4	14.2	0.10
No. 4	6	20	80	135.5	9.8	0.10
No. 5	6	20	80	152.2	16.0	0.06
Average	6	20	80	141.7	13.3	0.09

Compression tests are in principle conducted to ASTM D 3410 M-08 [9]. However, it is difficult to apply to the material to be used, because of dimension of the member. Therefore, compressive strength and compressive elastic modulus were determined from the experimental results using the compression test proposed by Yoon [10]. The results are given in table 2. Among the five specimens,

IOP Conf. Series: Materials Science and Engineering **372** (2018) 012040 doi:10.1088/1757-899X/372/1/012040

No. 1 and No. 2 specimens were not failed by fracture, so they were excluded from the average calculation.

The shear test is conducted according to ASTM D 5379 M-05 [11] and the test results are given in table 3.

Table 3. The shear test result in the longitudinal direction.

Specimen		Width (mm)	Length (mm)	Maximum displacement (mm)	Shear strength (MPa)	Shear modulus of elasticity (MPa)
Longitudinal Direction	No. 1	10.99	6.51	1.05	44.6	45.5
	No. 2	10.97	6.49	1.02	55.0	40.7
	No. 3	10.92	6.50	0.80	46.6	45.8
	No. 4	11.02	6.48	0.93	46.9	39.2
	No. 5	11.00	6.61	1.09	51.5	45.5
	Average	10.98	6.52	0.98	48.9	43.3
Transverse Direction	No. 1	10.99	6.51	3.06	67.3	51.4
	No. 2	10.97	6.49	1.47	75.4	54.8
	No. 3	10.92	6.50	1.38	74.9	55.5
	No. 4	11.02	6.48	1.32	70.6	55.6
	No. 5	11.00	6.61	1.34	69.1	53.9
	Average	10.98	6.52	1.71	71.4	54.2

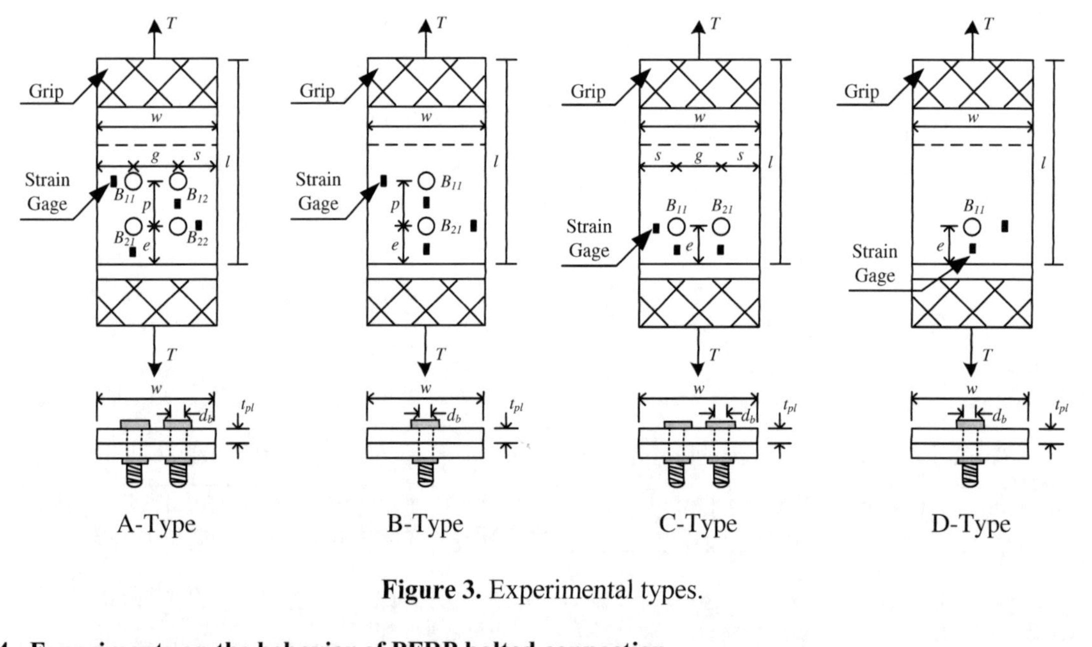

Figure 3. Experimental types.

4. Experiments on the behavior of PFRP bolted connection

The experimental types are shown in figure 3. Each type has a different number and arrangement of bolts. A-Type has a square arrangement of four bolts, and B-Type has double bolts arrangement vertically. C-Type has double bolts arranged horizontally, and D-Type has single bolt. The variable is the end distance to bolt diameter ratio (e/d_b). In this case, e/d_b is set to 3, 4, and 5, since the effect on the ultimate load of the joint is negligible even if the e/d_b value increases to 5 or more at the joint of single bolt [12]. The experiment results are given in table 4, and the failure modes are indicated by letters as bearing failure (B), shear-out failure (S), and cleavage failure (C).

Table 4. Experimental result on the behavior of PFRP bolted connection.

Specimen	e/d_b	Failure load (kN)	Failure mode			
			B_{11}	B_{12}	B_{21}	B_{22}
A-Type	3	43.6	B	S	S	S
	4	66.6	S	S	C	S
	5	66.2	C	B	B	B
B-Type	3	19.8	B	-	S	-
	4	36.2	S	-	None	-
	5	42.5	B	-	B	-
C-Type	3	17.4	S	S	-	-
	4	24.9	B	B	-	-
	5	26.3	S	S	-	-
D-Type	3	8.3	S	-	-	-
	4	11.7	C	-	-	-
	5	13.7	S	-	-	-

The experimental results show that the failure loads increase with increasing e/d_b 3 to 4 (about 55%). However, the failure loads are small enough to be ignored, in case of e/d_b 4 to 5. In addition, although the recommended geometric dimension is satisfied when e/d_b value is 3 [7], the fracture strength is very small compared with the case of 4 and 5, so it is considered that it is difficult to use in the practical design.

As e/d_b value increased, failure modes are changed from shear-out failure to bearing failure at A and B-Types. However, at C and D-Types, the obvious change is not observed. Therefore, additional experiments are needed to confirm whether the location of bolts affects to the change of failure mode.

5. Comparison of results obtained by FEA (finite element analysis) and experiment

To evaluate the stress distribution of the PFRP specimens with bolts and to predict the structural behavior of bolted connection, general purpose finite element analysis (FEA) program, ANSYS Ver. 11 [13] is used. With reference to the experiment mentioned above, the material assumed to be linear elastic and orthotropic. The tensile stress distribution in the axial direction is shown in figure 4.

Based on the test results, the tensile strength of the net section of the material was calculated using equations (1) and (2), when the specimen was broken.

$$\sigma_{net} = \frac{P_t}{A_{net}} \tag{1}$$

$$A_{net} = t_{pl}(w - nd_h) \tag{2}$$

In equations (1) and (2), σ_{net} is the stress in the net section of the specimen, P_t is the tensile load acting on the specimen, and A_{net} is the net cross section of the member excluding the bolt holes. t_{pl} is the thickness of the specimen, w is the width of the specimen, n is the number of bolts on the same horizontal line, and d_h is the diameter of bolt hole.

Comparison between experimental and FEA results for the strength at the net section is given in table 5. The reliability of A, B, C-Type is about 77%. However, D-Type has a different tendency (about 174%) from the previous experiment. It is necessary to increase reliability at the single bolt case through the additional experiment.

IOP Conf. Series: Materials Science and Engineering **372** (2018) 012040 doi:10.1088/1757-899X/372/1/012040

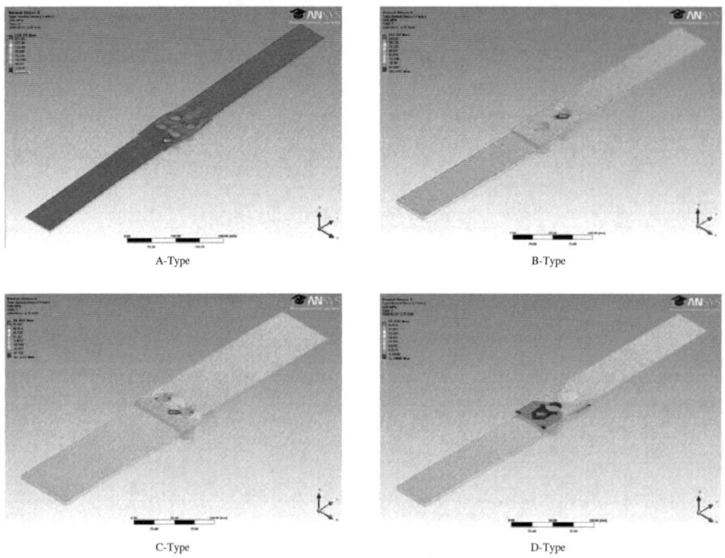

Figure 4. FEA result.

Table 5. Comparison of experiment results and FEA.

Specimen type	e/d_b	In-plane stress (MPa) Experimental result ①	FEA result ②	①/②
A-Type	3	134.6	202.83	0.66
	4	205.6	254.62	0.81
	5	204.3	256.09	0.80
B-Type	3	89.2	137.70	0.65
	4	163.1	195.46	0.83
	5	191.4	232.32	0.82
C-Type	3	53.7	64.31	0.84
	4	76.9	101.63	0.76
	5	81.2	105.30	0.77
D-Type	3	37.4	20.51	1.82
	4	52.7	32.79	1.61
	5	61.7	34.36	1.80

6. Conclusion

In this paper, the mechanical properties and structural behavior of bolted connections are investigated by experiment. The finite element analysis is also conducted on the bolt connected PFRP composite members. Following conclusions are made:

- Failure mode from the tensile test of bolted lap-joint in the PFRP were classified into shear-out, cleavage, and bearing failure with respect to e/d_b values.
- When e/d_b value of the joint decreases, shear-out failure mode tends to occur. As e/d_b value increases, cleavage and bearing failure modes tend to occur.
- The tensile strengths obtained by the experiments of the lap-joints are similar to those obtained by the finite element analysis at A, B, and C-Type specimens. However, D-Type, single bolt case, needs to increase reliability through the additional experiments. Although A, B, and C-Type specimens show similar trends, D-Type specimen shows no trend. Therefore, it is necessary to study further to establish the design criteria, in addition to these results.

From these results, we confirmed that the relation between the end distance to bolt diameter ratio

(e/d_b) and failure mode. In addition, the fracture strength in the net section fracture can be predictable using the finite element analysis. If the failure mode and strength can be predicted in advance, the net section tensile strength, shear-out failure strength, cleavage failure strength, and bearing failure strength of the bolted PFRP members can be determined and reflected in the structural design.

References

[1] Lee Y G, Park S Y, Park J S, Nam J H, An D J and Yoon S J 2011 Structural behavior of PFRP connection with single bolt *Proc 18th Int Conf Comp Material (ICCM18)* (Jeju Korea)
[2] Strongwell 2016 *Extren Design Manual* (Bristol Virginia)
[3] Abd El Naby S F M and Hollaway L 1993 The experimental behaviour of bolted joints in pultruded glass/polyester material part 2: Two bolt joints *Composites* **24** 397-401
[4] Prabhakaran R, Razzaq Z and Devara S 1996 Load and resistance factor design (LRFD) approach for bolted joints in pultruded composites *Composites: Part B* **27B** 351-360
[5] Choi J M, Chun H J and Byun J H 2005 Effects of various parameters on stress distribution around holes in mechanically fastened composite laminates *Composites Research* **18** 9-18
[6] Lee Y G 2009 The characteristics of structural behavior of bolted connection for the PFRP structural members (Seoul: Hongik University)
[7] Bank L C 2006 *Composites for Construction: Structural Design with FRP Materials* (New Jersey: Wiley & Sons) Chapter 15 pp 298-502
[8] Kim E H 2007 Investigation of pipe deflection behavior and prediction of longterm ring deflection of GFRP pipe buried underground hongik university research institute of science and technology *Research Report*
[9] ASTM D 3410 M-08 2008 *Standard Test Method for Compressive Properties of Polymer Matrix Composite Materials with Unsupported Gage Section by Shear Loading* American Society for Testing and Materials
[10] Yoon S J 1993 *Local Buckling of Pultruded I-Shape Columns* (Atlanta: Georgia Institute of Technology)
[11] ASTM D 5379 M-05 2005 *Standard Test Method for Shear Properties of Composite Materials by the V-notched Beam Method* American Society for Testing and Materials
[12] Rosner C N and Rizkalla S H 1995 Bolted connections for fiber reinforced composite structural members: experimental program *Journal of Materials in Civil Engineering* **7** 223-31
[13] ANSYS Inc. ANSYS® 2007 ANSYS Workbench 11 Cannonsberg Pennsylvania

IOP Conf. Series: Materials Science and Engineering **372** (2018) 012041 doi:10.1088/1757-899X/372/1/012041

Tensile and compressive modulus of elasticity of pultruded fiber-reinforced polymer composite materials

J H Lee[1], S H Kim[2], J K Park[3], W C Choi[2] and S J Yoon[1,4]

[1]Department of Civil Engineering, Hongik University, 94 Wausan-ro, Mapo-gu, Seoul 04066, Korea

[2]Department of Architectural Engineering, Gachon University, 1342 Seongnam-daero, Sujeong-gu, Gyeonggi-do 13120, Korea

[3]Daehan EnC, #1401 Techdong SKⓃTechno Park, 124 Sagimakgol-ro, Jungwon-gu, Seongnam-si, Gyeonggi-do 13207, Korea

E-mail: sjyoon@hongik.ac.kr

Abstract. Many researches focused on the mechanical properties of steel and concrete have been carried out for applications in the construction industry. However, in order to clarify the mechanical properties of pultruded fiber-reinforced polymer (PFRP) structural members for construction, testing is needed. Deriving the mechanical properties of PFRP structural members through testing is difficult, however, because some members cannot be tested easily due to their cross-section dimensions. This paper reports a part of studies that attempt to present conservative results in the case of members that cannot be tested reasonably. The authors obtained and compared experimental and theoretical modulus of elasticity values. If the mechanical properties of PFRP members can be predicted using reasonable and conservative values, then the structure can be designed economically and safely even in the early design stages. To this end, this paper proposes a strain energy approach as a conservative and convenient way to predict the mechanical properties of PFRP structural members. The strain energy data obtained can be used to predict the mechanical properties of PFRP members in the construction field.

1. Introduction

In order to use composite material because of its outstanding material properties as a structural material in the construction field, it is important to understand its mechanical properties. Composite material produced by pultrusion, referred to as 'pultruded fiber-reinforced polymer' (PFRP) composite, has different properties in the longitudinal and transverse directions of the fiber. To analyze and design such composite structures, mechanical properties tests, such as tension, compression, shear, and bending tests, should be conducted. However, some profile sections of structural members produced by pultrusion may not allow testing according to ASTM and ISO test methods, especially for the transverse direction, because the test methods suggested by ASTM and ISO limit the length of the specimens required for tensile tests. Thus, many PFRP composite members cannot be tested using standard test methods.

Research into the mechanical properties of steel has been carried out for practical applications in the construction industry. According to McCormac, the stress-strain curve of steel is assumed to be the same for members in tension and compression. However, this assumption does not hold for PFRP composite material. Recently, Kim *et al* measured and compared the tensile and compressive mechanical properties of PFRP composite members [1]. In this study, we compared the theoretical and experimental results of mechanical property tests of PFRP composite material for the conservative application of PFRP members in the construction field [1-4].

2. Theoretical predictions

The theoretical prediction method that is used to predict the mechanical properties of PFRP composite members applied in current construction is as follows. For the modulus of elasticity in the longitudinal direction (E_{11}), the 'rule of mixtures' is used to calculate the theoretical values of the PFRP composite properties. To calculate these values using the rule of mixtures, the volume ratio (v_f, v_m) and modulus of elasticity (E_f, E_m) of each component (fiber and matrix) are needed. Equation (1) is based on the rule of mixtures and is given by [5].

$$E_{11} = E_f v_f + E_m v_m \tag{1}$$

For the modulus of elasticity in the transverse direction (E_{22}), equation (2), proposed by Hopkins and Chamis, is used [5].

$$E_{22} = E_m\left(\left(1-\sqrt{v_f}\right) + \sqrt{v_f}\left(1-\sqrt{v_f}\left(1-E_m/E_f\right)\right)^{-1}\right)^{-1} \tag{2}$$

Equations (1) and (2) enable the prediction of the modulus of elasticity values of PFRP composite members. Table 1 presents the calculated results.

Table 1. Calculation results for prediction of mechanical properties of PFRP composite members (fiber volume ratio 41.19%).

Direction	E_f (GPa)	v_f (%)	E_m(GPa)	v_m (%)	Modulus of elasticity (GPa)
Longitudinal direction (E_{11})	72.5	41.19	4	58.81	32.215
Transverse direction(E_{22})					7.955
Direction	σ_f (MPa)	v_f (%)	σ_m (MPa)	v_m (%)	Ultimate strength (MPa)
Longitudinal direction (σ_{11})	3400	41.19	65	58.81	1.439×10^3
Transverse direction(σ_{22})					109.065

3. Mechanical properties of composite material

To carry out the tests under both tension and compression, test specimens with 5-mm thickness were fabricated. The specimen is made of E-glass fiber for the reinforcing material and unsaturated polyester for the matrix. The fiber's (including roving and fabric) volume ratio was 41.19 percent (weight ratio was 60%).

For the tests, 20 specimens (5 specimens for each test) were prepared. The tensile tests were carried out according to ASTM D 3039M-00 [6] and the compressive tests were carried out according to Yoon [7,8]. Table 2 presents the averaged test results and table 3 presents the specimen dimensions and stress-strain curves. The mechanical behavior of the pultruded composite material in the longitudinal direction is linear elastic and is mainly dependent on the characteristics of the fiber. However, the mechanical behavior of the pultruded composite in the transverse direction is nonlinear elastic and is largely dependent on the characteristics of polyester resin. The results obtained from

IOP Conf. Series: Materials Science and Engineering **372** (2018) 012041 doi:10.1088/1757-899X/372/1/012041

these tests were compared with the theoretically predicted results obtained following the method described in Section 2.

Table 2. Tensile and compressive test results.

Test	Ultimate strength (MPa)	Modulus of elasticity (GPa)	Poisson's ratio (ν_{12})
Tensile test in longitudinal direction	496.30	36.39	0.33
Tensile test in transverse direction	23.86	9.89	-
Compressive test in longitudinal direction	446.89	38.76	0.36
Compressive test in transverse direction	138.70	13.30	0.13

Table 3. Specimen dimensions and stress-strain curves.

Test	Specimen dimensions (length*width*thickness)	Stress-strain curve
Tensile test in longitudinal direction	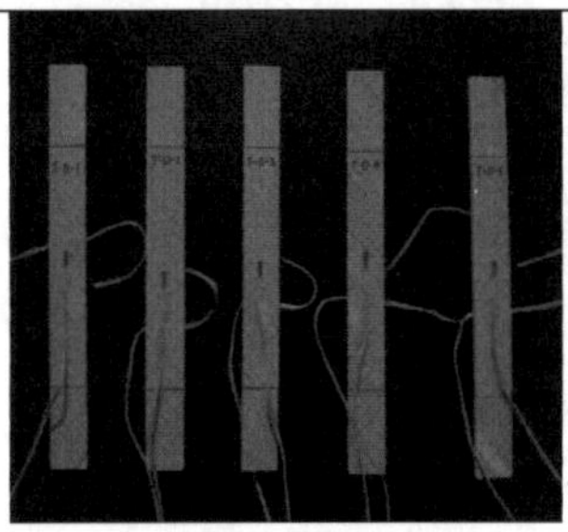 250 mm*25 mm*5 mm	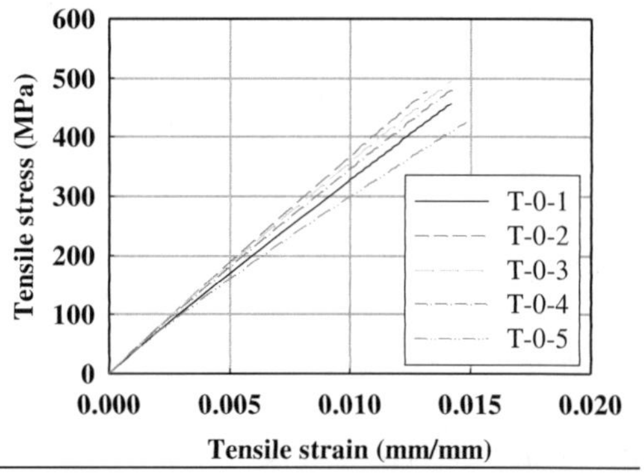
Tensile test in transverse direction	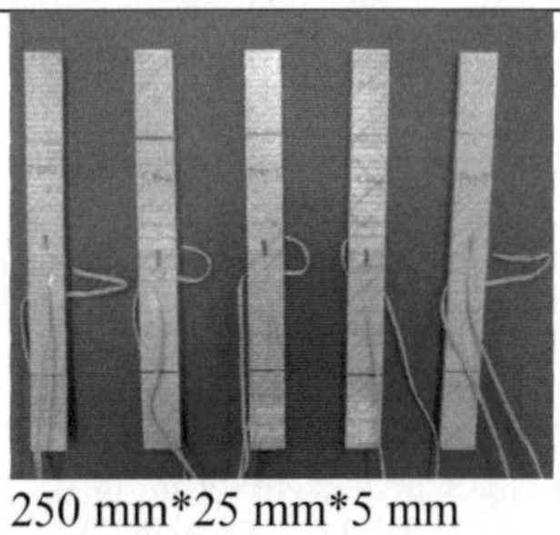 250 mm*25 mm*5 mm	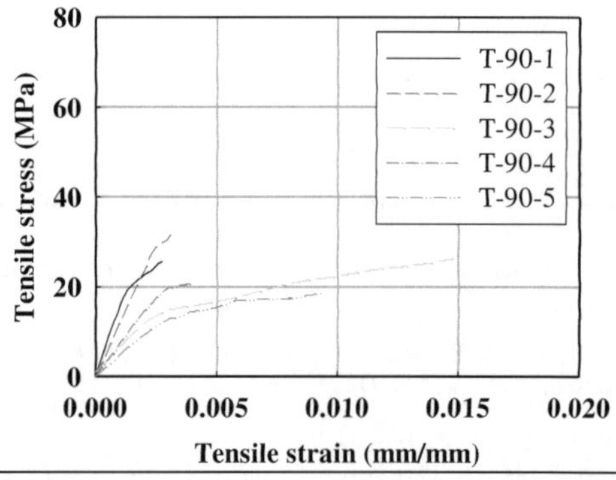

Compressive test in longitudinal direction

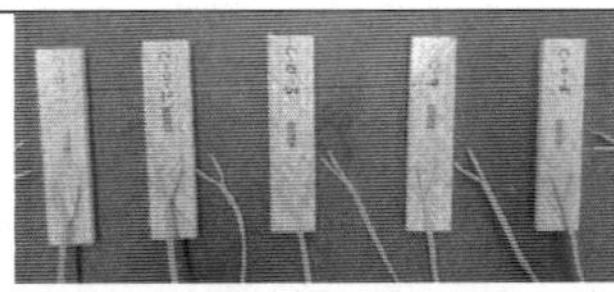

80 mm*20 mm*5 mm

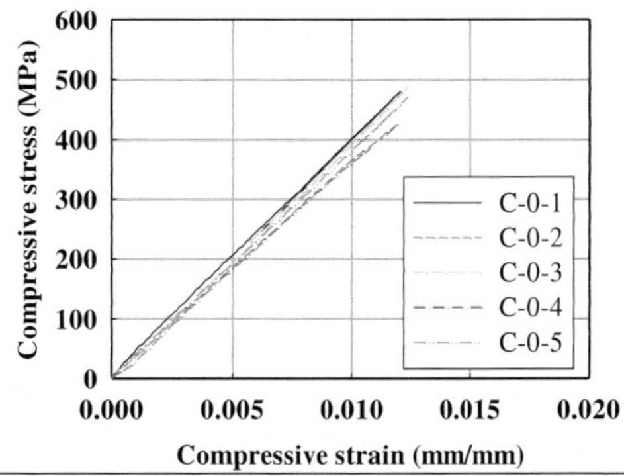

Compressive test in transverse direction

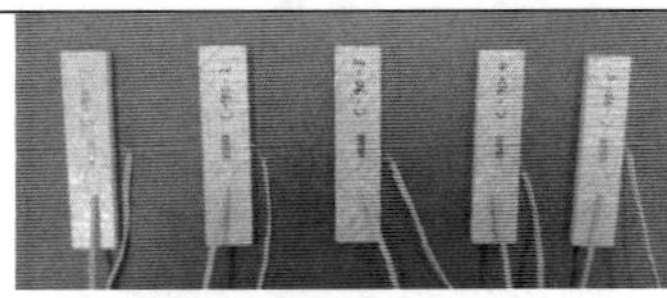

80 mm*20 mm*5 mm

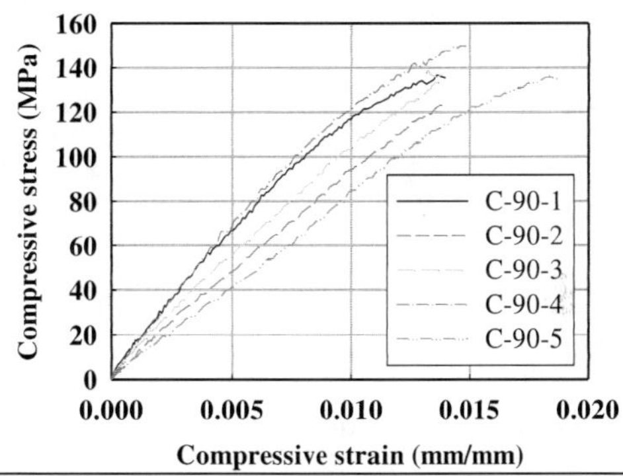

4. Comparison

It is necessary to confirm whether the tensile properties of the composite material can be substituted by the compressive properties. Table 4 provides the ratios of the mechanical properties obtained from the tests of the PFRP composite specimens. According to the ratios, the modulus of elasticity and the ultimate strength can be substituted for each other when they are in the fiber direction. However, because the difference is large in the transverse direction of the fiber, they cannot be substituted for each other when they are in the transverse direction.

Table 4. Ratios of mechanical properties.

Direction	Modulus of elasticity (GPa)		Ratio (①/②)
	Tension (①)	Compression (②)	
Fiber	E_{11t} 36.39	E_{11c} 38.76	0.94
Transverse	E_{22t} 9.89	E_{22c} 13.30	0.74
Direction	**Ultimate strength (MPa)**		**Ratio (①/②)**
	Tension (①)	Compression (②)	
Fiber	σ_{11t} 496.30	σ_{11c} 446.89	1.11
Transverse	σ_{22t} 23.86	σ_{22c} 138.70	0.17

The test results were compared with the theoretically predicted results. The mechanical properties of the test specimens were predicted using the rule of mixtures described in above and table 1. Figures

1(a) and 1(b) present a comparison of the results for the modulus of elasticity and ultimate strength, respectively. The theoretically calculated values are represented by the checkered bar in each graph.

Figure 1 shows that the modulus of elasticity obtained by tensile testing may lead to a conservative design and, thus, the theoretically predicted value may also result in a more conservative design. However, prediction of the ultimate strength is not possible; the test results under tension and/or compression are significantly different from each other. Hence, to predict the ultimate strength of the material under tension and compression, further study is needed.

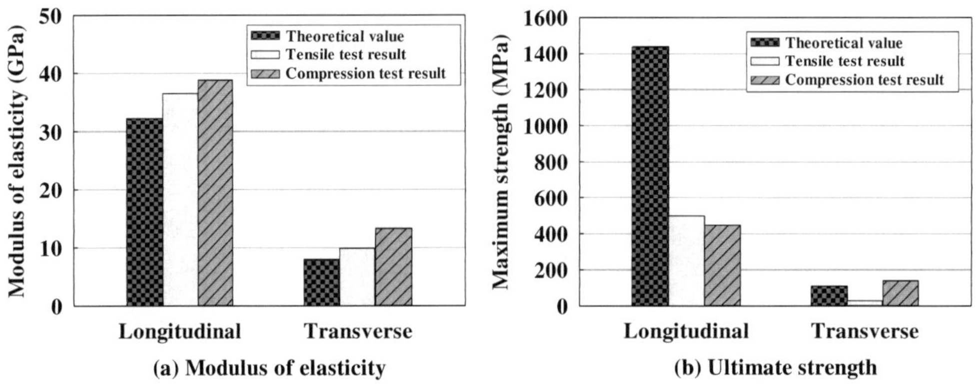

Figure 1. Comparison of tensile test results: (a) modulus of elasticity and (b) ultimate strength.

5. Suggested prediction method

According to Sakaguchi *et al* [9], the rule of mixtures is generally used only as an estimate of the envelope for the modulus of elasticity rather than as a numerically accurate predictor. The rule of mixtures provides accurate predictions when the difference between the modulus of elasticity values of the two phases is small. To solve the problem for applying the rule of mixtures, the Hashin-Shtrikman model, Voigt-Reuss/Angew model, phenomenological model, etc have been used to predict the mechanical properties in dental composites. However, for the construction field, no reasonable prediction model is available to determine the mechanical properties of PFRP composite material. If a prediction model could estimate the mechanical properties of PFRP composite material used in construction, it would also be helpful for predicting the mechanical properties of structural members. The theoretical calculation method described earlier cannot accurately predict the mechanical properties of specimens under tension and compression.

In this paper, strain energy constants are suggested to predict the mechanical properties of PFRP composite material. If the strain energy constants for the PFRP composite material are known, then the mechanical properties can be predicted with reasonable accuracy. The recommended strain energy constants that can be used to predict the mechanical properties of PFRP are given in equations (3), (4), and (5) [5,9].

$$av_f + bv_m = 1 \tag{3}$$

$$E_{11} = \left(a_1^2 v_f E_f{}^{-1} + b_1^2 v_m E_m{}^{-1}\right)^{-1} \tag{4}$$

$$E_{22} = a_2^2 E_f v_f + b_2^2 E_m v_m \tag{5}$$

where the strain energy constants, a, b, a_1, a_2, b_1, and b_2 can be expressed as a_{1t}, a_{2t}, b_{1t}, b_{2t}, a_{1c}, a_{2c}, b_{1c}, and b_{2c} in tension and compression.

For example, if E_{11} and E_{22} obtained from tests and E_f and E_m obtained from general material properties listed in any composite textbook are known, then the strain energy constants, a_{1t} and b_{1t}, may appear when equations (3) and (4) together are calculated, as given in table 5. These constants are suggested to predict the mechanical properties of PFRP (fiber volume ratio 41.19%). This method may be able to predict the mechanical properties of PFRP members for construction. Therefore, if data about the mechanical properties of composite materials used in the construction industry can be obtained, then a model that can predict the mechanical properties of PFRP composite for structural members can be derived.

Table 5. Strain energy constants for test specimens (fiber volume ratio 41.19%) (i= $\sqrt{-1}$).

Direction	Constants			
	Tension		Compression	
Longitudinal direction	a_{1t}	2.25-0.21i	a_{1c}	2.25-0.26i
	b_{1t}	0.12+0.15i	b_{1c}	0.12+0.18i
Transverse direction	a_{2t}	0.42	a_{2c}	0.56
	b_{2t}	1.41	b_{2c}	1.31

6. Conclusion

This study examined the tension and compression characteristics of PFRP composite material. The conclusions drawn from the study are as follows:

Modulus of elasticity values in the fiber direction (E_{11}), as obtained from tension and compression tests, can be used interchangeably in structural design. However, the modulus of elasticity values obtained from tension and compression tests in the transverse direction of the fiber (E_{22}) cannot be used interchangeably.

The theoretically predicted modulus of elasticity of a PFRP member may result in the conservative design of the member.

Ultimate strength predictions may not be feasible and require further study.

Acknowledgment

This work was supported by the National Research Foundation of Korea Grant, funded by the Korean Government (NRF-2017R1A6A3A01076282).

References

[1] Kim S J, Smith S T and Kemp M 2014 Characteristic values of pultruded fibre composite sections for structural design *23rd Australasian Conference on the Mechanics of Structures and Materials* (Lismore: ACMSM23) **1** 527-32

[2] McCormac J C and Csernak S F 2011 *Structural Steel Design 5th Edition* (New Jersey: Pearson Education)

[3] Lee J H 2018 Tension and compression characteristics of pultruded fiber reinforced polymer composite materials (Seoul: Hongik University)

[4] Yoo J H and Yoon S J 2017 Evaluation of safety factor on the buckling of pultruded glass fiber reinforced polyester composite plate *J of Korean Soc Adv Comp Struc* (Seoul: KOSACS) **8** 30-6

[5] Kim S J, Smith S T and Kemp M 2014 Characteristic values of pultruded fibre composite sections for structural design *23rd Australasian Conference on the Mechanics of Structures and Materials* (Lismore: ACMSM23) **1** 527-32

[6] Gibson R F 1994 *Principles of Composite Material Mechanics* (New York: McGraw-Hill)

[7] ASTM D 3039/D 3039M-00 2000 *Standard Test Method for Tensile Properties of Polymer Matrix Composite Materials* (West Conshohocken: ASTM International)

[8] Yoon S J 1993 Local buckling of pultruded I-shape columns (Atlanta: Georgia Institute of Technology)

[9] ASTM D 3410/D 3410 M-03 2003 *Standard Test Method for Compressive Properties of Polymer Matrix Composite Materials with Unsupported Gage Section by Shear Loading* (West Conshohocken: ASTM International)

[10] Sakaguchi R L, Wiltbank B D and Murchison C F 2004 Prediction of composite elastic modulus and polymerization shringkage by computational micromechanics *Dental Materials* (Amsterdam: Elsevier) **20** 397-401

An experimental study on the shear strength of FRP perfobond shear connector

S C Gwon[1], S H Kim[1], S J Yoon[2] and C W Choi[1,3]

[1]Department of Architectural Engineering, Gachon University, 1342 Seongnam daero, Sujeong-gu, Gyeonggi-do 13120, Republic of Korea
[2]Department of Civil Engineering, Hongik University, 94 Wausan-ro, Mapo-gu, Seoul 04066, Republic of Korea

E-mail: wchoi@gachon.ac.kr

Abstract. In this study, push-out tests were conducted to investigate shear behaviour of FRP perfobond shear connector. The parameters influencing shear capacity of FRP perfobond shear connector are concrete dowel effect, shear resistance effect of the laterally reinforced FRP re-bar, and frictional effect between shear connector and concrete. The specimens were designed to consider these parameters. The specimens coated with sand to increase frictional resistance between the FRP re-bar and concrete. Based on the test results and the parameters, new equation was suggested to predict shear strength of FRP perfobond shear connectors. The predicted results and the experimental results were compared to check the feasibility of prediction.

1. Introduction

Different deformation characteristics of concrete and FRP used in FRP-Concrete composite deck plate cause horizontal shear stress and slip at the interface of concrete and FRP when flexural deformation is occurred. Excessive slip leads to the destruction of the bonded surface and the independent behaviours of the two materials. It reduces the load carrying capacity of the FRP-Concrete composite deck. FRP perfobond shear connectors are widely used to induce efficient stress transmission and composite behaviour by resisting to slip. Studies have been extensively conducted to predict shear capacity of perfobond shear connectors [1-5].

In 1987, steel perfobond shear connector was firstly introduced to solve the fatigue problem of steel stud [1]. In 1992, research was conducted to compare shear capacity of steel stud and steel perfobond shear connectors [2]. In 2002, push-out tests to evaluate shear capacity of steel perfobond shear connectors were carried out and shear strength equation was proposed based on the test results [3]. After that, the spring model for FRP perfobond shear connector was proposed based on the pull-out test results and finite element analysis [4], and the slip modulus of FRP perfobond shear connector was proposed based on the push-out test and finite element analysis [5]. But, the research associated with the shear design equation for FRP perfobond shear connectors is relatively limited unlike steel perfobond shear connectors. Therefore, the equation predicting the shear strength of FRP perfobond shear connectors was proposed based on push-out tests of FRP perfobond shear connector and Medberry's equation [3].

2. Shear strength equation of the steel perfobond shear connector

Shear connectors are classified into flexible shear connectors and rigid shear connectors depending on the magnitude of the stiffness. Flexible shear connectors have longitudinal flexibility and low resistance to shear force. Rigid shear connectors have a higher resistance to shear force and less deformation in longitudinal direction than flexible shear connectors. The steel perfobond shear connector is a type of rigid shear connector and used at the interface of a composite members to maximize composite force between the two materials. Figure 1 shows mechanical behaviour of steel perfobond shear connector. The parameters affecting the shear capacity of the steel perfobond shear connectors are concrete dowel effect, shear resistance effect of the laterally reinforced re-bar, the frictional effect between shear connector and concrete, and the bearing effect from bottom concrete.

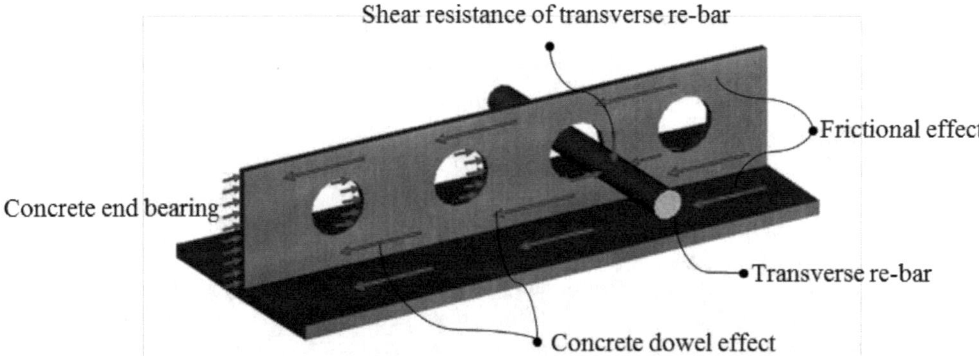

Figure 1. Mechanical behaviour of steel perfobond shear connector.

Based on these four parameters, previous research proposed the shear strength equation of steel perfobond shear connectors. Equation (1) shows the shear strength equation of steel perfobond shear connectors [3].

$$Q_{steel} = 0.75bh\sqrt{f'c} + 0.41b_f L_c + 1.66\pi\sqrt{f'c}\left(\frac{d}{2}\right)^2 + 0.9A_{vf}f_y \quad (1)$$

Where, Q_{steel} is the maximum shear resistance capacity of the steel perfobond shear connector, b is the depth of the slab, h is the height of the bottom slab, $f'c$ is the compressive strength of concrete, b_f is the width of the flange, L_c is the contact length between the flange and concrete of the steel girder, n is the number of holes, d is the diameter of the holes, A_{vf} is the cross-sectional area of the transverse reinforcements, and f_y is the yield strength of the steel re-bar.

First term in the equation (1) is the bearing effect from bottom concrete, 0.75 is obtained by the linear finite element analysis. Second term is the frictional resistance between concrete and connector, 0.41 which is average bond stress determined by the push-out tests. The third term is the dowel effect of concrete, 1.66 which is the average shear strength obtained from the experiment. The last term is the shear resistance effect of steel re-bar, 0.9 which is the reduction factor to consider 90% of the yield stress of steel re-bars.

3. Experiment

3.1. Tensile strength test on FRP re-bar

FRP re-bars were used to increase shear resistance of the FRP perfobond shear connector. Five specimens were fabricated in accordance with ASTM D 7205 [6]. Loading was applied by displacement control testing method with rate of loading of 5mm/min. The load and displacements for the specimens were measured by load cells and LVDTs. Figures 2 and 3 show the specimens and test set up, respectively.

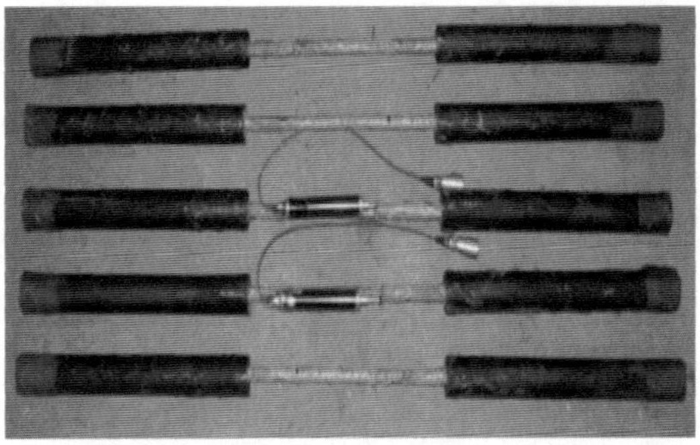

Figure 2. Specimens of FRP re-bars.

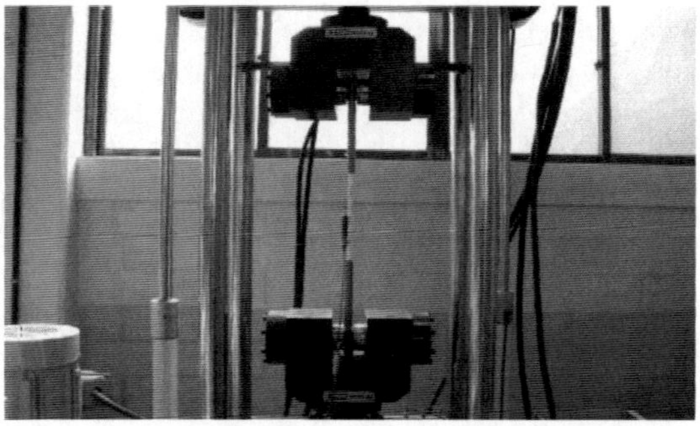

Figure 3. Test set up.

Figure 4 shows the stress-strain relation for the specimen No. 1. Specimen No. 1 showed brittle failure and linear elastic behaviour and all specimens showed similar behaviours. Table 1 shows the summary of tensile test results of all FRP re-bar specimens. The average tensile strength for the FRP re-bar specimens and the average elastic modulus of elasticity for the specimens are 737.9 MPa, 24.5 GPa, respectively.

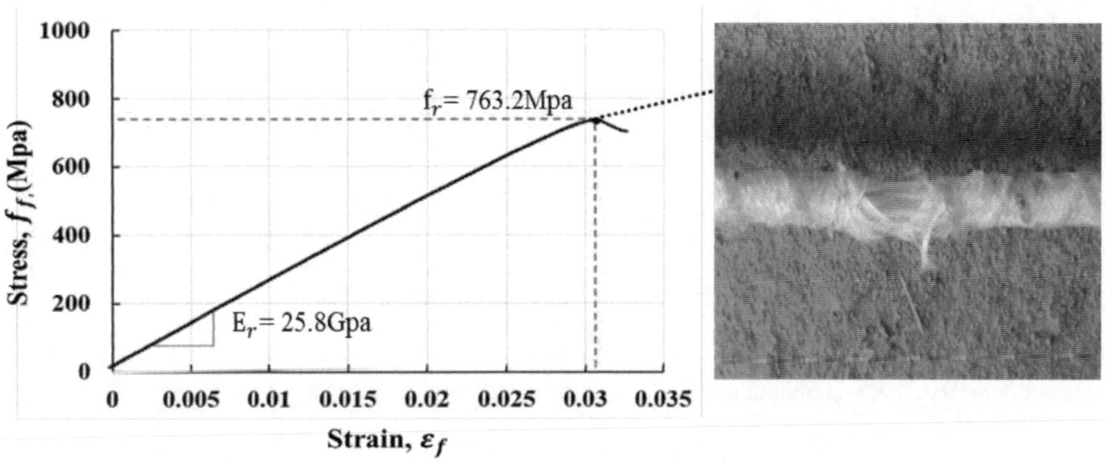

Figure 4. Stress strain relationship of specimen No. 1.

Table 1. FRP re-bar tensile test results.

Specimens	No. 1	No. 2	No. 3	No. 4	No. 5	Average
Tensile Strength (MPa)	763.2	656.9	763.3	752.1	753.8	737.9
Elastic Modulus (GPa)	25.8	17.4	26.3	29	23.8	24.5

Table 2. Compressive strength test results of concrete.

Curing (Days)	3 Days	7 Days	14 Days	23 Days
	33.7	41.5	44.8	43.9
Compressive strength (MPa)	31.4	22.6	45.3	51.2
	33.2	39.3	39.1	46.8
Average (MPa)	32.8	34.5	43.1	47.3

3.2. Compressive strength test on concrete

Concrete cylinder specimens for compressive strength tests were taken with respect to ages, 3 days, 7 days, 14 days and 23 days. Each three specimens were prepared with respect to ages and hence the total number of specimens are twelve. Table 2 shows test results for concrete compressive strengths. The average strength of concrete is 47.3 MPa.

3.3. Push-out test

3.3.1. Specimens. Three specimens coated with sand were fabricated to increase frictional resistance between concrete and FRP module. And the steel re-bars were placed in concrete to control the cracks caused by shrinkage. Figures 5 and 6 show the specimens used in push-out tests of FRP perfobond shear connector.

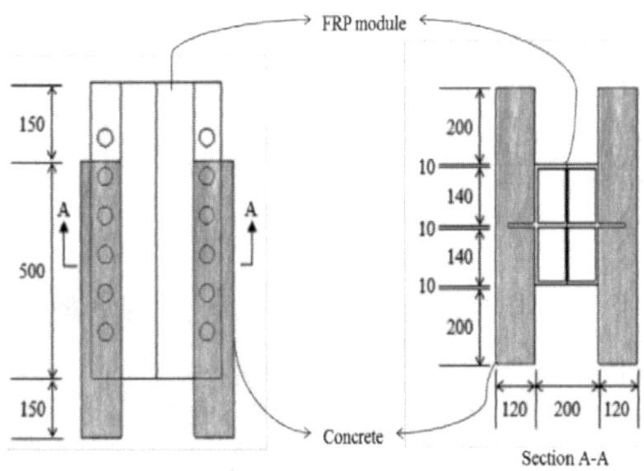

Figure 5. Specimens No. 1, No. 2.

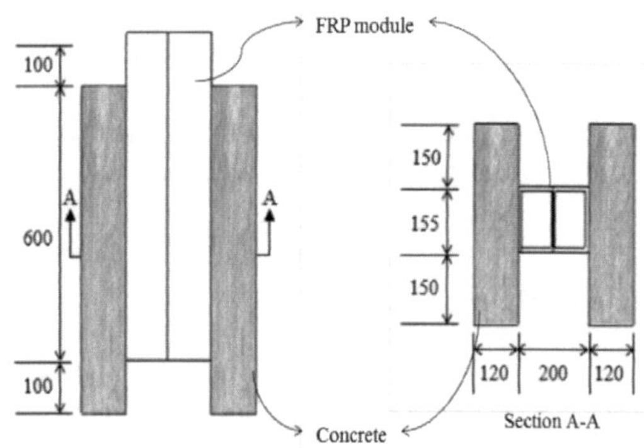

Figure 6. Specimen No. 3.

Figure 5 shows specimens No. 1 and No. 2. The specimens No. 1 and No. 2 were fabricated by pultrusion technique to make FRP shear connector and FRP module into one. The transmission of the shear force in specimens No. 1 and No. 2 is formed by transverse reinforcement placed at intuitional angles to the axis of the specimens. Figure 6 shows specimen No. 3 which is the FRP module without shear connector. It is manufactured to examine the bond strength of the sand coated surface.

3.3.2. Loading and measurements. Push-out tests for FRP perfobond shear connectors were carried out similarly with the test method conducted in steel perfobond shear connectors [3]. The load was applied through displacement control system at a rate of 0.2 mm/min using an actuator with the maximum capacity of 2940 kN. The relative displacements of the specimens were measured by the LVDTs installed at the both sides of specimens.

3.3.3. Test results. Figures 7 and 8 show the load-displacement curves and failure modes of specimens No. 1 and No. 2. At the early loading stage, the slopes of the load-displacement curves were increased with load value. After then, the fine cracks were occurred at the interface of FRP module and concrete. The slopes of the load-displacement curves were decreased at this stage. It is estimated that the fine cracks of specimens No. 1 and No. 2 were occurred at 1,200 kN and 1,500 kN, respectively. The specimens No. 1 and No. 2 were failed when the sand coated surfaces were debonded. At this stage, the relative displacements were 1.5mm. Since the failures were occurred at the bonded surface, it is estimated that the main parameter influencing the shear strength of specimens is the bond strength of sand coated surface.

Specimen No. 3 was designed to investigate only the bond strength of sand coated surface. The specimen No. 3 has smaller contact surface area between concrete and FRP module than specimens No. 1 and No. 2. Therefore, specimen No. 3 has smaller displacement and load at the failure than those of specimens No. 1 and No. 2. Table 3 shows summary of test results of the push-out tests.

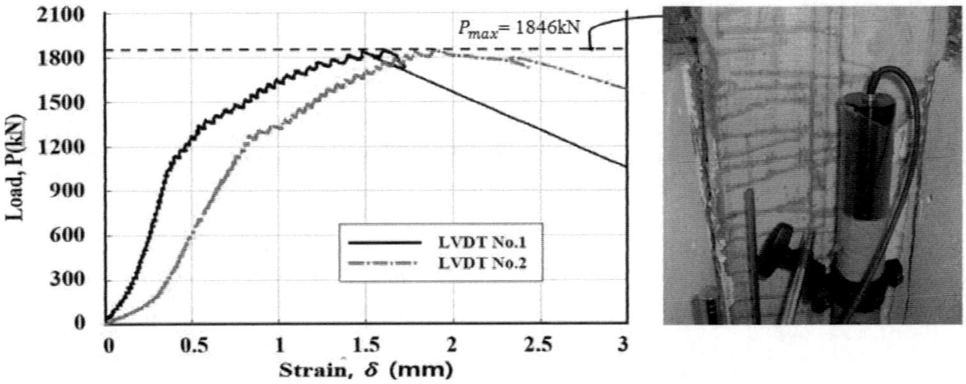

Figure 7. Test result of specimen No. 1.

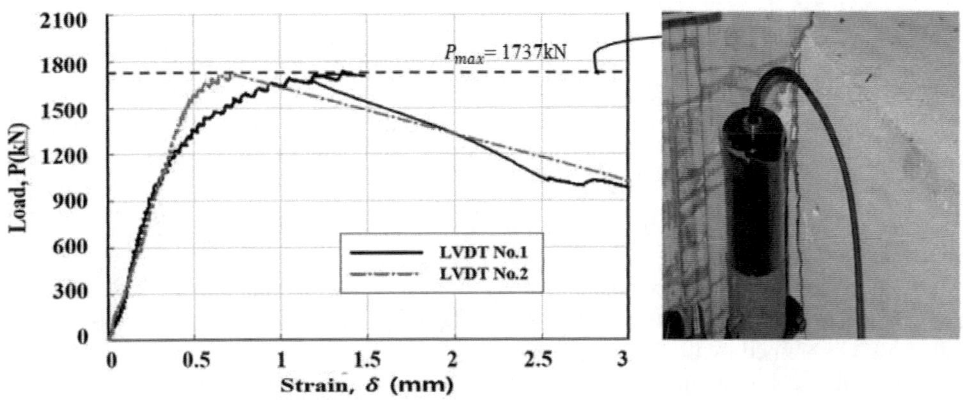

Figure 8. Test result of specimen No. 2.

Table 3. Push-out tests results.

Specimens		No. 1	No. 2	No. 3
P_{max} (kN)		1,846	1,737	341.57
Relative displacement (mm)	LVDT #1	1,610	1,360	0.041
	LVDT #2	1,895	0.715	0.090
d_{12}		0.285	0.645	0.320

P_{max} is the maximum load of the specimens when the bonded surface between concrete and FRP module was failed. d_{12} represents the interval of relative displacements measured from the LVDTs installed in the axial direction of each specimen. Table 3 shows that d_{12} is larger then the load is smaller. Therefore, the difference of the maximum load between specimen No. 1 and No. 2 seems to be attributed due to the moment induced by the eccentricity of loading.

4. Shear strength equation of the FRP perfobond shear connector

Shear strength evaluation equation of the FRP perfobond shear connectors was proposed based on the test results and equation (1). Since the FRP perfobond shear connector is manufactured with FRP module in one piece, the bearing effect between concrete and connectors does not occur.

The parameters affecting the shear capacity of the FRP perfobond shear connectors are concrete dowel action, shear resistance of the laterally reinforced FRP re-bar, the friction between shear connector and concrete. Based on these three parameters, shear strength equation of FRP perfobond shear connector was proposed as follow.

$$Q_{FRP} = 1.84 b_f L_c + 7.346 n\pi \sqrt{f'c} \left(\frac{d}{2}\right)^2 + 0.95 A_{vf} f_r \quad (2)$$

Where, Q_{FRP} is the maximum shear resistance of the FRP perfobond shear connector, f_r is the tensile strength of the FRP re-bar. 1.84 MPa in the first term is bond strength between concrete and FRP module of specimen No. 3. The second term is dowel effect of concrete. 7.347 in the second term is the spring factor based on the previous research [4]. The third term is the shear resistance of FRP re-bar. 0.95 in the third term is recommended as a reduction factor of FRP re-bar [7]. Table 4 shows the comparison of test results with predicted results by equation (2).

Table 4. Comparison test results with predicted results (equation (2)).

Specimens	Q_d	Q_r	Q_b	Q_{FRP}	Q_{exp}	Q_{FRP}/Q_{exp}
No.1	119.5	299.8	563	897.1	923	1.115
No.2					869	1.06

Where, Q_d is the concrete dowel effect, Q_r is shear resistance effect of FRP re-bar and Q_b is the frictional effect between shear connector and concrete. Q_{exp} is the value of test results converted into one shear connector. Comparing Q_{FRP} and Q_{exp}, the differences were in the range of +6% ~ +11.5% approximately.

5. Conclusion

In this study, push-out test were conducted to investigate the shear capacity of FRP perfobond shear connector. The shear strength equation was proposed based on the push-out test and previous equation model. The results obtained by the investigation are as follow:

- The parameters influencing the shear capacity of FRP perfobond shear connector are the bond strength of the sand coated surface, dowel action caused by hole in the shear connector, and shear strength of FRP re-bars.
- The main parameter influencing the shear capacity of FRP perfobond shear connector is bond strength of sand coated surface. Concrete dowel effect has little effect to the shear capacity.
- Comparing the predicted and tested results, the differences were in the range of +6%~11.5%. It found that the proposed shear strength equation estimates the shear strength of FRP perfobond shear connectors conservatively.

Acknowledgment
This work was supported by the Energy Research & Development of the Korea Institute of Energy Technology Evaluation and Planning (KETEP) grant funded by the Korea government Ministry of Trade, Industry & Energy (No. 20161120200190).

References

[1] Zellner W 1987 Recent designs of composite bridges and a new type of shear connector *Proc. of the ASCE/IABSE Engineering Foundation Conference on Composite Construction in Steel and Concrete* (New York: ASCE) pp 240-52
[2] Veldanada M R and Hosain M U 1992 Behaviour of perfobond rib shear connectors *Canadian Journal of Civil Engineering* (Ottawa: CJCE) **19** 1-10
[3] Medberry S B and Shahrooz B M 2002 Perfobond shear connector for composite construction *Engineering Journal* (Bangkok: EJ) **39** 2-12
[4] Cho J R, Park S Y, Cho K, Kim S T and Kim B S 2012 Pull-out test and discrete model of fibre-reinforced polymer perfobond rib shear connector *Canadian Jounral of Civil Engineering*

(Ottawa: CJCE) **39** 1311-20
[5] Zou X, Feng P and Wang J 2016 Perforated FRP ribs for shear connecting of FRP-concrete hybrid beams/decks *Composite structures* (Porto: Elsevier) **152** 267-76
[6] ASTM D 7205 2016 Standard test method for tensile properties of fiber reinforced polymer matrix composite bars *American Society for Testing Materials International* (West Conshohocken: ASTM International)
[7] ACI 440 2017 Guide for design and construction of externally bonded FRP systems for strengthening concrete structures *American Concrete Intitute* (Michigan: ACI)

IOP Conf. Series: Materials Science and Engineering **372** (2018) 012043 doi:10.1088/1757-899X/372/1/012043

Feasibility study of a brine boiling machine by solar energy

W Phayom[1,2]

[1]Energy Engineering, Faculty of Technology, Udon Thani Rajabhat University, Udon Thani, 41000, Thailand

E-mail: warachit@hotmail.com

Abstract. This study presented the technical and operational feasibility of brine boiling machine by using solar energy instead of firewood or husk for salt production. The solar salt brine boiling machine consisted of a boiling chamber with an enhanced thermal efficiency through use of a solar brine heater. The stainless steel solar salt brine boiling chamber had dimensions of 60 cm x 70 cm x 20 cm. The steel brine heater had dimensions of 70 cm x 80 cm x 20 cm. The tilt angle of both the boiling chamber and brine heater was 20 degrees from horizontal. The brine temperature in the reservoir tank was 42°C with a flow rate of 6.64 L/h discharging into the solar boiling machine. It was found that the thermal efficiency and overall efficiency of the solar salt brine boiling machine were 0.63 and 0.38, respectively at a solar irradiance of 787.6 W/m^2. The results shows that the potential of using solar energy for salt production system is feasible.

1. Introduction

The northeast of Thailand is home to the largest rock salt producers in the country. There are areas having the rock salt, salty groundwater, sandstone, or shale mixed with salt underground, creating an obstacle to agriculture [1]. For this reason, such saline areas have been used for salt production to generate income for farmers. Ban Dung District, Udon Thani Province, is an area in which salt can be produced. It has an area of approximately 800 hectares and is owned by 180 farmers. There are two processes for making salt. The first is the salt field method. The second is the salt boiling method. More than 250,000 tons of salt is produced here annually with a market value of approximately 100 to 150 million baht. As such, salt production is the major industry of Ban Dung District [2].

The preferred process to produce the salt is the boiling method, as it requires less area and labor. This is especially true in the rainy season when it is not possible to use the salt field method. Nevertheless, the boiling method requires costly fuel. In the past years, firewood was used as fuel. This is rarely done at present due to stricter regulations for use of forest resources. Currently, the preferred fuel is rice husks (figure 1). Burning rice husks generates greenhouse gases, dust pollution and heat. Additionally, rice husks have high transportation costs as there are delivered from remote areas. Additionally, transportation of rice husks consumes fossil fuels. Should the availability of rice husks decrease, the amount of salt produced will be reduced accordingly.

Solar energy can be considered an infinitely renewable energy source. Furthermore, it is a clean energy as it creates no pollution to the environment, unlike fossil energy. There have been numerous practical applications of solar energy. They employ either the use of energy from the sun's rays, or use of solar thermal energy. The intensity of solar irradiation in Thailand is high from January to April, approximately 20 MJ/m^2-day [3]. However, solar energy in Thailand creates very high levels of heat in comparison to the countries in the temperate zone, where the irradiation is more common [4]. This study presented the technical and operational feasibility of brine boiling machine by using solar energy instead of firewood or husk to produce salt in Northeastern Thailand.

Figure 1. An oven used for burning rice husks to generate heat for producing salt by the boiling method.

2. Experimental apparatus description

The solar salt brine boiling system in this study consisted of solar salt brine boiling chamber used with a solar brine heater or called flat plate collector to increase the efficiency of the system [5,6]. Details of the study are as follows.

2.1. Solar salt brine boiling system

The principle of solar salt brine boiling system is similar to solar distillation. Its purpose is to separate water from dissolved substances using heat [7] but its desired product is the salt in brine. In this research, brine water was mixed between water with salt (ratio of 2:1). The brine water was poured into a brine heater that was connected to a salt brine boiling chamber (figure 2) to increase the brine's temperature before releasing it into the salt brine boiling chamber. The temperature of brine in the brine heater was near the boiling point of water.

The heater increased the brine temperature before allowing it to flow into the solar salt brine boiling chamber. The brine heater was a solar collector consisting of 15 stainless steel tubes, each 600 mm long. Their inner diameter was 20 mm and they were placed in a 700 mm x 800 mm insulated box. The stainless steel tubes were connected and the total tube length was 9000 mm. The outer surfaces of the tubes were painted black and covered by a black metal sheet. The inside of the brine heater was insulated with fiberglass. The brine heater was covered with 3 mm thick transparent glass on the upper side. The brine heater was tilted at a 20° angle from horizontal which was the optimal annual tilt angle for Udon Thani [8,9]. The brine was fed to the heater from a reservoir into the stainless steel tubes from below and removed from the top after the heating process.

The photography of the solar salt brine boiling chamber is shown in figure 3. It was constructed from a stainless steel box with dimensions of 500 mm (width) 700 mm (length) and 70 mm (height), and its surface was black painted. It was insulated with fiberglass and ethylene rubber. The upper side was covered by a 3 mm thick sheet of transparent glass. The solar salt brine boiling chamber's surface was tilted 20° from the horizontal to allow condensing water to run down the condensate channel. The brine feed came from the brine heater and went into the distribution tube which has a longitudinal slot that was 50 mm wide with 0.5 mm diameter hole for each slot. The brine then ran through this slot onto the black absorption tray as a thin film [10]. Using the absorption tray is an easy method for collecting salt at the end of the process. Solar energy transferred heat to the absorption tray. The water evaporated and then condensed when it came in contact with the cool glass cover and flowed into the condensate channel. Only crystalline salt remained on the absorption tray.

Figure 2. The solar salt brine boiling machine showing its brine heater (A) and salt brine boiling chamber (B).

Figure 3. The experimental solar salt brine boiling chamber includes condensate channel (1), distribution tube (2) and black absorption tray (3).

The thermal efficiency (η_{th}) of the brine heater and the salt brine boiling chamber was investigated from the ration of the amount of power used per sq.m. ($\dot{Q}_{used}$) from the machine divided by the solar irradiance (I) as in equation (1). The overall efficiency (η) was the ratio of amount of evaporated brine (S_{evap}) compared to amount of brine taken in (S_{in}), as in equation (2).

$$\eta = \left(\frac{S_{evap}}{S_{in}} \times 100\%\right) \quad (1)$$

$$\eta_{th} = \frac{\dot{Q}_{used}}{I} \times 100\% \quad (2)$$

2.2. *Measuring devices*

Solar irradiance was measured using an EKO-MS-602 pyranometer with a HIOKI-LR-5041 voltage logger. These measurements were done in natural sunlight and the active surface of the specimen was within 5° being coplanar with the active surface of the reference device (Pyranometer) or θ angle. Air temperatures were measured in the salt brine boiling chamber. The brine temperatures were measured in the reservoir tank, at the outlet of the solar brine heater and at the inlet to the salt brine boiling chamber. The temperatures were measured using individually calibrated thermocouples (J type).

3. Experimental results

In this study, the flow rate of brine discharge into the solar salt brine boiling machine was 6.64 L/h. Experimental data was collected between the hours of 9:30 am to 2:30 pm. During this period, the weather was sunny with few clouds. Variation of the solar irradiance and the temperatures of the outdoor air, indoor air and brine in reservoir tank are shown in figure 4. Values of solar irradiance, temperature of outdoor air, indoor air and brine in reservoir tank varied from 619 W/m^2 to 858 W/m^2 (avg. of 787.6 W/m^2), 39°C to 50°C (avg. of 42.4°C), 34°C to 42°C (avg. of 37.8°C) and 37°C to 46°C (avg. of 42.0°C), respectively. The variation of brine temperature at the outlet of the brine heater and air temperature in stainless steel box of the salt brine boiling chamber were compared with brine temperature in reservoir tank and solar irradiance as shown in figure 5.

From figures 4 and 5 show that increasing solar irradiance led to an increase in all temperatures. Decreasing the solar irradiance produced the opposite effect. The air temperature in the salt brine boiling chamber was higher than 100°C between 10:30 am to 12:30 pm. At the same way, the brine temperature at the brine heater outlet was higher than 80°C after one half hour of operation until end of the experiment. It was double the brine temperature in the reservoir tank. Moreover the brine heater could produce brine at temperatures greater than 60°C (57°C to 104°C, average 88°C) throughout the day at flow rate of 6.64 L/h. The maximum temperature difference between inlet and outlet temperatures of the brine heater was 66°C at 11 am. The inlet and outlet temperatures were 38°C and 104°C, respectively. That means using the brine heater would increase the capacity the solar salt brine boiling machine, since the temperature difference between brine and its boiling was decreased. Thus brine flowed down the tray and crystallized immediately as shown in figure 6 and the resulting product is shown in figure 7.

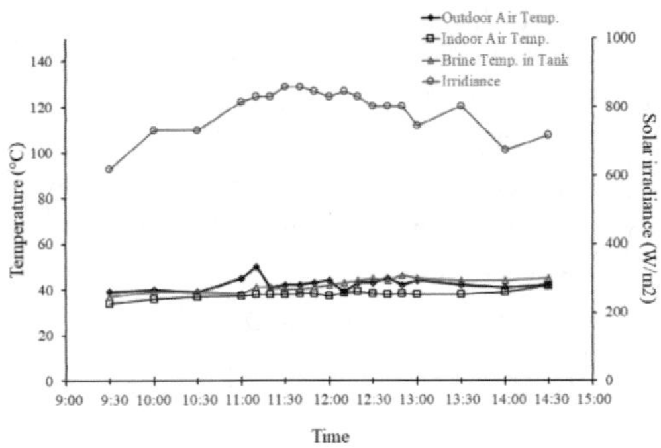

Figure 4. Solar irradiance and temperatures variations.

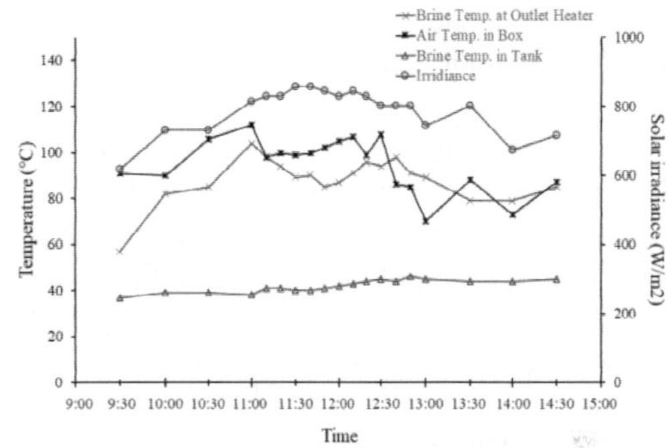

Figure 5. Comparison of brine temperature at outlet of brine heater and in the reservoir tank.

Figure 6. The solar salt brine boiling machine.

Figure 7. The sample of salt produced.

Table 1. The brine boiling machine efficiencies.

Period	kg of inlet brine	kg of outlet brine	kg of produced salt (kg of brine)	kg/m^2/day of produced salt	η
11:00 am-1:00 pm	13.28	6.40	2.29 (6.88)	-	0.52
09:30 am-11:00 am& 1:00 pm-2:30 pm	19.92	14.14	1.93 (5.78)	-	0.29
Whole day	33.2	20.54	4.22 (12.66)	12.06	0.38

Moreover brine boiling machine efficiency was pretty high at 0.52 around noon (11:00 am -1:00 pm) but it was rather low at 0.29 on morning (09:30 – 11:00 am) and afternoon (1:00 pm– 2:30 pm).

When consider all day, salt could be produced with 36.17 kg of brine per sq.m or 12.06 kg of salt per sq.m, the results are shown in table 1. Even the amount of salt in this study was less than a conventional salt boiling system (42 kg of salt per sq.m a day [11]) but solar energy was free.

For thermal efficiency (η_{th}), the brine heater had a η_{th} of 0.8 and it decreased to 0.63 when operated with the salt brine boiling chamber. This results might be due to increased surface area in the brine boiling chamber. Therefore, increasing brine temperature using solar energy before discharging brine to a conventional salt boiling system may be appropriate for a salt production system.

4. Conclusions

A solar salt production system in this study included a solar salt boiling chamber and solar brine heater. It was tested under the actual environmental conditions of Udon Thani Province, Thailand. The first step was to produce hot brine. To do this, brine flowed into the lower part of the solar brine heater and flowed out at the top. The second step was to produce salt. Hot brine flowed to an absorption tray in the solar salt brine boiling chamber, and then salt crystallized. The flow rate of brine was 6.64 L/h. The average brine temperature at outlet of the brine heater was 88°C. However brine temperature is a function of solar irradiance and time. The thermal efficiency and the overall efficiency of the solar salt brine boiling machine were 0.63 and 0.38, respectively at a solar irradiance of 787.6 W/m^2. Therefore, the feasibility of this study when using solar energy in a conventional salt production system is possible, however use the solar brine heater alone is appropriate.

Acknowledgments

The supporting funds for this work were provided by Udon Thani Rajabhat University and are gratefully acknowledged. The author wishes to thank Prof. Dr. Jeffrey C. Naff for his offering of helpful suggestions.

References

[1] Pannachat Moryadee 2013 *J. of Hum.&Soc. Sci., SRU.* **5(1)** 187-201
[2] Udon Thani Today's News Forum 2011 Flip Ban Dung salt filed to solar spa: waste salt field to salt spa-herb mud URL: http://www.udon-today.com/home/index. php/all-news/allnews */61-* -m-m-s, access on 15/09/2012, 2011
[3] Rapapate N and Göl Ö 2007 *RE&PQJ* **1** 638-44
[4] Ministry of Energy 2012 Department of alternative energy development and efficiency *Annual Report 2012* 91–5
[5] Sivakumar P, Christraj W, Sridharan M and Jayamalathi N 2012 *ARPN J. of Eng. & Appl. Sci.* **7** 45–9
[6] Kalogirou S A 2004 *Energ. & Comb. Sci.* **30** 231–95
[7] Aybar H S, Egelioglu F and Atikol U 2005 *Desalination* **180** 285–9
[8] Phayom W 2014 *Appl. Mech. & Mater.* **448-453** 1428–32
[9] Phayom W and Namahima A 2014 *Adv. Mater. Res.* **931-932** 1078-82
[10] Khalil M K and Abu-Abdou 1994 *Renew. Energ.* **4** 839–45
[11] Chaiyarak B 2012 *J. of Mek Soc.* **8** 71-94

IOP Conf. Series: Materials Science and Engineering **372** (2018) 012044 doi:10.1088/1757-899X/372/1/012044

Multi-objective optimization of process parameters of multi-step shaft formed with cross wedge rolling based on orthogonal test

S T Han[1,2], X D Shu[1,2,4], V Shchukin[3] and G Kozhevnikova[3]

[1]College of Mechanical Engineering and Mechanics, Ningbo University, Ningbo-315211, China
[2]Zhejiang Provincial Key Laboratory of Part Rolling Technology, Ningbo-315211, China
[3]Belarusian National Technical University, Minsk, Belarus

E-mail: shuxuedao@nbu.edu.cn

Abstract. In order to achieve reasonable process parameters in forming multi-step shaft by cross wedge rolling, the research studied the rolling-forming process multi-step shaft on the DEFORM-3D finite element software. The interactive orthogonal experiment was used to study the effect of the eight parameters, the first section shrinkage rate φ1, the first forming angle α1, the first spreading angle β1, the first spreading length L1, the second section shrinkage rate φ2, the second forming angle α2, the second spreading angle β2 and the second spreading length L2, on the quality of shaft end and the microstructure uniformity. By using the fuzzy mathematics comprehensive evaluation method and the extreme difference analysis, the influence degree of the process parameters on the quality of the multi-step shaft is obtained: β2>φ2>L1>α1>β1>φ1>α2>L2. The results of the study can provide guidance for obtaining multi-stepped shaft with high mechanical properties and achieving near net forming without stub bar in cross wedge rolling.

1. Introduction

Cross wedge rolling is one of new technologies forming of shaft parts, compared with traditional cutting and forging parts, it can form stepped shaft parts that are close to the final shape with a little loss of material. It is an important part of advanced manufacturing technology with the advantages of remarkable lumber and high efficiency in production [1]. In recent years, the transportation industry has been developing rapidly all over the world. As the core part of the driveline, multi-step shafts have high requirements for the macro size accuracy and comprehensive mechanical properties; however, shafts have large deformation and complex metal flow and deformation mechanism during forming process of cross wedge rolling. It is easy to appear the phenomenon of mixed crystal and coarse grain, which affects the mechanical properties of products [2-4]. In order to achieve near net forming and good mechanical properties, it was researched by the finite element method in this paper that process parameters influence the end equality and microstructure uniformity of multi-step shaft. The influence of eight main technological parameters on the concave and microcosmic grain sizes was measured. According to the importance of the measurement index, the measured data were weighted and synthetically judged through the method of combining the orthogonal experiment method and the fuzzy mathematics comprehensive evaluation [5,6], the influence degree of different factors and different levels of each factor on the comprehensive value were determined through the range analysis. The best process parameters combination of multi-step shaft was found, which validity was verified. Therefore, the results of this study can provide guidance for the production of high comprehensive cross wedge rolling products with high comprehensive performance.

2. The establishment of the finite element model

The stepped shaft is the main component of the automobile transmission, and it has a high requirement for the forming precision and the mechanical performance. In this paper, the numerical simulation of the cross wedge rolling process is simulated by the DEFORM-3D finite element software, taking the three-step shaft cross wedge rolling as an example. The shape of the part will be as shown in figure 1(a). When the above model is imported to software DEFORM-3D, 1/2 of the model is taken to build the symmetrical shaft and the symmetry constraint is applied to save the computation time, reduce the storage space, and keep the symmetry of the model [7,8]. The rigid plastic finite element model established in this paper will be as shown in figure 1(b). The simulation is meshed by relative meshes, which number is 50000, the rolled material is 42CrMo alloy steel, the billets diameter is 50mm, and the rolling temperature is 1050°C. The heat exchange coefficient is set to be 11 W/(mm^{2}· K), and the convective heat transfer coefficient with air 0.02 W/(mm^{2}· K). The initial grain size is set to be 150um in the finite element simulation according to the same experimental result of initial grain size [9,10].

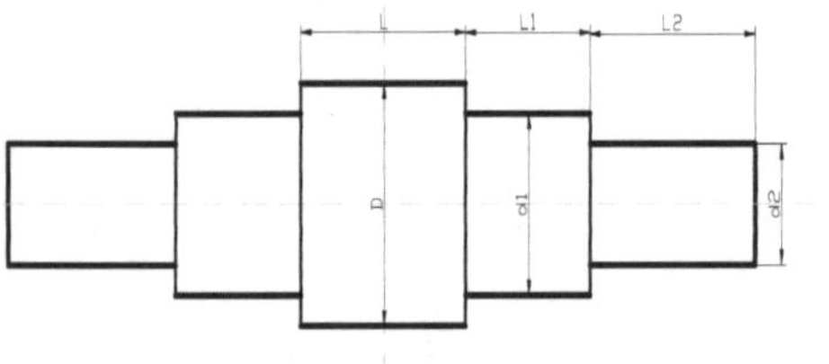

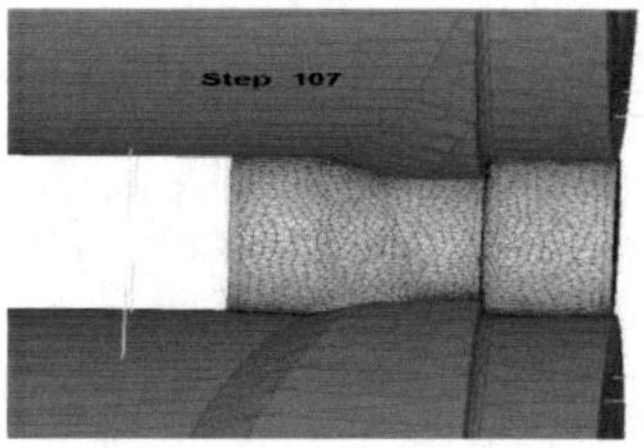

(a) Step-shaft part (b) Finite element model of cross wedge rolling

Figure 1. Step-shaft part and finite element model.

3. Experimental design

3.1. Selection of process parameters

The three-step shaft passes through two times cross wedge rolling plastic forming process, and there are many technological parameters that affect the forming quality. This paper mainly selects 8 model processing parameters as follows: the first section shrinkage rate φ1, the first forming angle α1, the first spreading angle β1, the first spreading length L1, the second section shrinkage rate φ2, the second forming angle α2, the second spreading angle β2 and the second spreading length L2, as shown in figure 2.

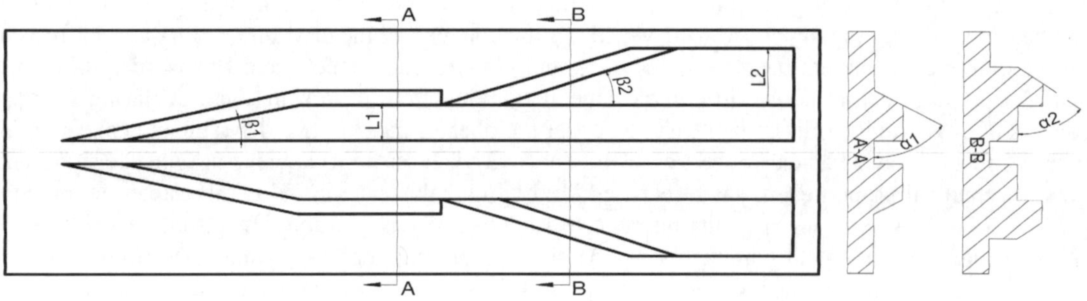

Figure 2. Cross wedge rolling die.

Table 1. Level of the process parameters and factors.

Level	first section shrinkage rate φ1	second section shrinkage rate φ2	the first spreading length L1	second spreading length L2	first forming angle α1	second forming angle α2	the first spreading angle β1	second spreading angle β2
	A	B	C	D	E	F	G	H
1	36	60	20	20	28	24	5	5
2	41	65	30	30	30	26	7	7
3	46.5	70	40	40	32	28	9	9

Table 2. Experimental scheme and experimental result table.

Test number	Influencing factor level setting													test results		
	A	B	A×B (C×D)	B×A	C	E	F	G	D	H	Empty column	Empty column	C×D	Average grain size/um	Concave size/mm	Overall rating
1	2	1	2	1	1	1	3	3	2	3	3	2	3	47.20	17.40	8.44
2	1	3	3	2	1	2	2	1	2	3	3	1	2	46.70	20.64	7.15
3	3	2	1	3	1	3	1	2	2	3	3	3	1	47.50	20.10	7.33
4	1	1	1	1	1	1	1	1	1	1	1	1	1	45.90	20.10	7.64
5	2	3	1	1	2	3	2	1	3	1	3	3	3	27.85	30.71	3.00
6	3	2	1	2	2	1	2	3	2	2	1	1	3	40.00	21.10	7.74
7	2	2	3	2	2	1	3	1	1	3	2	3	1	63.60	22.29	4.17
8	3	3	2	2	1	2	3	2	1	1	1	3	3	45.20	23.72	4.82
9	2	2	3	3	1	3	2	3	1	1	1	2	2	47.00	19.65	7.67
10	3	3	2	3	3	1	2	1	1	2	3	2	1	36.30	24.80	5.45
11	1	1	1	3	2	2	2	2	1	3	2	2	3	46.03	16.78	8.78
12	2	1	2	3	2	2	1	1	2	2	1	3	2	33.10	22.43	7.04
13	1	3	3	3	3	1	1	3	2	1	2	3	3	58.60	28.46	0.31
14	2	3	1	2	1	2	1	3	3	2	2	2	1	29.22	20.86	8.31
15	1	2	2	3	1	3	3	1	3	2	2	1	3	62.00	23.23	3.29
16	1	2	2	1	3	2	2	3	3	3	1	3	1	47.90	17.80	8.22
17	3	1	3	1	1	1	2	2	3	2	2	3	2	44.93	19.22	8.23
18	1	1	1	2	3	3	3	3	1	2	3	3	2	42.80	17.05	9.20
19	3	1	3	2	3	3	1	1	3	3	1	2	3	56.50	16.44	7.29
20	1	3	3	1	2	3	3	2	2	2	1	2	1	40.23	23.20	5.90
21	3	3	2	1	2	3	1	3	1	3	2	1	2	60.00	21.27	5.09
22	2	1	2	2	3	3	2	2	2	1	2	1	1	42.20	22.60	6.25
23	2	3	1	3	3	1	3	2	3	3	1	1	2	33.98	20.52	8.52
24	1	2	2	2	2	1	1	2	3	1	3	2	2	52.00	27.40	1.27
25	3	2	1	1	3	2	3	1	2	1	2	2	2	33.53	24.24	5.36
26	3	1	3	3	2	2	3	3	3	1	3	1	1	37.06	22.30	6.36
27	2	2	3	1	3	2	1	2	1	2	3	1	3	40.00	20.00	8.46

The level of the process parameters and factors are shown in table 1, In the design of orthogonal test, this paper, in addition to the above eight factors, also investigate the interactions between the first section shrinkage rate and the second section shrinkage rate, the first spreading length and the second spreading length. The corresponding experimental design and results are shown in table 2.

3.2. Measurement indicators

In this paper, the effect of different process parameters on the end quality and microstructure uniformity during the cross wedge rolling process is mainly investigated. For the end quality, it is mainly measured by the axial length of the concave, and the microstructure uniformity is mainly measured by the average grain size.

3.2.1. Calculation method of concave value. As one of the measuring indexes, the concave directly reflects the quality of the end of the workpiece, so it is particularly important to measure the size of the concave effectively. This article takes the mean value measured by multiple measurements as a measure, specific methods as shown in figure 3: 25 nodes are evenly selected on the outermost ring of the end section of the end of the rolling piece. Then the central point of the concave and 24 points evenly distributed on the circular boundary which takes the central point as a centre point and approximately 0.2 mm as a radius are selected. The value of the single side concave of the rolled piece is the average axial coordinate value of the 25 nodes of the outer ring of the workpiece minus the average axial coordinates of the 25 nodes of the central region. The calculation formula is as follows:

$$Y = \frac{1}{25}\sum_{i=1}^{25} Si - \frac{1}{25}\sum_{i=1}^{25} Sj \quad (1)$$

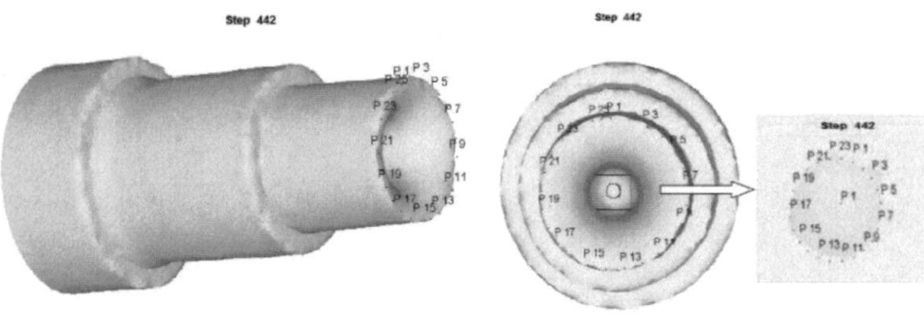

Figure 3. distribution of feature points.

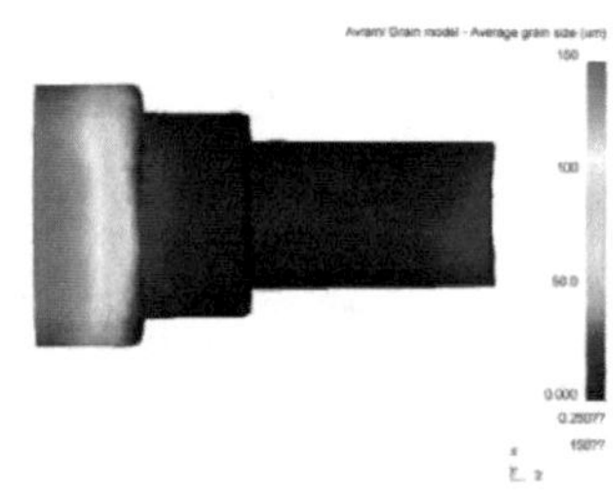

Figure 4. The microcosmic average grain size.

3.2.2. Microstructure uniformity. The uniformity of microstructure mainly includes the uniformity of grain distribution and the size of grain size [11]. The grain refinement in the process of multi-step rolling is mainly concentrated in the forming area. The microcosmic average grain size of the step shaft is shown in figure 4. minimum grain size of second-steps and third-steps metal of shaft after rolling process are 0.2 μm, and the first-step metal of shaft has not been rolled, although due to the axial deformation of penetration, part of the grain was refined, but the overall grain size is still maintained at more than 140 μm. Based on the above microcosmic grain distribution characteristics and taking into account the measurability, we choose the average grain size as a measure of microstructure uniformity.

3.3. Data processing

In the simulation experiments, the two selected indicators have great difference in dimension, and they also have different emphasis on the importance of products. It is impossible to directly superimpose the two sets of data as an evaluation index. Therefore, according to the industry standard and the actual production experience, four indexes intervals are divided combined with the statistical method of fuzzy mathematics as shown in table 3, The index values of each interval are mapped to the [0,1] numerical interval respectively through the mapping function of the S type distribution, according to the difference of the important degree of each factor, the weighted comprehensive score is used for the mapping values of each index. In this paper, the weighted values of the size of the concave and the average grain are b1=7, b2=3, respectively, as shown in formulas (2) and (3).

Table 3. Value of each parameter of the function.

measuring indexes.	Index values		
	a	b	c
Average grain size/um	63.5	45	31.8
Concave size/mm	30	23	16

$$\text{S type mapping function } Y_{ij} = \begin{cases} 1 & x_{ij} < c \\ 1-2\left(\dfrac{x_{ij}-c}{a-c}\right)^2 & c \le x_{ij} < b \\ 2\left(\dfrac{x_{ij}-a}{a-c}\right)^2 & b \le x_{ij} < a \\ 0 & x_{ij} \ge a \end{cases} \tag{2}$$

$$\text{the weighted comprehensive score } Y = b_1 Y_{i1} + b_2 Y_{i2} \tag{3}$$

In the formula: i is the test number, j is the number of measuring indexes.

In order to get the relationship between the various factors and the comprehensive index, to find out the influence laws and trends of the factors on the quality of the rolled pieces, the data were analysed by range analysis. Through range analysis of experimental data, the greater the range value, the greater the impact of the factor on the comprehensive index, the more critical the factor, and vice versa. The range values and effects of various factors are shown in table 4.

Table 4. Range analysis table.

projects	A	B	A×B (D×C)	B×A	C	E	F	G	D	H	C×D		
T1	51.7	69.2	65.88	60.34	62.8	51.7	52.7	50.3	61.2	42.6	59.63	64.8	60.4
T2	61.8	53.5	49.86	56.19	49.3	64.4	62.4	59.5	55.5	63.6	59.51	49.7	58.4
T3	57.6	48.5	55.53	54.74	59.0	55.0	56.0	61.3	54.4	64.9	52.13	56.6	52.3
t1	5.8	7.7	7.3	6.7	7.0	5.8	5.9	5.6	6.8	4.7	6.6	7.2	6.7
t2	6.9	6.0	5.5	6.2	5.5	7.2	6.9	6.6	6.2	7.1	6.6	5.5	6.5
t3	6.4	5.4	6.2	6.1	6.6	6.1	6.2	6.8	6.1	7.2	5.8	6.3	5.8
Excellent level	2	1	1	1	1	2	2	3	1	3	1	1	1
R	10.1	20.7	10.8		13.5	12.7	9.7	10.9	6.8	22.3	11.7	15.0	8.2
The order of primary and secondary					H>B>C>E> C×D>G>A×B>A>F>D								

3.4. Result analysis

According to the range analysis, it can be clearly seen that the second spreading angle H and the second section shrinkage rate B have great influence on the comprehensive index. the second spreading length D and the second forming angle F have little influence on the comprehensive index. The important influence degree of each factor from large to small is H>B>C>E>C×D>G>A×B>A>F>D, The combination of the best factor level of the model is A2B1C1D1E2F2G3H3. From the horizontal impact trend diagram of figure 5, it can be seen that the changing trend of H is bigger, and the changing ranges of D and F are relatively small. The horizontal combination of the maximum values are the best process parameters. There is no corresponding combination in the test scheme of table 2, so it must be tested by experiments.

Through DEFORM-3D finite element numerical simulation, under the optimized process level condition, namelyA2B1C1D1E2F2G3H3, the average grain size is 34.5 μm, the average concave value is 17.79 μm, the comprehensive score is 9.724. In comparison with table 2, it is known that the scheme is the best process plan.

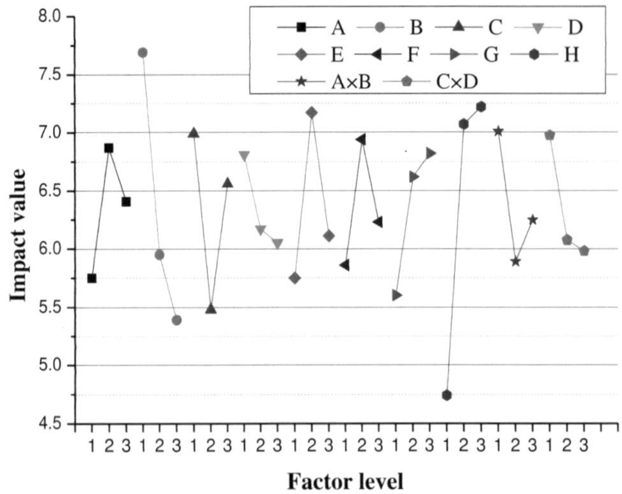

Figure 5. Tendency chart of various factors at different levels.

4. Conclusion

- Process parameters are the key factors affecting the quality of cross wedge rolling products. Through the reasonable design of the technological parameters of the cross wedge rolling die, the end quality and microstructure uniformity of the rolled piece will be improved effectively, thus the comprehensive performance of the product has be improved.
- With the comprehensive index as the evaluation standard, the main and secondary relationship between the factors of the process parameters and the rolling quality of the multi- stepped shaft is in turn as follows: the second spreading angle H, the second section shrinkage rate B, the first spreading length C, the first forming angle E, the first spreading angle G, the first section shrinkage rate A, the second forming angle F, the second spreading length D, namely $\beta2>\varphi2>L1>\alpha1>\beta1>\varphi1>\alpha2>L2$, the best combination is A2B1C1D1E2F2G3H3.

Acknowledgments

The Project is supported by the National Natural Science Foundation of China (Grant No. 51475247), the Natural Science Foundation of Zhejiang Province (Grant No. LZ17E050001) and the Science and Technology Project of Zhejiang Province (Grant No. 2016C31018).

References

[1] Wei J, Shu X, Tian D, *et al* 2017 Study in shaft end forming quality of closed-open cross wedge rolling shaft using a wedge block *Int J Adv Manuf Technol* **93** 1095-105
[2] Yan B, Shu X D and Hu Z H 2008 Development and status of study on microstructure evolvement regularity of workpiece of cross wedge rolling *China Metall* **2** 8-10
[3] Pater Z 2010 Development of cross-wedge rolling theory and technology steel research international *Steel Res Int.* **25** 32
[4] Zhou Z R, Shu X D, *et al* 2016 Effects of process parameters on microstructure of workpiece in closed-open joint rolling *Hot Work Technol* **45** 111-4
[5] Xia J M, Zhang L and Wen L 2016 Multi-objective optimization on stamping process parameters based on orthogonal test *Forg Stamping Technol* **41** 33-6
[6] Pan B S, Gong H L and Liu H 2007 Optimized design of the injection molding process for multi-object based on orthogonal experimental method *J Zhejiang Univ Technol* **35** 308-12

[7] Wang M-T, *et al* 2013 Numerical simulation and modeling of hot deformation microstructure evolution of a non-quenched and tempered steel in cross wedge rolling *Trans Mat Heat Treat* **2** 34
[8] Pater Z 2005 Finite element analysis of cross wedge rolling *J Mater Process Tech* **2** 173
[9] Chen M S, Lin Y C and Ma X S 2012 The kinetics of dynamic recrystallization of 42CrMo steel *J Mater Sci Eng A* **556** 260-6
[10] Wei Y L, Shu X D, Han S T, *et al* 2017 Analysis of microstructure evolution during different stages of closed-open cross wedge rolling *Int J Adv Manuf Technol* **95** 1975-88
[11] Zhang N, Wang B Y and Lin J G 2012 Effect of cross wedge rolling on the microstructure of GH4169 alloy *Int J Adv Manuf Technol.* **19** 836-42

IOP Conf. Series: Materials Science and Engineering **372** (2018) 012045 doi:10.1088/1757-899X/372/1/012045

The three-dimensional simulation analysis of dynamic response on perforated strings

M F Li[1,2], H F Liu[1], Y H Dou[1], L H Cao[1] and Y X Liu[1]

[1]Northwestern Polytechnical University, School of Mechanical Engineering, Xi'an, China

E-mail: limf88@126.com

Abstract. It analyzes the dynamic response and stresses of perforating tubular string to detonating impact load in oil-gas well in ANSYS, obtains the response of vibration displacement, velocity and acceleration of perforating tubularstring caused by detonating impact load, finds the influence of the length and wall thickness of perforating tubular string to working stresses. The result shows that:when the detonating impact load exerts the perforating tubular string with compressive and tensile axial force alternatively;the vibration displacement, velocity and acceleration of perfora-ting tubular string change periodically at same cycle;the closer to the perforating gun, the larger the amplitude of vi-bration velocity and acceleration;the closer to the packer the smaller the vibration displacement, the larger the work-ing equivalent stress of perforating tubular string;the longer or the thicker the perforating tubular string, the smaller the working equivalent stress and the higher the strength safety. Therefore, it uses the damping tube between packer and perforating gun as well as thick walled tubing to increase the strength safety of perforating tubular string.

1. Introduction

In the HTHP deep wells, such as KS101 and WC1 which belong to CNPC, the fracture of the center pipe of the string of the packer which used to complete testing and perforation has occurred [1], and the accident of the strings broken off well was also occurred in a development well in the northeast of Sinopec [2]. With the view of understanding the impact of perforation detonation load on perforation string and downhole tool safety, engineers have carried out theoretical and experimental research. After nearly ten years of research, the Schlumberger has developed the optimization design technology of the perforated string. Through the prediction of the perforation detonation load and the strength safety of the string, the perforation parameters and the perforation string configuration are optimized to prevent the damage of the perforation string and the downhole tools [3]. Drawing lessons from the thought of Schlumberger, the relationship between the structure of the well and the length of the string and the stress strength of the string is analyzed in the case of the perforated string of a HTHP ultra deep well [4-6].

The influence of perforation on the strength of the string mainly lies in the vibration of the string which caused by perforation detonation, furthermore, the response of the string to perforation impact load. Due to the complexity of this problem, the current research mainly rely on finite element numerical analysis and experimental study [7,8]. Considering the inclination and azimuth, the finite element analysis model of horizontal well perforation was established, meanwhile, the relationship between vertical displacement and velocity of perimeter along the perimeter was analyzed [9], Puls Frac perforation engineering software and ANSYS finite element analysis software were used to analyze the implicit-explicit dynamic behavior of perforated pipe string. It was found that the stress of the pipe string at the lower end of confined packers was the highest when perforation was found. According to the existing research foundation, finite element model of perforation string is established by finite element analysis software ANSYS in this paper. The dynamic response and stress intensity of perforation string are analyzed. The vibration displacement, velocity and acceleration of perforation pipe string the response law of equivalent stress to perforation impact load, the effect of perforation length and wall thickness on the stress intensity of the string, and providing guidance for optimizing the configuration of perforation string.

2. Perforated section pipeline finite element model establishment and loading

To analyze the transient dynamic response of the string under the impact of fast changing impact load, the finite element analysis software ANSYS Workbench transient dynamic analysis module is applied to the finite element analysis of the perforated string. As shown in figure 1, the 20-m-long perforated section of the packer under the 88.9 mm × 9.52 mm P110 tubing-mounted perforating string is taken as the research object. The tubing material has a yield strength of 758 MPa, an elastic modulus of 206 GPa, a Poisson's ratio of 0.3 and the density of 7.85 g/cm^3. As shown in figure 2, the casing is a homogenous casing, the tubing can be regarded as an elastomer. Vibrates occurs under the combined action of perforating shock waves and percussive fluid pressure pulsation, the eight-node hexahedron entity unit is used to mesh the pipe string by using the mapping meshing method. The 20 m long pipe string is divided into 24,160 units with a total of 169,200 nodes.

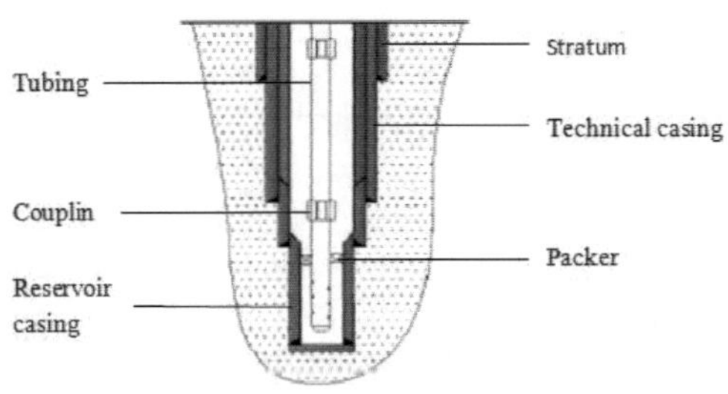

Figure 1. Schematic diagram of tubing conveyed perforating string.

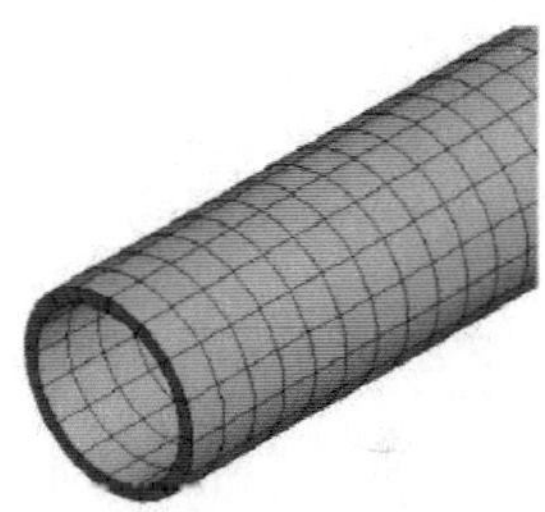

Figure 2. Local diagram of finite element model of the perforated strings.

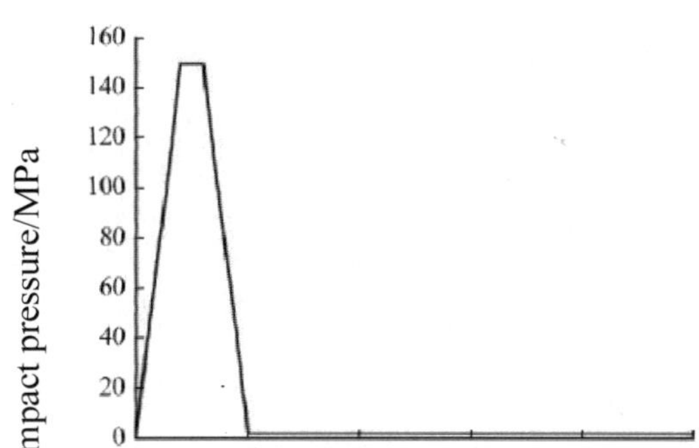

Figure 3. Axial shock load by perforating.

The simplified perforation impact load-time curve is shown in figure 3 [7], the peak pressure is 150 MPa and the loading time is 50 ms. The impact load that ia shown in figure 3 is applied to the finite element analysis model of perforated pipe string which is shown in figure 2, and the finite element analysis of dynamic response of perforated pipe string can be used to obtain the vibration displacement, velocity and acceleration of perforation string. In addition, we can know the dynamic response of perforation pipe string to perforation impact load from the law of equivalent stress changing with time.

3. Dynamic response analysis column vibration displacement perforations

Figure 4 is the curve of the displacement-time of the pipe string at 5 m, 10 m and 20 m from the packer. It can be seen from the diagram that the vibration displacement of string at different locations varies periodically with time, similarly to the sine curve, and the cycle is the same, which is about 16 ms. The peak vibration displacement of the distance packer which located at 20 m is 22 m, the peak vibration displacement of the distance packer is 13 mm, and the peak vibration displacement is 8 mm at the distance packer 5 m, which indicates that the closer the distance packer is, the smaller the vibration displacement of the string is.

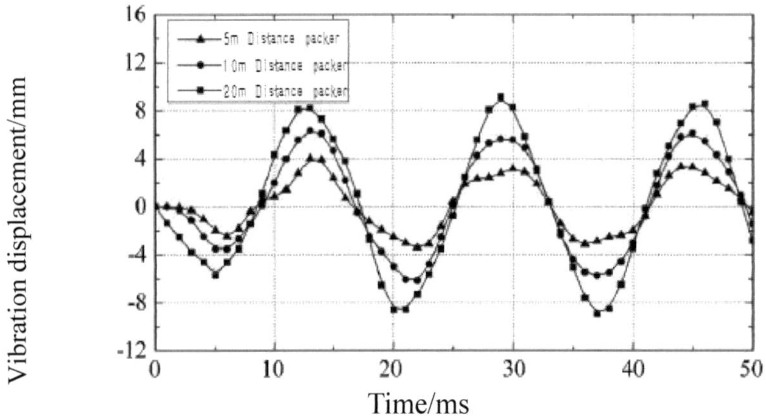

Figure 4. Vibration displacement of perforating string at different positions.

4. Dynamic response analysis of vibration velocity of perforated pipe string

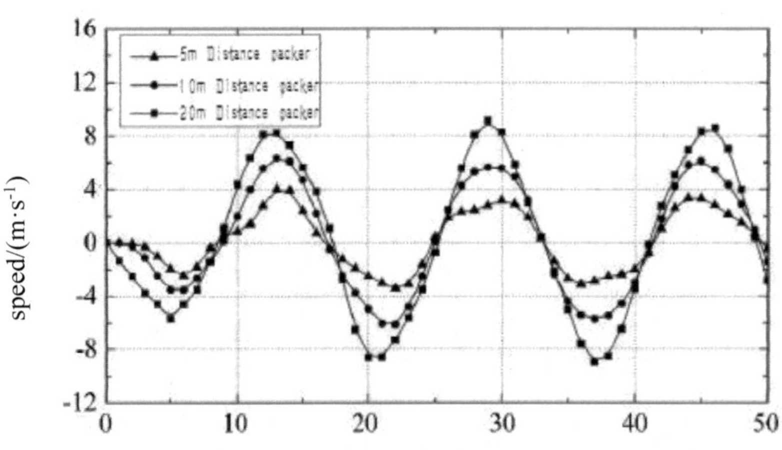

Figure 5. Vibration velocities of perforating string at different points.

Figure 5 is a curve of the vibration velocity from 5 to 10 and 20 m away from the packer. It can be seen that the vibration velocity of the column varies with time periodically with the same period. The peak vibration velocities of the columns at 5 m, 10 m and 20 m were 4 m/s, 6 m/s and 9 m/s respectively, that is,the farther the distance packer is, the greater the amplitude of the vibration velocity.

5. Dynamic response analysis column vibration acceleration perforations

Figure 6 shows the curve of vibration acceleration of a string at 5 m, 10 m and 20 m from the packer with time. It can be seen from the diagram that the vibration acceleration of the pipe string at different positions alos changes periodically with time, and the cycle is the same. The peak vibration acceleration at 20 m, 20 m, 5 m distance packers are respectively 3500 m/s^2 (about 350 g), 2200 m/s^2 and 1600 m/s^2, that is, the farther the distance from the packer ,the great the vibration acceleration.

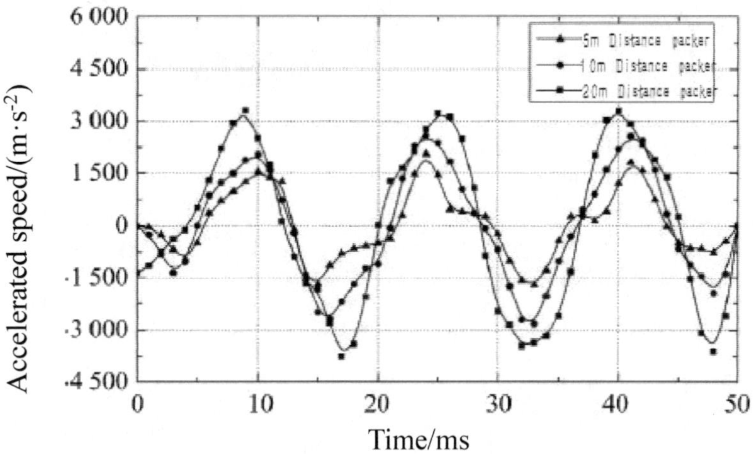

Figure 6. Vibration velocities of perforating string at different points.

6. Dynamic response analysis of equivalent stress of perforated pipe string

Figures 7 and 8 show the equivalent stress contour of the string at the packer at 6 ms and 10 ms respectively. It can be seen that the stress wave has not been transmitted to the packer at 6 ms and the equivalent stress of the string at the packer. The maximum equivalent stress is about 182 MPa. With the propagation of percussion shock waves, the equivalent stress of the column at the packer gradually increases and reaches 441 MPa at 15 ms. In addition, it can be seen that the equivalent stress distribution on the pipe section is not uniform, and the maximum equivalent stress always appears at the outer wall of the pipe string. Figure 9 shows the curves of the maximum equivalent stress of the packer processing column and the perforating section with time.

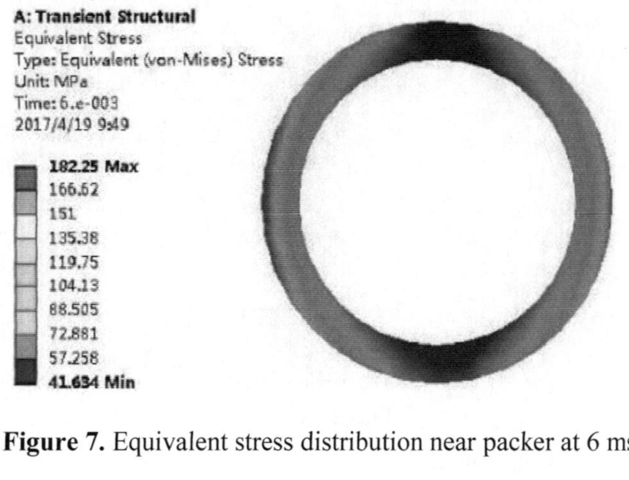

Figure 7. Equivalent stress distribution near packer at 6 ms.

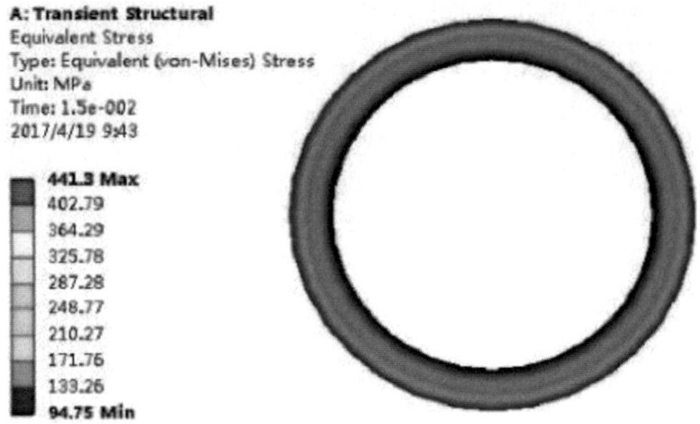

Figure 8. Equivalent stress distribution near packer at 15 ms.

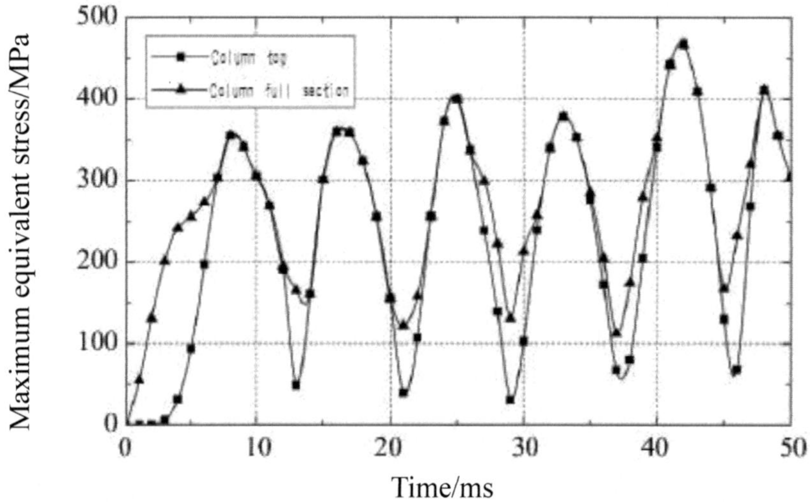

Figure 9. The maximum equivalent stress of string versus time.

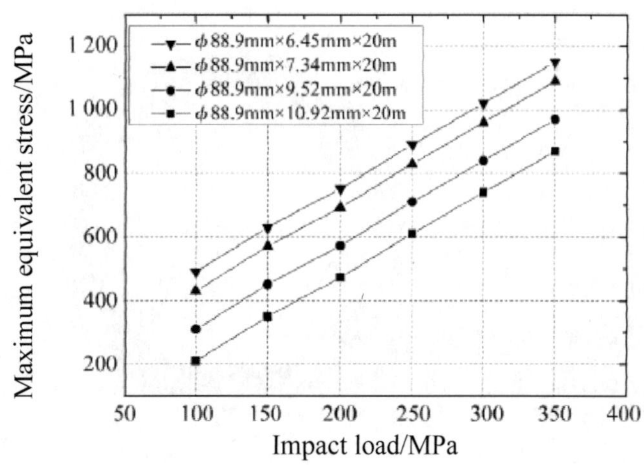

Figure 10. Maximum equivalent stress of columns with perforations of different wall thicknesses under different amplitude perforation impact loads.

7. Influence of length and wall thickness of perforated pipe string on stress intensity

Using the above-mentioned finite element analysis model and method, the impact load amplitude of perforation is assumed to be 100, 150, 200, 250, 300 and 350 MPa respectively. Observing the effect of the length and wall thickness of the perforation section on the maximum equivalent stress is investigated, as shown in table 1 and figure 10.

Table 1. Under different amplitude impact load of different lengths perforation section of the maximum equivalent stress.

Tubing length/m	**Impact load**/MPa					
	100	150	200	250	300	350
20	248	398	498	677	798	857
30	217	369	469	649	769	826
40	203	357	456	637	758	812
50	195	351	448	629	752	805

As can be seen from table 1 and figure 10, the maximum equivalent stress is positively correlated with the amplitude of perforating impact load. The longer the column of perforation or the thicker the wall, the smaller the maximum equivalent stress is. Therefore, setting the shock-absorbing tubing between the perforating gun and the packer (increasing the length of the perforating section), or increasing the thickness of the perforating section can improve the strength as well as safety of the perforating section's string.

8. Conclusion

- Under the percussive detonation impact load, the pipe string under perforation is subjected to alternating compressive and tensile impact loads. The vibration displacement, vibration velocity and vibration acceleration around the string periodically change with time as well as the period. The farther away from the packer (the closer to the detonation source of the perforation), the greater the amplitude of the vibration displacement, the vibration velocity and the vibration acceleration is, the peak of the vibration acceleration of the string caused by perforation impact can reach hundreds of gravitational accelerations, and go through a dynamic load.
- The closer to the packer, the smaller the displacement amplitude of the string is, but the larger the equivalent stress is, the equivalent stress of the packer at the beginning of the perforation is smaller, in addition, as the time increases, the propagation of the detonation wave, the equivalent stress of the column at the packer gradually increases, and the column at the packer is the dangerous point of stress in the column of the perforation section. As a result, perforation strings often experience plastic buckling near the packer even cause damage to the packer core tube.
- Under the same perforation impact load, the longer the column of perforating section or the thicker the wall of the column, the smaller the maximum equivalent stress is and the safer the column. Therefore, it is possible to provide a damping tube between the perforating gun and the packer (to increase the length of the perforating section string), or to increase the wall thickness of the perforating section column to improve the strength and safety of the perforating section string.

Acknowledgments

The research was funded by National Natural Science Fund for Study on pulsating pressure and dynamic response of tubular string during perforating in high pressure high temperature deep well (51374171) and Research Project of Comprehensive Perforation Optimization of Unconventional Oil and Gas Reservoirs Based on the Strength and Safety of Oil Casing Columns (16JK1606) and National Natural Science Fund for Study on stress of gas production string in directional well based on energy method

(2014JQ7289).

References

[1] Dou Y H, Xu H J, Jiang X H, *et al* 2007 Perforation test co-packers center tube damage analysis *China Petroleum Machinery* **35** 113-5
[2] Peng X Z and Huang Q S 2010 On the perforating section of the pipe tubing trip reason *West Prospecting Project* **23** 88-90
[3] Bauman C, Barnard K, Lu A, *et al* 2013 Prediction and re-duction of perforating gunshock loads *International Petro-leum Technology Conference* (Beijing: IPTC Programme Committee)
[4] Dou Y H, Li M F, Zhang F X, *et al* 2012 Influence of well structure on strength and safety of tubing string in perforation section *China Petroleum Machinery* **40** 27-9
[5] Lian Z H, Lin T J, Liu J, *et al* 2006 Finite element analysis of perforation vibration of horizontal well tubing string *Oil Drilling & Production Technology* **28** 56-9
[6] Chen H B, Tang K, Ren G H, *et al* 2010 Dynamic analysis of ultra-deep well perforation pipe string *Logging Technology* **34** 487-91
[7] Cai L Z, Zhao W, Xue S F, *et al* 2015 Dynamic response analysis of tubular structure during perforation *Oilfield Machinery* **44** 26-30
[8] Zhou H F 2014 *Dynamic Response of Well Perforating Column to Detonation Impact Load* (Beijing: Beijing Institute of Technology)
[9] Zhang Q, Zhang M L, Zhou Z H, *et al* 2015 Transmission process of underground perforation shock wave and pipe string movement *Journal of Yangtze University* (Natural Science Edition) **12** 45-7

Cracking mechanism of shale cracks during fracturing

X J Zhao[1,2,5], Q Zhan[1,2], H Fan[2], H B Zhao[3] and F J An[4]

[1]School of Aeronautics, Northwestern Polytechnical University, Xi'an, Shanxi, 710072, China
[2]School of Electronic Engineering, Xi'an Shiyou University, Xi'an, Shanxi, 710065, China
[3]Xi'an Center of Geological Survey, Shanxi Xi'an 710054, China
[4]Dongying Shengfeng Safety Service Corporation, Shandong Dongying 257000, China

E-mail: zhaoxsyu@qq.com; zhqu@xsyu.edu.cn; 723515507@qq.com; 315695573@qq.com; 94500732@qq.com

Abstract. In this paper, we set up a model for calculating the shale fracture pressure on the basis of Huang's model by the theory of elastic-plastic mechanics, rock mechanics and the application of the maximum tensile stress criterion, which takes into account such factors as the crustal stress field, chemical field, temperature field, tectonic stress field, the porosity of shale and seepage of drilling fluid and so on. Combined with the experimental data of field fracturing and the experimental results of three axis compression of shale core with different water contents, the results show that the error between the present study and the measured value is 3.85%, so the present study can provide technical support for drilling engineering.

1. Introduction

The shale formation is mainly composed of illite and mixed layer clay minerals, the horizontal bedding, crack and micro crack development of shale are obvious [1]. When the shale meets water, it can easily hydrate [2-4], the mechanical strength of it is reduced, the expansion stress is produced, and the pore pressure of the rock is increased [5]. Furthermore, water molecules in drilling fluids flow through the formation pores, then the additional stress field yields around the well wall [6], which leads to the decrease of the effective circumferential stress of the well. Besides, rock wall temperature changes can also cause wellbore instability, especially for the shale with high elastic modulus and low Poisson's ratio [7]. During the process of drilling fluid circulation, the thermal expansion of the surrounding rock of the shaft wall is not allowed to expand freely due to the limitation of wellbore liquid pressure and the surrounding rock of the shaft wall, so that the expansion stress will generate, which is the thermal stress affecting the shale fracture pressure [8].

At present, experts have proposed various methods for predicting formation fracturing pressure. Huang's model [9] is mainly for general reservoirs. It considers the relationship between overburden pressure and depth of rock strata, the influence of stress concentration on wellbore, the effect of tectonic stress of underground nonuniform distribution and the strength of strata and so on. Therefore, it can be applied to different deformations. Yan *et al* [10] established a fracture pressure model considering hydration based on Huang's model, which improved the prediction accuracy, but did not consider effect of perforation and natural fractures on the cracking of hydraulic cracks. Li *et al* [7] established a calculation model of fracture pressure on sandstone stratum coupling with temperature. Deng *et al* [11] established a model for calculating the fracture pressure considering the influence of high temperature and seepage, the prediction accuracy increased by 7%. However, this model is only fit for sandstone.

However, the fracture pressure model of shale considering the additional stress fields caused by the crustal stress field, chemical field, temperature field, tectonic stress field, the porosity of shale and seepage of drilling fluid is still absent. This paper is based on Hang's model, considering all these factors and established a new model for calculating the fracture pressure of shale.

1.1. Influence of seepage and porosity on crustal stress field

On the infinite plane, a circular hole receives uniform internal pressure, at the same time, the infinite distance in the plane is subjected to two horizontal stresses (σ_1 and σ_2) and overlying pressure in the vertical direction. Considering that the rock is a small deformation elastic body, the principle of linear superposition is applicable to it. The stress state of the wall rock of the wellbore can be obtained by studying the stress contribution of the stress components to the wall rock of the wellbore, and then superimposing them. It is assumed that the strata are homogeneous isotropic and linear elastic porous materials, the rocks around the borehole are in a state of plane strain [2]. The stress model of the wellbore wall is shown in figure 1. R is the maximum borehole radius, m; r is the radius of borehole, m; θ is the polar angle of the reverse clockwise rotation of the maximum radial direction in the radial direction.

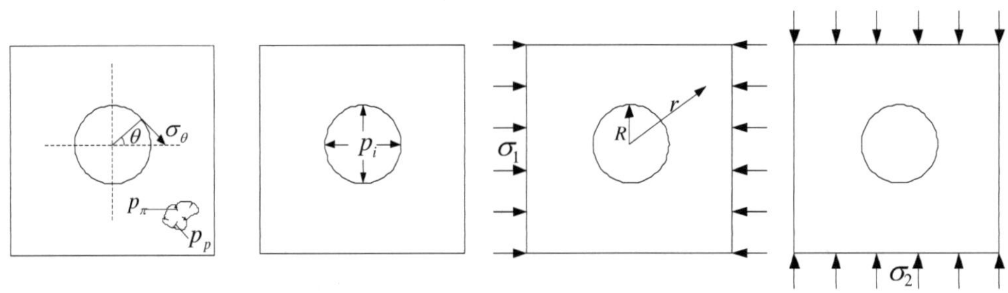

Figure 1. Stress state around wellbore.

The wall rock of the borehole wall is porous medium, so the flow of fluid satisfies Darcy's law, the radial flow of drilling fluid filtrate in the stratum will generate additional stress field around wellbore rocks. Combined with drilling fluid column pressure and crustal stress, the distribution of the circumferential stress in the crustal stress field around the shaft wall in a vertical well is [2].

$$\sigma_\theta = -\frac{R^2}{r^2} p_i + \frac{(\sigma_1+\sigma_2)}{2}\left(1+\frac{R^2}{r^2}\right) - \frac{(\sigma_1-\sigma_2)}{2}\left(1+\frac{3R^4}{r^4}\right)\cos 2\theta + \delta\left[\frac{\alpha'(1-2\mu)}{2(1-\mu)}\left(1+\frac{R^2}{r^2}\right) - \phi\right](p_i - p_p) \quad (1)$$

σ_θ is the circumferential stress, MPa; p_i is the pressure of the liquid column of the drilling fluid, MPa; p_p is the pore pressure of the original formation, MPa; α' is the effective stress coefficient; μ is a Poisson ratio; σ_1 is the maximum horizontal crustal stress, MPa; σ_2 is the minimum horizontal crustal stress, MPa; δ is the permeability coefficient; ϕ is the porosity of shale.

When $\delta=1$, the wall of the well is permeable. When $r=R$, the circumferential stress on the surface of the wellbore is

$$\sigma_\theta = -p_i + (1-2\cos 2\theta)\sigma_1 + (1+2\cos 2\theta)\sigma_2 + \delta\left[\frac{\alpha'(1-2\mu)}{2(1-\mu)} - \phi\right](p_i - p_p) \tag{2}$$

The formation cracks produced by the fluid in the well are mostly vertical, the reason is that the tensile strength of the tangent stress exceed the tensile strength of the rock from compression to tension. When $\theta = 0^\circ$ or $\theta = 180^\circ$, σ_θ is the minimum value.

$$\sigma_\theta = 3\sigma_2 - \sigma_1 - p_i + \left[\frac{\alpha'(1-2\mu)}{2(1-\mu)} - \phi\right](p_i - p_p) \tag{3}$$

1.2. Effect of chemical field on crustal stress field
When water-based drilling fluid encounters shale, under the action of water pressure difference, chemical potential difference and the pressure difference between the drilling fluid pressure and pore pressure, water molecules flow into the pores between the cracks and particles, they cause the transfer of water molecules and ions, at this time, hydration occurs [2]. Under certain conditions, the shaft wall can be considered as a semi permeable membrane, and the two sides are drilling fluid system and stratigraphic fluid system. On this basis, Chenevert and Pernot [12] put forward the calculation model of hydration stress. The calculation formula is

$$p_\pi = I_m \frac{R'T}{\overline{V}} \ln \frac{(A_w)_m}{(A_w)_{sh}} \times 10^{-6}. \tag{4}$$

p_π is hydration stress, MPa; I_m is the membrane permeable efficiency; R' is gas constant; T is the absolute temperature; $\overline{V}$ is the partial molar volume of pure water; $(A_w)_m$ is the activity of the formation liquid entering the stratum; $(A_w)_{sh}$ is the activity of water in shale.

It is assumed that the structure of shale can make the fluid in the pores circulate in them and form the uniform pore pressure to the skeleton of the rock, so the effective stress is equal to the positive stress minus the pore pressure and the hydration stress. Therefore, the effective circumferential stress is as follows:

$$\sigma_\theta' = 3\sigma_2 - \sigma_1 - p_i + \left[\frac{\alpha'(1-2\mu)}{2(1-\mu)} - \phi\right](p_i - p_p) - p_p - p_\pi. \tag{5}$$

1.3. Influence of temperature on crustal stress field
Boas [13] and Maury [14] put forward the idea that the thermal pressure caused by the temperature change of wellbore rock can cause the instability of the wellbore wall. The effect of temperature should be considered when the vertical depth of the open hole is more than 1 000 m. It is proved that the stress of 0.4 MPa is produced by 1 °C increases in the temperature of the medium hardness rock [14]. According to the generalized Hoek's law and thermoelastic theory, the circumferential stress caused by the variation of the wellbore and the formation temperature is as follows:

$$\sigma_{\theta t} = \frac{E\alpha_m}{3(1-\mu)}\left[\frac{1}{r^2}\int_R^r T_f(r,t)rdr + T_f(r,t)\right]. \tag{6}$$

In the upper formula, $T_f(r,t) = T(r,t) - T_0$. $T_f(r,t)$ is rock temperature change field around well; $\sigma_{\theta t}$ is the temperature variable stress; T_0 is the original formation temperature, °C; α_m is the thermal expansion coefficient of rock volume; E is the elastic modulus of rock, GPa; t is the time, h.

When $r = R$, formula (6) can be written as:

$$\sigma_{\theta t} = \frac{E\alpha_m}{3(1-\mu)}[T_w - T_0]. \tag{7}$$

T_w is the temperature of the wellbore, T_0 is the original stratum temperature.

1.4. Influence of tectonic stress field

According to the stress superposition criterion of small pore elastic deformation, under the influence of in-situ stress, seepage, temperature field and chemical field, the effective circumferential stress of the stress field around the borehole can be expressed as:

$$\sigma_\theta^z = 3\sigma_2 - \sigma_1 - p_i + \left[\frac{\alpha'(1-2\mu)}{2(1-\mu)} - \phi\right](p_i - p_p) - p_p - p_\pi + \frac{E\alpha_m}{3(1-\mu)}[T_w - T_0] \tag{8}$$

According to the maximum tensile stress criterion, the condition of stratum cracking is

$$\sigma_\theta^z = -S_t. \tag{9}$$

S_t is the tensile strength of rock, MPa. When the condition is satisfied, p_i is the fracture pressure p_F.

If $K = \frac{\alpha'(1-2\mu)}{2(1-\mu)} - \phi$, formula (8) can be written as

$$p_F = \frac{1}{1-K}(3\sigma_2 - \sigma_1 + S_t - (1+K)p_p - p_\pi + \frac{E\alpha_m}{3(1-\mu)}[T_w - T_0]). \tag{10}$$

The main stress in the vertical direction is called the overlying stress σ_3, which is produced by the weight of the overlying strata and varies with the depth of the rock strata. The effective overlying pressure $\sigma_3^{'}$ is:

$$\sigma_3^{'} = \sigma_3 - p_p, \text{ or } \sigma_3^{'} = S - p_p. \tag{11}$$

According to the view of Huang [6], if tectonic pressure exists, the effective stresses added to the two horizontal main directions in general are not equal. They are expressed as

$$\sigma_1^{'} = \frac{\mu}{1-\mu}\sigma_3^{'} + \alpha\sigma_3^{'} + p_p, \quad \sigma_2^{'} = \frac{\mu}{1-\mu}\sigma_3^{'} + \beta\sigma_3^{'} + p_p. \tag{12}$$

Where $\sigma_1^{'}$ is the maximum horizontal effective stress, MPa; $\sigma_2^{'}$ is the minimum horizontal effective stress, MPa.

Submitting formula (12) into the formula (10), the shale stratum fracture pressure model is got.

$$p_F = \frac{1}{1-K}(S_t - p_\pi + \frac{E\alpha_m}{3(1-\mu)}[T_w - T_0]) + p_p + \frac{1}{1-K}(\frac{2\mu}{1-\mu} + 3\beta - \alpha)\sigma_3^{'}. \tag{13}$$

Where p_F is the stratum fracture pressure, MPa; α, β are stress coefficients of geological structure.

2. Example calculation of shale fracture pressure

According to the crustal stress and measured fracture pressure parameters of oilfield [6,11,12,14], the basic parameters are selected as shown in tables 1 and 2.

Table 1. Conventional parameters of shale formation.

Membrane permeable efficiency I_m	Gas constant R'	Absolute temperature T / K	Partial molar volume of pure water $\overline{V}$	the activity of the formation liquid entering the stratum $(A_w)_m / m^3 mol^{-1}$
0.1	8.314	363	1.8×10^{-5}	0.78
the activity of water in shale $(A_w)_{sh}$	Shale porosity ϕ	Poisson ratio μ	Structural stress coefficient α ; β	Effective stress coefficient α'
0.915	0.1811	0.25	1.4; 0.46	0.7

Table 2. Ground stress and fracture pressure model in oil field.

Oil field	Well No.	Average depth of well /m	Measured fracture pressure p_f /MPa	Stratum pressure p_p /MPa	Tensile strength S_t/MPa	Effective overlying stress σ_3' /MPa	Predicted rupture pressure /Mpa				
							Eaton model	Anderson model	Exlog model	Huang's model	Model of this paper
QY	Y2#	1782.7	39.00	14.80	4.9	24.51	24.79	34.80	36.07	39.20	39.96
	Y4#	1700.9	39.20	14.21	4.9	23.13	23.65	33.08	34.28	37.51	38.43
	Y14#	1609.9	36.36	13.43	4.9	21.75	22.30	31.17	32.30	35.63	36.70
	Y13#	1582	36.55	12.64	4.9	21.15	21.27	29.90	31.00	34.38	35.51
	Y11#	1495.8	34.30	12.54	4.9	19.26	20.40	28.25	29.26	32.76	34.12
HCH	C1#	1050.8	24.01	8.72	3.92	13.10	14.06	19.41	20.10	23.07	25.06
	C33#	1127.2	25.77	9.41	3.92	14.11	15.16	20.92	21.65	24.56	26.43
MLB	186#	1452.7	31.85	12.05	3.92	18.84	19.74	27.42	28.40	30.97	32.30
	93#	1446.7	33.81	12.05	3.92	18.71	19.68	27.32	28.29	30.86	32.22
	188#	1432	33.12	12.15	3.92	18.94	19.87	27.61	28.60	31.15	32.48
MLZ	98#	1345.4	32.83	11.27	3.92	17.08	18.24	25.21	26.10	28.78	30.32
	207#	1447.4	33.81	12.05	4.9	18.73	19.69	27.33	28.31	31.86	33.27
	256#	1388	30.28	11.56	3.92	17.82	18.84	26.10	27.03	29.66	31.12
	209#	1595.7	37.24	13.33	4.9	20.77	21.81	30.27	31.35	34.76	35.93
	303#	1501.5	33.03	12.54	4.9	19.38	20.45	28.36	29.37	32.87	34.20
MLN	39#	1377	29.20	11.47	3.92	17.68	18.68	25.89	26.81	29.46	30.93
	75#	1361.5	29.01	11.37	3.92	17.45	18.49	25.61	26.52	29.18	30.67
	84#	1423	31.56	11.86	3.92	18.40	19.36	26.87	27.83	30.43	31.81

According to the formula (13), the calculation results of the fracture pressure of shale are shown in figure 2.

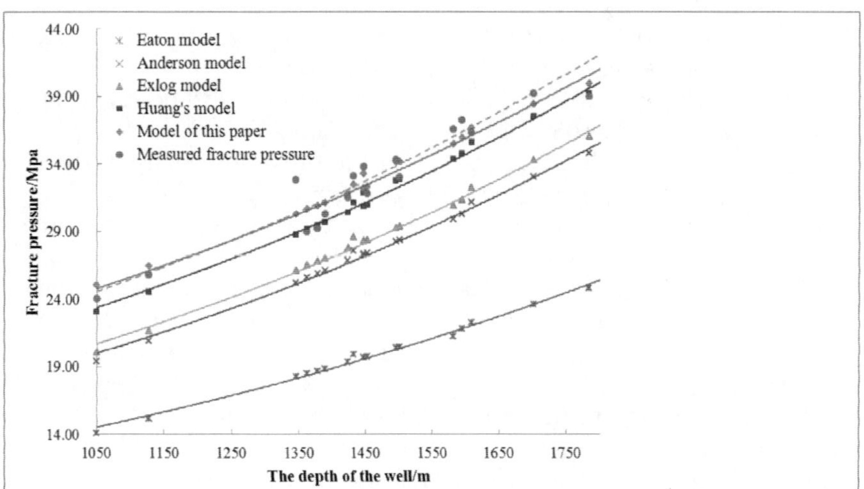

Figure 2. The relation curve of fracture pressure and depth of well.

The calculation model is compared with other classic models as shown in figure 2. It can be seen that the fracture pressure of shale gradually rises with the rise of the well depth. The shale fracture pressure calculated by the present study is higher than that of other models at the same well depth. The reason is that the model of this paper not only considers the effect of these factors: the relationship between the overlying stress, the depth of the rock layer, stress concentration in wellbore wall, the effect of tectonic stress on underground nonuniform distribution and rock stratum strength, but also considers the additional stresses caused by percolation of drilling fluid, the hydration stress, the thermal stress caused by the temperature difference in the wellbore. The results show that the prediction error of the model of this paper is 3.85% compared with the measured results. The predicted results of this model are in good agreement with the actual fracture pressure. So the density of the equivalent drilling fluid calculated is more practical.

3. Summary

- Combined with the theory of elastic mechanics and rock mechanics, the application of the maximum tensile stress criterion, the present study established a model of shale fracture pressure. Based on Huang's model, the model of the present study has considered the additional influence of seepage of drilling fluid, hydration stress, shale porosity, temperature variable stress and tectonic stress and so on. Compared with the measured results, the error of this model is 3.85%, so the calculation accuracy is improved.
- Based on the stress superposition principle of small deformations on elastic material with pores, the present study made the linear superposition of crustal stress, tectonic stress, additional stresses caused by seepage and temperature and hydration stress. However, the interaction between them has not been considered yet, which needs to be studied in the subsequent work.

Acknowledgment
The present work is supported by National Natural Science Foundation of China (Grant No. 51674200, 51704233 and 51704237), and the special research project of Shanxi Provincial Department of education (Brittleness analysis of drilling rig system based on improved extension theory and FAHP).

References

[1] Liu X J, *et al* 1997 Study on the stability of well wall in mudstone strata *Natural Gas Industry* **17** 45-8
[2] Qu Z and Wang P 2016 *Creep Damage Instability Study of Shale* (Science Press, Beijing) 30-45
[3] Wang P, Qu Z, Huang H, *et al* 2016 Creep experimental study of brittle shale triaxial state under aqueous *Science Technology & Engineering* **16** 66-71
[4] Wang P, Qu Z and Huang H 2015 Experimental study of the effect of formation water salinity on creep laws of the hard brittle shale *Petroleum Drilling Techniques* **43** 63-8
[5] Wen K M, Chen P, Li X Q, *et al* 2015 The Influence between shale hydration and rock stress distribution relationship *Science Technology & Engineering* **15** 43-6
[6] Huang R 1982 Initiation and propagation of fractures in hydraulic fracturing *Petroleum Exploration and Development* **1982** 65-77
[7] Li S G, Deng J G, Yu B H, *et al* 2005 Formation fracture pressure calculation in high temperatures wells *Chinese Journal of Rock Mechanics & Engineering* **24** 5669-73
[8] Maury V and Guenot A 1995 Practical advantages of mud cooling systems for drilling *SPE 25732* **1995** 42-8
[9] Huang R 1984 A model for predicting formation fracture pressure *Journal of the University of Petroleum China* **1984** 16-28
[10] Yan X, Hu Y Q, Li N, *et al* 2015 Calculation model of breakdown pressure in shale formation *Lithologic Reservoirs* **27** 109-13
[11] Deng J 2009 Formation fracture pressure prediction method in high temperature and high

pressure formations *Petroleum Drilling Techniques* **37** 43-6
[12] Chenevert M E and Pernot V 1998 Control of shale swelling pressures using inhibitive water-base muds *SPE 49263* 27-30
[13] Boas B M V 1992 Temperature profile of a fluid flowing within a well *SPE 21133* 439-46
[14] Maury V and Guenot A 1995 Practical advantages of mud cooling systems for drilling *SPE 25732* 42-8

IOP Conf. Series: Materials Science and Engineering **372** (2018) 012047 doi:10.1088/1757-899X/372/1/012047

Lightweight design of automobile frame based on magnesium alloy

R Lyu [1,3,a], X Jiang [1,b], O Minoru [2,c] and D Y Ju [1,d]
[1]Department of Electronic Science and Engineering, Saitama Institute of Technology University, Japan
[2]Manufacturing research center, Saitama Institute of Technology, Japan

E-mail: [a]lr518ever@gmail.com, [b]xin88886@gmail.com, [c]ootake.minoru@sit.ac.jp, [d]dyju@sit.ac.jp

Abstract. The structural performance and lightweighting of car base frame design is a challenging task due to all the performance targets that must be satisfied. In this paper, three kinds of materials (iron, aluminum and magnesium alloy) replacement along with section design optimization strategy is proposed to develop a lightweight car frame structure to satisfy the tensile and safety while reducing weight. Two kinds of cross-sections are considered as the design variables. Using Ansys static structure, the design optimization problem is solved, comparing the results of each step, structure of the base flame is optimized for lightweight.

1. Introduction

In recent years, The measures of reducing CO_2 and the development of EV vehicle has been developed rapidly. In addition to the aging society and adapt to the problem of the elderly and children ikuyo user ultra compact and lightweight electric vehicle.

According to the research [1,2], the fuel consumption can be reduced by 6%-8% and the emission volume can be reduced by 5%-6%, which is 10% of the vehicle weight loss. Therefore, vehicle lightweight is an important measure to save energy and reduce pollution. Magnesium alloys have the advantages of high specific strength, strong energy absorption, good formability and recovery, and have a wide application in the field of automobile. [3]

In order to reduce the weight of the car, this study uses a new type of magnesium alloy material applied to a car chassis structure, without changing the original structure, by changing the car frame girder section and materials, so that achieving the purpose of reducing the weight of the frame along with contenting the stress intensity of magnesium alloy frame.

2. Analysis methods and theories

The choice of the test lightweight material has to be made in the context of the design requirements and the load spectrum imposed to the structure. The lightweight potential has to be shown within constraints given by the material itself, the useable designed space and the manufacturability of real parts. An overall cost estimation should not be reduced to the material itself, and it should also consider production, assembly, tooling, maintenance and other cost. Accordingly, it should reflect its proper share on the total cost of ownership to the customer.

In this research, there is a goal of designing an automobile base frame lighter than a commercial frame. Design variables for weight reduction are material constants such as material density and plate thickness. Therefore, the optimal design of this time is to make it safer and lightweight. In the optimum design, static structural analysis is carried out while adjusting the parameters and thickness of the material, and the parameter which becomes even better by using lighter weight material and thickness while comparing with the weight of the base frame we predicted and performed the following static structural analysis.

In this research, in order to guarantee the safety within the elastic range of the base frame of the automobile, it is determined by two discrimination criteria and the optimum design is performed.

The material strength is determined by using the Tresca hypothesis (maximum shear stress theory) which is often used for material dynamics [4-7]. That is, it is safe if the difference between the maximum principal stress and the minimum principal stress is smaller than the tensile strength of the material.

$$\sigma_s = 1/2\ (\sigma_1 - \sigma_2) < \sigma_{max} \tag{1}$$

Here, σ is the main shear stress, σ_{max} is the maximum shear stress, σ_1 is the maximum principal stress, and σ_2 is the minimum principal stress.

Depending on the Mises yielding condition when plastic deformation occurs, the homologous stress (Mises stress) can be expressed by the following equation.

$$\bar{\sigma} = \frac{1}{\sqrt{2}}[(\sigma_1 - \sigma_2)^2 + (\sigma_2 - \sigma_3)^2 + (\sigma_3 - \sigma_1)^2]^{1/2} \tag{2}$$

For safety discrimination using static structural analysis within the elastic deformation range, the Mises stress of the structure is smaller than the yield stress σ_y at the time of occurrence of plastic deformation, $\bar{\sigma} \le \sigma_y$.

3. The analysis process

3.1. Material editing and loading

In this paper, three kinds of materials are used for the analysis and calculation of automobiles (table 1). In the whole base frame of gravity condition, the 40% percent of the loading is in the first half of the frame, and remaining 60% loading is acting on the rest of the frame, and the supporting force in the hub link is loading chassis upward, making the car balance.

Table 1. The mechanical properties of Aluminum , Steel and Magnesium.

	SPFH540	Aluminum 6061-T6	Magnesium AZ91
Density (kg/m³)	7850	2700	1830
Coefficient of elasticity (GPa)	210	69	45
Poisson ratio	0.3	0.33	0.35
Yield strength (MPa)	355	276	160
Tensile strength (MPa)	540	310	340

After selecting the material, the stress analysis of the model is carried out, figure 1. Firstly, the load of the different parts of the model is restrained after the freedom of the end face is restrained.

Figure 1. Boundary condition of Base frame model.

3.2. Meshing and analysis

In the choice of materials and loading after the needs of the model grid, then choose the size of the grid, to the small number of mesh complex (30~40 mm), the simple model can be set up a relatively large number of grid (40~70 mm). In this paper, we use 50 mm mesh to simplify the calculation,as shown in figure 2. After the grid is divided, the analysis and calculation can be carried out.

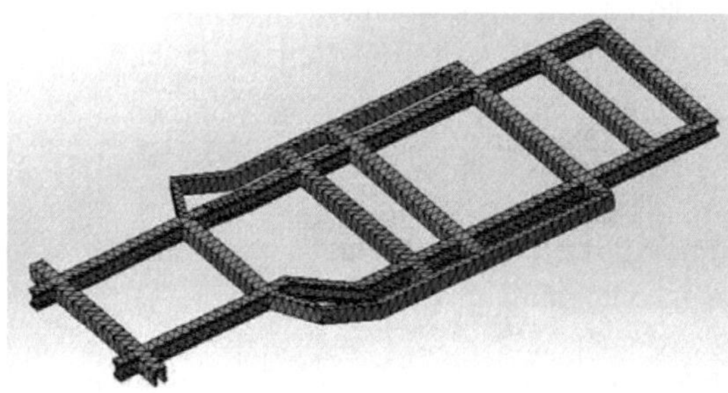

Figure 2. Mesh of Base frame model.

4. Example analysis of car model

In this study, the market is the target frame to design a lightweight vehicle base frame. Therefore, the optimal design of safety while satisfying the lightweight. In the optimum design, structural analysis, static, while adjusting the cross-sectional shape and material parameter, structure and material adopted by lightweight design method, and in addition, the structural static analysis, the parameters such as Von-Mises stress and deformation are predicted and are shown in figures 3 through 5.

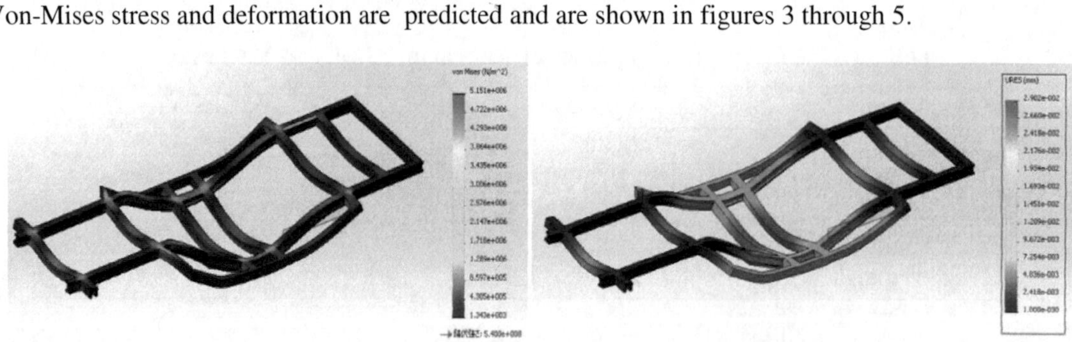

(a) (b)

Figure 3. The calculation results of Steel base frame. (a) Von-Mises and (b) Deformation quantity.

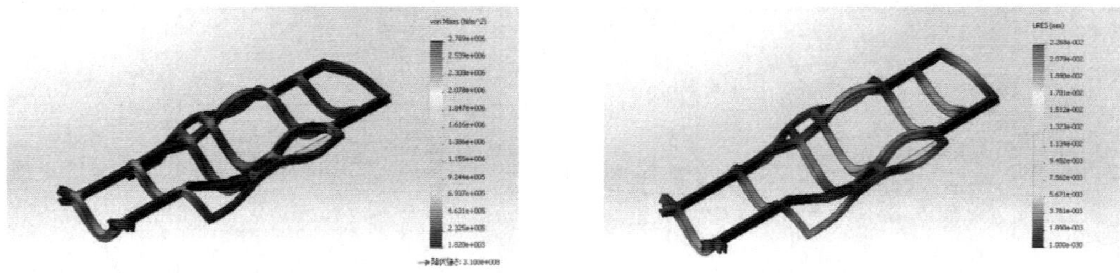

(a) (b)

Figure 4. The calculation results of Aluminum base frame. (a) Von-Mises and (b) Deformation quantity.

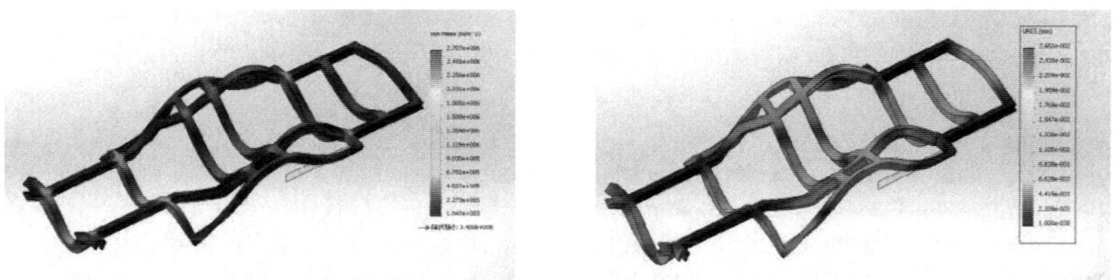

(a) (b)

Figure 5. The calculation results of Magnesium base frame. (a) Von-Mises and (b) Deformation quantity.

Table 2. Simulation result.

	Square Type					I Type				
	H	b	Fe (355 MPa)	Al (276 MPa)	Mg (160 MPa)	H	b	Fe (355 MPa)	Al (276 MPa)	Mg (160 MPa)
Weight（kg）	80 mm	5 mm	260.41	89.57	60.71	80 mm	4 mm	247.82	85.24	57.77
ΔL_{max}(mm)			2.05	5.923	9.022			2.431	7.157	10.91
$(\sigma_1-\sigma_3)_{max}$(MPa)			379.8	369.9	368.8			241.4	235	234
Weight		6 mm	272.8	93.83	63.59		5 mm	256.93	88.37	59.9
ΔL_{max}(mm)			1.788	5.248	7.993			2.088	6.142	9.366
$(\sigma_1-\sigma_3)_{max}$(MPa)			283.1	275	273.8			214.5	209.2	208.8
Weight		7 mm	284.75	97.94	66.38		6 mm	265.83	91.43	6197
ΔL_{max}(mm)			1.551	4.548	6.927			1.842	5.416	8.258
$(\sigma_1-\sigma_3)_{max}$(MPa)			275.1	267.1	266.1			200.3	193.9	192.5
Weight	90 mm	5 mm	267.87	92.13	62.45	90 mm	4 mm	252.05	86.69	58.76
ΔL_{max}(mm)			1.697	4.988	7.601			2.15	6.325	9.638
$(\sigma_1-\sigma_3)_{max}$(MPa)			476.1	463.5	462.2			238.7	231.4	230
Weight		6 mm	281.33	96.76	65.58		5 mm	261.7	90.01	61.01
ΔL_{max}(mm)			1.342	3.931	5.986			1.825	5.367	80184
$(\sigma_1-\sigma_3)_{max}$(MPa)			251.8	247.8	249.5			201	194.5	193.1
Weight		7 mm	294.35	101.24	68.62		6 mm	270.7	93.23	63.35
ΔL_{max}(mm)			1.427	4.187	6.376			1.596	4688	7.148
$(\sigma_1-\sigma_3)_{max}$(MPa)			202.3	196.9	196.7			162.8	157	155

Weight			275	84.4	64.3			265.48	91.31	61.89
ΔL_{max}(mm)		5 mm	1.446	4.24	6.47		4 mm	1.781	5.227	7.961
$(\sigma_1-\sigma_3)_{max}$(MPa)			425.3	414.1	413			187.4	180.9	178.9
Weight			290	99.7	67.6			266.47	91.65	62.12
ΔL_{max}(mm)	100 mm	6 mm	1.257	3.688	5.616	100 mm	5 mm	1.698	4.983	7.589
$(\sigma_1-\sigma_3)_{max}$(MPa)			210.1	203.7	202.7			172.7	166.4	164.6
Weight			304	105	70.9			276.45	95.08	64.45
ΔL_{max}(mm)		7 mm	1.131	3.312	5.042		6 mm	1.448	4.246	6.468
$(\sigma_1-\sigma_3)_{max}$(MPa)			181	174.9	173.8			158.7	152.8	151.2

Under the above conditions, the model of the base frame is analyzed and calculated, and the shape variables, maximum and minimum stresses of the model are obtained (figures 3-5). Through the analysis and calculation of the chassis with different sectional area, the stress, deformation and weight ratio are obtained (table 2), and achieved the purpose of optimizing the chassis.

5. Conclusion

In this paper, a new method in analysing automobile base frame were carried out using the Ansys Workbench. The results can be obtained, when the section is I type (H=100, b=5), the light weight of the car frame is realized under the condition of satisfying the stress.And the results show that Magnesium alloy frame is 70% lighter than iron frame,40% lighter than aluminium frame, and the lightweight design was achieved.

References

[1] Ma M T and Ma L X 2008 Application and prospective technology of aluminum alloy in automotive lightweight *New Material Industry* **2008** 43-50

[2] Yin G F, Yin X Y and Hiu X B 2006 *SolidWorks API Example* (Beijing: Machinery Industry Press)

[3] Qian L 2016 Study on the optimum design of automotive frame using FEM 2015 *Saitama Insitutute OF Technology 15th Research Forum*

[4] Li G, Zhang C P and Li L 2000 *Visual Basic 6.0 Foundations and Examples of Chinese Programming* (Beijing: Electronics Industry Press)

[5] Tan Q Q, Luo Y J and Wang J 2008 *Implementation of Solidworks Embedded Plug-in Based on VB.NET* (Computer and Network)

[6] Liang C C and Le G N 2012 Lightweight optimization of bus frame structure considering rollover safety *The sustainable City VII* **2** 1185-96

[7] Chiandusssi G, Gaviglio I and Ibba A 2004 Topology optimization of an automotive component without final volume constraint specigication *Advances in Engineering Software* **35** 609-17

IOP Conf. Series: Materials Science and Engineering **372** (2018) 012048 doi:10.1088/1757-899X/372/1/012048

Lightweight design and analysis of automobile wheel based on bending and radial loads

X Jiang[1], R Lyu[1], Y Fukushima[1], M Otake[2] and D Y Ju[1,3]

[1]Graduate School of Engineering, Saitama Institute of Technology, Japan
[2]Advanced Science Research Laboratory, Saitama Institute of Technology, Japan

Email: dyju@sit.ac.jp

Abstract. Lightweighting of automobile vehicle is a significant application trends, using magnesium alloy wheels is a valuable way. This article discusses design of a new model of automobile wheel. Then bending test and radial test finite element model were established. Considering three different materials namely magnesium alloy, aluminium alloy and steel, the stress and strain performances of each material can be obtained. Through evaluating and analyzing model in bending test and radial test, we obtained the reasonable and superior results for magnesium alloy wheel. The results of the equivalent stress and deformation were compared, the magnesium alloy wheel practicality has been confirmed. This research predicts the reliability of the structural design, some valuable references are provided for the design and development of magnesium alloy wheel.

1. Introduction

As the automotive industry is increasingly demanding on energy saving and environmental protection, people are taking more attention on the lightweight design of automobiles [1]. Wheel is one of the most important parts of a vehicle. It determines a number of parameters such as security, stability, performance and fuel economy. To ensure energy efficiency, the wheels must be as lightweight as possible. Magnesium alloys are considered one of the most promising materials in the 21st century, with low density, high specific strength and good casting properties [2].

In the modern design, it is important to improve the efficiency of development and reduce the number of tests. To achieve these goals, computer aided engineering (CAE) is a useful tool for the improvement and development of automotive wheels. B Anusha *et al* [3], introduced the results of the four-wheel strength analysis using four different wheel shapes, including straight, sloping, y-shaped and honeycombed shapes. The analysis variables include different materials and different models of the load. Recently, casting and forging techniques have been improved, and the corrosion resistance of magnesium has also been improved.

In this research, a kind of wheel structure is designed. The finite element model of wheels are established based on the radial test and bending test. Compared the finite element results of an aluminium alloy wheel, magnesium alloy wheel and steel wheel. The rationality and superiority of the designed magnesium alloy wheel are obtained.

2. Design of automobile wheel

When designing wheel, two major factors must be considered, such as safety and engineering standards. The wheel mainly consists of rim and spokes [4,5]. The shape and size of the wheel is

IOP Conf. Series: Materials Science and Engineering **372** (2018) 012048 doi:10.1088/1757-899X/372/1/012048

specified in standard according to JATMA. In this research, the market is the target to design a lightweight vehicle wheel. Therefore, the design is lightweight while satisfying the safety. The specifications of the wheel: rim diameter 355 mm, rim width is 165 mm, offset ET 45 mm, pitch centre diameter PCD 100 mm, hub centre diameter 59 mm.

Orthogonal experimental design is a common technique for optimization. It is based on probability theory, quantitative statistics and practical experience. The test scheme of the quasi-orthogonal table is arranged, and the results are calculated and analyzed. It is a design method to find the optimal test scheme quickly. The advantage of orthogonal design is that it greatly reduces the test. The frequency of inspection improves the working efficiency. Using orthogonal experimental design can be more efficient for wheel design and analysis. The three dimensional model of the wheel is shown in figure 1 below.

Figure 1. Wheel modle.

The wheel optimization flowchart is shown in figure 2 below. Using optimization, we can achieve our target.

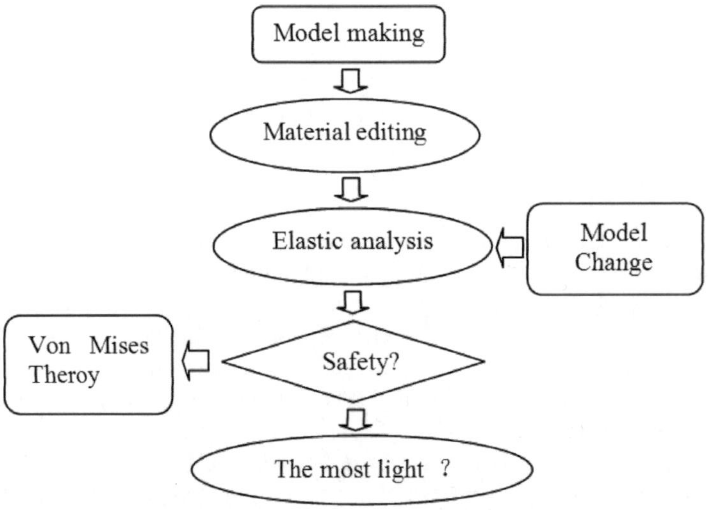

Figure 2. Optimization flowchart.

3. Numerical modelling of the wheel

In modern design, using finite element analysis can help us determine the strength of the wheel in advance and reduce the test times and cost [6]. Bending load while driving and static load while vehicle stops is two working conditions of the wheel, those two working conditions should be considered seriously when designing wheel [7,8]. The mesh will be applied on the total surface of the wheel with equal element size.

In this research, the gross weight of the vehicle is about 1175 kg, load on each wheel is 2937.5 N. Three kinds of materials are used for the analysis and calculation of wheel as table 1 lists. According to the Mises yield condition, our Mises stress results can be considered and compared in the basic strength theory (mainly Mohr strength theory). On this basic, we can ensure the safety of the wheel in the elastic range.

Table 1. The mechanical properties of Aluminum, Steel and Magnesium.

	SPFH540	Aluminum 6061-T6	Magnesium AZ91
Density (kg/m³)	7850	2700	1830
Modulus of Elasticity (GPa)	0.21	0.069	0.045
Poisson ratio	0.3	0.33	0.35
Yield strength (GPa)	0.355	0.276	0.16

3.1. Bending endurance of the wheel
The bending endurance of the wheel can show the wheel strength under bending load. The bending moment M is calculated by using below equation:

$$\vec{M} = (\mu R + d)\vec{F} \times S \tag{1}$$

In the equation, $\vec{M}$ is bending moment ($N \cdot m$), μ is friction coefficient between tires and the road set as 0.7, R is tire static load radius (m), d is offset of wheel (m), $\vec{F}$ is maximum rated load (N), S is strength coefficient set as 1.6. We can obtain bending moment is 1168.9 $N \cdot m$.

3.2. Radial endurance of the wheel
The radial endurance is intended to detect the wheel performance when the total load of the vehicle compresses the wheel radially. The radial load Fr shall be determined from the equation:

$$\vec{F}_r = K \times \vec{F} \tag{2}$$

In the equation, $\vec{F}_r$ is radial load (N), $\vec{F}$ is maximum rated load (N), K is coefficient according to the industrial standards set as 2.25. We can obtain radial load is 6609.4 N.

In this research, using Stearns J wheel and tire contact research results, the force on the magnesium alloy wheel from the tire can be replaced by the radial force directly on the wheel to simplify the modeling. The pressure W_r, W and W_0 is given by the following equation:

$$W = b\int_{-\theta_0}^{\theta_0} W_r r_b d\theta, \quad W_r = W_0 \cos\left(\frac{\pi\theta}{2\theta_0}\right), \quad W_0 = \frac{W\pi}{4br_b\theta_0} \tag{3}$$

In the equation, W is radial load on the wheel, b is width of the bead seat, r_b is radius of the bead seat, θ_0 is the maximum deflection angle of radial load. In this way we can get the pressure loaded in the wheel inner ring is 0.45 MPa, the pressure loaded on the rim of the wheel is 0.785 MPa.

The load of test model is shown in figure 3 below. Setting the length of the loading axis as 1000mm and the force is 1168.9 N.

IOP Conf. Series: Materials Science and Engineering **372** (2018) 012048 doi:10.1088/1757-899X/372/1/012048

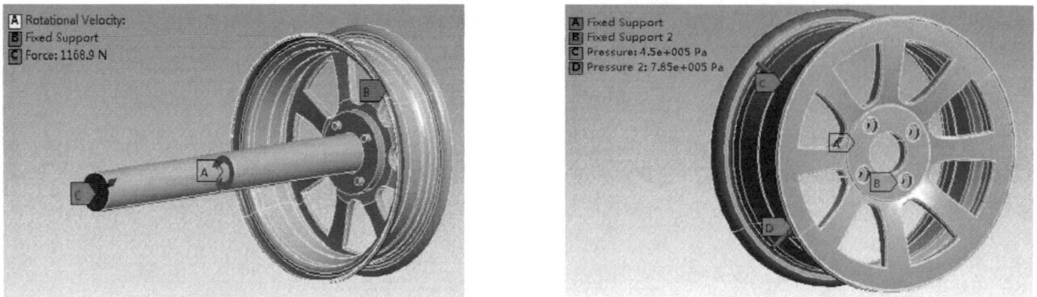

(a) (b)

Figure 3. Bending endurance test and radial endurance test model. (a) Bending endurance test model, and (b) Radial endurance test model.

4. Results and discussions

Based on above loading conditions, the model of the wheel is analyzed and calculated. The equivalent stress and total deformation were determined in ANSYS and presented in figures 4-9.

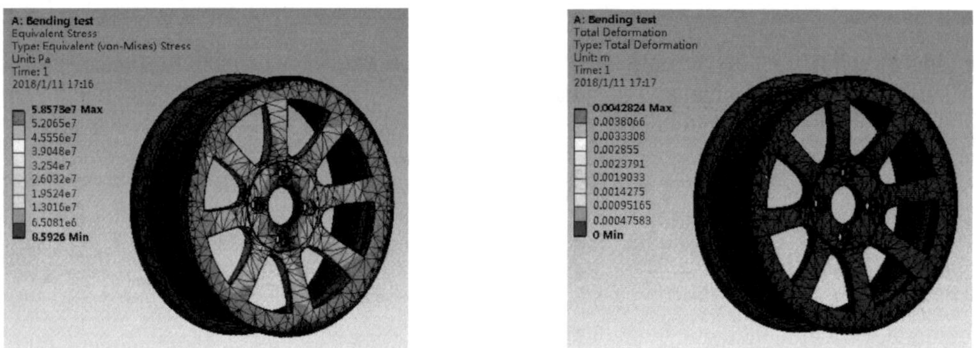

Figure 4. Equivalent stress and deformation of AZ91 magnesium alloy wheel.

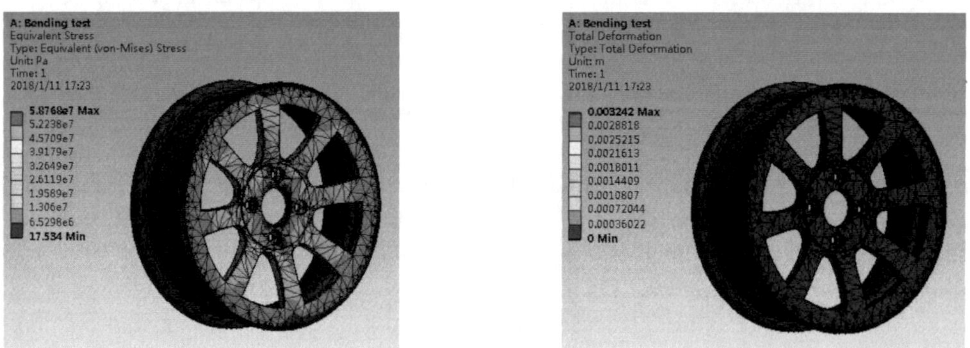

Figure 5. Equivalent stress and deformation of Aluminum alloy wheel.

IOP Conf. Series: Materials Science and Engineering **372** (2018) 012048 doi:10.1088/1757-899X/372/1/012048

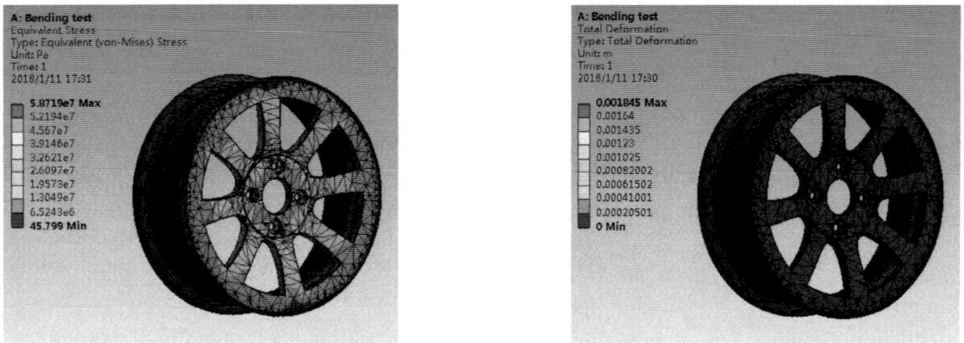

Figure 6. Equivalent stress and deformation of Steel wheel.

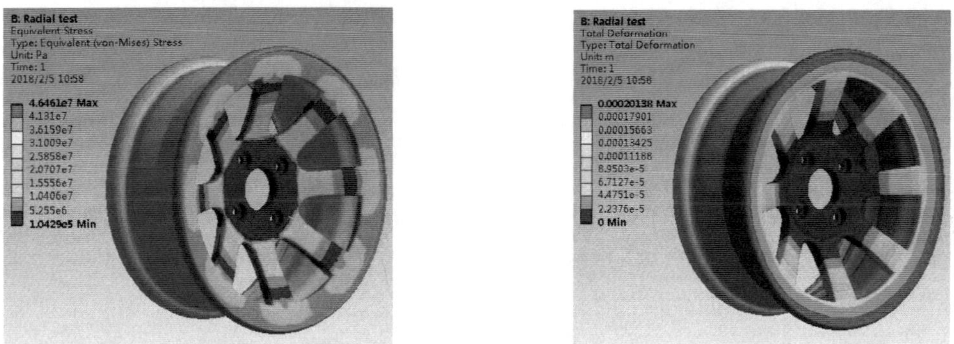

Figure 7. Equivalent Stress and deformation of AZ91 magnesium alloy wheel.

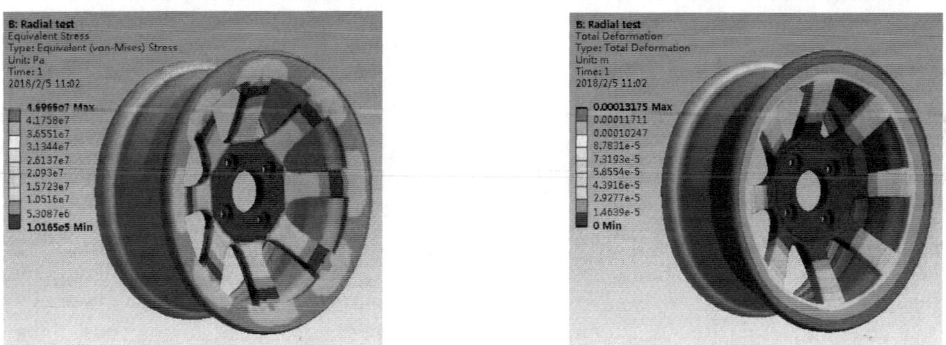

Figure 8. Equivalent Stress and deformation of Aluminum alloy wheel.

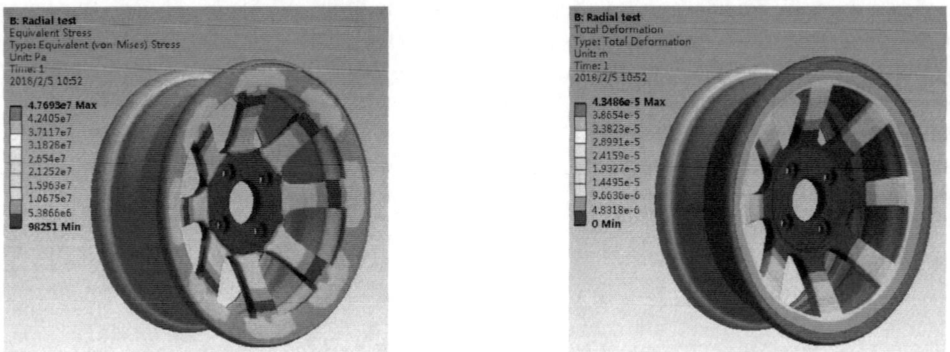

Figure 9. Equivalent Stress and deformation of Steel wheel.

4.1. Bending endurance results of the wheel
From the above analysis, the results under bending endurance test such as equivalent stress and deformation are obtained.

For the magnesium alloy and aluminum alloy wheel maximum equivalent stress is concentrated on spokes shown in above figures, steel wheel maximum equivalent stress also concentrated on spokes, which is less than the yield stress of each material suggested. Strain of each material wheel is acceptable according to the principle of elastic-plastic deformation.

4.2. Radial endurance results of the wheel
From the above analysis, the results under radial endurance test such as equivalent stress and deformation are obtained.

The magnesium alloy wheel and aluminum alloy maximum equivalent stress is concentrated on rim shown in above figures, steel wheel maximum equivalent stress also concentrated on rim, which is less than the yield stress of each material suggested. Each material wheel with acceptable strain.

Table 2. Comparison table.

Results	Parameters	AZ91	6061-T6	SPFH540
Bending endurance results	Weight(Kg)	3.82	5.63	16.37
	Equivalent Stress(Mpa)	58.57	58.76	58.71
	Deformation(mm)	4.28	3.24	1.85
Radial endurance results	Weight(Kg)	3.82	5.63	16.37
	Equivalent Stress(Mpa)	46.46	46.96	47.69
	Deformation(mm)	0.20	0.13	0.04

After the finite element analysis and orthogonal experimental design. From the above table 2, after material optimization, contrast about different materials, considering the equivalent stress, deformation and weight, we can get the most suitable result. The quality of magnesium alloy is about 30% lighter than aluminum and 70% lighter than steel, meets the lightweight design concept. In this research, the wheel model was improved, magnesium alloy wheel has the least equivalent stress, which is less than the yield stress of the material suggested, the displacement is acceptable, satisfying lightweight wheel design.

5. Conclusion
The finite element analysis has been carried out on the wheel using ANSYS software. The design wheel meets the bending load and static load conditions and therefore the design is safe. By comparing results of different materials, az91 magnesium alloy wheel can be subjected to less stress value. In terms of stress, our new wheels have a better performance; in terms of strain, although the strain result is within an acceptable range, we will focus on improving it. Hence, magnesium alloy is more feasible to be used in wheel than other materials for the lightweight design.

Acknowledgment
This research receives ongoing support from the Next-Generation vehicle Project at Saitama Institute of Technology.

References
[1] Prem J, Raghupathi P and Kalaiyarasan A 2016 Analysis of magnesium alloy wheel for four wheeler IJRSR **7** 13126-30
[2] Kumar D S, Sasanka C T, Ravindra K and Suman K N S 2015 Magnesium and its alloys in

automotive applications – A review AJMST **4** 12-30
[3] Srikanta B A and Veeraraju P 2014 Geometrical and Material optimization of alloy wheel for four wheeler IJRI **1** 1401-2
[4] Ganesh S and Periyasamy D P 2014 Design and analysis of spiral wheel rim for four wheeler IJES **3** 29-37
[5] Theja M S and Krishna M V 2013 Structural and fatigue analysis of two wheeler lighter weight alloy wheel IOSR-JMCE **8** 35-45
[6] Torgal S and Mishra S 2012 Stress Analysis of wheel rim IJMER **1** 34-7
[7] Wang L M, Chen Y F, Wang C Z and Wang Q Z 2011 Fatigue life analysis of aluminum wheels by simulation of rotary fatigue test SV-JME **57** 31-9
[8] Adigio E M and Nangi E O 2014 Computer aided design and simulation of radial fatigue test of automobile rim using ANSYS *IOSR-JMCE* **11** 68-73

IOP Conf. Series: Materials Science and Engineering **372** (2018) 012049 doi:10.1088/1757-899X/372/1/012049

Homogeneous and heterogeneous micro-structuring of austenitic stainless steels by the low temperature plasma nitriding

T Aizawa[1,2] and S-I Yoshihara[1]

[1]Shibaura Institute of Technology, 3-9-14 Shibaura, Tokyo 108-8548, Japan

E-mail: taizawa@sic.shibaura-it.ac.jp

Abstract. The austenitic stainless steels have been widely utilized as a structural component and member as well as a die and mold substrate for stamping. AISI316 dies and molds require for the surface treatment to accommodate the sufficient hardness and wear resistance to them. In addition, the candidate treatment methods must be free from toxicity, energy consumption and inefficiency. The low temperature plasma nitriding process has become one of the most promising methods to make solid-solution hardening by the nitrogen super-saturation. In the present paper, the high density RF/DC plasma nitriding process was applied to form the uniform nitrided layer in the AISI316 matrix and to describe the essential mechanism of inner nitriding in this low temperature nitriding process. In case of the nitrided AISI316 at 673 K for 14.4ks, the nitrided layer thickness became 60 μm with the surface hardness of 1700 HV and the surface nitrogen content of 7 mass %. This inner nitriding process is governed by the synergetic interrelation among the nitrogen super-saturation, the lattice expansion, the phase transformation, the plastic straining, the microstructure refinement and the acceleration of nitrogen diffusion. As far as this interrelation is sustained during the nitriding process, the original austenitic microstructure is homogeneously nitrided to have fine grains with the average size of 0.1 μm and the high crystallographic misorientation angles and to have two phase ($\gamma + \alpha$') structures with the plateau of nitrogen content by 5 mass%. Once this interrelation does not work anymore, the homogeneous microstructure changed itself to the heterogeneous one. The plastic straining took place in the selected coarse grains; they were partially refined into subgrains. This plastic localization accompanied the localized phase transformation.

1. Introduction

High nitrogen stainless steels (HNS) have been developed to save the nickel resource and to improve the corrosion and wear resistance in practice [1]. Even at present, the maximum nitrogen solubility in these steels is still limited by 1 mass% [2]. High nitrogen concentration becomes an engineering issue to advance the steps of HNS research. On the other hand, the plasma nitriding has been also utilized as one of the most feasible surface treatments to massive surface treatment [3]. In case of the conventional plasma nitriding at higher temperature than 723 K, iron and chromium nitrides are inevitably synthesized as a precipitate to form the fragile white layer. A complete nitrogen diffusing zone with the sufficient nitrided layer thickness and hardness is difficult to form by those processing in practice.

The low temperature plasma nitriding has been highlighted to make nitrogen super-saturation into the stainless steels and chromium based binary alloys without formation of iron and chromium nitrides even at 673 K in [4-6]. In particular, both the austenitic and martensitic stainless steels were efficiently

IOP Conf. Series: Materials Science and Engineering **372** (2018) 012049 doi:10.1088/1757-899X/372/1/012049

nitrided at 673 K to have thick nitrided layer up to 80 μm and to have high nitrogen solubility up to 10 mass% after [6]. Those studies stated that both γ-phase and α'-phase lattices were significantly expanded by this nitrogen super-saturation into lattices and that phase transformation from γ- to α'-phases and from α'- to γ-phases was driven by this lattice expansion. In the recent works on the low temperature plasma nitrided AISI304, AISI316, and AISI420 stainless steels, the microstructure was also modified by this nitrogen super-saturation in [7-9]. The average grain size of 9 μm in the bare martensitic stainless steels was refined down to 0.1 μm or less than, by the high density plasma nitriding at 673 K. The similar grain size refinement also took place in the austenitic stainless steels even for the plasma nitrided AISI304 at 673 K. In addition, once the original γ-phase expanded to γ_N-phase with occupation of nitrogen solute atoms into vacancy sites in the fcc-structured lattices, the highly nitrogen super-saturated AISI316 was modified into the two phase stainless steels with the γ_N-phase and its transformed α'-phase.

In the present paper, the inner nitriding processing at the lower temperature is precisely analyzed to describe the nitriding mechanism by the nitrogen super-saturation. The synergetic interrelation is thought to drive this low temperature plasma nitriding among the nitrogen super-saturation, the γ-lattice expansion, the phase transformation, the plastic straining, the microstructure refinement and the enhancement of nitrogen diffusion processes. AISI316 austenitic stainless steels are employed as a test-piece to experimentally describe these two inner nitriding modes at 673 K; i.e., homogeneous and heterogeneous nitriding processes. Each process is experimentally defined to characterize its feature and to discuss the driving mechanism.

2. Experimental procedure

2.1. High density low temperature plasma nitriding system

The present plasma nitriding system consists of the vacuum chamber, the evacuation system, the DC-RF generators working in the frequency of 2 MHz, the gas supply of N_2 and H_2, and, the heating unit located under the cathode plate as illustrated in figure 1.

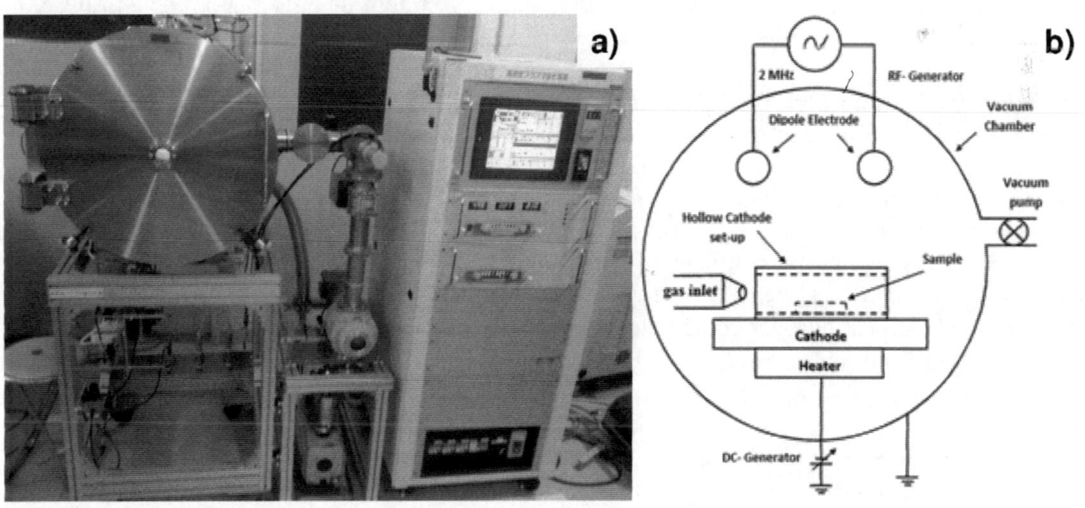

Figure 1. High density RF/DC plasma nitriding system, (a) Outlook of system and (b) System diagram.

The hollow cathode device is utilized to confine the RF-ignited nitrogen- and hydrogen plasmas and to have higher ion and electron densities than those in the standard nitriding process. The maximum nitrogen ion density reached to 3×10^{17} ions/m^3, which is equivalent to that by the electron beam assisted plasma nitriding in low pressure in [10].

In the present experimental set-up, the specimen was placed and fixed in the electrically conductive hollow, which was connected to the DC-bias. The nitrogen and hydrogen mixed gas was supplied from the either inlet of hollow. After placing the sample in the hollow cathode setup, the chamber was evacuated down to 0.5 Pa in pressure, and, then filled with the nitrogen gas with the flow rate of 160 ml/min until the pressure was constant by 300 Pa before heating up to the specified holding temperature. Next, the sample was pre-sputtered for 1.8 ks to remove any oxide passive layer from its surfaces. This pre-sputtering process operated by using the DC discharge of 350 V in pure nitrogen gas by 60 Pa. After pre-sputtering, it was plasma nitrided for 14.4 ks or four hours at the holding temperature of 673 K (or 400°C) by 70 Pa in pressure. The mixture of nitrogen and hydrogen gasses was provided to have the ratio of 160 ml/min for nitrogen and 30 ml/min for hydrogen. After the nitriding process, the sample was cooled in the vacuum chamber to minimize the possibility of surface contamination.

2.2. Preparation of samples
The austenitic stainless steel type AISI316 plate with the size of 20 x 40 x 2 mm^3 were employed as a substrate after heat treatment at 833 K for qualification of grains. Its chemical compositions are: [C] = 0.08 mass%, [Si] = 1.00 mass%, [Mn] = 2.00 mass%, [P] < 0.045 mass%, [S] < 0.030 mass%, [Ni] = 12.0 mass%, [Cr] = 17.0 mass%, and [Mo] = 2.5 mass% for iron in balance. Both sample surfaces were mirror-polished and cleaned by the ultrasonic cleaner before plasma nitriding.

2.3. Measurement and observation
The nitrided specimens were analyzed by the X-ray diffractometer (XRD; Rigaku SmartLab), SEM (Scanning Electron Microscope; HITACHI SU-70), and, EDX (Electron Dispersive X-ray spectroscopy) were utilized for material characterization from the surface down to the specified depth. EBSD (Electron Back-Scattering Diffraction) was also used to analyze the microstructure refinement, the phase transformation and the plastic straining. The inverse pole figure was utilized to describe the crystalline structure. The phase mapping was performed to understand the phase distributions from the surface to the depth. The KAM (Kernel Average Misorientation) was also used to represent the intergranular misorientations among the neighboring grains. Hardness depth profile was measured on the cross-sections by the microhardness testing apparatus (Mitsutoyo HM-200) with the load of 50 g or 0.5 N in every 10 μm in depth, respectively.

3. Experimental results and discussion
AISI316 specimens are nitrided at 673 K for 14.4 ks to describe the difference between the homogeneous and heterogeneous nitriding processes. EDX is utilized to investigate the nitrogen content distribution from the surface to the depth (d) of the nitrided AISI316 plates at 673 K. As shown in Fig. 2, this depth profile consists of two sections; e.g., high nitrogen layer (HNL) and low nitrogen layer (LNL). The nitrogen content is kept nearly constant by [N] = 4 to 5 mass% in HNL; [N] becomes higher to be 6 mass% at the vicinity of surface. This nitrogen content plateau ranges from the surface to the depth of d = 60 μm at 673 K. The nitrogen content decreases across the border between HNL and LNL; e.g., [N] = 4 mass% at d = 60 μm decreases down to [N] = 0.8 mass% at d = 80 μm. This high nitrogen concentration both in HNL and LNL proves that the γ-lattice in AISI316 is super-saturated by nitrogen solutes.

In the conventional plasma nitriding, the nitrogen solute content decreases exponentially from the maximum nitrogen content of 0.1 to 0.2 mass% at the surface toward the nitriding front end, as reported in [11]. No plateau with constant nitrogen content was observed in the nitrogen depth profile. This implies that the inner nitriding process at 673 K is governed not only by the nitrogen diffusion but also the nitrogen super-saturation process.

This structural change of γ-lattices by the nitrogen super-saturation is analyzed by XRD analysis. Figure 3 shows the XRD diagrams for the nitrided AISI316 plates at 673 K. The original γ-phase peaks for (111) and (200) lattice planes shifted to the lower 2θ angles. The original fcc-structured lattice expands by the nitrogen super-saturation. This expanded γ-phase (γ_N) lattices are detected in figure 3 as

γ_N (111) and γ_N (200) in correspondence to γ (111) and γ (200), respectively. Besides for these two γ-phase peak shifts, a new peak was detected at $2\theta = 44°$ in figure 3. This corresponds to the α'-phase for (110). This suggests that γ_N to α' phase transformation takes place together with the nitrogen super-saturation.

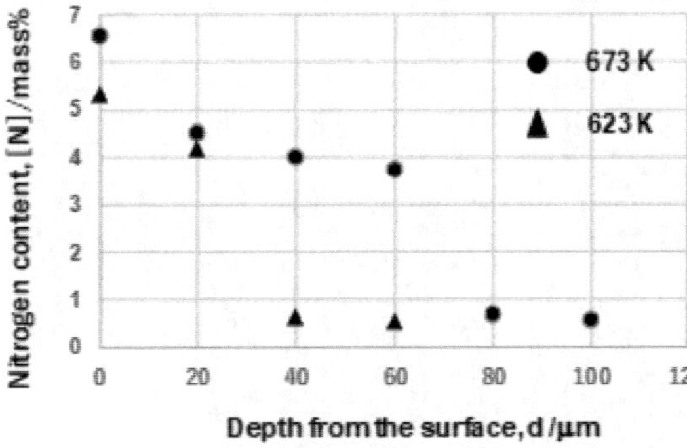

Figure 2. Nitrogen content depth profiles analyzed by EDX for the nitrided AISI316 plates at 673 K.

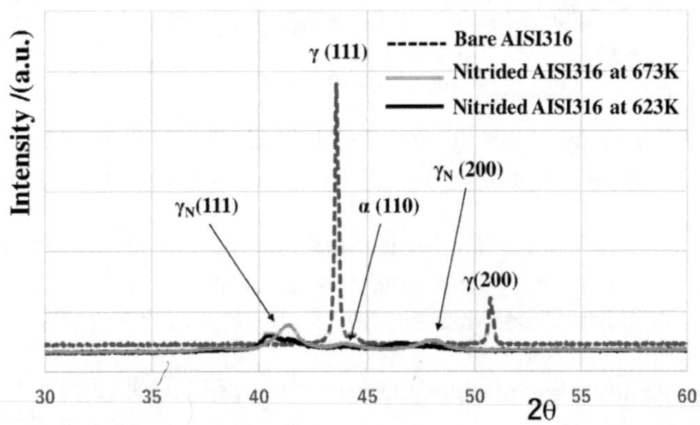

Figure 3. XRD analysis of the nitrided AISI316 plates at 673 K and 623 K as well as the bare plate.

Figure 4 depicts the KAM distribution, the phase mapping and the inverse pole figure in the normal axis on the cross-section of the nitrided AISI316 at 673 K. KAM represents the equivalent plastic straining state of materials after [12]. In the absence of plastic strains, the misorientation angle between the adjacent grains becomes zero; while this KAM increases with increasing the equivalent plastic strains. The plastic straining takes place homogeneously with high KAM in HNL for d < 60 μm at 673 K. This homogeneous KAM distribution implies that the plastic strains are uniformly induced to compensate the strain incompatibility between the nitrogen unsaturated and saturated lattices. In fact, the highly plastic strained region is neighbouring to the phase-transformed one, as shown in Figs. 4 a) and b). The surrounding regions around the phase-transformed zone with high elastic strains are plastically strained to drive the microstructure refinement. Fine distribution of the expanded γ-phase and the transformed α'-phase in this layer reveals that the phase transformation also advances uniformly in the depth.

The inverse pole figures in the normal direction for the nitrided AISI316 at 673 K is shown in figure 4(c). The original coarse grains of 10 to 20 μm are completely refined to sub-micro meter sized grains or nearly down to 0.1 μm, which is equivalent to the spatial resolution in the present EBSD analysis. In addition, each refined grain has a random crystallographic orientation. This uniform KAM distribution, fine distribution of mixed two (γ, α')-phases and refined grains with random orientation from the surface to the depth of 60 μm, drives this homogeneous inner nitriding.

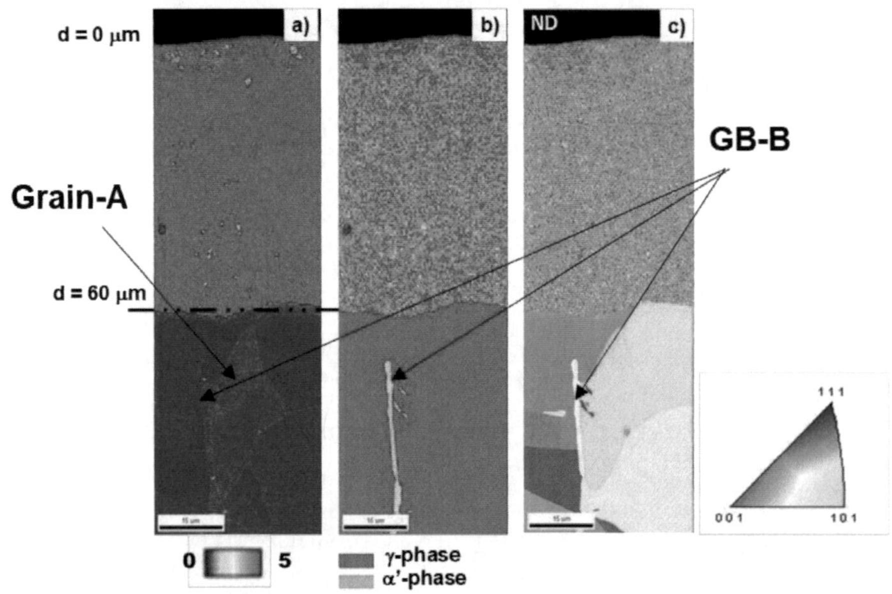

Figure 4. EBSD analysis results of the nitrided AISI316 at 673 K. (a) KAM distribution, (b) Phase mapping, and (c) Inverse pole figure in the normal axis.

As shown in figure 2, not only this HNL layer for 0 < d < 60 μm but also the LNL for d > 60 μm is nitrogen super-saturated since the nitrogen solute content is measured to be 4 mass% at d = 60 μm. However, KAM distribution, phase mapping and IPF-profile are uniform in the range of 0 < d < 60 μm while they are heterogeneous below d = 60 μm in Fig. 4.

The plastic straining locally takes place even in the LNL for d > 60 μm as shown in figure 4(a). In particular, many slip-lines by plastic straining penetrate through the grain-A and along the grain boundary (GB) - B. In figure 4(b), the expanded γ-phase transformed to α'-phase in part of this grain-A and at this GB-B. This selective plastic straining and phase transformation implies that the nitrogen diffuses through the grain-A and along the GB-B and that the grains and grain boundaries at the vicinity of grain-A and GB-B are only nitrogen super saturated. In fact, microstructure is also modified locally around the grain-A and GB-B.

This change in Fig. 4 from homogeneous nitriding mode to heterogeneous nitriding one at d = 60 μm, proves that there is a critical process to control the inner nitriding at 673 K. Since nitrogen super-saturation advances across the border at d = 60 μm in Fig. 2, the nitrogen diffusion process triggers this difference. The homogeneous nitriding is accommodated by the nitrogen diffusion paths through the grain boundaries of fine grains and the α'-phase lattices for d < 60 μm. These nitrogen diffusion is localized to the limited grains and grain boundary at d = 60 μm; the homogeneous nitriding turns to be heterogeneous for d > 60 μm. As discussed in [8, 13-14], the synergetic interrelation among the nitrogen super-saturation, the γ-lattice expansion, the lattice elastic straining, the phase transformation, the plastic straining, the microstructure refinement and the nitrogen diffusion, works simultaneously; then, the homogeneous nitriding process is sustained into the depth of matrix. Once the nitrogen diffusion process is locked, each constituent process in this interrelation is forced to localize and result in the heterogeneous nitriding.

4. Conclusion

The low temperature plasma nitriding of stainless steels works as a method to make surface hardening and nitrogen super-saturation in the austenitic stainless steels. AISI316 stainless steels are nitrided at 673 K for 14.4 ks to describe this hardening and nitriding processes. Under the homogeneous nitriding, the nitrogen content becomes constant by 5 mass% till the end of nitriding front end, both the γ_N-phase and α'-phase are homogeneously formed with fine mixture of two phases, and, the intense plastic straining is induced to make microstructure refinement with the reduction of grain size down to 0.1 μm. This homogeneously nitrided layer has high hardness; e.g., 1700 HV at the surface and 1200 HV at d = 60 μm. This homogeneous nitriding turns to be heterogeneous across this border at d = 60 μm. This is because the nitrogen diffusion process is localized to penetrate through the selected grains and grain boundaries into the further depth of matrix. Hence, the total nitriding process is first driven by the heterogeneous nitriding and followed by the homogeneous nitriding with uniform plastic straining, fine phase transformation and microstructure refinement.

Acknowledgments

The authors would like to express their gratitude to Mr. A. Farghali (SIT) for his help in experiments. This study was financially supported by the METI-project on the supporting industries in 2017 and 2018.

References

[1] Stein G, Menzel J and Dorr H 1988 *High Nitrogen Steels HNS88* J Foct and A Hendry (Eds.) (London: Maney Publishing) p 32

[2] Berns H and Kühl A 2004 Reduction in wear of seapage pump through solution nitriding *Wear* **246** 16-20

[3] Aizawa T and Sugita Y 2013 Plasma nitriding of tool steels *Res. Rep. SIT* **57** 1-10

[4] Dong H 2010 S-phase surface engineering of Fe-Cr, Co-Cr and Ni-Cr alloys *Int. Mater. Reviews* **55** 65-98

[5] Kim S K, Yoo J S, Priest J M and Fewell M P 2003 Characteristics of martensitic stainless steel nitrided in a low-pressure RF-plasma *Surf. Coat. Technol.* **163-164** 380-5

[6] Katoh T and Aizawa T 2016 Low temperature high density plasma nitriding of stainless steel molds for stamping of oxide glasses *Manufacturing Review* **3** 1-6

[7] Farghali A, Aizawa T and Suzuki Y 2017 High nitrogen concentration on the fine grained austenitic stainless steels *Proc. 11th SEATUC Conference* OS6-4 1-6

[8] Aizawa T and Yoshino T 2018 Plastic straining for microstructure refinement in stainless steels by low temperature plasma nitriding *Proc. 12th SEATUC Conference* (in press)

[9] Fraghali A and Aizawa T 2017 Phase transformation induced by high nitrogen content solid solution in the martensitic stainless steels *Mater. Trans.* **58** 697-700

[10] Yamakawa K 2017 Introduction to electron beam enhanced plasmas *Proc. 7th Advanced Tribo-Processing Seminar* 21-33

[11] Hiraoka Y and Inoue K 2010 Prediction of nitrogen distribution in steels after plasma nitriding *J. Denki-Seiko* **86** 15-24

[12] Kamaya M, Wilkinson A J and Titchmarsh J M 2006 Quantification of plastic strain of stainless steel and nickel alloy by electron backscatter diffraction *Acta Materialia* **54** 539-48

[13] Aizawa T 2017 Micro-fabrication of textured stainless steels via the nitrogen super-saturation process *Proc. 1st Engineering Innovations and Technology Conference-2017* (in press)

[14] Farghali A and Aizawa T 2018 Homogenous and heterogeneous structuring processes in the nitrided austenitic stainless steels below 673 K *Mater. Sci. Engg. B* (in press)